N. Clauer S. Chaudhuri

Clays in Crustal Environments

Isotope Dating and Tracing

With 169 Figures

Springer-Verlag
Berlin Heidelberg New York
London Paris Tokyo
Hong Kong Barcelona
Budapest

Dr. Norbert Clauer
Research Director
Centre de Geochimie de la Surface
1, rue Blessig
67084 Strasbourg
France

Professor Dr. Sambhu Chaudhuri
Department of Geology
Kansas State University
Manhattan, KS 66506
USA

ISBN 3-540-58151-0 Springer-Verlag Berlin Heidelberg New York
ISBN 0-387-58151-0 Springer-Verlag New York Berlin Heidelberg

Library of Congress Cataloging-in-Publication Data. Clauer, Norbert. Clays in crustal environments: isotope dating and tracing/Norbert Clauer, Sambhu Chaudhuri. p. cm. Includes bibliographical references. ISBN 3-540-58151-0. – ISBN 0-387-58151-0 1. Clay minerals. 2. Isotope geology. I. Chaudhuri, Sambhu, 1938- . II. Title. QE389.625.C514 1995 549\.6 – dc20 94-35344

Typesetting: Best-set Typesetter Ltd., Hong Kong

SPIN: 10467864 32/3130/SPS – 5 4 3 2 1 0 – Printed on acid-free paper

Preface

Clay minerals have been well recognized as one of the most important components of the Earth's shallow crustal materials. These minerals, together with others of similar structure, have been investigated with varied success for their use as stratigraphic markers, tectonic indicators, paleocurrent or paleohydrologic tracers, and paleoclimatic indicators. They have been considered as prime regulators for a wide variety of chemical processes both on the surface and in the subsurface within the shallow part of the Earth's crust. The different interests in the same material have a common basis of knowing when and how the minerals respond to changes in the environment in which they form or occur. Isotope studies of these minerals have shed much light on these common quests.

The history of isotope research on clay minerals is about 35 years old. In the early years a great deal was published in the field of dating of clay minerals. Some results appeared enigmatic and helped to maintain continued interest in this line of research; others produced results of dubious value. The expansion of isotopic research during the early years proceeded randomly and was based on assumptions that could not always be corroborated. Although the specific goal of defining the time of deposition could not be easily achieved, the results showed that many clay minerals in many modern sediments and ancient sedimentary rocks are detrital in origin, indicating that dating of the small amount of the authigenic components is not simple or straightforward. New evidence from increasing isotopic research, especially on weathered profiles on the continent and deep-sea sediments from DSDP Sites, expanded and modified ideas about clay genesis in continental soils and modern ocean sediments. Diagenetic changes of clay minerals became a major focus in many isotopic investigations of clay minerals in latter years. New insights were gained about diagenetic formations of clay minerals, but with this increasing evidence came also fresh debates about time and mechanism of formation of diagenetic clay minerals in deeply buried sediments. In recent years, the horizon of isotope research on clay minerals has greatly expanded to search for solutions about clay mineral formation in shallow crustal environments by following different lines of approach. New ground for research in the field of isotope geochemistry on clay minerals has emerged as a result of increased attention to finding solutions to problems related to climate, waste disposal, and pollution.

The amount of material that has appeared in the literature is vast and is rapidly growing. The territory that has been covered through isotope analyses of clay minerals is wide; much has yet to be explored systematically. Reexamination of some old problems is needed in light of recent evidence. Studies of clay minerals in shallow crustal regions have shown that isotope signatures of clay minerals are often intimately connected to those of their immediate precursors. These carryover traits are not weaknesses but strengths, as they elucidate the understanding of many different cyclic processes within the shallow crust of the Earth.

Isotope geochemistry of clay minerals has become a subject of considerable diversity. To cover everything in a single text is an extremely difficult, if not impossible, task. Omissions of certain areas and unequal treatments of topics that have been covered, if they are found to have occurred, are not intentional. Such weaknesses may have resulted from difficulties in dealing adequately with all the developments that have come during the last 20 years or so. A major purpose of this book is to acquaint the reader with various isotopic approaches, to gain an understanding of the physical and chemical dynamics in the evolution of clay minerals in shallow crustal environments, while being exposed to a wide range of factual knowledge. We hope that the reader will become familiar with the state of the art in the field of clay isotope geochemistry. Another major purpose of the book is to stimulate progress in isotope research on clay minerals. Although the book is organized in a way that each chapter can be considered as being nearly self-sufficient with introduction and summary, the texts are presented in a sequence from isotopic inheritance by clay minerals in weathering environments, through isotopic inheritance and modification of clay minerals in depositional and diagenetic environments, to isotopic inheritance and modification of clay and mica-type minerals in widely varied low-grade metamorphic conditions. A chapter at the end of the book highlights some areas of research with potential benefits from isotope analyses of clay minerals.

The major impetus of isotope research on clay minerals began with the pioneering work led by Patrick M. Hurley and his associates at the Massachussetts Institute of Technology and William Compston and his collaborators at the Australian

National University in the late 1950s and the early 1960s. Most of their works focused on K-Ar and Rb-Sr geochronologies of clay minerals. About this time, Gunter Faure, at the Massachussetts Institute of Technology and at the Ohio State University, contributed to the understanding of water and clay sediment interactions at surficial conditions in fluvial, lacustrine, and marine environments. The early 1970s saw the emergence of stable isotope research on clay minerals, pioneered primarily by Samuel M. Savin and Samuel Epstein at the California Institute of Technology. The widespread interest in isotope studies on clay minerals grew not only because of the works of those active in this particular field of geochemistry, but also due to major stimuli that were provided by those who were not specialized in isotope geochemistry, but had the foresight to recognize the high potential of the use of isotopes as tracers in studies of clay minerals. Their conviction that isotopic signals are major clues to the origin of clay minerals led them to encourage many to follow this line of research. Dr. Georges Millot, a scholar of international stature in the field of clay mineralogy, became such an inspiring leader by establishing in 1963 an isotopic laboratory specialized in studies of superficial environments, at the Research Center of the French National Research Council (CNRS) in Strasbourg, France.

The book is dedicated to the memory of Dr. Georges Millot, whose vision on the origin of clay minerals and leadership as Director of the Centre de Sédimentologie et Géochimie de la Surface have contributed in a major way to developments in the field of the isotope geochemistry of clay minerals. We would also like to take the opportunity to thank friends and colleagues who were generous with their time in helping us in the task of conceiving and completing this book. Ray E. Ferrel Jr. (Louisiana State University), Larry A. Frakes (University of Adelaide), Hélène Paquet (Centre de Géochimie de la Surface) and Peter Stille (Centre de Géochimie de la Surface) made many valuable suggestions for the composition of the text. Gunter Faure (Ohio State University), François Gauthier-Lafaye (Centre de Géochimie de la Surface), and James R. O'Neil (University of Michigan) were prime interlocutors during the conception, gestation and completion of the work. Claude Hammel generously spent many days in drafting all the figures. Finally, our very deep appreciation goes to both Michèle and Nupur for their patience with our unending endeavour in completing the manuscript.

Strasbourg, France
Manhattan, KS 66506, USA
January 1995

N. Clauer
S. Chaudhuri

Contents

Chapter 1

An Introduction to Clay Minerals and Isotope Geochemistry

Clays are probably the most important component of surficial materials of the Earth. Because clays have been widely used by mankind since High Antiquity, they have been subject of a wide variety of studies in a number of different disciplines such as geology, soil science, mining, engineering, and biology. Readers interested in the various applications of clays are referred to the book by Robertson (1961).

1 Fundamentals of Clay Mineralogy

Clay minerals occur in very small particle size, so that precise mineralogical descriptions were not possible until the X-ray technique for atomic structural analyses came into use, based on the principles outlined by Pauling (1930). The pioneer studies on clays are about 100 years old, and describe primarily the chemical compositions of these fine materials. Until the 1940s, clay minerals were considered as essentially amorphous material. Increased understanding of the structural details came with the progress of X-ray technology. Additional knowledge of the morphology and crystallography of the clay minerals came through the use of electron microscopies (scanning and transmission) which were introduced during the 1960–1970s. These technological developments produced quantum progress in knowledge about the evolution of clay minerals in the upper crustal environment. Thermodynamic applications to clay mineral studies have produced further insight to the stability domains of clay minerals in different upper crustal environments. The frontier of knowledge about clay mineral genesis has been vastly extended by the application of isotope geochemistry to studies on clay minerals, which began more than 30 years ago.

1.1 Definition

Clay minerals have been variously defined, serving the interests of each specific discipline. We accept here the term commonly used by geologists, which emphasizes both the specific chemical and physical attributes of the minerals. The basic chemical composition of such minerals is hydrous aluminum silicate. They have a layered structure, high malleability when wet, high affinity for water, high swelling property, high ion exchange capacity, are soft when dry and small, less than 2 μm (<2 μm) in size, with large surface to volume ratios.

1.2 Basic Structural Units and Layer Types

Clay minerals are phyllosilicates or sheet silicates with stacking of octahedral and tetrahedral sheets which are the two basic building blocks making up the basic structural units. In the tetrahedral sheet, each of the three basal oxygens of a tetrahedron is linked to a neighboring tetrahedron, while the fourth or apical oxygen of all tetrahedra points in the same direction above or below the plane of basal oxygens. The apical oxygens form a hexagonal to pseudohexagonal ring. The tetrahedral cations are commonly Si and Al, but may occasionally be Fe^{3+} and Cr. In the octahedral sheet, each octahedron is linked to its neighbors by shared edges. The cation of each octahedron is co-ordinated by a combination of OH and O ions in the sheets. Ideally, with OH anions co-ordinating the octahedral cation, the octahedral sheet forms either a brucite-like structure called trioctahedral structure in which all cation positions are filled, or a gibbsite-like structure called dioctahedral structure in which two-thirds of the cation positions are filled. Because of the size limit, the cation positions are filled by such common medium-sized ions as Al, Fe^{3+}, Fe^{2+}, Mg, Mn and Li, and less commonly by Zn, Ni, Cu, Co, and others with similar ion size.

The clay mineral structures, based on the combination of tetrahedral and octahedral sheets, are of 1:1, 2:1, and 2:1:1 types. In the 1:1 types consisting of one octahedral sheet and one tetrahedral sheet, the plane of apical oxygens of the tetrahedral sheet is superimposed on the OH plane of the octahedral sheet (Fig. 1.1). The common plane of junction consists of both oxygen and unshared OH. The unshared OH occurs at the center of each six-fold ring defined by the apical tetrahedral oxygens. The 2:1 type is the stacking of a octahedral sheet sandwiched between two tetrahedral sheets, so that their planes of apical oxygens face toward each other. The common planes between the two tetrahedral sheets and the octahedral sheet consist of sharing oxygens and unshared OHs, each unshared OH being at the center of tetrahedral apical oxygen hexagonal rings (Fig. 1.2). The 2:1:1 structure consists of a 2:1 layer arrangement with an additional octahedral sheet between the 2:1 layers.

The thickness of each structural unit depends not only on the number of sheets involved, but also on the compositions of tetrahedral and octahedral sheets and the charge compensations as a result of unbalanced substitutions

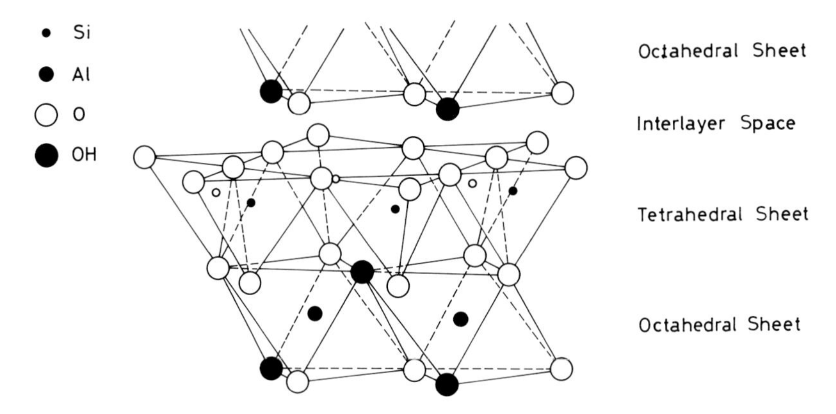

Fig. 1.1. Schematic representation of the crystal structure of a 1:1 type clay mineral showing the relations between tetrahedral and octahedral layers

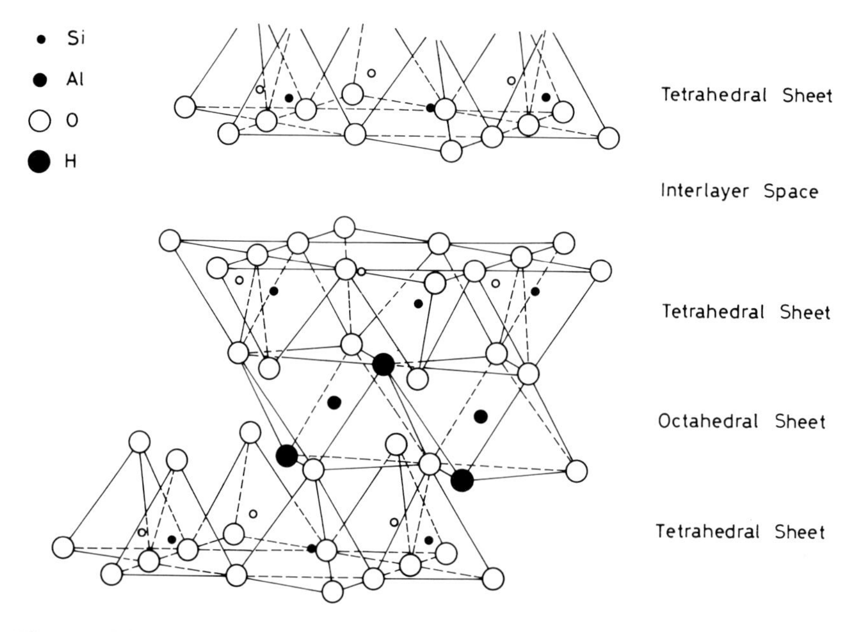

Fig. 1.2. Schematic representation of the crystal structure of a 2:1 type clay mineral showing the relations between tetrahedral and octahedral layers, and interlayer spaces

of cations in the sheets. The net compensating charge may be located either in the interlayer site as in 2:1 layer arrangements or in the additional octahedral sheet in 2:1:1 layer arrangements. The most common interlayer cations are Ca, Mg, H, Na, K, NH_4. Interlayer sites might also be occupied by Fe and Al in a hydroxy form.

A detailed account of the chemistry and crystallography of clay minerals can be found in Brindley and Brown (1984). X-ray diffraction data are used for mineralogical identification of the minerals, but such data of clay material should be carefully analyzed, as the X-ray diagrams may be influenced by: (1) the small size of the crystals causing decrease in intensity of the signals, (2) defects in the structures by distortions of the layers due to substitutions, (3) disorder in the stacking of the layers within a crystallite, and (4) stacking of different types of layers producing mixed-layers phases. Basic studies on X-rays have also shown that for particles larger than 100 Å, the X-ray reflections are only dependent on the crystalline structure of the particles (Guinier 1964). Clay particles are at the limit of this critical size, and perturbations might be noticeable on the X-ray charts due to the size.

To differentiate among different types of clay minerals, standard treatments are routinely applied to all samples for X-ray diffractometric analyses. One treatment consists of ethylene-glycol vapors, which might occupy the interlayer positions of some types of clay minerals. These ethylene-glycol molecules will modify the basal spacing between the layers. Another treatment consists of hydrazine monohydrate vapors which might also be added as a component between weakly bonded layers of some types of clay minerals. Addition of these molecules will also change the basal spacing of these specific clay minerals. A heat treatment of about 490 °C is also used routinely to dehydrate the clay minerals which naturally incorporate water molecules in their interlayer positions.

1.3 Classification of Clay Minerals

Clay minerals are characterized by the type of layer structure, the composition of the sheets that make up the layer structure, the layer charge, and the composition of the interlayer material (Brindley and Brown 1984). A general scheme of the nomenclature of clay minerals is given in Table 1.1 (Bailey 1980). Chemical data on these different types of clay minerals will also be given along with their presentation and additional information is available in the very complete compilation of Weaver and Pollard (1973), as well as in a more recent publication edited by Newman (1987).

1.3.1 The Serpentine-Kaolin Group

The serpentine-kaolin group is characterized by a 1:1 layer type without layer charge, and it has a [001] layer spacing of about 7 Å. The serpentine subgroup, which is of trioctahedral composition, will not be discussed, because the constitutive minerals are not of interest for the purpose of this book. The kaolin subgroup, on the other hand, is a common constituent of sedimentary rocks and is relevant to our scope.

Table 1.1. Classification of hydrous phyllosilicates. (Bailey 1980)

Layer type	Interlayer	Group	Subgroup
1:1	None or H_2O only	Serpentine + kaolinite (x ± 0)	Serpentines, kaolins
2:1	None	Talc + pyrophyllite (x ± 0)	Talcs, pyrophyllites
	Hydrated exchangeable cations	Smectite (x ± 0.2–0.6)	Saponites, montmorillonites
	Hydrated exchangeable cations	Vermiculite (x ± 0.6–0.9)	Trioctahedral vermiculites, dioctahedral vermiculites
	Nonhydrated cations	True mica (x ± 0.5–1)	Trioctahedral true micas, dioctahedral true micas
	Nonhydrated cations	Brittle mica (x ± 2.0)	Trioctahedral brittle micas dioctahedral brittle micas
	Hydroxide sheet	Chlorite (x variable)	Trioctahedral chlorites dioctahedral chlorites

x = Charge per formula unit.

Kaolin minerals consist of a Si-O tetrahedral sheet and an Al dioctahedral sheet. As the former one is larger than the latter one, a distortion of the tetrahedral sheet occurs so that its basal dimensions nearly equal that of the octahedral sheet to accommodate the fit between the hexagonal networks of the tetrahedral and the octahedral sheets. Each 1:1 layer is bonded to the adjacent layers by hydrogen bonds between the OH ions in one of the planes of an octahedral sheet in one layer and the basal O ions of the tetrahedral sheet in the next layer. The [001] spacings of the 1:1 kaolin layers range from 7.1 to 7.3 Å, depending on stacking imperfections. Stacking sequences of layers with different interlayer shifts and different locations of the vacant octahedral cation site result in different polytypes of kaolin minerals (Bailey 1963, 1988a). The kaolin subgroup consists of kaolinite, dickite, nacrite, and halloysite minerals.

Kaolinite and dickite have the same interlayer sheet but different vacant octahedral cation locations in the successive layers. In the former, the vacant octahedral cation site is the same in successive layers, whereas in the latter the site alternates at every third layer which tends to create a two-layer structure. Nacrite results from a six-layer stacking sequence as alternate layers are rotated by 180°. The vacant cation sites rotate by 60° and successive layers are shifted by the same amount along the a-axis. Its polymorphic form is of 6R type with a chemical composition identical to that of kaolinite. The nacrite particles often appear irregular and even rounded.

The crystallinity of kaolin minerals is determined by the regularity of the displacement of one layer with respect to another, which is also a measure of the degree of stacking order. Thus in well-crystallized kaolinite, nacrite, and dickite, the stacking is regular; but in poorly crystallized kaolin minerals, the randomness in stacking along the b-direction could be further compounded by the nonrepetitiveness in the position of the vacant octahedral cation site in successive layers.

The X-ray reflections of the [001] planes of well-crystallized kaolinite are at 7.16 Å, and the high order reflections occur successively at 3.57, 2.34 Å, etc. The [001] spacing of poorly crystalline kaolinite is slightly higher than that of a well-crystallized one. An ideal chemical formula for kaolinite is $Si_4Al_4O_{10}(OH)_8$. However, chemical analyses of kaolinite samples have frequently indicated presence of small amounts of Fe, Ti, Ca, Mg, K, and other elements in trace amounts. In some instances, the Si/Al ratios have been found to differ from 1 (Weaver 1968), and the deviations from the ideal formula are often produced by different types and amounts of impurities in the samples. The cation exchange capacity (CEC) of kaolinite minerals commonly ranges from about 3 to 15 meq/100 g. The low values are probably representative of pure kaolinite, and the increase could be due to impurities. Much of the CEC of pure kaolinite is owed to broken bonds at the edges of the crystals, occurring as hexagonal to pseudohexagonal plates or flakes.

Kaolinite, a common product of weathering in soils, is often present as a detrital component in coastal and off-shore sediments and also in sedimentary rocks, especially sandstones. It might also occur together with authigenic dickite in deeply buried sandy sequences. Dickite is coarse-grained and better crystallized relative to kaolinite, and it commonly occurs as high-temperature kaolin minerals, in more bulky booklet-forming crystals. than kaolinite. The good crystallinity and relatively coarse size of dickite have afforded valuable information on structural and chemical aspects of kaolin minerals. Poorly crystalline kaolinite is common in fire clays, flint clays, and ball clays.

Halloysite is a highly disordered kaolin mineral which occurs in two forms: a hydrous form with an ideal chemical composition of $Si_4Al_4O_{10}(OH)_8.4H_2O$ and a basal spacing of about 10.1 Å, and a dehydrated form with a chemical composition, $Si_4Al_4O_{10}(OH)_8$, similar to that of kaolinite, and a basal spacing of about 7.2 Å. The 10 Å halloysite contains one molecular layer of water between the 1:1 layers and may be converted to the 7 Å form at temperatures less than 70 °C. The CEC of halloysite minerals is generally slightly higher than that of kaolinite and ranges from about 5 to 50 meq/100 g. Halloysite is frequently associated with amorphous to poorly crystalline allophane and imogolite in soil materials, where it occurs as fine-grained rolled spirally tubes. It can be complexed with organic matters which enter the structure as flat molecules. Because of its fine-grained disordered structure, halloysite is chemically more impure than other kaolin minerals.

1.3.2 The Illite Group

The illite group has been studied most extensively because of its ubiquitous presence as clay-sized 10 Å mica-type minerals in sedimentary rocks. The term "illite" was first introduced by Grim et al. (1937) as a general term for the mica-type clay constituent of argillaceous sediments. The stucture of illite consists of negatively charged 2:1 layers that are "keyed" together, commonly by large interlayer cations which balance the net charge deficiency in the 2:1 layers. Most of the natural illites are dioctahedral with Al being the highly dominant cation and minor amounts of Mg and Fe ions. Illite differs from ideal muscovite by having less substitution of Al for Si in the tetrahedral sheet and a small substitution of Al by Mg and Fe^{3+} in the octahedral sheet. The total charge deficiency in the illite 2:1 layers is the sum of a large negative charge in the tetrahedral sheet and a small negative charge in the octahedral sheet. The charge deficiency in the 2:1 layer is balanced primarily by K ions in the interlayer site. A representative structural chemical formula for common illites is $Si_{6.40}\ Al_{1.60}(Al_{3.70}Fe^{3+}{}_{0.10}Mg_{0.20})K_{1.80}O_{20}(OH)_4$. In an ideal illite, the charge deficiency is about 0.9 per unit cell, but in many natural illitic materials the interlayer charge ranges from 0.6 to 0.8 per unit cell, owing primarily to many natural illites having some amounts of expandable layer. The properties of pure illite have been determined by extrapolation of various physical parameters such as CEC versus expandability and surface area, and chemical parameters such as K content versus percent expandable layers and tetrahedral Al content, or the ratio of illite/smectite mixed-layer minerals to pure illite. The CEC of many natural illites vary between 10 and 40 meq/100 g. An ideal illite with "zero" expandable layers is Al-rich dioctahedral in composition and is thus similar to muscovite in the octahedral composition; but some natural illites with nearly zero expandable layers contain much more Mg and Fe in the dioctahedral sheet than a near-"ideal" illite. These Mg-rich illites are chemically more similar to phengite, which has a total charge close to that of muscovite and a representative structural chemical formula of $Si_{6.78}Al_{1.22}(Al_{2.86}Fe^{3+}{}_{0.10}Fe^{2+}{}_{0.18}Mg_{0.10})K_{1.74}Na_{0.14}Ca_{0.06}O_{20}(OH)_4$. Phengite commonly occurs in low to intermediate-grade metamorphic rocks, but the mineral can also be found in some very late stage or highly evolved diagenetic rocks.

Mica-type clay minerals in sedimentary environments are mostly of dioctahedral type. Reported occurrences of trioctahedral illites are few and they have been observed in some evaporites and soil clays. Although K is the most dominant cation in most illites, other cations may be equally important in some types of illites. NH_4 illites have been reported in some organic-rich black shales. The chemistry of these illites is poorly known, but their NH_{4+} contents make up nearly 50% of the interlayer cations (Daniels and Altaner 1990).

Glauconites and celadonites are common dioctahedral illite-like authigenic minerals forming in ocean sediments. As compared to illite, these minerals have higher contents of Fe and Mg in the octahedral sheet. An average structural chemical formula of glauconite is $Si_{7.38}Al_{0.62}(Al_{0.92}Fe^{3+}{}_{1.98}Fe^{2+}{}_{0.36}Mg_{0.80})K_{1.32}Na_{0.12}Ca_{0.14}O_{20}(OH)_4$(Weaver and Pollard 1973). On the average, about 60% of the dioctahedral positions are filled with Fe. The total charge deficiency in the "ideal" glauconite is about the same as in "pure" illite but, unlike that in illite, much of the charge deficiency in the majority of glauconites is located in the octahedral sheet. In some instances, tetrahedral charge deficiency has been found to be greater than octahedral charge deficiency. As in illite, varied amounts of expandable layers occur in glauconite minerals and the K content is negatively correlated to the amount of expandable layers; but at the same percentage of expandable layers between the two minerals, illite has a slightly higher K content than glauconite. Illite and glauconite are set apart by their chemical compositions, suggesting lack of genetic relationship between them (Fig. 1.3; Velde and Odin 1975). The CEC of glauconite with extremely low amount of expandable layer is less than 15 meq/100 g. Celadonite is also a Fe-Mg-rich dioctahedral 10 Å mineral, but con-

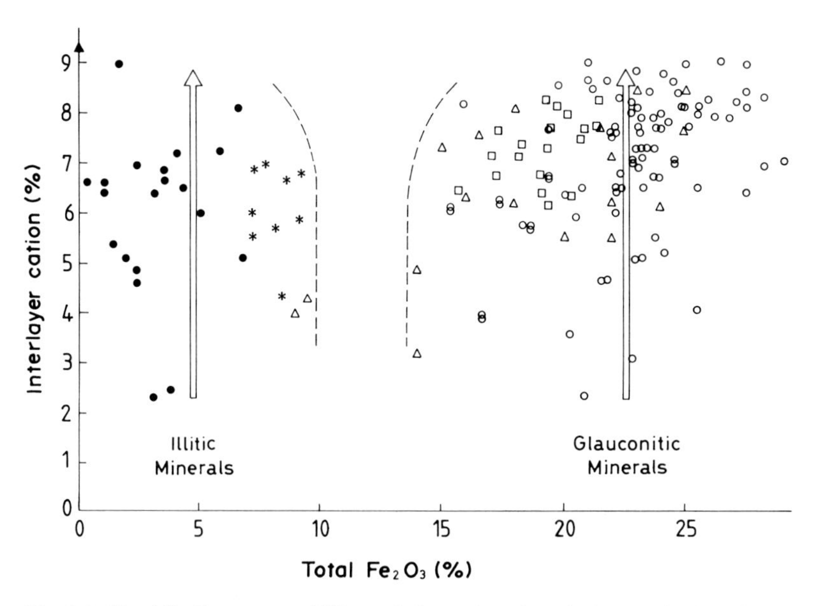

Fig. 1.3. Total Fe_2O_3 content of illite and glauconite minerals. No continuity exists between the two mineral types. (Velde and Odin 1975)

taining more Mg and less Fe than glauconite, the Fe/Mg ratio being about one-half that in glauconite. A representative structural chemical formula is $Si_{7.62}Al_{0.38}(Al_{0.98}Fe^{3+}{}_{1.44}Fe^{2+}{}_{0.42}Mg_{1.2})K_{1.28}Na_{0.26}Ca_{0.12}O_{20}(OH)_4$. Celadonite minerals are common products of many submarine hydrothermal alteration of basalts.

Glauconite occurs in marine rocks of all geologic ages and its origin has been variously explained (McRea 1972). The early theory about the formation of glauconite advocated a process of coprecipitation of Mg-, Fe-, Al-, and Si-gels which subsequently absorbed K (see discussion by Valeton 1958). Support of this early theory later lost ground upon the introduction of a widely accepted theory of formation of glauconite from degraded 2:1 layer silicate minerals which were subjected to diagenetic replacement of Al by Fe and concomitant uptake of K (Burst 1958a, b; Hower 1961). A more recent popular theory of glauconitization was proposed by Odin and Matter (1981), who advocated a two-stage process: authigenesis of Fe-smectite minerals in pores of different substrates followed by the formation of glauconitic micas through recrystallization of the Fe-smectite minerals accompanied by dissolution of the substrate. According to this theory, the glauconite pellets seem to have started their evolution as granular, rounded objects built around detrital clay particles and other mineral phases such as oxides. The primary mineral phases may have been variously modified by interaction with fluids migrating through the sediments. Odin (1975) believes that a major role in this progressive glauconitization process during sedimentation time is a partial chemical isolation of the grain interior with respect to sea water, resulting in a microenvironment which is different from surrounding marine environment. This concept was mainly based on the fact that grains of different sizes are differently evolved: the small ones hardly having any trace of glauconitization, and the very large grains having only partial glauconitization at their surfaces. The existence of isolated microenvironments during glauconitization is further supported by the fact that, while pellets occurring in shales or silts are glauconitized, the associated clay matrix is never.

A problem of definition of glauconitic minerals, especially evident by the contrast in the use between the English and the French literature, has existed for many years. In this book we have adopted Odin's (Odin and Létolle 1980; Odin and Matter 1981) use of the word *glauconite*. According to Odin, the term glauconite should be restricted to the dark green, pelletal or not, Fe-rich with K_2O contents higher than 6%, mica-type mineral of marine origin, while the word *glaucony(-ies)* should be used to designate all the Fe-rich but K-poor (<6% K_2O), light green to greenish, yellow pellets with mica-like structure. The glaucony pellets then represent a continuum between a glauconite end-member and a K-depleted smectite end-member.

Identification and characterization of mica-type sedimentary minerals have been adequately covered in a number of books and articles (Grim 1968; Millot 1970; Velde 1977, 1985; Chamley 1989). The X-ray diffraction patterns of air-dried smears of pure illite minerals on glass slides give [001] reflections generally ranging from 9.98 to 10.02 Å. The [001] reflections of NH_4 illite is about 10.2 Å. These variations in [001] spacing are related primarily to combined effects of varied octahedral and interlayer compositions. The higher order reflections are repeated at regular multiples at about 4.45, 3.35, 2.56 Å. Illite with Al-rich octahedral sheet has a [002] intensity that is about 0.25 that of the [001] intensity, whereas that with Fe-rich octahedral sheet has the [002]/[001] intensity ratio much less than 0.25. In all mica-type minerals, the d-spacing of the [001] reflections remain unchanged by solvation with ethylene-glycol or hydrazine monohydrate solutions and by heat treatment to about 500 °C. The dioctahedral nature is determined by the position of the [060] X-ray reflection at about 1.50 Å obtained from randomly oriented powdered samples. X-ray diagrams of powdered glauconitic samples show a clear dependence of the shape and the position of the [001] reflection on the K_2O content of the minerals (Odin and Dodson 1982). The [001] peak is broad at about 14 Å when the K_2O content is below 3%, but the peak sharpens and moves toward 10 Å when the K_2O content increases to about 8.5% (Fig. 1.4). Furthermore, the [003] peak is surrounded by two other peaks whose intensities increase as the K_2O content increases.

Smear slides may be used to determine the "crystallinity index" of the illite minerals. Weaver (1958, 1961) was apparently the first who observed that increasing burial produced a sharpening of the illite [001] peak profile. He described this change by the "sharpness ratio". Kubler (1964, 1966) utilized this property as a convenient criterion to define the positions of drill-core samples of sedimentary rocks relative to the window of hydrocarbon or gas generation, and proposed the concept of "illite crystallinity", which is a measure of the width at half peak of the reflection [001] obtained from the untreated X-ray diagrams. Alternative expressions are the measurement of the width of the [001] peak in angle (Kisch 1980) and the "Srodon index" which is a calibration of the amount of illite layers in an illite/smectite mixed-layer mineral based on the location of the [0010] reflection peak (Srodon 1984). Many agree that temperature is the main factor which controls the illite crystallinity, but the fluid composition and the porosity and permeability properties are also important factors (see Kisch 1983). The measure of the illite crystallinity is affected by the size of the crystallites, chemistry and origin of the illite, occurrence of certain associated species such as paragonite, and composition and amount of mixed-layer minerals present. Ross (1968) and Weber (1972) showed that peak broadening occurs in very fine fractions (<1 μm). Srodon (1984) maintained that the <0.2 μm size fractions would give an index that would be independent of sample composition and grain size. Weaver and Broekstra (1984) convincingly demonstrated that grain size dependence disappears at the top of the epizone (greenschist facies; Fig. 1.5). Direct comparison of values obtained in different laboratories should be considered with care, because the values of crystallinity index are biased by the instrumental conditions.

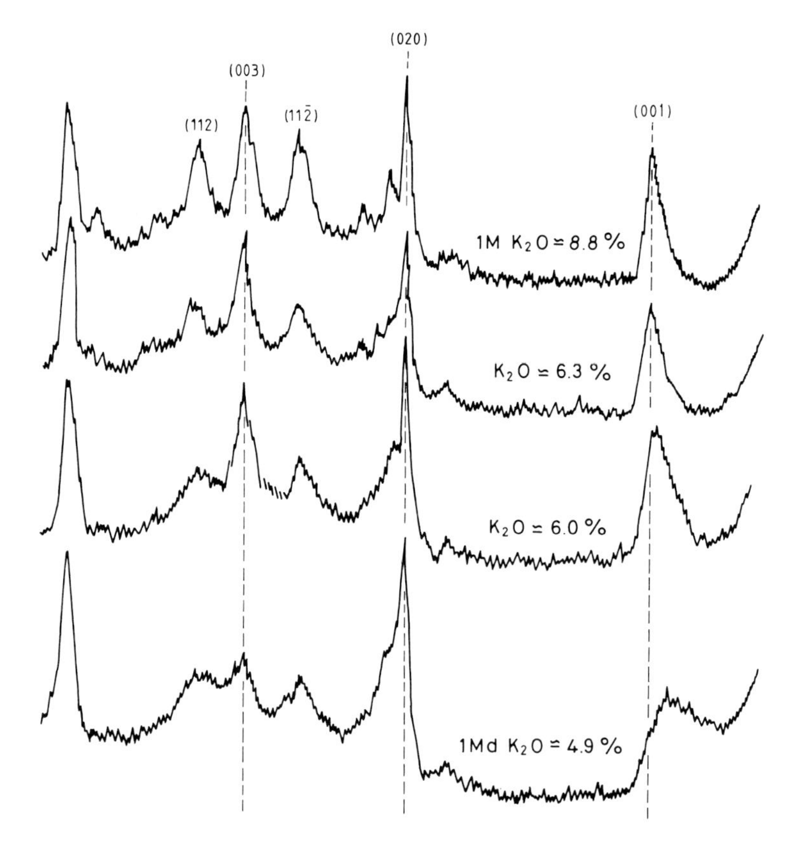

Fig. 1.4. Relatonship between the shape of the X-ray peaks and the K_2O contents of glauconitic minerals. (Odin 1975)

Illite-like minerals that are found in sedimentary rocks can have different structures arising from different stacking sequences of successive layers. Four polytypes have been described among the illite-like minerals: 1Md, 1M, 2M, and 3T. Polytypes of illite-like minerals are determined from X-ray powder patterns. The diagnostic [hkl] reflections of the different polytypes are given in Bailey (1988a and b). Temperature and chemical composition are major factors which control the stacking sequences. Illites commonly occur as 1M and 2M types, whereas glauconite and celadonite as 1Md and 1M types and phengite as all four types. The 1Md and 1M illites are the low-temperature species and are generally believed to be epigenetic or diagenetic in origin. The 2M type is a higher temperature form (above 250 °C) and is known to be associated with metamorphosed or hydrothermally altered rocks, as suggested by Yoder and Eugster (1955), and Velde (1965). Hence, the occurrence of 2M polytype in sedimentary rocks can be interpreted as a detrital component. The 3T polytypes are uncommon: an occurrence has

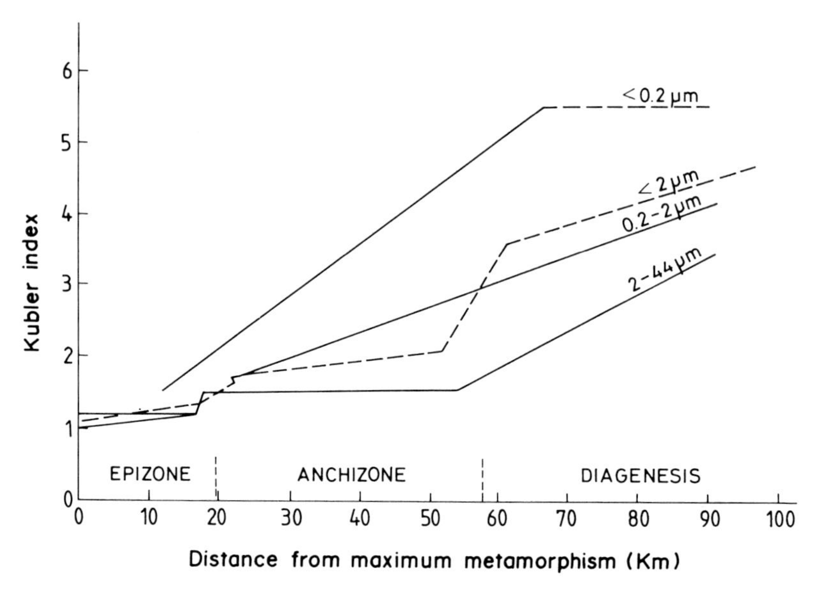

Fig. 1.5. Size-dependent variation in the Kubler index, as a measure of the crystallinity index of illite (see text) relative to the intensity of recrystallization. (Weaver and Broekstra 1984)

been described in host sedimentary rocks for uranium deposits in Canada where illite was considered to be related to a hydrothermal event that produced the uranium deposits (Ey 1984).

1.3.3 The Smectite Group

Smectite, together with illite, constitutes the most dominant clay mineral component in recent continental and marine sediments. The minerals belonging to the smectite group have a 2:1 layer with a range of negative charges between 0.2 and 0.6 per unit cell, averaging at about 0.35 per unit cell (Table 1.1). The charge deficiency results from combined isomorphous substitutions in octahedral and tetrahedral sheets. The net charge deficiency is balanced by interlayer cations with one to two water layers at about 50–70% relative humidities. The interlayer water constitutes an ice-like hexagonal structure. Smectites have varied CEC between 80 and 150 meq/100 g. The CEC of smectites is about ten times higher than that of the other clay minerals (Van Olphen and Fripiat 1979). About four-fifths of the CEC relates to the exchangeable interlayer cation and the rest to bonds at the edges of the crystallites. Smectite minerals are also known for their ability to dehydrate and rehydrate repeatedly in continental weathering and soil environments because of fluctuating moisture conditions.

Dominant exchangeable cations for smectites are Ca^{2+}, Mg^{2+}, Na^{+}, and H^{+} ions differing in the hydration energy, which in part accounts for the variations in the [001] spacings. In normal laboratory conditions, Na and K-smectites have [001] spacings of about 12.5 Å, whereas Ca and Mg-smectites have 14 to 15.5 Å. The basal spacing of smectite minerals is very sensitive to ethylene-glycol and heating treatments. The solvation with ethylene-glycol produces a shift of the 14 Å [001] peak to about 17 Å. The heating at about 500 °C expels the water molecules from the interlayer position and reduces the basic interlayer distance to about 10 Å or even slightly less (Fig. 1.6). Smectites are known as both dioctahedral and trioctahedral types. Common dioctahedral types are montmorillonite, beidellite, and nontronite, whereas common trioctahedral smectites are saponite, hectorite, and stevensite (Table 1.2). Dioctahedral types can be differentiated from trioctahedral types by the location of the [060] peak on an X-ray powder diagram, the former having the reflection at 1.49 to 1.52 Å, the latter at 1.53 to 1.57 Å.

Ross and Hendricks (1945), Grim (1968), Schultz (1969), and Weaver and Pollard (1973) provided a large collection of chemical data on smectite minerals. In dioctahedral sheets, Mg often occupies about 10 to 30% of the cation positions, but both Al and Fe can have a very wide range of values in

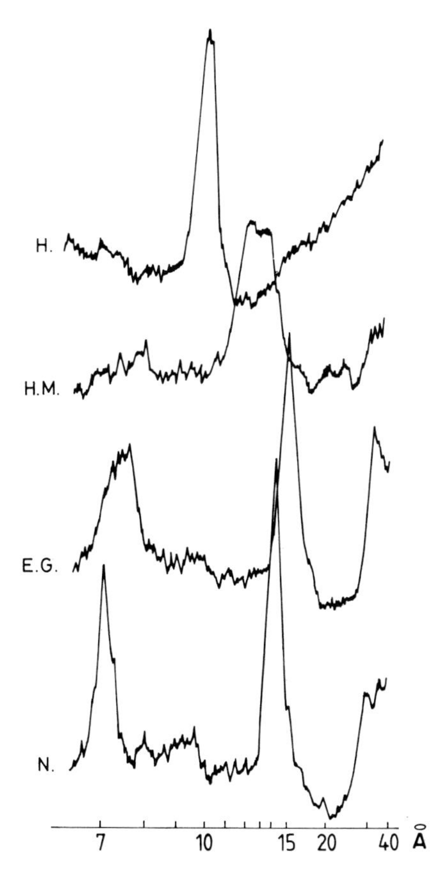

Fig. 1.6. X-ray patterns of smectite: untreated (*N.*), ethylene-glycol treated (*E.G.*), hydrazine-monohydrate treated (*H.M.*), and heated at 500 °C (*H.*)

Table 1.2. Different types of smectites

Types	Octahedral occupancy	Groups
Dioctahedral smectites	Al dioctahedral type	Montmorillonite (octahedral charge)
		Beidellite (tetrahedral charge)
	Fe dioctahedral type	Nontronite (tetrahedral charge)
Trioctahedral smectites	Al trioctahedral type	Saponite (octa. and tetra. charges)
	Li and F trioctahedral type	Hectorite (octahedral charge)
	Mg (Li) trioctahedral type	Stevensite (octahedral charge)
Smectites with uncommon elements	Li di-trioctahedral type	Swinefordite
	Cr	Volkhonskoite
	Zn	Sauconite

the dioctahedral sheets. Montmorillonite and beidellite are the Al-rich end-members, whereas nontronite is the Fe-rich counterpart. A major distinction between montmorillonite and beidellite is that the origin of the layer charge in the former is primarily in the octahedral sheet due to substitution of Mg and also some Fe^{2+} for Al and Fe^{3+}, and that in the latter is primarily in the tetrahedral sheet by replacement of Al for Si. A small amount, less than 5% on the average, of tetrahedral substitution of Al for Si may also occur in montmorillonite, in which the layer charge commonly varies between 0.4 and 1.0, averaging at about 0.7 per unit cell. The average layer charge in beidellite may be somewhat higher than that in montmorillonite. Greene-Kelly's (1953, 1955) Li-test may be used to distinguish between montmorillonite and beidellite. When saturated with Li and dried afterwards at about 350 °C for about 12 h, beidellite, but not montmorillonite, will expand to about 17 Å under glycol treatment. The nonexpansion in the montmorillonite may be explained by balancing of the octahedral charge by Li ions which entered the vacant sites of the octahedral sheet. Representative structural chemical formulas are $Si_8(Al_{3.34}Mg_{0.66})Na_{0.66}(OH)_4O_{20}$ for montmorillonite and $Si_{7.28}Al_{0.72}(Al_{2.92}Fe^{3+}{}_{1.00}Mg_{0.08})Na_{0.72}(OH)_4O_{20}$ for beidellite. Nontronite has Fe^{3+}-enriched dioctahedral sheets with an average layer charge about the same as in montmorillonite or beidellite, but the principal charge deficiency may be located in either the octahedral or the tetrahedral sheet. A representative formula of nontronite, which is often produced from alteration of basalts at oceanic ridges, may be given as $Si_{7.34}Al_{0.66}(Fe^{3+}{}_{3.22}Al_{0.78})K_{0.66}(OH)_4O_{20}$.

Saponite, hectorite, and stevensite constitute the most common Mg-rich trioctahedral smectites whose tetrahedral sheets are essentially made up of Si cations. In stevensite, the charge deficiency can be due to less than complete filling of the octahedral cation sites by Mg, or to isomorphous substitutions of Mg by low charge ions, or to both; but in hectorite, the charge deficiency is related to substantial replacement of Li for Mg in the trioctahedral sheet. A formula for stevensite may be given as $Si_8(Mg_{4.84})Na_{0.32}(OH)_4O_{20}$, whereas that for hectorite as $Si_8(Mg_{5.34}Li_{0.66})Na_{0.66}(OH)_4O_{20}$. Unlike hectorite and stevensite, saponite has an appreciable amount, as much as 14%, of substitution of Al for Si in the tetrahedral sheet, which is the primary cause for its charge deficiency. In some saponite samples, nearly half of the Mg in the trioctahedral sheet is replaced by Fe^{2+} ions. In relation to the location of charge deficiency, saponite may be considered as the trioctahedral analogue of beidellite. An example of a saponite chemical formula might be given by $Si_{7.34}Al_{0.66}(Mg_6)Na_{0.66}(OH)_4O_{20}$.

1.3.4 The Vermiculite Group

The vermiculite group consists of 2:1 expandable minerals and is differentiated from smectite group mainly on the basis of a higher layer charge, tentatively set as greater than 1.2 per unit cell. The higher charge leads to a spontaneous rehydration in normal atmospheric conditions following dehydration at less than 300 °C, and acceptance of only one layer of ethylene-glycol molecules in the interlayer space. Like smectite, vermiculite can be both dioctahedral and trioctahedral. Barshad and Kishk (1970) found that CEC of soil vermiculites are generally between 100 and 150 meq/100 g and could even be more than 200 meq/100 g for a total charge deficiency of about 0.8 per unit cell. The charge deficiency can be primarily in either the tetrahedral sheet or the octahedral sheet. Part of the tetrahedral negative charge in some soil vermiculites may be balanced by excess positive charge in the octahedral sheet. Al, Fe^{3+}, and Mg are the dominant octahedral cations, whereas Na, Ca, and K are the common interlayer cations. Some Al and Fe occurring in hydroxy forms may also occupy the interlayer sites, thereby causing a decrease in CEC and expandability of the layers.

The vermiculite minerals are mainly alteration products of biotites or chlorites and occur especially in weathering profiles and soils, but they may also result from alteration of illite. They could represent transitional phases, as progressive removal of interlayered water molecules leads to the development of a series of less hydrated stable phases. XRD determinations of smear slides show a basic [001] reflection at 14.3 Å which is shifted toward 15–16 Å by ethylene-glycol saturation. Heating at 300 °C shifts the 14 Å peak to about 11.6 Å, whereas heating at 500 °C shifts the 14 Å peak to about 10 Å (Walker 1961).

1.3.5 The Chlorite Group

The chlorite group is composed of negatively charged 2:1 mica-like units which are interlayered with positively charged octahedral sheets leading to a

2:1:1 structure. The octahedral sheets in both the 2:1 layer and the interlayer position consist predominantly of Fe^{2+} and Mg ions. The positive charge in the octahedral sheet is due to replacement of divalent cations by Al, Fe^{3+} and other high charge ions, and the negative charge in the 2:1 layer is primarily due to substitution of Al for Si in the tetrahedral sheet. The CEC of the chlorite minerals varies between 10 and 40 meq/100 g. Macroscopic chlorites are often known to have trioctahedral sheets and less commonly dioctahedral sheets. In soil chlorites, the interlayer octahedral sheets are imperfectly developed with presence of Al and Fe^{3+} hydroxy interlayers. Beside electrostatic bonds between the positively charged interlayer of the octahedral sheet and the negatively charged 2:1 layer, additional bonding comes from H bond between the basal oxygens of two 2:1 layers and the hydroxyls of the interlayer octahedral sheet adjacent to the two 2:1 layers.

The basic [001] reflection ranges from about 14 to 14.4 Å depending in part on the composition of the unit cell. The higher-order reflections occur at multiple integer of [001] reflections. The chlorite reflections are not affected by ethylene-glycol treatment, and the [001] spacing shifts only by a small distance when the sample is heated to about 500 °C. Difficulty arises in recognizing presence of kaolinite mixed with chlorite; but this may be overcome by treatment with hydrazine-monohydrate which causes a shift in the kaolinite [001] reflection from 7 to 10 Å, but no shift in the chlorite [002] reflection at 7 Å. Chlorite is generally soluble in moderately strong HCl, but not the kaolinite. A careful X-ray reading may also help to distinguish the [002] kaolinite reflection at 3.56 Å from [004] reflection of chlorite at 3.50 to 3.55 Å. As low-temperature chlorite often contains imperfect "brucite" sheets, the [001] reflections may slightly expand in ethylene-glycol treatment, but after the heat treatment at about 500 °C, the [001] spacing of the heated sample differs very little from that of the untreated one. Bailey and Brown (1962) recognized several different polytypes of chlorite, but the lack of pure chlorite sample in sedimentary materials makes identification of the polytype nearly impossible. A representative structural chemical formula of a Mg-rich chlorite is $(Si_{2.94}Al_{1.06})(Mg_{8.88}Al_{2.50}Fe^{2+}{}_{0.46}X_{0.16})O_{20}(OH)_{16}$.

Precise description of chemical compositions of chlorite minerals in soils and sedimentary rocks is difficult because the minerals seldom occur in these rocks as a single clay mineral phase. X-ray data may be used to infer relative abundances between Mg and Fe contents. Fe-rich chlorites have low intensity for odd order reflections, while Mg-rich chlorites have nearly the same intensity between the odd and even order reflections. Chemical and structural information about chlorites has been obtained primarily from analyses of high temperature macroscopic forms associated with metamorphic and hydrothermal rocks. Chlorites, although common in sediments and sedimentary rocks, have low abundances in these materials ($<15\%$). Neither the origin nor the composition of the clay-sized chlorites is well understood. Undoubtedly, some of them are detrital, but evidence also exists that chlorites in sediments can be diagenetic. These authigenic chlorites can range from a Mg-rich component (clinochlore) to a Fe-rich one (thuringite). Chlorite content has been found to increase in deep-burial diagenetic environments, possibly as a relict product of the conversion from smectite to illite. This late diagenetic chlorite appears to be Fe-rich, but Weaver et al. (1984) have shown that Mg contents in the chlorites tend to increase with increasing temperature. Although some of these late diagenetic chlorites appear to be trioctahedral in both the 2:1 unit and the interlayer site, chlorites with dioctahedral 2:1 layer and trioctahedral interlayer have been found in some K-bentonite beds, as reported by Weaver (1959). Chlorites in soils often have incomplete "brucite" interlayers as a result of varied degradation of the interlayer octahedral sheet. These less than perfect chlorites probably form from the precipitation of Mg, Fe, or Al hydroxy layers in the interlayer spaces of expandable smectite and vermiculite minerals. Several experimental studies have demonstrated precipitation of such hydroxy layers in the interlayer positions of expandable 2:1 layer minerals (Jackson 1963). At higher temperatures, under deep burial conditions, these hydroxy layers can be reorganized to brucite-like layers to form chloritic minerals. Many low temperature chlorites occur as mixed-layer or interstratified clay minerals.

1.3.6 The Mixed-Layer Group

The mixed-layer minerals, also called interstratified minerals, result from combination of two or more kinds of clay mineral species along the c-direction. They constitute one of the most abundant groups, if not the most abundant, of clay minerals in sedimentary materials, and were first described by Gruner (1934). Although stacking of different layer silicates can theoretically integrate a large number of different interstratified minerals, natural occurrences of such minerals are limited to only a few types of minerals. Mixed-layer clay minerals consisting of three or more components may not be uncommon, but detection of such complex combinations is difficult by the X-ray diffraction methods routinely used for identification of clay minerals.

Detailed description of stacking sequences requires knowledge of the proportion of each component in the bulk composition and the probability of each interface type in the mixed-layer crystallites. Two contrasting views have been presented as regard to the manifestation of different types of interstratification. By considering the most common interstratification in the form of illite/smectite mixed-layer minerals, Reynolds (1967) and Reynolds and Hower (1970) explained the mixed-layering in terms of neighbor-to-neighbor interactions among interlayers of the crystallites which are in general of the order of 100 nm in the a- and b-dimensions and of less than 10 nm in the c-direction. These crystallites are considered to consist of about 1 nm-thick (=10 Å) silicate layers containing K-bearing anhydrous illite interlayers and hydrous smectite interlayers. The neighbor-to-neighbor relations among the interlayers describe the degree of order ranging from random interstratification through partially ordered interstratification to perfectly ordered interstratification of the crystallites.

A different view of illite/smectite interstratification was presented by Nadeau et al. (1984, 1985), who, based on studies of dispersed illite/smectite particles by transmission electron microscopy, suggested that the illite/smectite mixed-layer minerals are composed of "fundamental particles" that have internal interlayers of illitic composition and are bound by hydrous interfaces which mimic smectitic interlayers. Following this model, two 2:1 silicate layers interlayered by K yield a fundamental particle of about 2 nm thick (=20 Å) and the stacking of such particles with layers of water molecules between them leads to IS-ordered illite/smectite clay minerals with about 50% illite. Similarly, three 2:1 silicate layers with two internal K-interlayers yield a fundamental particle of about 3 nm thick (=30 Å) and the hydrous interfacing of such particles with about 67% illite can be described as IIS-ordered illite/smectite clay minerals. An interfacing of 4 nm-thick (=40 Å) fundamental particles, each consisting of four 2:1 silicate layers with three internal K-interlayers, result in ISII-ordered illite/smectite minerals with about 75% illite.

Random to partially ordered sequences are the most common naturally occurring mixed-layer clay minerals. The random sequences are recognized by irrational series of X-ray reflections with the low angle reflection being at a position between the [001] reflections of the two components. The precise position of the low angle reflection and its intensity are determined by the d-spacings of the two components, the fraction of each component in the mixture, the degree of regularity in the stacking sequence, the structure factor, and the size of the mixed-layer particles. Common examples of random interstratifications are illite/smectite, glauconies (Fe-equivalents of illite/smectite mixtures), smectite/chlorite, and chlorite/vermiculite, of which illite/smectite is the most abundant. Kaolinite/smectite interlayerings are rare. Several different methods have been developed for the X-ray identification of the components and their proportions in the mixed-layer minerals (MacEwan et al. 1961; Reynolds 1980, 1985; Srodon 1980, 1981; Mossmann 1991).

Regular interstratified (perfectly ordered) minerals have a rational series of X-ray reflections with the spacing of the [001] reflection being equal to the sum of the d-spacings of each component. Examples of some common regular interstratified clay minerals are allevardite (illite/smectite) or rectorite (illite/ or mica/montmorillonite), and corrensite (trioctahedral chlorite/smectite) or tosudite (dioctahedral chlorite/smectite). Corrensite has a [001] reflection between 28 and 29 Å which shifts to about 30–31 Å upon ethylene glycolation and to about 24 Å upon heating at about 500 °C. Corrensite has been found as a weathered product of basalts and also occurs in evaporite, hydrothermal, and diagenetic deposits. Rectorite has been found in hydrothermal deposits.

Because of their common occurrences in a wide variety of shallow crustal environments, the illite/smectite mixed-layer minerals have become the subject of many mineralogical and chemical studies during the past 35 years. Srodon (1980, 1984) has described a satisfactory method for the identification of the illite/smectite mixed-layer minerals in terms of the regularity in the stacking sequence and the fraction of each component in the mixture. Srodon and Eberl (1984) examined the chemical data of several illite/smectite mixed-layer minerals in bentonites from different stratigraphic units and geographic locations. They noted that K contents and relative amounts of tetrahedral Al increase with the number of illite layers. However, the total charge, which varied from 1.06 to 1.64 per silicate layer, remained unaffected by the number of illite layers possibly due to variations in the composition and layer charge of the precursor smectite layers. The illitization process can be quite complex as evident from the results of a study by Huff and Turkmenoglu (1981), who reported a higher octahedral charge of about 0.98 per silicate layer relative to a tetrahedral charge of about 0.52 per silicate layer and also a high Mg content in the octahedral composition, suggesting the importance of the parental material and pore fluid composition during the process. Srodon and Eberl (1984) have shown that the concentration of K in the illite layer is varied depending on the type of interstratification. The concentration of K is about 1.1 per illite layer in randomly mixed material with more than 50% expandable layers, whereas it is nearly 2 per illite layer in the highly ordered ones. The authors determined that the end-member illite derived from the alteration of smectite should have a K average of about 1.5 per illite layer and a cation exchange capacity of about 15 meq/100 g. Velde and Bruzewitz (1981) concluded that the smectite in the illite/smectite mixed layers is usually of montmorillonitic composition with layer charges varying from about 0.6 to 1.3 per unit cell. Examining the relationship between the layer charge and the amount of fixed K in the illite/smectite mixed layers in bentonite deposits, Srodon et al. (1986) noted that illite layers in these minerals have two distinct amounts of charge deficiency. Gaultier and Mamy (1978) and later Eberl et al. (1986) reported that repeated wetting and drying of illite/smectite mixed layers, in the presence of K-rich environments, can produce illitization of high-charged smectite layers.

Bethke and Altaner (1986) summarized the relationship between expandability and order of stacking in illite/smectite mixed-layer minerals. They concluded that the mixed-layer minerals with more than 45% expandable layers are randomly interstratified, those with about 30 to 45% expandable layers are both randomly and orderly interstratified, and those with less than 30% expandable layers are only orderly interstratified. The change from randomly to orderly interstratified illite/smectite minerals remains a widely open question. Compositions of both solid and fluid, temperature, and time are considered to be the major factors in the course, together with the rate of illitization of smectitic layers. Knowledge of these factors is important in the reconstruction of history of sedimentary basins. Keller et al. (1986) recognized textural differences between the illite/smectite mixed layers with less than 60–70% illite layers and those with more than 60–70% illite layers, which is often the interval for the formation of the ordered type. Pollastro (1985) and Whitney and Northrop (1987) also produced evidence of textural changes in the illite/smectite minerals at advanced illitization stage.

The reaction mechanism during illitization remains poorly understood. Based on results of an isotopic experimental investigation, Whitney and Northrop (1988) suggested that illitization is governed by two modes of reaction mechanism. When the illite/smectite mixed-layer minerals consist of 60% or more expandable layers, the process of illitization is thought to begin to occur along a transformation reaction path which involves resetting of about 65% of the oxygen in each layer. When the process has advanced to the point of an occurrence of less than 30% expandable layers, the process seems to be dominated by a neoformation reaction path in which the illite layers are isotopically 100% reset.

1.3.7 The Sepiolite-Palygorskite Group

Palygorskite and sepiolite are 2:1 layer silicates with channel-like structures resulting from alternate bonds of 2:1 units as the directions of the apices of the tetrahedra are periodically reversed along the b-axis. The channels created are filled with loosely bound water molecules and also structural bound water molecules which attach to the octahedral layers. Palygorskite and sepiolite differ in size of the unit structure. They, especially palygorskite minerals, have been known for years, as Brongniart (1807) made the first description as "asbestos-like forms".

Sepiolite contains nine octahedral-cation positions filled primarily by Mg. Palygorskite has five octahedral-cation positions filled by both Al and Mg. Sepiolite gives a major [001] peak at 12 Å, whereas palygorskite has a major reflection at 10.4 Å. The ideal structural formula of sepiolite following Nagy-Bradley's model is $Si_{24}Mg_{12}O_{60}(OH)_{12}(OH_2)_8 12H_2O$, and that of palygorskite $Si_{16}Mg_{10}O_{40}(OH)_4(OH_2)_8 8H_2O$. In palygorskite, the octahedral Al/Mg ratio is near unity, unlike in sheet structures where this ratio is either higher or lower than 1. Both types of minerals yield similar values for their CEC which is between 20 and 300 meq/100 g. Palygorskite has been found in marine coastal areas, as well as in deep marine sediments. Like sepiolite, it has also been reported in sediments of alkali lakes and in soils of semiarid to arid regions.

Martin-Vivaldi and Cano-Ruiz (1956) suggested that palygorskite could derive from smectite as the former occupies the region of discontinuity in the octahedral sheet between the dioctahedral montmorillonite and the trioctahedral saponite. The idea was that the structural motif of these minerals is energetically favorable for accommodation of certain ratios of octahedral cations to vacancies in a regular pattern. Later studies by MacKenzie et al. (1984) and Paquet et al. (1987) showed that the octahedral compositions of the smectite and fibrous clay overlap considerably. This observation is of interest in view of the fact that smectite in sedimentary environments, or illite in diagenetic and hydrothermal environments, can also occur as laths which represent an energetically favorable crystal form.

1.4 Thermodynamic Considerations for Clay Minerals

Major questions that are frequently addressed are whether a particular clay mineral is in equilibrium with solutions in any given soil environment, or stable in a given depositional or burial environment. Clay minerals formed in continental environments often find their way into marine environments. The fate of terrigenous clay minerals in marine environments has long been a central question in studies about the origin of clay minerals in sedimentary environments. Logarithmic chemical activity diagrams have proved very useful in interpreting both the controls of the solution chemistry on equilibrium clay mineral assemblages in geologic environments and the stability relations among clay minerals. Much disagreement exists about the characteristics of mixed-layer phases, such as illite/smectite mixed-layer clays, whether they behave as polyphase aggregates or solid solutions in geochemical processes. Activity diagrams generated from thermodynamic data have provided comprehensive understanding of the chemical behavior of mixed-layer clays in nature. Garrels and Christ (1965) have discussed in detail the procedures involved in constructing such diagrams for clay minerals. Aagaard and Helgeson (1983) and several others have recommended some modifications about slopes and intercepts of the stability field boundaries in activity diagrams because of the nonstoichiometric compositions for the clay minerals in natural systems. Despite the limitation that standard molal Gibbs free energies of formation for nonstoichiometric clay minerals have to be considered from values for clay minerals with rather idealized stoichiometry, which makes the activity diagrams open to question for true representation of natural materials, these diagrams demonstrate close correspondence between predicted phase relations and those that are observed in sediments at surface or moderately buried conditions.

Helgeson et al. (1970) presented an analysis of stable clay mineral assemblages in different natural systems from a consideration of equilibrium phase relations in the system K_2O-Al_2O_3-SiO_2-H_2O and Na_2O-Al_2O_3-SiO_2-H_2O at 0 °C and 1 atm (Figs. 1.7 and 1.8). Because of the range in compositions for the surface sea waters, the deep ocean waters and the sediment pore waters both in deep oceans and shelf areas, the assemblage of mica (illite)-kaolinite-Na-montmorillonite-quartz is in equilibrium in some parts

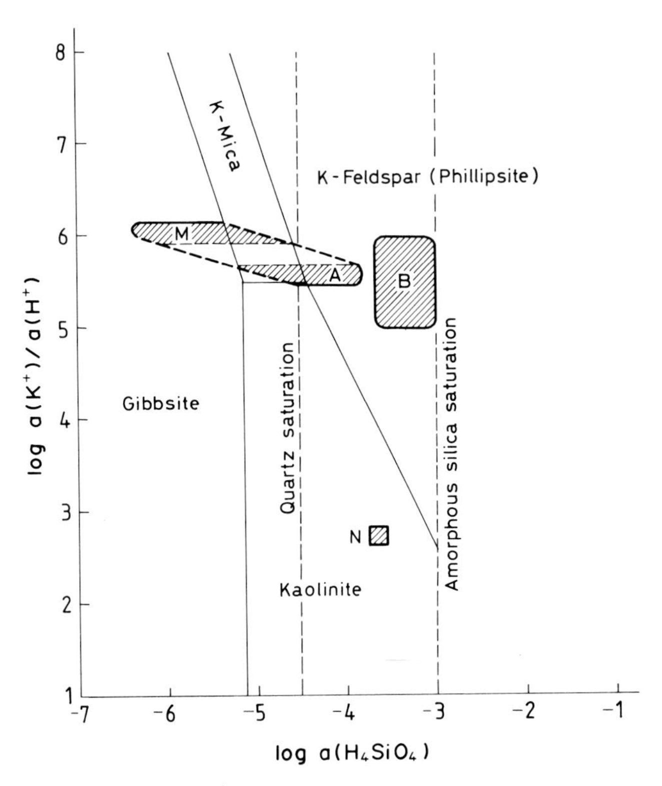

Fig. 1.7. Logarithmic activity diagram depicting equilibrium phase relations among clay minerals and other silicate minerals and an aqueous phase at 0 °C, 1 atm, and unit activity of water. The *shaded areas labeled M and A* represent the compositional range of surface and deep ocean waters, respectively. The area *B* represents *the compositional range of some sediment pore waters in deep sea and shelf areas*. The area *N* represents *the average composition of world streams*. (Helgeson et al. 1969)

of the ocean, whereas in other parts the assemblages of illite-Na-montmorillonite-K-feldspar (phillipsite)-quartz and of kaolinite-Na-montmorillonite-K-feldspar may be in equilibrium. Where sea water is enriched in dissolved silica, the equilibrium assemblage will consist of K-feldspar (phillipsite) and Na-montmorillonite. In areas of the ocean where high silica activity is due to the presence of submarine volcanics or to local river water contributions, detrital kaolinite or illite will react with high-silica sea water to produce smectite or zeolite. Interstitial waters in deep sea sediments may experience increased silica concentration due to dissolutions of diatoms incorporated in the sediments. As a result, kaolinite and K-mica or illite may react with pore water to form smectite. If the silica content is very high, zeolites may form at the expense of clay minerals. Helgeson and Mackenzie (1970) contended, using activity diagrams for stabilities of different clay minerals, that much of the detrital clay materials presently delivered into the ocean has a miner-

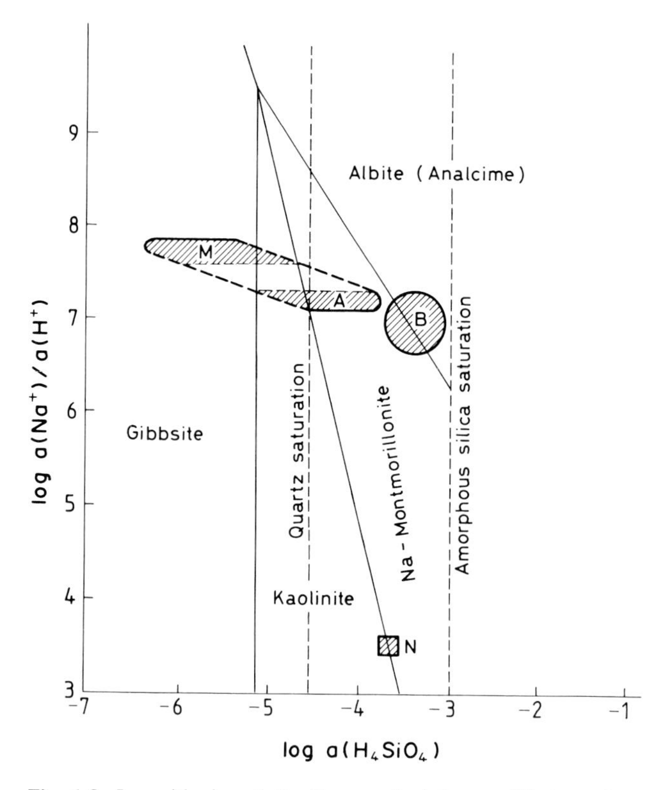

Fig. 1.8. Logarithmic activity diagram depicting equilibrium phase relations among clay minerals and other silicate minerals and an aqueous phase at 0 °C, 1 atm, and unit activity of water. For the significance of the *shaded areas*, see Fig. 1.7. (Helgeson et al. 1969)

alogic composition which is in equilibrium with waters in various parts of the ocean basin, suggesting that without any appreciable amount of reaction the clays should essentially pass through the ocean water before being buried beneath the sea floor (Fig. 1.9).

The relative stability and the formation of minerals in weathering environments has been widely debated. Much of the disagreement has come from lack of understanding of the relative importance of controls on variations in physical and chemical conditions in these weathering environments. The use of logarithmic activity diagrams has served as a convenient means of charting the course of reactions in response to variations in the chemical environments within weathering profiles. Kittrick (1971) and Rai and Lindsay (1975) stressed that Al, instead of being commonly regarded as an inert or inactive component, should be regarded as a mobile component in soil environments. They recommended the use of Al as one of the variables or part of a dependent variable in the construction of relative stability diagrams for different phyllosilicates. Clay mineral distributions in poorly drained

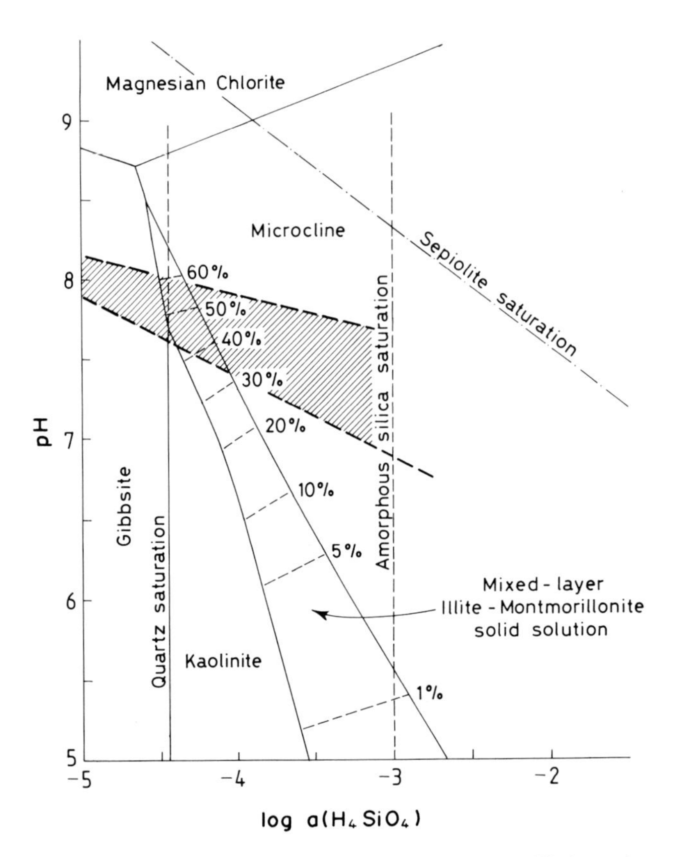

Fig. 1.9. Logarithmic activity diagram depicting equilibrium phase relations among aluminosilicates and sea water at 0 °C, 1 atm, and unit activity of water. The activities of Na^+, K^+, Ca^{2+}, and Mg^{2+} used in constructing the diagram are that of sea water. The *hatched area* represents the compositional range of sea water at this temperature, and the *dot-dash lines* indicate the composition of sea water saturated with quartz, amorphous silica, and sepiolite, respectively. The *dashed contours* designate the composition (in percent illite) of the mixed-layer illite/montmorillonite solid solution in equilibrium with sea water. (Helgeson and Mackenzie 1970)

soils are known to be different from those in well-drained soils. The difference is clearly explained in terms of the levels of activities of different chemical species. For example, the levels of activities of H_4SiO_4 and other soluble species are low in well-drained soils, because of the high solubility of these species relative to Al and Fe. By contrast, the activities of the silica and other soluble cations are maintained at high levels in poorly drained soils. As portrayed in Fig. 1.10A, for a poorly drained soil with pH of 6 and high activity of silica, the relative stability of secondary minerals increases in the order: chlorite, halloysite, gibbsite, illite, dickite, beidellite, kaolinite, and montmorillonite. At the same pH but with very low silica activity, as in

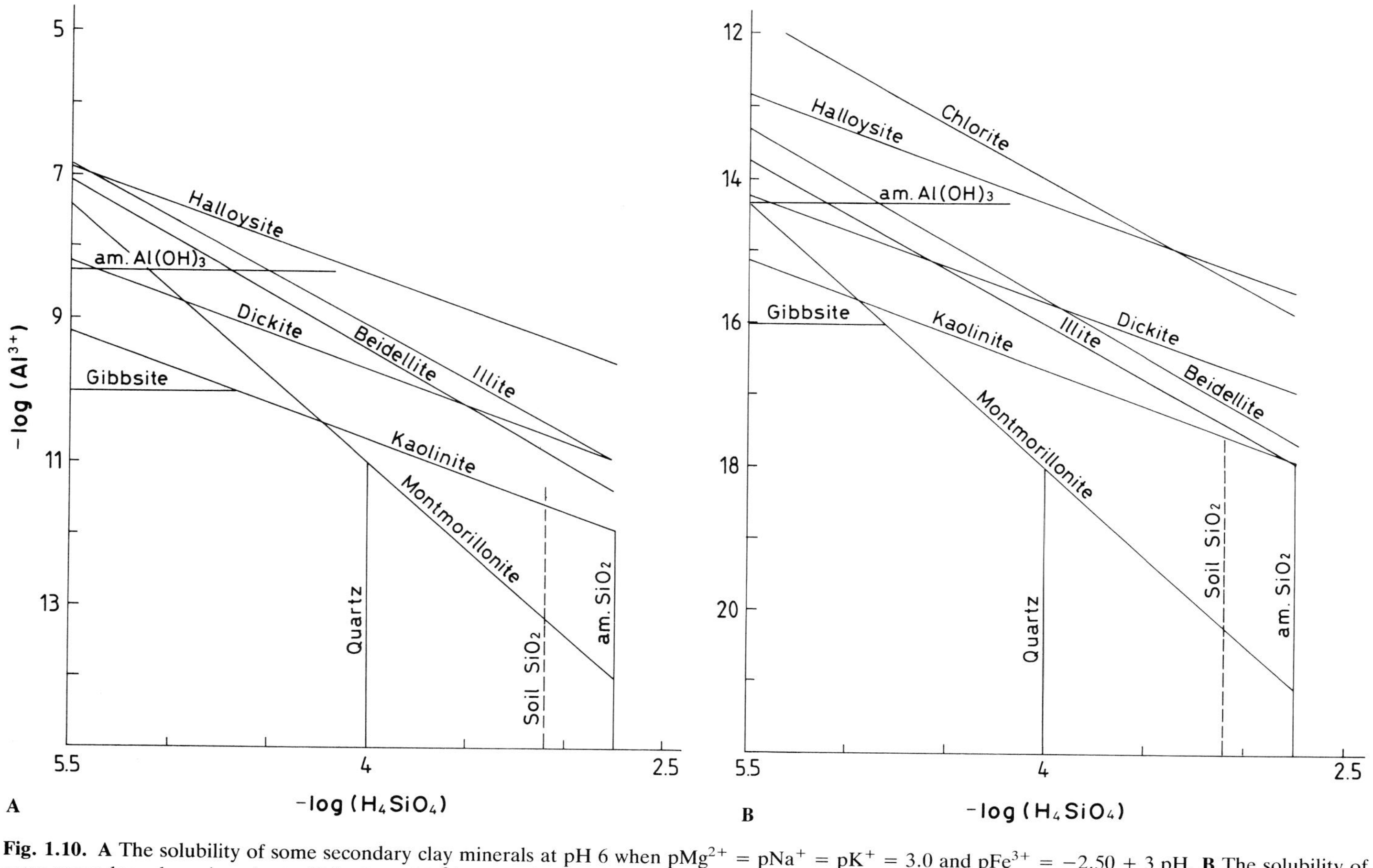

Fig. 1.10. A The solubility of some secondary clay minerals at pH 6 when $pMg^{2+} = pNa^{+} = pK^{+} = 3.0$ and $pFe^{3+} = -2.50 + 3$ pH. **B** The solubility of some secondary clay minerals at pH 8 when $pMg^{2+} = pNa^{+} = pK^{+} = 3.0$ and $pFe^{3+} = -2.50 + 3$ pH. (Rai and Lindsay 1975)

a well-drained soil, the mineral stability increases in the order: chlorite, halloysite, illite, beidellite, montmorillonite, and gibbsite. The effect of change from an acidic to an alkaline soil environment on the relative stability of clay minerals also becomes readily apparent by the use of activity diagrams. For example, at pH 8 and high silica activity, chlorite becomes relatively more stable than halloysite and bedeillite becomes more stable than illite, as the relative stability of secondary minerals increases in the order: halloysite, chlorite, gibbsite, dickite, beidellite, illite, kaolinite, and montmorillonite (Fig. 1.10B). Hence the logarithmic activity diagrams constructed from thermodynamic data provide a fundamentally sound basis for understanding the relative stability and transformation of clay minerals in soil environments.

1.5 Clay Separation and Characterization

We are not overstating our claim that a conventional crushing using combined steps of a jaw-crusher and a disc-grinder or a ball-mill crusher produces an overcrushing of most of the larger grains of a rock, creating artificially clay-sized particles which become intimately mixed with the natural clay component of the rock. Therefore, the mass of clay separated from such a crushed sample becomes variously contaminated by the presence of fine fragments of relatively coarser detrital grains composed of such minerals as feldspar, muscovite, and biotite, among many others. The presence of even a small amount of these detrital minerals in the clay sample can greatly influence the isotopic date of the fraction. The effects of these contaminations on the isotopic data of clay fractions have been recently documented by Liewig et al. (1987a), who strongly recommended the use of a gentle sample disaggregation method of repetitive freezing-thawing rock pieces of centimeter size placed in distilled water.

Separation and purification of clay fractions may be improved by the use of a specific chemical reagent which causes no significant damage to the isotopic system of the minerals. A leaching of clay by dilute HCl or HAc is usually performed when carbonate minerals are present. The relative impacts of the use of such reagents on the isotopic composition of clays will be discussed later, but, in general, a careful use of acids is definitely advantageous to the goals of the isotopic investigations. Leaching by H_2O_2 has been widely used to remove organic matter from sedimentary rocks. Although Douglas and Fiessinger (1971) have shown that the intensity of X-ray reflections of clay particles decreases dramatically after H_2O_2 treatment, we believe that this decrease in intensity is not a sign of clay degradation, but rather an indication of clay disorganization causing no significant change in the isotopic data, because the organic matter seems to bind the elementary clay particles together.

Clay fractions are appropriately identified and quantified in terms of abundance of the different minerals in the isolated fraction using X-ray diffraction technique. Additional information such as crystallinity index, polymorphic type, and population of the octahedral sheets can also be obtained by X-ray analyses. Knowledge of the morphology of the clay particles by electron microscopic data is also necessary in any study of clay material, because this information helps to differentiate generations and types of clay minerals in the analyzed materials. All these mineralogic data should be integrated with both major and trace element data, whenever possible. The Sm-Nd isotope method applied on clays should also be completed with analyses of the abundances of the rare-earth elements.

2 Principles of Isotope Geochemistry

Since the pioneer studies on glauconies by Cormier (1956) and Wasserburg et al. (1956), many different isotopic methods have proved useful in studies of clay minerals. Rb-Sr, K-Ar, Sm-Nd, oxygen, and hydrogen analyses are now being routinely carried out on such minerals, whereas new grounds are being explored for $^{40}Ar/^{39}Ar$ and Pb-Pb isotope compositions in clays. The basic principles in the different isotopic methods are discussed here, additional details and geochemical behaviors of the elements being given in several publications (Hoefs 1980; Faure 1986; DePaolo 1988; Geyh and Schleicher 1990).

2.1 Fundamentals of Isotope Geochemistry

The total number of different nuclei on Earth and the surrounding atmosphere and stratosphere is about 1700, of which only 260 are stable. Those nuclei containing the same number of protons and neutrons are most stable. The unstable nuclei decay spontaneously to reach a stable configuration. These unstable nuclei are called radioactive isotopes and their spontaneous transformation is radioactivity. The decay of a nucleus is accompanied by emissions of particles, such as α, β, γ and positrons, and heat. Spontaneous fission of nuclei can also occur for heavy isotopes. In *Isotope Geology*, we are mainly interested in that limited number of naturally occurring radioactive isotopes, and their daughter products, whose half-lives are comparable to the age of the Earth.

The rate of decay of a nucleus of a radioactive species is proportional to the number of its atoms (N) present at a given moment t, expressed by:

$$-dN/dt = \lambda N,$$

where λ is the decay constant. The integration of the above equation gives:

$$t = 1/\lambda \ln(1 + D/N),$$

where D is the number of daughter isotopes formed at t. This equation may be used to obtain a meaningful age for a geologic material, provided it has met some strict requirements under the mathematical formulation. A rock or a mineral which is to be dated must have behaved since its formation as a closed system against gain or loss of both the radioactive or parent and the radiogenic or daughter isotopes. Furthermore, the amounts of both parent and daughter isotopes and the decay constant should be precisely known. The amount of daughter isotopes present at time $t = 0$ should also be known so that the appropriate amount of the daughter isotope generated in the system since its formation can be precisely calculated.

Natural fractionation of isotopes is not detectable for nuclei having masses over about 40; but for nuclei of low atomic number elements such as H, C, O, and S, natural processes induce significant amount of isotopic fractionation, causing wide variations in these isotopic compositions among terrestrial materials. The fractionations are caused by kinetic effects which are commonly associated with evaporation, diffusion, and "vital" effects, and also by thermodynamic equilibrium effect resulting from varied bond energies among the isotopes.

2.2 Radiogenic Isotope Geochemistry

2.2.1 The Rb-Sr Method

The Rb-Sr method is based on the change in the amount of ^{87}Sr due to the natural decay of ^{87}Rb to ^{87}Sr accompanied by ß emission. Natural Sr has four isotopes: ^{84}Sr, ^{86}Sr, ^{87}Sr, and ^{88}Sr, whereas Rb has two isotopes ^{85}Rb and ^{87}Rb which is radioactive. The decay constant of ^{87}Rb is $1.42 \times 10^{-11}\,a^{-1}$. The Sr isotopic composition of a natural material is expressed by the ^{87}Sr/^{86}Sr ratio. For the dating purpose, the ^{87}Rb/^{86}Sr and the ^{87}Sr/^{86}Sr ratios are determined for each sample. The ^{87}Rb/^{86}Sr ratio is a measure of the Rb/Sr ratio of each sample. The basic equation for age determination is as follows:

$$(^{87}Sr/^{86}Sr)_{total} = (^{87}Sr/^{86}Sr)_{initial} + {}^{87}Rb/^{86}Sr\,(e^{\lambda t} - 1).$$

Knowledge of the initial ^{87}Sr/^{86}Sr ratio is helpful to determine the age of each sample. Because such knowledge cannot be easily gained for geologic systems, the difficulty in dating may be overcome by the use of an "isochron" method. The use of this method consists in plotting analytical points in a rectangular diagram with the Y-axis representing the ^{87}Sr/^{86}Sr ratios and the X-axis the ^{87}Rb/^{86}Sr ratios. Cogenetic samples with the same initial ^{87}Sr/^{86}Sr but different ^{87}Rb/^{86}Sr or Rb/Sr ratio during their formation, and with the history of maintaining closed system to both Rb and Sr since the formation, will have their analytical points displayed along a line in such a X-Y diagram. This line is called an isochron (Fig. 1.11), and its slope is equal to ($e\lambda^t - 1$) from which t is calculated. A minimum number of three to four samples with sufficient spread in the Rb/Sr ratios are needed for a well-defined isochron line.

A line in an isochron diagram is not necessarily an isochron, as it can also result from mixing in varied amounts of two components which are isotopically and chemically heterogeneous. In such a case of mixing, the line is a "mixing line" and the slope has no geological meaning. The problem involving sedimentary materials is therefore to gather criteria that will be helpful to distinguish between a mixing line and an isochron line.

The $^{87}Sr/^{86}Sr$ ratio at the time of formation of a clay mineral phase may be used to depict the origin of the fluid genetically associated with the phase, because the Sr isotopic compositions of common upper crustal minerals and fluids may be defined within broad limits. When the material analyzed contains some Rb and is old enough to have accumulated radiogenic ^{87}Sr, a correction for radiogenic growth of ^{87}Sr has to be made to determine the initial Sr isotopic ratio of the considered material. This correction, however, is not needed, either for any material formed in Recent times or for any material of Phanerozoic age when its Rb/Sr ratio is less than 0.01, such as in many carbonate and sulfate minerals.

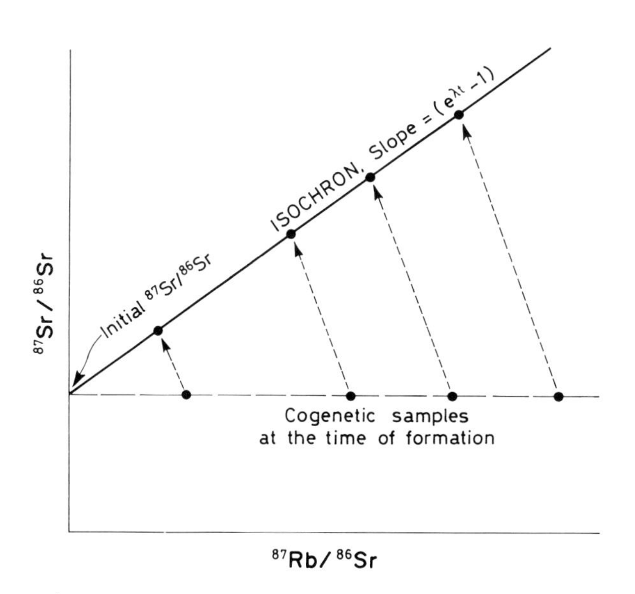

Fig. 1.11. An illustration of a Rb-Sr isochron diagram

2.2.2 The K-Ar Method

The K-Ar method is based on the natural decay of ^{40}K to ^{40}Ar accompanied by ß emission and electronic capture. However, approximately 89% of ^{40}K decays to ^{40}Ca with ß emission. Although this dominant decay scheme can be potentially another useful isotopic method for dating geological materials, the natural abundance of ^{40}Ca is such (96.98%) that it causes extreme difficulty in measuring with high precision the growth of radiogenic ^{40}Ca in most terrestrial materials. This K-Ca method is now applicable to a very limited number of minerals which are highly depleted in Ca, but slightly enriched in K. The K-Ar method, on the other hand, is widely applied to almost any type of minerals and rocks. For the conventional K-Ar method, two splits of the same sample are needed to determine K and Ar separately, the former by atomic absorption, flame spectrometry, or isotope dilution, and the latter by mass spectrometry following its extraction by fusion of a sample under vacuum. As required for all isotopic methods of dating, a closed-system condition of the material is essential. This is especially crucial for the K-Ar method of dating, because Ar, being a noble gas, is not strongly bonded to other atoms or ions in a mineral phase, rendering most minerals highly susceptible to Ar loss under increase in temperature.

As ^{40}K decays to both ^{40}Ar and ^{40}Ca, each branch of the decay system has its decay constant separate from the other. The decay constants are 0.581×10^{-10} and 4.962×10^{-10} a^{-1} for the decay by electron capture and by ß emission, respectively (Steiger and Jäger 1977). The total decay constant of ^{40}K, λ, is equal to $(0.581 + 4.962) \times 10^{-10}$ a^{-1}. The equation of the K-Ar age is:

$$t = 1/\lambda\ (^{40}Ar^{*}\lambda_{total}/^{40}K\ \lambda_{capt.\ elec.}) + 1.$$

The value of the initial $^{40}Ar/^{36}Ar$ ratio during the growth of a mineral has to be known. In most cases, we assume that it is identical to that of the atmosphere at 295.5 (Nier 1950). In general, the amounts of initial Ar are very low in K-rich clay minerals and hence the uncertainty in the value of the initial Ar isotopic ratio has no major influence on the age calculation for very old rocks or minerals. The initial $^{40}Ar/^{36}Ar$ ratio for clay minerals can be higher than 295.5, but the consequences of such high initials can be corrected using the isochron technique, the isochron being defined by $^{40}K/^{36}Ar$ as the abscissa and $^{40}Ar/^{36}Ar$ as the ordinate. A fraction of the non-radiogenic Ar in any mineral could be due to contamination of present-day atmospheric Ar which is adsorbed by the mineral particles during preparation and purification of the samples. The effect of adsorbed Ar is negligible for most plutonic minerals when they are preheated under vacuum before Ar extraction, but this adsorbed Ar can be a problem with fine particles such as clay particles. Adsorption of large amounts of atmospheric Ar decreases the accurracy of the technical data, and an initial $^{40}Ar/^{36}Ar$ different from that of the present-day atmosphere can induce mixing lines in isochron diagrams. The isotopic age of the analyzed minerals can subsequently be modified by way of an increase or a decrease.

Evaluations of the K-Ar isochron technique were given by McDougall et al. (1969), Shaffiqullah and Damon (1974), and Hayatsu and Carmichael (1977). In most examples of clay-isotope dating, the isochron approach is not needed for old minerals which contain high amounts of K_2O, leading to high $^{40}K/^{36}Ar$ and high $^{40}Ar/^{36}Ar$ ratios (Clauer et al. 1985b). The uncertainty of the regression calculation is then too high to present any advantage relative to the conventional age calculation. In that case, the average of the individual K-Ar apparent ages, assuming an initial $^{40}Ar/^{36}Ar$ of 295.5, may be compared to the age calculated by regression. Some studies reported isochrons with abnormally low or even negative values for the intercepts (Bonhomme et al. 1978; Langley 1978). Obviously, the initial $^{40}Ar/^{36}Ar$ of a mineral cannot be negative, and lines with such intercept values can only be mixing lines with meaningless slopes. Hunziker (1986) observed that fine fractions ($<0.6\,\mu m$) of sediments define a straight line in the $^{40}Ar/^{36}Ar$ and $^{40}K/^{36}Ar$ coordinates, probably describing an isochron, whereas the coarse fractions ($>0.6\,\mu m$) increased in their $^{40}Ar/^{36}Ar$ ratios with hardly any increase in the $^{40}K/^{36}Ar$ ratios (Fig. 1.12). This trend for the coarse fractions must be related to mixing of various amounts of detrital components with varied $^{40}Ar/^{36}Ar$ ratios, but similar K/Ar ratios. Clearly, a linear trend with a negative intercept can result from mixing of the coarse fractions having a limited range in the $^{40}Ar/^{36}Ar$ ratio with a fine fraction having relatively much lower $^{40}Ar/^{36}Ar$ and K/Ar ratios (Fig. 1.12). High initial Ar isotopic values can occur in the case of clay minerals that generated in diagenetic conditions with the $^{40}Ar/^{36}Ar$ ratios of the environment being significantly above the atmospheric $^{40}Ar/^{36}Ar$ value. The $^{40}Ar/^{36}Ar$ ratio of the environment can indeed be higher than that of the atmosphere, as suggested by the evidence that the $^{40}Ar/^{36}Ar$ ratios of some natural gases in sedimentary rocks were found to range from about 600 to about 1800 (N. Clauer, unpubl. data).

Harper (1970) presented theoretical examples by which the ^{40}Ar-^{40}K data yield a negative intercept of ^{40}Ar. He suggested that: (1) these negative values had to be interpreted as consequences of a ^{40}Ar loss from cogenetic crystalline phases irrespective of their K contents, and (2) the slope of the line represents the age of the mineral crystallization. It is very difficult to consider that any mineral phase contains less than 0% ^{40}Ar during crystallization or recrystallization. A more rational cause for a line in a ^{40}Ar vs. ^{40}K diagram to yield a negative value for the intercept is a rotation due, for instance, to an artificial mixing of inhomogeneous materials as discussed above (Fig. 1.13A), or to a differential loss of ^{40}Ar by clay minerals differing in the K_2O contents and hence in their Ar retentivity, during a recrystallization event (Fig. 1.13B). If the event of disturbance is strong enough to completely recrystallize all mineral phases, the linear trend of the data

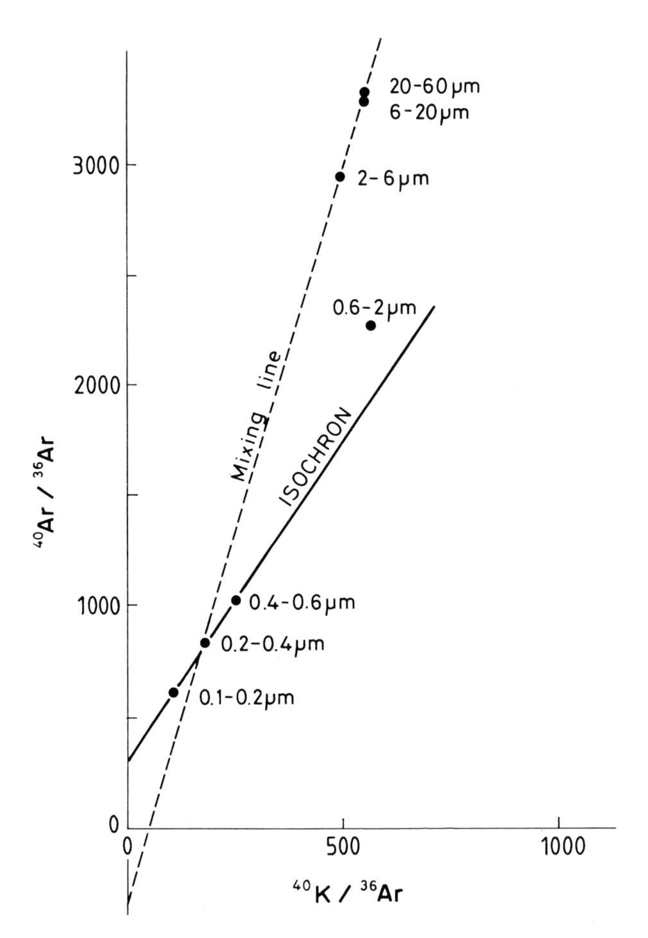

Fig. 1.12. K-Ar isochron plot for particles of different size fractions of the same sample. Data points of the fine fractions define an isochron, whereas mixing between coarse and fine fractions gives a line with a negative intercept. (Hunziker 1986)

points should pass through 0 or yield a positive value for its intercept in a ^{40}Ar vs. ^{40}K diagram.

2.2.3 The ^{40}Ar/^{39}Ar Method

The ^{40}Ar/^{39}Ar technique is a recently extended variation of the ^{40}K/^{39}Ar technique (Sigurgeirsson 1962; Merrihue and Turner 1966), but this technique has so far been of limited use in studies of sedimentary and low-grade metamorphic rocks. The principle is that ^{39}K decays to ^{39}Ar by irradiation under fast neutrons in a reactor. The technique theoretically has the advantages of having the measurements of both radioactive parent and radiogenic daughter isotopes made from the same aliquot of the sample.

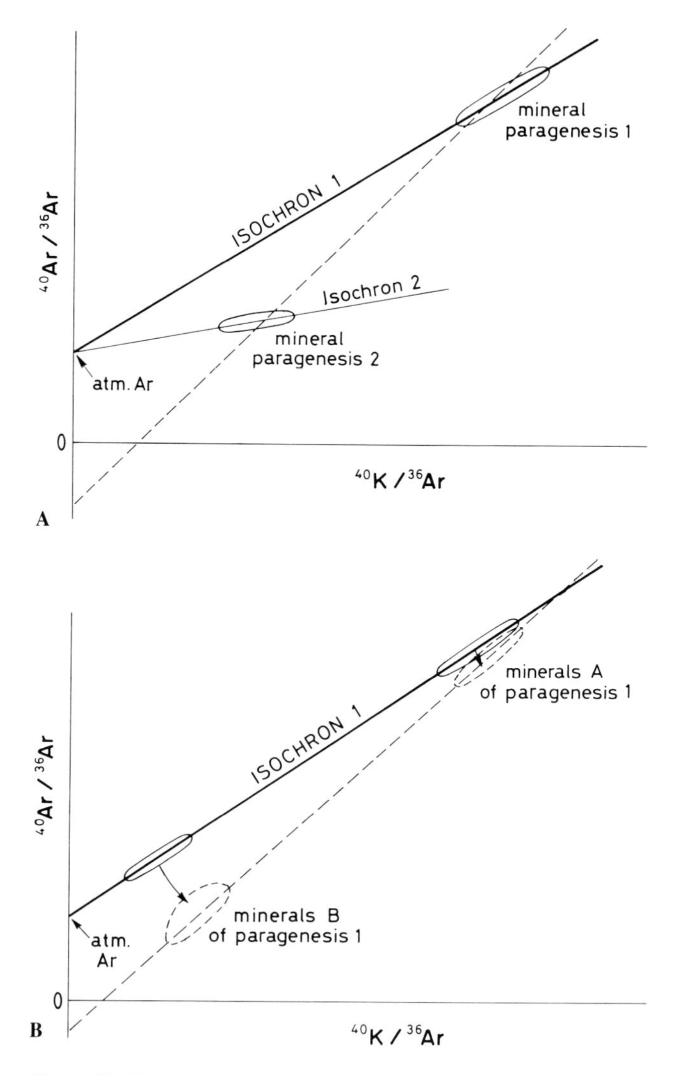

Fig. 1.13. Theoretical mixing lines with negative intercepts in K-Ar isochron diagrams. **A** The *line* results from mixing of minerals belonging to two parageneses; **B** the *line* results from differential loss of radiogenic ^{40}Ar from minerals having different Ar retention capabilities depending on their crystallinity. *atm. Ar* stands for atmospheric Ar

This avoids the uncertainty from sample heterogeneity, the radioactive and radiogenic isotopes being measured simultaneously by a single isotopic determination of the Ar, with a high precision by using the most abundant ^{39}K isotope (^{39}K = 93.26%, while ^{40}K represents only 0.01%). The simultaneous measurement of the parent and daughter isotopes of the same powder split allows analysis by stepwise progressive heating, leading to a separation of the Ar released at low temperature from that released at higher temperatures. However, the analyses are not without some inconveniences and uncertainties which include the effect of recoil on the retention of Ar, the knowledge of the flux of neutrons, and the correction necessary to apply for the production of ^{39}Ar from the neutron irradiation of ^{39}Ca. The amount of Ca in the analyzed minerals has, therefore, to be known precisely. Recoil effects can be especially significant when the particles are small, because the isotopes which are impacted by the neutrons are moved in the lattices and can even be expelled out of the structure.

Kunk and Brusewitz (1987) have precisely documented the problem of ^{39}Ar recoil effect on an illite/smectite mixed-layer mineral placed in a quartz vial before irradiation. Before starting any step heating, they measured that about 50% of ^{39}Ar was extracted from mineral structure by neutron activity (Fig. 1.14). The step-heating apparent ages gave subsequently high values between 700 and 800 Ma. The ^{40}Ar/^{39}Ar total gas age including the ^{39}Ar released into the vial during irradiation gave about 389 Ma which is slightly above the K-Ar conventional value at 340 Ma. This difference suggests that ^{39}Ar may be lost in varied amounts from the clay minerals due to recoil effect. Glauconies, for instance, seem to suffer loss of ^{39}Ar due to the recoil effect (Yanase et al. 1975; Brereton et al. 1976; Foland et al. 1984; Klay and Jessberger 1984). Structural alterations of glauconies attendant with irradiation in a reactor may explain the loss of ^{39}Ar.

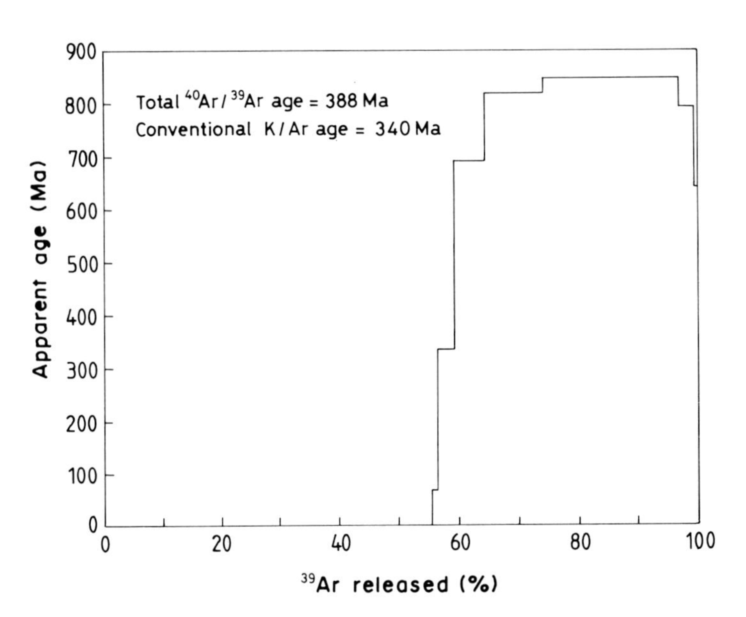

Fig. 1.14. ^{40}Ar/^{39}Ar incremental release age spectrum of an illite/smectite mixed-layer mineral placed in a quartz vial before irradiation. The first 55% of released ^{39}Ar are due to recoil effect. (Kunk and Brusewitz 1987)

The recoil-related loss of ^{39}Ar may not be the rule for all clay minerals, because Burghele et al. (1984), Hunziker et al. (1986), Kligfield et al. (1986), Reuter and Dallmeyer (1987a, b), and Dallmeyer et al. (1989) reported cases of similar conventional K-Ar and $^{40}Ar/^{39}Ar$ ages for different size fractions of illitic micas from very-low to low-grade metamorphic domains. Reuter and Dallmeyer (1989), for instance, documented the recoil effect on the redistribution of ^{39}Ar from high-K bearing illite phases to low-K bearing chlorite or albite phases (Fig. 1.15). This recoil and redistribution effect seems to happen preferentially in particles with large surface areas and poor grain edges (Fig. 1.16). Therefore, the potential application of the $^{40}Ar/^{39}Ar$ technique to well-crystallized, diagenetic illite for dating purpose rests on clear understanding of the behavior of large particles (in the 10 μm range) during irradiation, inasmuch as these clay particles might have crystallized in a temperature range above that of the irradiation conditions. Additional studies regarding the recoil-related loss of radiogenic ^{40}Ar by clay minerals are warranted. Systematic studies like those of Foland et al. (1984), Hunziker et al. (1986), and Reuter and Dallmeyer (1987a, b) should be pursued before any expectation of routine application of the $^{40}Ar/^{39}Ar$ method on sediments can materialize.

A potential of the $^{40}Ar/^{39}Ar$ method also lies in the prospect of in situ measurements of grains by coupling the mass-spectrometer to a laser microprobe (York et al. 1981; Sutter and Hartung 1984). Some results of laser $^{40}Ar/^{39}Ar$ dating of illite associated with uranium deposits in northern Saskatchewan are already available (Bray et al. 1987). However, the in situ $^{40}Ar/^{39}Ar$ data are more scattered than the corresponding K-Ar data from extracted clay particles for reasons which still have to be understood.

2.2.4 The Sm-Nd Method

Rare-earth elements (REE) represent a group of elements having very similar chemical properties due to the very similar nature of electronic configurations of these elements. All REE have the same +3 oxidation state, except Ce and Eu, which can also have +4 and +2 oxidation states, respectively. These differences suggest that Ce and Eu can be more mobile than the other REE, or depending on the Eh of the environment, that they are able to be scavenged by some specific mineral phases. Despite the close similarity in the behavior of the REE with +3 oxidation states, they can be fractionated from each other by combined effects of their complexing properties with different organic and inorganic ligands and structural properties of the minerals.

Sedimentary rocks commonly have low Sm/Nd ratios, averaging at about 0.21 for shales, 0.20 for graywackes, 0.23 for sandstones and limestones, 0.34 for organic-rich sediments (Fleet 1984; McLennan 1989). The various processes controlling mobilization and fractionation of the REE in

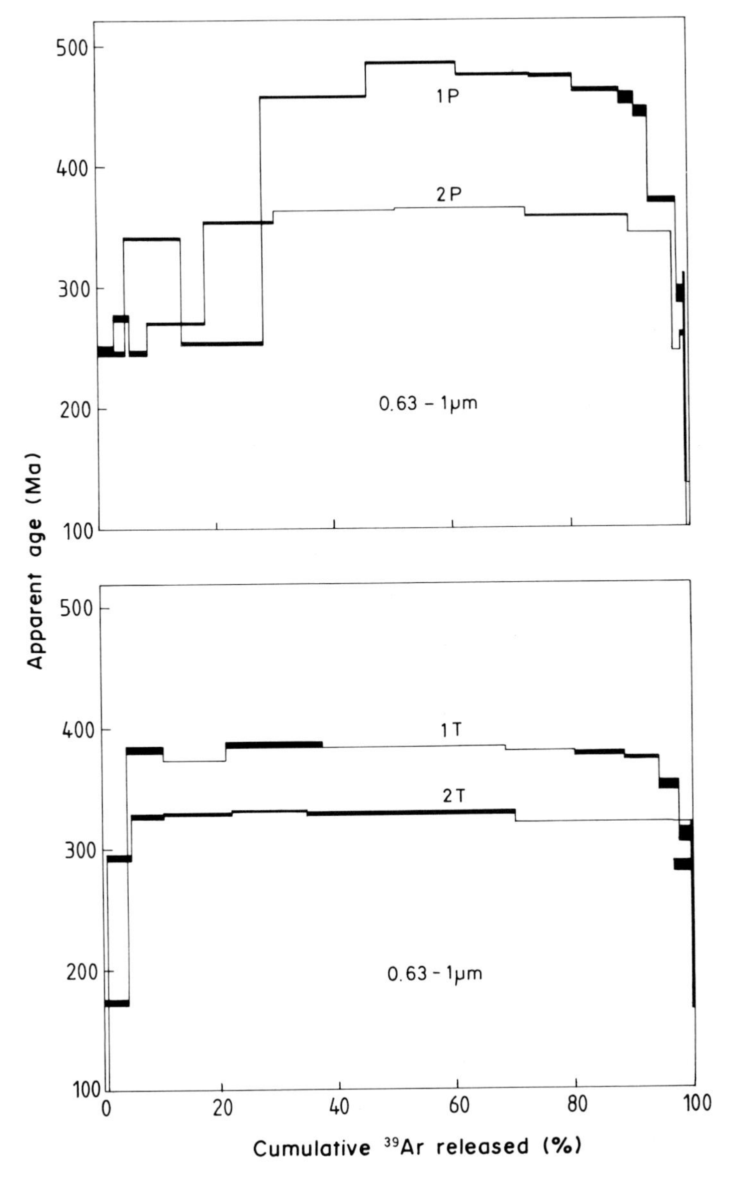

Fig. 1.15. $^{40}Ar/^{39}Ar$ incremental release spectra of illite-rich size fractions. *Top* The two size fractions are representative of an anchizonal (1 P) and an epizonal metapelite (2 P); *Bottom* the two size fractions are representative of an anchizonal (1 T) and an epizonal metatuff (2 T). (Reuter and Dallmeyer 1989)

sedimentary environments have been discussed in a number of articles (Bonnot-Courtois 1981; McLennan 1989; Chaudhuri et al. 1992a). Surficial inorganic and organic processes are known to have major impacts on the fractionations of the REE (Balashov et al. 1964; Ronov et al. 1967; Cullers

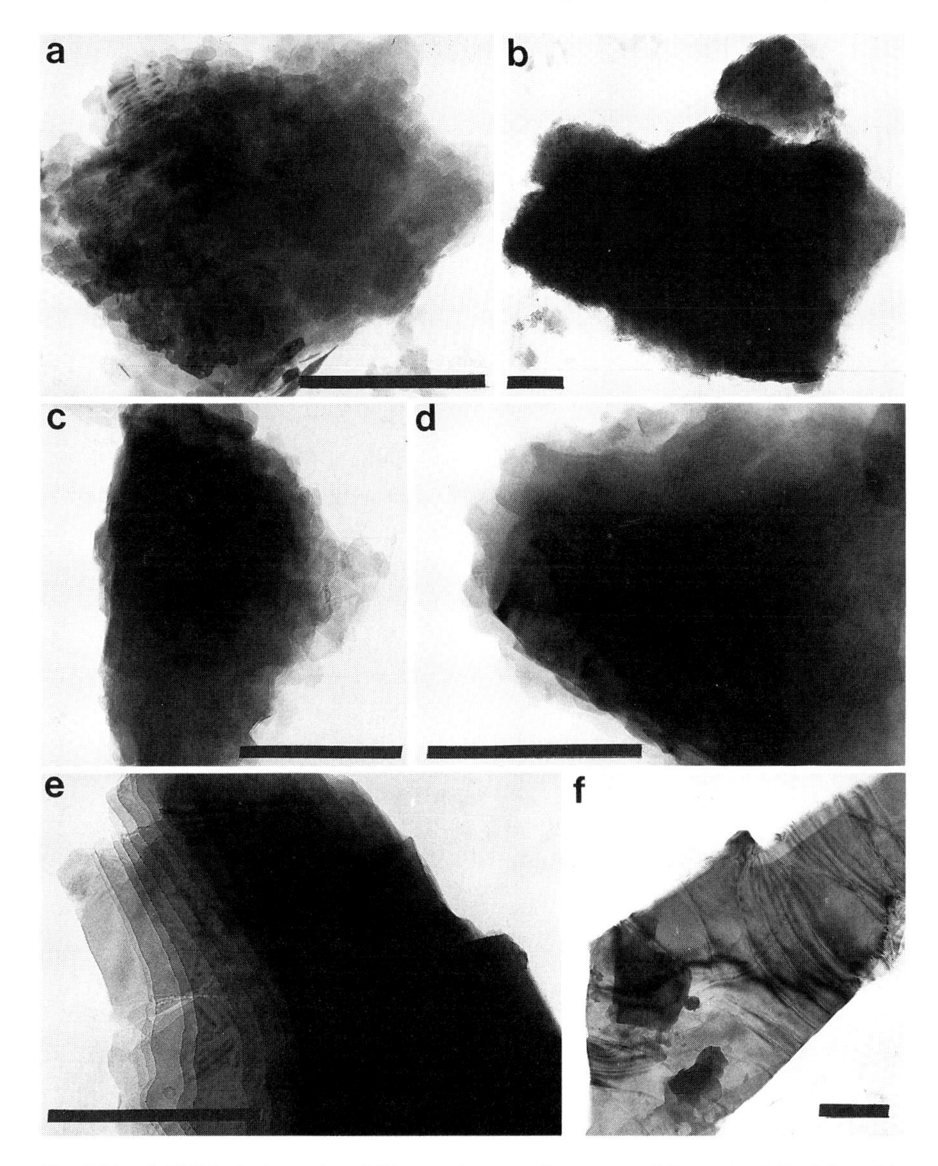

Fig. 1.16a–f. TEM photographs of illite- and muscovite-type particles. **a** to **d** Particles with poorly defined edges that are typical of detrital illite minerals. **e**, **f** Particles with straight edges which belong to authigenic well-crystallized illite to muscovite-type particles. *Bars* 0.5 μm

et al. 1975; Nesbitt 1979). The diagenetic process appears to have little effect on the fractionation of these elements in the bulk clay samples (Chaudhuri and Cullers 1979). The effect of fractionation produced mostly at the surface is often lost in the bulk sediments or sedimentary rocks as a result of mechanical mixing of the various components in the depositional basin.

Samarium has seven natural isotopes of which ^{147}Sm and ^{148}Sm are radioactive. As the half-life of ^{148}Sm is too long to be useful for dating terrestrial materials, the decay of ^{147}Sm to ^{143}Nd with α emission and a half-life of 1.06×10^{11} years is used for isotopic dating of rocks and minerals. Neodymium has seven isotopes, and the Nd isotopic composition of any material is expressed by its $^{143}Nd/^{144}Nd$ ratio. Because of the very long half-life, the Sm-Nd method is well suited for dating very old materials.

The equation used for age determination by the Sm-Nd method is expressed as:

$$(^{143}Nd/^{144}Nd)_{total} = (^{143}Nd/^{144}Nd)_{initial} + (^{147}Sm/^{144}Nd)(e\lambda^{t} - 1).$$

An isochron for cogenetic samples may be constructed by using the $^{143}Nd/^{144}Nd$ ratio as the ordinate and the $^{147}Sm/^{144}Nd$ ratio as the abscissa. The age is calculated from the slope of the line defined by the data points of the samples analyzed. As natural materials commonly have a narrow range in the Sm/Nd ratios, an isochron for cogenetic clay minerals has yet to be demonstrated. Hence, model ages are often calculated to understand the genetic history of sedimentary materials. The calculation of model ages requires the assumption that the Nd isotopic composition of a crustal material derived from a reference source that had a linear growth in the $^{143}Nd/^{144}Nd$ ratio during the last 4.5 Ga. The chondritic reservoir, also called CHUR (Chondritic Uniform Reservoir), is often used as the reference source (DePaolo and Wasserburg 1976). The equation for a model age is given as:

$$t = 1/\lambda \ln(^{143}Nd/^{144}Nd)_{sample} - (^{143}Nd/^{144}Nd)_{CHUR}/(^{147}Sm/^{144}Nd)_{sample} - (^{147}Sm/^{144}Nd)_{CHUR} + 1.$$

The present-day $(^{143}Nd/^{144}Nd)_{CHUR}$ and $(^{147}Sm/^{144}Nd)_{CHUR}$ ratios are 0.512638 and 0.1967, respectively (Wasserburg et al. 1981). The Nd isotopic composition of a sample at any time t may also be expressed in terms of per mill difference from the isotopic value of CHUR, at the time, which is given as:

$$\varepsilon_{Nd} = (^{143}Nd/^{144}Nd)_{sample} - (^{143}Nd/^{144}Nd)_{CHUR}/(^{143}Nd/^{144}Nd)_{CHUR} \times 10^{4}.$$

2.2.5 The U-Th-Pb Method

Lead is widely distributed throughout the Earth and occurs not only as the radiogenic daughter product of U and Th (U-Th-Pb method), but also forms its own minerals from which U and Th might be excluded. The isotopic compositions can therefore vary widely from highly radiogenic Pb in very old U, Th-bearing minerals to the common Pb in galena that have low U/Pb and Th/Pb ratios. Both U and Pb occur as trace elements in most common rocks and minerals. The abundance of U and Pb in clays are 1.5–4.0 and $<$10–140 µg/g, respectively (Wedepohl 1978).

Minerals may contain three radioactive isotopes of uranium: ^{238}U, ^{235}U, and ^{234}U and one radioactive isotope of thorium, ^{232}Th. The decays of ^{238}U, ^{235}U, and ^{232}Th to the stable isotopes of ^{206}Pb, ^{207}Pb, and ^{208}Pb, respectively, are occurring through a long chain of disintegrations, ^{234}U being an intermediate daughter product in the decay chain of ^{238}U. The time-dependent growths of Pb isotopes from radioactive decays of the U and Th isotopes are expressed as:

$$(^{206}Pb/^{204}Pb)_{total} = (^{206}Pb/^{204}Pb)_{initial} + (^{238}U/^{204}Pb)(e^{xt} - 1),$$
$$(^{207}Pb/^{204}Pb)_{total} = (^{207}Pb/^{204}Pb)_{initial} + (^{235}U/^{204}Pb)(e^{yt} - 1),$$
$$(^{208}Pb/^{204}Pb)_{total} = (^{208}Pb/^{204}Pb)_{initial} + (^{232}Th/^{204}Pb)(e^{zt} - 1),$$

where x, y, and z are the decay constants of ^{238}U ($1.55125 \times 10^{-10}\,a^{-1}$), ^{235}U ($9.8485 \times 10^{-10}\,a^{-1}$), and ^{232}Th ($4.9475 \times 10^{-10}\,a^{-1}$), respectively. The equations for the growths of ^{207}Pb and ^{206}Pb may be combined to express the ^{207}Pb-^{206}Pb equation:

$$(^{206}Pb/^{204}Pb)_{total}/(^{206}Pb/^{204}Pb)_{total} = (^{235}U/^{238}U)(e^{yt} - 1)/(e^{xt} - 1),$$

where the ratio of $^{235}U/^{238}U$ is a constant with the value of 1/137.88 for terrestrial materials with normal U isotopic composition.

The dating of rocks and minerals by the individual U-Pb isotope method often gives highly discordant ages, commonly as a result of differential losses of radiogenic Pb during the geologic history of the materials; but a reliable age may be obtained by the use of the combined ^{207}Pb-^{206}Pb data. The age of a suite of cogenetic minerals with the same initial Pb isotopic composition may be theoretically determined from the slope of a Pb-Pb isochron defined by the $^{207}Pb/^{204}Pb$ ratio as the ordinate and the $^{206}Pb/^{204}Pb$ ratio as the abscissa. The age will not be affected by either loss or gain of U or loss of Pb by the clay minerals due to weathering or any other crustal processes in very recent geologic time.

Clay minerals which are nearly depleted in U but enriched in Pb may also be analyzed for a model age of crystallization, using a model with single or multiple stage evolution of Pb. A single-stage model assumes that the isotopic compositions of the common Pb, such as that in the Pb-sulfide minerals, evolved in different environments with different U/Pb and Th/Pb ratios from the same primordial isotopic value at 4.55×10^9 a until the Pb is separated from the different sources to form Pb minerals. The Pb isotopic compositions of different samples of minerals or rocks would then define an isochron in the $^{207}Pb/^{204}Pb$ and $^{206}Pb/^{204}Pb$ coordinates, and an age can be calculated from slope of the line. The two-stage evolution model of Stacey and Kramers (1975) is a frequently used multi-stage model. It assumes that the Pb evolved first from primordial isotopic value between 4.55 and 3.7 Ga in a reservoir with a $^{238}U/^{204}Pb$ ratio of 7.19 and a $^{232}Th/^{204}Pb$ ratio of 32.21, and then, as a result of differentiation at 3.7 Ga, in a reservoir with a $^{238}U/^{204}Pb$ ratio of 9.735 and a $^{232}Th/^{204}Pb$ ratio of 36.937 until the different types of Pb were separated from these sources. An isochron defined by the

$^{207}Pb/^{204}Pb$ and $^{206}Pb/^{204}Pb$ data relate to the time elapsed since the Pb in a sample was isolated from the second reservoir.

Very large uncertainties exist as to whether or not clay minerals have the ability to meet the requirement that their U-Pb systems remained closed to both U and Pb since crystallization. Clay minerals are well known for their high ion-exchange capacity, and this inherent property alone casts a large shadow on the usefulness of the Pb-isotope method to studies of clay genesis. Pb-Pb isotopic data on sedimentary materials are still too few to assess the full potential for an extended application.

2.3 Stable Isotope Geochemistry

Isotopic variations among stable isotopes of light elements (those with atomic numbers less than 40) in natural materials arise from kinetic and equilibrium isotopic effects. Common examples of kinetic isotope effects are those which result from different translation velocities of molecules with different isotopic masses, as in diffusion, where lighter molecules preferentially diffuse out relative to heavier molecules from a system or where lighter molecules escape preferentially over heavier molecules during evaporation. Kinetic isotope effects can be significant also in the case of bond dissociation by bacterial reactions. The equilibrium isotopic effects are produced by variations on bond energy associated with thermodynamic equilibrium in a chemical reaction. In many low-temperature processes, kinetic isotope effects are important controls on the variations in isotopic abundance, and the magnitude of kinetic isotopic effects can be significantly higher than that of equilibrium isotopic effects.

The equilibrium isotopic effect comes into play when isotopic substitution, such as the substitution of a light isotope by a heavy isotope, occurs, producing major changes in the vibrational levels and the "zero point energy"; but differences in the energy between isotopic species are considerably diminished at high temperatures. A given molecule with light isotopes has a higher vibrational frequency, and hence a higher zero-point energy, than a similar molecule with heavier isotopes. A consequence is that the bonds of lighter isotopes are weaker than those of heavier isotopes, making the molecules with lighter isotopes more reactive. In equilibrium isotope exchange reactions, the energy change is the difference in the zero point energy between two molecular species and, consequently, the energy change is much smaller in comparison to free energy change in a chemical reaction. The magnitude of equilibrium isotope effects may be expressed in terms of isotope fractionation factor between the phases involved in thermodynamic equilibrium. The fractionation factor, α, between phases A and B can be stated as:

$$\alpha_{A-B} = (R_A/R_B),$$

where R_A is the ratio of heavy to light isotopes of an element in phase A, that is $^{18}O/^{16}O$ or D/H in phase A, and R_B is the same isotopic ratio in phase B. In general, the fractionation factor and the thermodynamic equilibrium constant for the isotopic exchange reaction between two phases, where isotopes are randomly distributed among all sites in the molecules, are related by:

$$\alpha = K^{1/n},$$

n being the number of atoms exchanged in the expressed reaction and K the equilibrium constant for the isotope exchange reaction. Customarily, the exchange reactions are expressed for a single atom exchange so that α becomes equal to K. This equilibrium constant for the isotope exchange reaction is temperature-dependent, like all equilibrium constants. At temperatures down to as low as 0 °C, $\ln \alpha$ is proportional to $1/T^2$ where T is the absolute temperature. The relationship between α and T is commonly expressed as:

$$1000 \ln \alpha = A(10^6/T^2) + B,$$

where A and B are constants. The stable isotope composition of a substance is expressed with greater facility by comparison with the abundance of a reference material than by using the absolute isotope ratio which is always very small. Hence the isotopic abundance of a sample is commonly expressed by delta (δ) values in per mill given as:

$$\delta_{sample} = R_{sample} - R_{standard}/R_{standard} \times 1000,$$

where R is the ratio of the heavy to light isotopes. Standard material commonly used for both oxygen and hydrogen isotope analyses is Standard Mean Ocean Water (SMOW), and for carbon isotope determinations belemnite powder from Cretaceous PeeDee Formation (PDB). The equilibrium isotope fractionation factor between two phases, α_{X-Y}, and their isotopic abundance values, δ_X and δ_Y, bear the following relationship:

$$\alpha_{X-Y} = 1000 + d_X/1000 + d_Y.$$

As α in most cases are close to unity, $1000 \ln \alpha_{X-Y}$ is approximately equal to $(\delta_X - \delta_Y)$.

Although temperature is a very dominant factor which affects vibrational energies or partition functions of atoms in substances, and hence the fractionation factors in the isotopic exchange between two phases, chemical compositions of crystal structures of solids can also have varied degrees of influence on the fractionation factors. High vibrational frequencies correspond to those bonds with low atomic masses and high ionic potential and, hence, to lower the free energy in the isotopic exchange reaction, heavy isotopes are preferentially incorporated by the substance. An Al-rich layer silicate mineral should theoretically have greater preference for heavy isotopes of oxygen than an identical silicate mineral with Mg ions because of the higher measured ionic potential of Al. This compositional effect is illustrated by the study of Lawrence and Taylor (1971), who reported that Fe-rich smectites were depleted in ^{18}O by about 2–3 per mill to Fe-poor smectites. Also, O'Neil et al. (1969) found that $CaCO_3$ is enriched in ^{18}O by about 1 per mill relative to $SrCO_3$ and by 3 per mill relative to $BaCO_3$ at 25 °C. Similarly, Tarutani et al. (1969) noted that Mg-rich $CaCO_3$ is enriched in ^{18}O relative to Mg-poor $CaCO_3$ by about 0.6 for each mole percent $MgCO_3$ in $CaCO_3$ formed at 25 °C. These examples illustrate that the fractionation of oxygen isotopes is dependent on the composition of cations to which the oxygens are linked. The hydrogen isotopes bonded to oxygen in the cation-oxygen link is also fractionated due to change in the vibrational frequencies associated with OH. Taylor and Epstein (1966) and Brigham and O'Neil (1985) demonstrated that the hydrogen isotope properties are significantly affected by the presence of Fe in octahedral sites in layer silicate minerals, as the deuterium content of octahedral OH tends to increase with increase in the Mg/Fe or Al/Fe ratios of these minerals.

Aside from chemical bonding, crystal structure may also have some influence on the fractionation of the oxygen and deuterium isotopes. In general, well-ordered and close-packed structures preferentially concentrate heavy isotopes. O'Neil (1968) found ^{18}O fractionations by about 3.0 per mill and D fractionations by about 19.5 per mill between ice and water at 0 °C. Much smaller oxygen isotope fractionations have been found in many natural materials which have different structural forms. Tarutani et al. (1969) noted a small ^{18}O fractionation of 0.5–0.6 per mill between calcite and aragonite at 25 °C. Murata et al. (1977) reported small amounts of oxygen isotope fractionation among various diagenetic silica minerals, such as biogenic opal, cristobalite, and cryptocrystalline quartz. Thus, isotope fractionation effects due to crystal structure factor are generally small as compared to those due to differences in chemical composition.

The equilibrium oxygen and hydrogen isotope fractionation factors between different clay minerals and water at different temperatures have been summarized by Kyser (1987). The dependence between oxygen and hydrogen isotope fractionations for illite and other clay minerals and surficial temperature are illustrated in Fig. 1.17. Based on the present-day geothermal gradient and an interpolation between the illite-water and smectite-water fractionations, $\delta^{18}O$ values of pore waters at various depths can be calculated, using the following fractionation equations:

$$1000 \ln \alpha_{\text{illite-water}} = 2.43 \times 10^6 T^{-2} - 4.82$$
$$1000 \ln \alpha_{\text{smectite-water}} = 2.67 \times 10^6 T^{-2} - 4.82.$$

During D/H fractionation between clay minerals and water, the hydroxyls of the minerals are depleted in D relative to water, and the degree of depletion increases with decreasing temperature. D/H fractionation for smectite-water between 0 and 100 °C has been determined empirically.

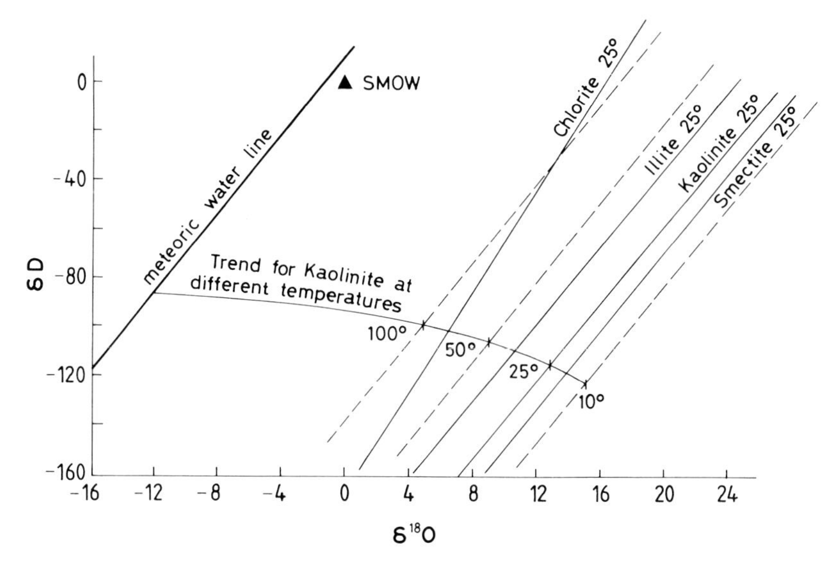

Fig. 1.17. Oxygen and hydrogen isotope compositions of different clay minerals relative to that of meteoric water under equilibrium conditions at 25 °C. The *dashed lines* illustrate the oxygen and hydrogen isotope compositions of kaolinite relative to that of meteoric water at different temperatures. The *curved line* indicates the trend in the changes of the isotope compositions of kaolinite with changes in temperature relative to a fixed isotope composition of meteoric water. (Kyser 1987)

3 Specific Aspects of Clay Isotope Geochemistry

Appropriate interpretation of isotopic results of clay minerals requires some familiarity with the ease by which the different isotopes may be removed from clay minerals during their natural history since formation, or during their laboratory treatment, in connection with the isotopic analyses. Any selective removal of an isotope, either a parent or a daughter one, from clay minerals due to a natural or a laboratory cause must be clearly identified to provide the correct interpretation of the isotope data. A blanket application of a popular belief may not be altogether justified in a particular situation and may even lead to the failure of recognition of many different subtleties in the natural process or processes during the growth history of the clay minerals. The discussion which follows addresses specific areas concerning whether or not clay minerals meet the isotopic requirements of a closed system condition in response to forces of natural, post-deposition conditions, or laboratory conditions.

3.1 Retentivity of Radiogenic Argon

Since the beginning of K-Ar isotopic studies on clay minerals, many have raised the question about the use of K-Ar data on clay minerals for dating purposes, because their small size might render them incapable of acting as closed systems especially with respect to Ar, which is vulnerable to diffusive loss. Similar skepticism has surfaced about the retention of radiogenic Sr by clay minerals. The ground for skepticism was largely created by results of preliminary studies made on glauconitic materials in the late 1950s and the early 1960s. These studies found that the isotopic "ages" of glauconitic materials are often lower than the stratigraphic age (Hurley et al. 1960). An explanation that has been commonly conveyed for these low ages is that Ar is readily lost from the minerals. We are discussing the merits of such an explanation in light of several experimental studies.

3.1.1 The Problem of Ar Diffusion

Considering that the rates of a dissolution reaction of a glauconitic mineral will be different if the K ions have different coordinations in the mineral structure, Thompson and Hower (1973) conducted an experiment of leaching glauconite minerals with 0.5 N HCl to test the hypothesis of linking low ages to varied K coordinations in the mineral structure, and hence to varied degrees of retention of radiogenic ^{40}Ar by the mineral. A congruent dissolution of a mineral follows a first-order reaction which is characterized by a linear trend between the time and the log normal of the fraction of the original amount of an element still remaining in the mineral structure. The slope of the line gives the rate constant. An incongruent dissolution of a mineral would be marked by two or more different first-order reactions, each having a constant characteristic rate. In the simplest case of two first-order reactions for an incongruent dissolution of a mineral, the two linear trends in the relationship between the log normal of the fraction of the original amount remaining and the time for the two first-order reactions will be manifested as a composite curve, the early part of which being dominated by the relatively more rapid dissolution component (Fig. 1.18).

Thompson and Hower (1973) observed from their experimental study that K from glauconites with low expandabilities is removed at three different rates, indicating that it is present in three different coordinations in the structure of these minerals. Thompson and Hower noted that 75 to 98% of K in all the samples were removed at the lowest rate, K_1 (Fig. 1.19). The authors attributed this K to the position of 12-fold coordinations in the mica-like layers. The K that was removed with the highest dissolution rate, K_3, was considered to have come from either the hydrated smectitic layer or the surface of crystallites. The K with the intermediate removal rate, K_2, was thought to have come from the hydrated layers. However, unlike the one

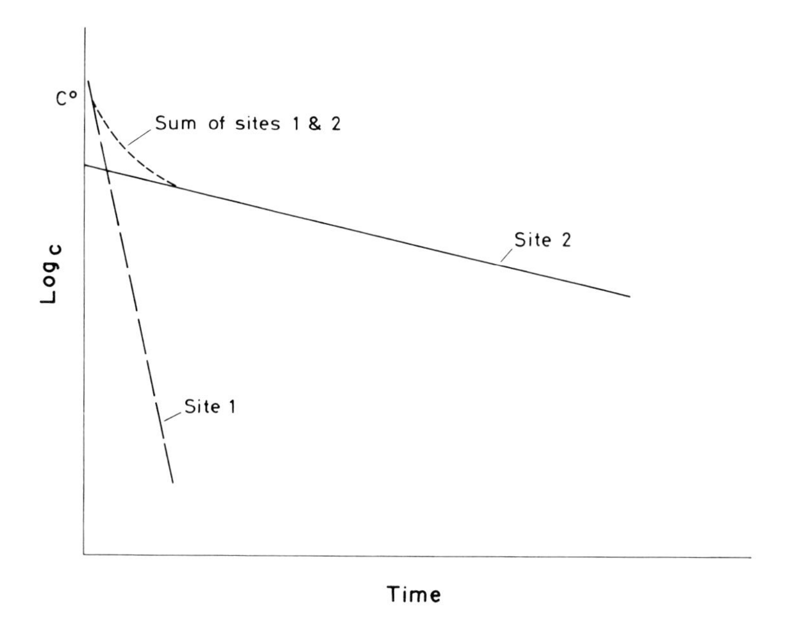

Fig. 1.18. Log-normal concentration of K following different times of leaching of a mineral. The trend reflects different solubility rates of K due to different locations in the mineral. (Thompson and Hower 1973)

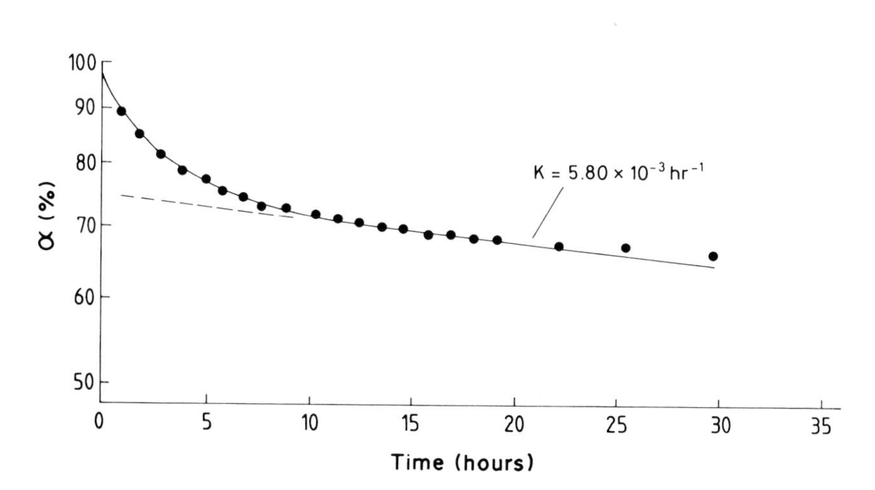

Fig. 1.19. Experimental data showing the lowest rate of K removal out of glauconite minerals, α being the amount of K remaining in the mineral structures relative to time. (Thompson and Hower 1973)

with the highest removal rate, K_3, the K with the intermediate removal rate was thought to have come from the hexagonal holes in the basal plane of the silica tetrahedral layers surrounded by six oxygens. Based on the experimental data, the authors devised a correction factor which may be applied to compensate for the loss of the daughter isotope in the calculation of the ages.

Aronson and Douthitt (1986) performed an experiment to test the Thompson-Hower (1973) model of different coordinations for K in mixed-layer mica-like minerals. The experiment involved measurements of K contents and K-Ar ages of a sample of illite/smectite mixed-layer mineral as it is progressively dissolved in HNO_3 at two different concentrations. The acid dissolution curves for K indicated two different rates of reaction, a slow rate, K_1, and a rapid rate, K_2. The dissolution curve was dependent on the acid normality, but independent of the acid type (Fig. 1.20). The extrapolation of the linear asymptote of the dissolution curve back to time t

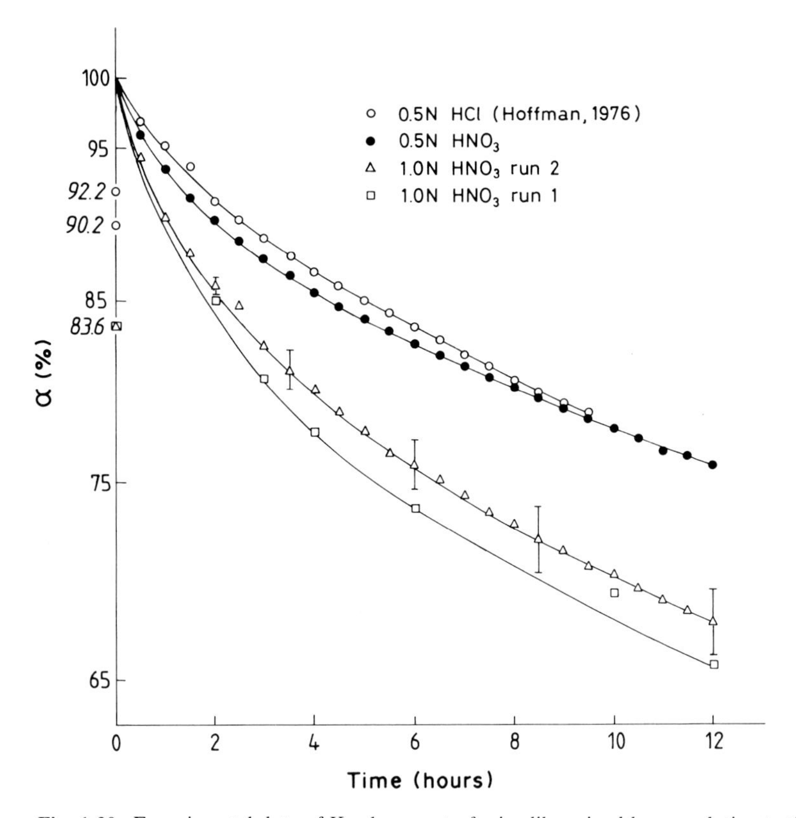

Fig. 1.20. Experimental data of K release out of mica-like mixed-layers relative to time. The K dissolution curves are dependent on acid normality but not on acid type. (Aronson and Douthitt 1986)

= 0 indicated that about 91% of the K belong to the slow dissolution mode upon reaction with 0.5 N HNO_3, whereas about 84% of the K belong to the slow reaction category when reacted with 1.0 N HNO_3. Although the acid strengths varied, the slopes of the slow rate dissolution curves were nearly identical at the two different acid strengths. Aronson and Douthitt further suggested that the slow reaction component is a composite of two subcomponents with different reaction rates, the slower one of the two containing as much as 83% K. Hence, the results of the study on the illite/smectite mixed-layer clay by Aronson and Douthitt (1986) are similar to those on glauconites by Thompson and Hower (1973), signifying that these mica-like minerals have inherent structural inhomogeneities for K retention and hence for Ar retention.

The data on the acid-treated residues in Aronson and Douthitt's (1986) study indicated that K_2O decreases exponentially from the initial value of about 5.5% to a near constant value of about 4.8% at a somewhat earlier period with 1.0 N than with 0.5 N acid dissolution, and that ^{40}Ar decreased similarly but with considerable scatter (Fig. 1.21). Aronson and Douthitt

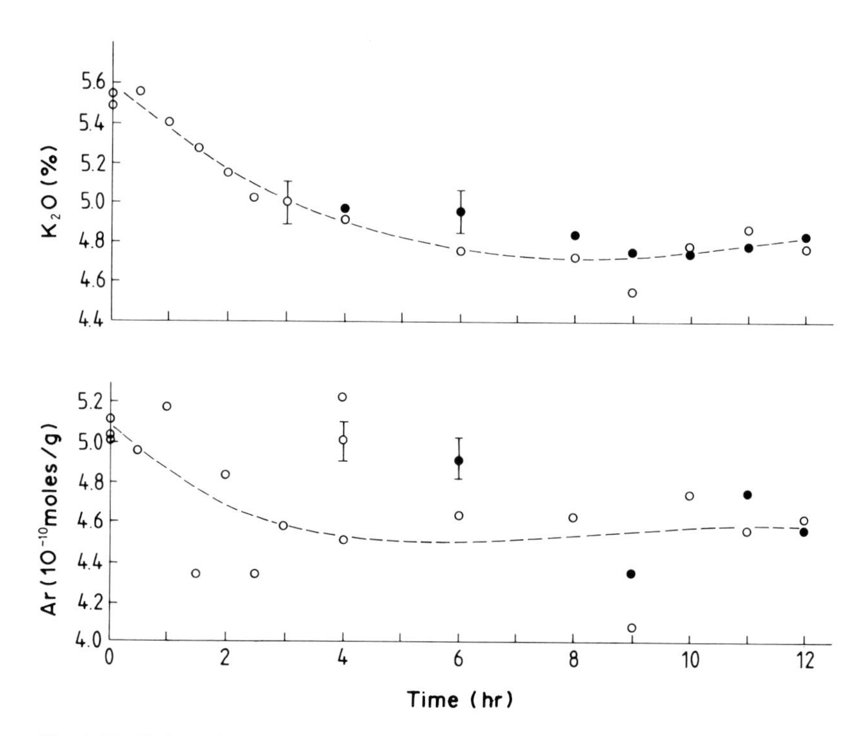

Fig. 1.21. K_2O and Ar contents of clay residues after experimental dissolution in acid. The *open circles* correspond to 1 N acid treatment, and the *full circles* to 0.5 N acid treatment. The *two curves* are reasonably similar, indicating no preferential Ar loss. (Aronson and Douthitt 1986)

calculated the K-Ar ages for the solid residues after acid treatments which remained approximately constant. These ages do not conform to the model ages based on the Thompson-Hower hypothesis of no Ar retention at K_2 site and the internal age agreement also argued for no preferential loss of Ar. Aronson and Douthitt claimed that site K_1 was as retentive as site K_2 for the Ar. They further suggested that the site for rate K_1 in the mixed-layer illite/smectite is similar to that for the mica, whereas the site for rate K_2, which also retains Ar, is one of the partly converted interlayer in the evolutionary transformation process from smectite to illite.

3.1.2 The Effect of Temperature on Ar Diffusion

To provide an understanding of the diffusion of Ar from glauconitic minerals, different investigators conducted laboratory studies on these minerals subjected to varied conditions of temperature and pressure. Amirkhanov et al. (1961) and Evernden et al. (1961) concluded that loss of Ar from glauconite can occur at temperatures as low as about 100 °C, but most of the Ar is lost at temperatures exceeding about 200 °C. Evernden et al. also found a small change in the K-Ar age at temperatures less than about 220 °C, but they observed a major decrease in the K-Ar age above 200 °C. Hurley et al. (1960) observed no change in the K-Ar age after having heated glauconitic samples for about 24 h under atmospheric pressure at a temperature of about 110 °C. Several others also concluded from studies of Ar loss under vacuum that glauconitic minerals only begin to lose Ar around 200 °C (Polevaya et al. 1961; Sardarov 1963; Ghosh 1972; Odin 1975; Odin and Bonhomme 1982). The rate of Ar diffusion is much slower for well-crystallized minerals and under confining water pressure than under vacuum (Odin et al. 1977). Thus the different studies demonstrate that K-Ar ages remain essentially unchanged between 110 and 200 °C. Zimmermann and Odin (1982) heated five glauconitic fractions with varied chemical compositions between 200 and 1100 °C under vacuum, and observed that the richer the glauconitic material in K, the better its retention of radiogenic ^{40}Ar, and that Ar loss was accompanied by structural disorganization.

The experimental studies on Ar loss from illite and smectite are relatively very few. The study of Odin and Bonhomme (1982) has clearly shown that following the initial bakeout between 80 and 110 °C under vacuum, the Ar content for smectite and illite heated to about 250 °C remains constant. This Ar retention behavior of either illite or smectite is similar to that of glauconite. These different experimental studies show that continuous significant diffusion of Ar from phyllosilicate minerals under normal sedimentary conditions is most unlikely. Any Ar diffusion from these minerals under sedimentary conditions must, therefore, be related to recrystallization processes.

3.1.3 The Effect of Induced Irradiation on Ar Diffusion by Recoil

Reuter and Dallmeyer (1989) have compared the K-Ar and $^{40}Ar/^{39}Ar$ dates of different mica-rich size fractions of metapelites and metatuffs from Rheinisches Schiefergebirge in Germany. The two sets of dates from middle anchizonal metapelite are slightly but significantly different: the $^{40}Ar/^{39}Ar$ total-gas dates of the 0.4–0.6, 0.6–1, 1–2, and 2–6 μm size fractions are systematically higher than the K-Ar values of the corresponding fractions. This could result from a recoil of ^{39}Ar which, however, seems not to be grain-size-dependent, as the coarsest 2–6.3 μm size fraction displayed the largest difference from about 400 to about 430 Ma (Fig. 1.22). The equivalent size fractions of the upper anchizonal metapelite yielded K-Ar and $^{40}Ar/^{39}Ar$ dates which are much more concordant, the largest difference of about 20 Ma being determined again on the coarsest 2–6.3 and 6.3–20 μm fractions (Fig. 1.22). The upper anchizonal sample underwent a higher recrystallization degree, the decrease of the surface to volume ratio being partly induced by the variable morphology of the grain edges.

Reuter and Dallmeyer (1989) explained that both recoil and redistribution of ^{39}Ar seem to have occurred preferentially in particles with large surface areas and poorly defined grain edges, which relate to poorly crys-

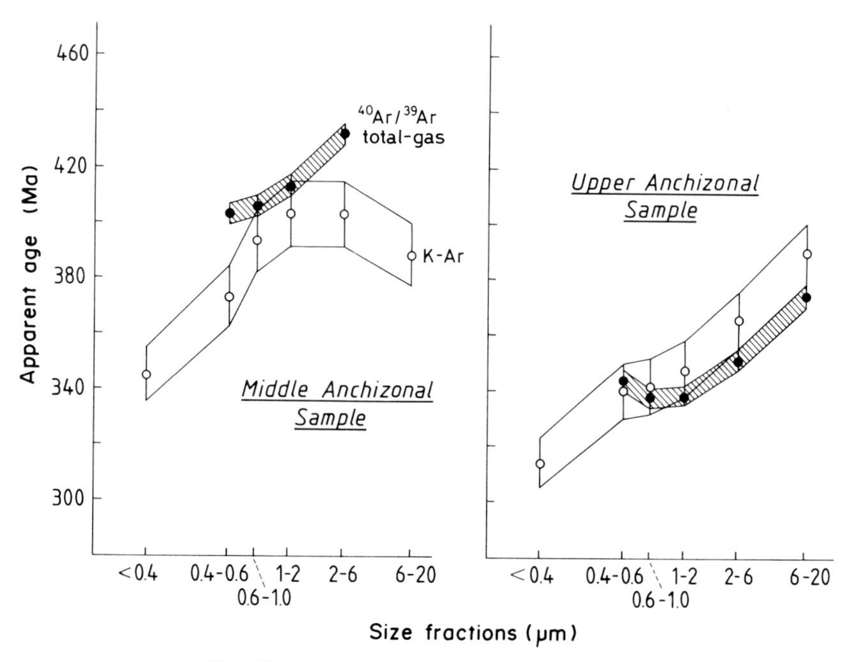

Fig. 1.22. K-Ar and $^{40}Ar/^{39}Ar$ isotopic data of different size fractions of a middle and an upper anchizonal metapelite from Rheinisches Schiefergebirge, Germany. (Reuter and Dallmeyer 1989)

tallized minerals, relative to well-crystallized ones with straight edges. This interpretation is supported by the plateau ages (Fig. 1.15) which are rather complicated for the middle anchizone sample, but tend towards plateau-type spectra in the upper anchizone sample. Additional thoughts about the behavior of the ^{39}Ar recoil and ^{40}Ar diffusion might be suggested on the basis of these data. As discussed above, the $^{40}Ar/^{39}Ar$ total ages of the size fractions of the middle anchizonal metapelite are systematically higher than their corresponding K-Ar dates, whereas the $^{40}Ar/^{39}Ar$ total ages of the different size fractions of the upper anchizonal metapelite are almost always lower than their corresponding K-Ar ages. The problem is, at this point, to consider if these age differences could have been induced by the irradiation procedure producing mainly ^{39}Ar recoil in the case of the middle anchizonal sample and mainly ^{40}Ar diffusion in the case of the upper anchizonal sample. In other words, it could be that ^{39}Ar recoil and ^{40}Ar diffusion happened in all analyzed size fractions, with recoil more important than diffusion in the middle anchizonal material and diffusion more important than recoil in the upper anchizonal material.

3.2 Effects of Mechanical Treatments

Ultrasonic treatment is the most commonly used method for mechanical purging of impurities associated with sedimentary clay minerals. The K-Ar systematics of the ultrasonically disaggregated materials have been the subject of some studies. Obradovich (1965) observed from the analyses of various fractions of ultrasonically disaggregated glauconitic minerals that the loss of K following ultrasonic treatment ranged from about 0 to as much about 20%, the higher amounts of loss being in general associated with the low K-bearing minerals; but regardless of the amounts of K loss, the K-Ar ages were not altered. Obradovich concluded that ultrasonic treatment essentially fractured the minerals without affecting the K-Ar date. Odin and Rex (1982) confirmed Obradovich's conclusion about very little K being lost from the high K-bearing glauconites that are ultrasonically disaggregated. Because highly evolved coarse-grained glauconites have the highest fidelity in retaining the memory of K-Ar characteristics since the time of formation of the minerals, Odin and Rex recommended only a moderate period (a few minutes) of ultrasonic treatment.

3.3 Effects of Chemical Treatments

Leaching by acids (hydrochloric or acetic acid of low strength, 1 N or lower) or some other reagents (ammonium acetate, ammonium chloride, amberlite ion exchange, etc.) first came into use for the purpose of removal of interfering soluble minerals or elements from clay minerals to be analyzed for the isotopic studies. The leachings have been thought to have simulaneously caused removal of both parent and daughter isotopes from easily exhangeable sites (surface, edges, lattice defects, expandable layers, etc.) of clay minerals, potentially producing unknown effects on the isotopic memory of the minerals. The relative effectiveness of different reagents commonly used for the leaching and the merits of these different modes of leaching on the isotopic analyses of clay minerals have been discussed by Clauer et al. (1993).

The determination of the CEC of clay particles might be of interest in connection with leaching experiments, as the results can give some information on the surface properties of the clay material, the principle of the technique being to replace all exchangeable cations of a mineral by protons, the overall values ranging from 3–15 meq/100 g for kaolinite to about 150 meq/100 g for smectite and vermiculite. The very high CEC of vermiculite and smectite is of some importance for their Rb, Sr, Sm, Nd, or Pb contents, as most of these elements may be adsorbed on the particles and not trapped in the lattices. The meaning of the isotopic compositions of these elements might, consequently, be very different: the adsorbed elements being more representative of an ancient or recent aqueous environment, while the internal elements being more representative of the time related decay.

3.3.1 Effects on K-Ar Analyses

Treatment of glauconite with hydrochloric acid of low strength has been commonly known to release a significant amount of K to the solution. By leaching glauconite with 0.5 N HCl, Thompson and Hower (1973) made the first extensive attempt to determine the structural sites for K in a layered silicate mineral of sedimentary origin. They concluded, based on the rate of K release, that K is located in three different crystallographic sites and that the fraction of the most soluble K increases with increase in the proportion of expandable layers in glauconite. The results of Thompson and Hower suggest differential mobility of Ar in glauconite material. However, this study did not demonstrate to what extents both K and Ar are removed relative to each other as a result of the acid leaching. Aronson and Douthitt (1986) acid-leached a sample of illite/smectite mixed layer and also found three different subsites for K in the mineral. Based on the evidence of internal K-Ar age agreement, the authors preferred no preferential loss of radiogenic ^{40}Ar. More recently, Clauer et al. (1993) analyzed two different size fractions of a sample of diagenetic illite treated with various reagents which included 1 N HCl, 1 N NH_4Cl, cation exchange resin, NH_4-EDTA, acetone, and humic acid extracts. They found no significant preferential removal of radiogenic ^{40}Ar (Fig. 1.23), and recommended the use of low strength HCl for the purpose of leaching.

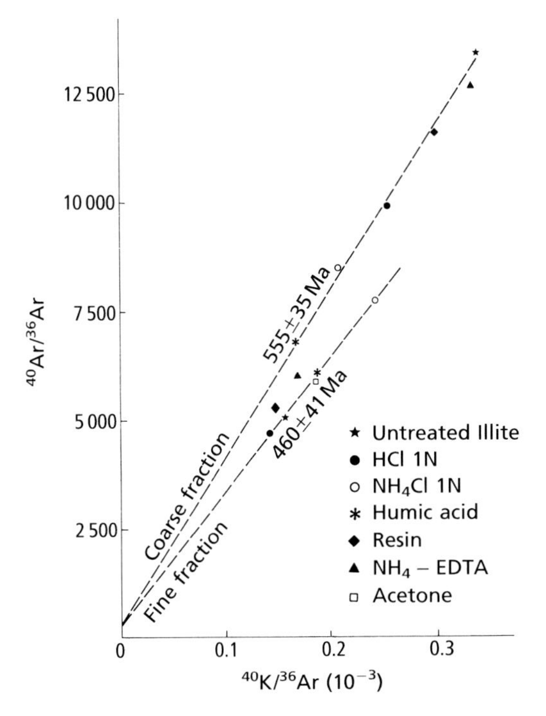

Fig. 1.23. K-Ar data of diagenetic illite size fractions experimentally leached with different reagents. No preferential loss is visible, but the data points of the fine fraction treated with NH_4-EDTA and resin fall off the lower line. For discussion, see text. (Clauer et al. 1993)

3.3.2 Effects on Rb-Sr Analyses

The initial attempts of Rb-Sr analyses on clay sediments leached by hydrochloric acid focused on removal of the carbonate component as impurity (Bofinger et al. 1968, 1970). These initial studies commonly showed increases in both the Rb/Sr and the $^{87}Sr/^{86}Sr$ ratios, primarily due to removal of large amounts of Sr by dissolution of the carbonate phase. The acid-leached clays tended to yield a less scattered Rb-Sr isochron diagram, and the resultant age seemed to be lower than the age of the corresponding untreated samples. Leachings by dilute hydrochloric acid were thought to remove not only carbonate impurities, but also loosely held Rb and Sr from surface of the clay particles (Clauer 1976). Some studies provided Sr isotopic ratios of leachates which were about 1 to 2 parts per thousand higher than that of the marine Sr. This elevation in the Sr isotopic values for the leachates hinted preferential acid leaching of radiogenic Sr from clay minerals (Clauer 1979a).

Chaudhuri and Brookins (1979) determined the effects of different concentrations of HCl (0.1 to 4.0 M) on the Rb-Sr isotopic compositions of four different standard illitic and smectitic minerals. They reported that the leaching increased both the Rb/Sr and $^{87}Sr/^{86}Sr$ ratios of the clays and also found that the Rb-Sr model ages of the different minerals increased following the acid leaching. The increases were apparently produced by removal of Sr low in its $^{87}Sr/^{86}Sr$ ratio. The authors found highly variable $^{87}Sr/^{86}Sr$ ratios for the leachates, ranging from 0.7070 to 0.7095 for leachates from smectites, and from 0.7124 to 0.7277 for leachates from illites. The leaching experiments obviously did not produce preferential extraction of radiogenic ^{87}Sr as the authors did not observe significant increases in the Sr isotopic ratios when they increased the concentration of the acid.

Kralik (1984) compared the Rb-Sr data of a <0.5 µm fraction of a Fithian illite leached differently with 1 N HCl, cation exchange resin and NH_4-EDTA. He observed various amounts of radiogenic ^{87}Sr in the leachates, but no preferential removal of ^{87}Sr from mineral structures and therefore recommended the use of amberlite resin to dissolve away carbonate minerals associated with clays and also to remove loosely bound Rb and Sr from the clays. However, a comparable study by Clauer et al. (1993) on another type of illite emphasized the difficulties of using this technique because of potential losses of the very finest particles during analytical procedure (Fig. 1.24).

Keppens et al. (in prep.) studied the effects of ultrasonic treatments on glauconitic aggregats put into different reagents. They found that the

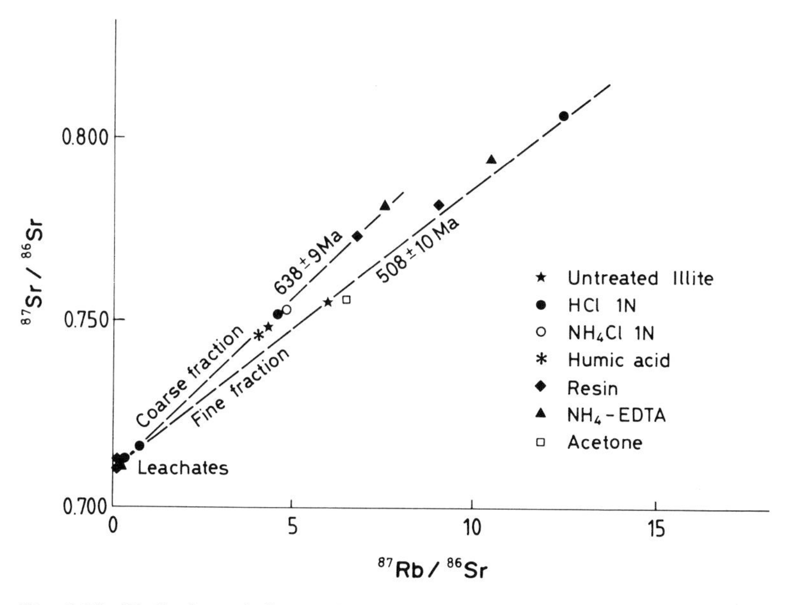

Fig. 1.24. Rb-Sr data of diagenetic illite size fractions experimentally leached with different reagents. (Clauer et al. 1993)

ultrasonic treatments removed Fe-rich, and probably also Al-rich oxy-hydroxides, together with carbonate phases which are dissolved by the reagents. These removals did not affect the K-Ar ages, but significantly modified the Rb-Sr ages, suggesting that these accessory mineral phases might contain Sr with $^{87}Sr/^{86}Sr$ ratios which may be different from that of the initial Sr of the glauconitic mineral phases at the time of their crystallization.

Morton and Long (1980) compared the Rb-Sr isotopic results of several different untreated glauconitic samples with that of the same samples which were treated differently by HCl, NH_4OAc, HOAc buffered with NH_4OAc, and NH_4-EDTA. They recommended the use of NH_4OAc for the purpose of isotopic dating, because the leaching seemed to remove all loosely bound Rb and Sr from the minerals, signifying that the Rb and Sr in the NH_4OAc-treated residual glauconitic materials conformed to the condition of a closed system. Although several other isotopic studies followed the NH_4OAc treatment (Grant et al. 1984; Morton 1985b; Ohr et al. 1991), the claim of benefit in using NH_4OAc by Morton and Long (1980) was exaggerated in light of the existing hard data.

A presumed cause for the use of NH_4OAc is that the $^{87}Sr/86$ Sr ratio of the leachate from a NH_4OAc treatment is lower than that from a HCl treatment or from other treatments. This claim is mostly subjective, because the use of NH_4OAc fails to remove such impurities as apatite, calcite, or dolomite, which may be present together with the clay material in the fractions to be analyzed. The inadequacy of the NH_4OAc treatment in removing the exchangeable Sr has been demonstrated by Grant et al. (1984). They observed that following the treatment of a sample of glauconite in chloride brine, the NH_4OAc leaching was unable to remove all the Sr that was adsorbed by the glauconite from the brine. X-ray evidence showed that no phase change in the assemblage occurred during the entire experimental stage, leading Grant et al. to conclude that some Sr could have remained as a difficultly exchangeable ion in the expandable layers. In other words, the purging of the impurities seems either far from being complete or altogether ineffective by the NH_4OAc treatment.

The claim that high $^{87}Sr/^{86}Sr$ ratios for leachates derived from HCl treatment is indicative of selective leaching of radiogenic ^{87}Sr is a highly superficial conclusion. Clauer et al. (1992a) and Clauer et al. (1993) have provided strong evidence for the lack of selective leaching of radiogenic ^{87}Sr from glauconite and illite minerals when treated with 1 N HCl. They also showed that the Rb-Sr isotopic data of an illite sample leached with HCl acid are concordant with that of the same sample treated with other reagents such as NH_4OAc, NH_4Cl, amberlite resin, etc. (Fig. 1.24). An additional proof that 1 N HCl leaching is not preferentially extracting radiogenic ^{87}Sr is given by the results of leachings made by N. Clauer (unpubl. data) on aliquots of Estonian illites which were leached separately with 1 N NH_4OAc by Gorokhov et al. (1994). The data, which are presented in Table 1.3,

Table 1.3. Leaching experiments on the same aliquots of diagenetic illites. (Data from Gorokhov et al. 1994; N. Clauer, unpubl. data)

Samples	Initial $^{87}Sr/^{86}Sr$ ratio ($\pm 2\sigma$)	Age in Ma ($\pm 2\sigma$)
77 NC74 <0.1 μm		
1 N HCl	0.7111 ± 0.0002	508 ± 10
1 N NH_4OAc	0.7111 ± 0.0004	505 ± 10
77 NC80 <0.1 μm		
1 N HCl	0.7115 ± 0.0002	499 ± 10
1 N NH_4OAc	0.7102 ± 0.0008	521 ± 12
807/8 <0.1 μm		
1 N HCl	0.7133 ± 0.0008	501 ± 10
1 N NH_4OAc	0.7122 ± 0.0005	512 ± 14
77 NC102 <0.1 μm		
1 N HCl	0.7141 ± 0.0003	465 ± 8
1 N NH_4OAc	0.7134 ± 0.0007	466 ± 9
77 NC106 <0.1 μm		
1 N HCl	0.7120 ± 0.0002	481 ± 8
1 N NH_4OAc	0.7109 ± 0.0005	481 ± 8

Sample numbers are those of Gorokhov et al.; initial $^{87}Sr/^{86}Sr$ ratios are derived from the isochron intercepts; each calculation includes the untreated aliquot, its leachate, and residue.

clearly show that the untreated aliquot, the leachate, and the residue for each sample plot along a line, each gives an age and an initial $^{87}Sr/^{86}Sr$ ratio which agree with that of the others.

The high ratio for a leachate from HCl treatment on clay minerals could reflect congruent dissolution of a fraction of the clay mass, as Bath (1977) showed that the congruent dissolution of a clay phase, and not a selective leaching of radiogenic ^{87}Sr, may be responsible for an elevated Sr isotopic ratio of the dissolved phase. He found covariation between Rb and ^{87}Sr contents in the leachate from the treatment of clay minerals in different chloride solutions at temperatures between 315 and 360 °C (Fig. 1.25). Therefore, the claim that NH_4OAc treatment causes no selective leaching, but the HCl treatment does, is yet to be backed by hard evidence.

A reasonable agreement between the calculated isotopic dates and the presumed stratigraphic ages of some leached glauconites has also been advanced to argue for the use of the NH_4OAc treatment (Morton and Long 1980). The data of Morton and Long have also shown that the isotopic dates of some glauconites following HCl treatments are nearly identical to that of the same glauconites following NH_4OAc treatments, regardless of whether the dates are in agreement with the stratigraphic ages or not. Furthermore, the claims of Morton and Long might have been biased to some degree by experimental error in the determination of the concentration of Sr in the glauconitic minerals. For example, they reported that a glauconite

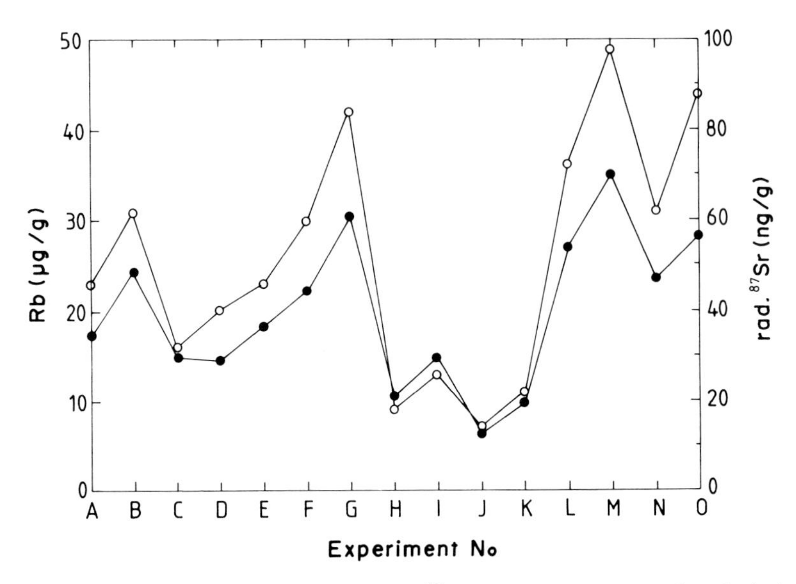

Fig. 1.25. Amounts of Rb and radiogenic ^{87}Sr extracted from clay minerals during different experimental reactions with chloride solutions at different temperatures. No preferential loss of radiogenic ^{87}Sr is visible. (Bath 1977)

sample from the Houy Formation in the Llano Uplift region in Texas had a Sr content of 20.7 μg/g with an $^{87}Sr/^{86}Sr$ ratio of 0.807 for the untreated sample and a Sr content of 6.16 μg/g with a Sr isotopic ratio of 1.3464 for the acid-leached residue. They did not report the Sr isotopic ratio for the leachate, but a mass calculation based on these data shows that the leached Sr, which in this case is about 14.54 μg, should have a calculated $^{87}Sr/^{86}Sr$ ratio of about 0.669. As this figure is totally unrealistic, we believe that the error was more with the concentrations of Sr rather than with the Sr isotopic values. The data of Grant et al. (1984) also cast doubt on the validity of the claim that NH_4OAc-treated samples are better suited for the Rb-Sr analyses. These authors found an isotopic date of 437 Ma for a NH_4OAc-treated sample of glauconite from an Upper Cambrian Bonneterre unit. The isotopic date of the NH_4OAc-treated sample was found to be higher than the 415 Ma for the corresponding 0.1 N HCl-treated sample. However, this data cannot be used to justify the claim that the NH_4OAc-treated sample gives an apparently realistic age, because an isotopic mass balance calculation shows that the data are questionable. Grant et al. reported that the Sr content was about 5.53 μg/g and the Sr isotopic ratio about 1.4982 for the untreated Upper Cambrian glauconite sample, that the Sr content was about 4.36 μg/g and the Sr isotopic ratio of 1.7371 for the NH_4OAc-treated sample, and that the Sr content was about 4.25 μg/g and the Sr isotopic ratio about 1.8085 for the HCl-treated sample. The isotopic mass balance calculations show that the calculated $^{87}Sr/^{86}Sr$ values for the NH_4OAc leachate and the HCl leachate are 0.608 and 0.468, respectively. These calculated values are, of course, unrealistic, and the low calculated Sr isotopic date probably reflects an experimental error which is most likely attributable to the determination of the Sr content of the glauconite samples. Thus, the evidence which may be advanced to argue the case for the advantages of using NH_4OAc treatment and the disadvantages of using HCl acid should be carefully reviewed. The recent work by Clauer et al. (1993) has clearly shown that 1 N HCl treatment may have more advantages than NH_4OAc treatment and that the high Sr isotopic ratio for the HCl leachate is not a convincing argument that a sample treated in HCl yields an age which is less reliable than the age obtained from the same sample treated in NH_4OAc or any other reagent.

3.3.3 Effects on Stable Isotope Analyses

Three kinds of water are recognized in clay minerals. These comprise waters that are adsorbed on the surface of the particles, those that are held in the interlayer positions, and those that are part of the sheet or structural frameworks. Adsorbed and interlayer waters exchange their isotopes with ambient waters at extremely rapid rates by a few seconds to days. On the other hand, the rates of isotopic exchange between structural and ambient waters for clay minerals are generally slow and are dependent on the temperature and the types of clay minerals. Laboratory experiments spanning a period of 2 years have shown that oxygen isotopic exchange between the structural oxygen of either well-crystallized illite or kaolinite and the oxygen of ambient water can be considered to be nearly insignificant over a period of a few million years at temperatures of about 100 °C or less, but certainly is considerable at temperatures of about 300 °C (James and Baker 1976; O'Neil and Kharaka 1976). Unlike the rate of oxygen-isotope exchange, that of hydrogen-isotope exchange between the structural hydrogen of these clay minerals and the hydrogen of ambient water is relatively rapid at temperatures as low as 60 °C. O'Neil and Kharaka (1976) concluded that hydrogen isotope exchange takes place by a mechanism of proton exchange that is completely independent of oxygen isotope exchange. Their experiments have also shown that the isotopic exchanges for smectitic clay minerals with their ambient waters are rapid and significant relative to that of either illite or kaolinite with their ambient waters. The evidence gathered from these experiments may be used to suggest that the oxygen-isotope exchange between the structural oxygen of clay minerals and the oxygen of the ambient water may not occur for tens of million years in most burial diagenetic environments. Therefore, structural oxygen of illitic and kaolinitic minerals in many diagenetic environments may be analyzed to characterize the growth history of the clay minerals by describing the composition of their ambient fluids and their temperature of formation, facilitating the depiction of tectonics and paleohydrology of the sedimentary basin in which these minerals evolved.

3.4 Significance of the Leachates

Strontium trapped by the clay minerals during their crystallization and the radiogenic ^{87}Sr formed by decay of the radioactive ^{87}Rb are well fixed in the structures of the clay minerals. Variable amounts of Sr are also adsorbed on surfaces of the particles or located in easily exchangeable sites at the edges of the inter-layer positions, or even located in inter-particulate sites. Knowledge of the Sr isotopic signature of this exchangeable Sr may be useful to characterize the isotopic composition of the fluid during deposition or diagenesis, and hence makes it possible to constrain the environment of crystallization of a clay mineral (Clauer 1976; Clauer et al. 1990, 1992b). Also, this Sr potentially may be used to gain an insight to types of water that might have been involved in "interlamellar" or "interparticulate" domains of clay particles during their growth (Tessier 1984; Touret 1988; and others).

Leachings of clay minerals by reagents facilitate, to some degree, separation of two, or more, components of the same element held in structurally different sites. Separate analyses of these differently located elements can be done, but care must be exercized in the experimental steps, because uncontrolled steps may lead to extraction of radiogenic ^{87}Sr from clay structure, as has been reported by Faure and Barrett (1973) and Clauer (1976, 1979a). The carefulness of the leachings is reflected in the alignment of the data points of the leachate, the untreated sample, and the residue in an isochron co-ordinate. Failure to achieve such alignment is a result of uncontrolled practice during the experiments and leads to an unsound interpretation of the results. To our best knowledge, the scatter of the data often occurs as a result of loss of some amounts of isotopically heterogeneous clay particles during the experimental procedure in connection with the leaching (Clauer et al. 1993).

4 Summary

Clays are important records of major physical and chemical modifications within the shallow part of the Earth's crust. Several different naturally occurring isotopes commonly present in the clays may be analyzed to depict the physical and chemical conditions attendant with their formation and accumulation in the different crustal sectors. Detailed information on structure and chemical compositions of the clays is needed for well-constrained interpretations about their formation history based on evidence from the isotopic data.

Structures of common clay minerals or clays, all of which belong to the family of phyllosilicates, are of sheet or fibrous types and consist of differently combined octahedral and tetrahedral sheets forming 1:1 layer (kaolin), 2:1 layer (smectite, illite, vermiculite, glauconite, palygorskite-sepiolite), 2:1:1 layer (chlorite), or a combination of these layers (mixed layer). The layer units are made essentially of oxygen, hydroxyl, silicon, aluminum, and magnesium or iron or both. Depending upon the charge of the layer arising from substitutions in the units, interlayer sites may be occupied by a wide variety of cations such as sodium, potassium, calcium, strontium, rare-earth elements, lead, barium, among many others, which differ in their hydration energies. Various chemical activity diagrams, commonly using systems K_2O-Al_2O_3-SiO_2-H_2O and Na_2O-Al_2O_3-SiO_2-H_2O, have proved very useful in predicting the stability relations among groups of clay minerals occurring in different parts of the upper crust.

Naturally occurring isotope tracers that have been most commonly analyzed in studies of clay minerals include ^{40}Ar, ^{87}Sr, and ^{143}Nd radiogenic isotopes and ^{18}O and D stable isotopes. The abundances of the radiogenic isotopes in the crustal materials are varied due to generations of these isotopes from disintegration of their respective radioactive parents. Ideally, the abundances of these isotopes in clay minerals reflect both the isotopic composition at the time the minerals formed and also the amounts generated from the radioactive disintegration since the formation of the minerals. The natural variations in the abundances of the stable isotopes of oxygen and hydrogen in a given type of clay mineral are dependent on the temperature and the isotopic composition of genetically associated fluids, potentially allowing these isotope measurements to be used for reconstruction of a wide range of physical and chemical parameters that existed during the growth of the minerals. The reconstruction of primary history of any group of clay minerals is possible provided no isotope exchange or disturbance occurred since the formation of the minerals.

Chapter 2

Isotope Geochemistry of Clay Minerals in Continental Weathering Environments

The surface of the continent is constantly modified both by major magmatic and tectonic forces that change the shape and the elevation of the landmass, and by denudation effected by varied climatic or atmospheric, hydrologic, biologic, and gravitational processes, resulting in transport of the continental products for ultimate deposition in the oceans. The exogenic cycle of the Earth consists of denudation of surface materials, temporary storage of part of the materials in various reservoirs within the continent, accumulation of the bulk of the materials in ocean basins, and the conversion of sediment into sedimentary rocks. Weathering is the beginning of the exogenic cycle, initiated by chemical alteration and physical breakdown of rocks and minerals at or near the surface of the Earth. The process of alteration and breakdown is then followed by various processes of denudation.

The surface environment of the Earth consists of four major reservoirs; lithosphere, hydrosphere, biosphere, and atmosphere. Weathering of rocks and minerals is driven by combined forces of atmosphere, biosphere, and hydrosphere. The alteration of the surface material is accompanied by varying degrees of dissolution and in situ precipitation of authigenic products, of which clay minerals are a major component. Silicate minerals which make up the bulk of the Earth's crust are the major precursors of authigenic clay minerals formed in the weathering environment. Common rock-forming silicate minerals weather at variable rates. The relative stabilities among the silicate minerals under weathering conditions have been generalized by Goldich (1938), and numerous subsequent studies have shown that this generalization remains largely valid. Climate, topography, vegetation, degree of leaching, redox potential, chelation, and solution ion activities are important factors in the stability of these minerals in the weathering environment (Huang and Keller 1970; Goode 1974). A complex set of variables is involved in chemical weathering, but for the sake of simplicity, a weathering system is driven mainly by four variables: mineral or rock, water, dissolved gases, and dissolved organic constituents. Rates of chemical weathering of rocks and minerals have been estimated from laboratory experiments (e.g., Lasaga 1984), thermodynamic and kinetic aspects (Del Nero 1992) and geochemical considerations of transfers of solutes by percolating solutions in soils or by rivers, as well as considerations of the composition of the weathering residue relative to that of the parent material (Tardy 1969; Velbel 1985; Dethier 1986). According to these estimates, saprolites form at present in many localities at a rate ranging from about 30 to 37 mm per 1000 years. Lasaga (1984) calculated the mean lifetime of a 1-mm crystal of various common rock-forming silicate minerals, based on the dissolution rate of the minerals at a pH of 8 and a temperature of 25 °C; the estimated mean lifetime ranged from about 34 million years for quartz to about 2.7 million years for mica, to about 0.52 million years for K-feldspar, to about 80 000 years for albite, to about 112 years for anorthite.

Bedrock may be covered by regolith derived from either the underlying rocks or sources elsewhere unrelated to the underlying bedrock. Our focus is on the soil or regolith which is genetically linked to the underlying bedrock because such a mantle of unconsolidated material, or such a profile of soil affords an opportunity to compare the chemical and isotopic compositions of authigenic minerals with that of their precursors, enabling us to determine the physical and chemical environments during the growth of the clay minerals. Furthermore, knowledge of the isotopic compositions for different clay minerals at this initial stage of the exogenic cycle is quite important for any interpretation of chemical evolution of clay minerals at subsequent stages of their evolution.

The discussion in this chapter will focus on changes in isotopic geochemistry of some major silicate minerals under progressive weathering. These minerals are of interest because they are the most common precursors of clay minerals formed in weathering environments. We will also discuss the isotopic systems of authigenic clay minerals formed in weathering environments. The collective evidence provides insight into the mechanism of clay mineral formation in weathering environments.

1 Clay Authigenesis in Soil Profiles

Weathering of geologic materials comprises both destructive and constructive processes. Minerals such as quartz and muscovite are primarily disintegrated, producing smaller particles, whereas other common silicate minerals are decomposed to form a wide variety of authigenic phases frequently consisting of oxides of Fe and Al (goethite, limonite, hematite, gibbsite, and others), carbonates of Ca and Mg (calcite, dolomite, and others), and clay minerals (kaolinite, halloysite, smectite, vermiculite, and others). Of these authigenic phases, clay minerals and calcite have received most of the attention for isotopic investigations. The oxide phases have yet to be thoroughly investigated, Fe-oxide phases being of much interest especially for Sm-Nd isotopic investigations.

1.1 Isotope Redistribution in Silicate Precursors During Weathering

To understand clearly the isotopic systematics of authigenic clay minerals from weathering profiles, it is instructive to examine first the effects of weathering on the isotope geochemistry of those silicate minerals from which clay minerals are commonly formed in weathering environments. This understanding of the parent materials is crucial to gain information on the distribution of the isotopes in the soil components and associated fluids.

Because of their importance to geochronologic investigations, biotite and feldspars in weathered rocks have become subjects of several Rb-Sr and K-Ar isotopic studies (Zartman 1964; Marvin et al. 1965; Goldich and Gast 1966; Fullagar and Bottino 1969; Mitchell et al. 1988). The early studies concluded that weathering causes a decrease in the age of biotite, as well as microcline, as a result of preferential leaching of the radiogenic isotope (^{87}Sr, ^{40}Ar) relative to the parent radioactive isotope (^{87}Rb, ^{40}K). They also established that the decrease in the age of microcline due to weathering is smaller than that of biotite. However, none of these studies provided any convincing argument explaining the difference in the ages between the two minerals.

1.1.1 The Weathering of Biotite

A variety of experimental studies has been performed on biotite minerals to understand structural and compositional changes, especially the process of K leaching, which occur in these minerals during weathering. The oxidation of Fe^{2+} to Fe^{3+}, lessening the required amount of the interlayer charge of biotite, may explain the leaching of K (Robert 1970). The degree of K removal was found to change with increased leaching, because the removal of K induces a contraction of the lattice along the b-dimension (Burns and White 1963). Contraction has the effect of increased bracing of the remaining interstitial cations within the lattice. This effect "stabilizes", at least temporarily, the K during the weathering process. The composition of the octahedral layers, as mentioned above, and the orientation of the proton of the OH radical in the structure, appear also to influence the bond strength of K and other interlayer cations (Basset 1960). Characteristics of K release from biotite and other micaceous minerals seem also to be dependent on the presence of other alkali elements. Bashour and Carlson (1984) noted from results of an experiment on ion exchange that Rb is able to replace K until the content of Rb in the minerals reaches about 1% of the exchangeable sites.

The effect of oxidation of Fe on the oxygen isotope composition of biotite was studied experimentally by Komarneni et al. (1985), who observed that oxidation caused a decrease in the $\delta^{18}O$ value. O'Neil and Kharaka (1976) investigated experimentally the oxygen and hydrogen isotopic exchanges between clay minerals and water at temperatures higher than 100 °C. The authors concluded that, at low temperature, oxygen isotopic exchange is negligible (less than 3%), but the hydrogen isotopic exchange varied between about 7% for illite and about 28% for smectite. Yeh and Savin (1976) found that ^{18}O exchange is detectable when particle sizes are smaller than a few tenths of a micron. Hence, both experimental studies and field data suggest that the structural protons and, to a much smaller degree, oxygen atoms are quite "active" during the weathering of biotites.

Experimental studies investigating the effect of weathering on isotopic ages of biotite minerals are very few. Kulp and Engels (1963) examined the relationships between K-Ar dates of biotites and loss of K through replacement by Ca and Mg ion exchange producing vermiculites. They noted that the K-Ar dates remain unchanged by removal of as much as 50% of K and that the dates are lowered by only 10% when the K removal amounts to as much as 80%. Their study suggested that both the K and the Ar are removed simultaneously and proportionately during the alteration, so that the K-Ar date of the residual unaltered material remains essentially unchanged.

Mitchell and Taka (1984) provided a mathematical basis for the patterns of K and Ar losses and subsequent changes in the K-Ar dates of biotites, as observed in natural weathering profiles or experimental studies. The authors derived a theoretical relationship between the fraction of ^{40}K lost (α) and the fraction of radiogenic ^{40}Ar lost (β) at the time of disturbance of the K-Ar system due to weathering. This relationship, which takes into account the amount of ^{40}K remaining and the accumulation of radiogenic ^{40}Ar since the disturbance, is expressed by:

$$\beta = (e^{\lambda t_0} - e^{\lambda t^*})/(e^{\lambda t_0} - e^{\lambda t_2}) + \alpha(e^{\lambda t^*} - e^{\lambda t_2})/(e^{\lambda t_0} - e^{\lambda t_2}),$$

where t_0 is the formation age, t_2 the disturbance (weathering) age, and t^* the measured conventional age. Based on the above mathematical relationship, they constructed theoretical patterns of K and Ar losses attendant with decreases in ages to explain experimental and natural data on weathering of biotites as well as muscovites (Fig. 2.1). The figure makes it evident that the K-Ar dates are hardly changed when the K loss is less than 20%, but the K-Ar dates decrease by the preferential loss of Ar relative to K when the K loss is more than 20%. The effects of K removal on the structures of mica have been reported by Burns and White (1963), Farmer et al. (1971), and Walker (1979). Hence, Mitchell and Taka suggested that the highly preferential loss of Ar for the mica minerals which have lost more than 20% K is probably due to two factors. One of the causes is that after an initial loss of a fraction of the original amount of K, the stability of the residual K is increased, and the other is that the Ar loss is facilitated by a structural b-dimension increase of the mica minerals following a large amount of K removal. Mitchell and Taka hypothesized three stages of weathering: (1) at the initial stage, octahedral modification occurs due to Fe^{2+} oxidation,

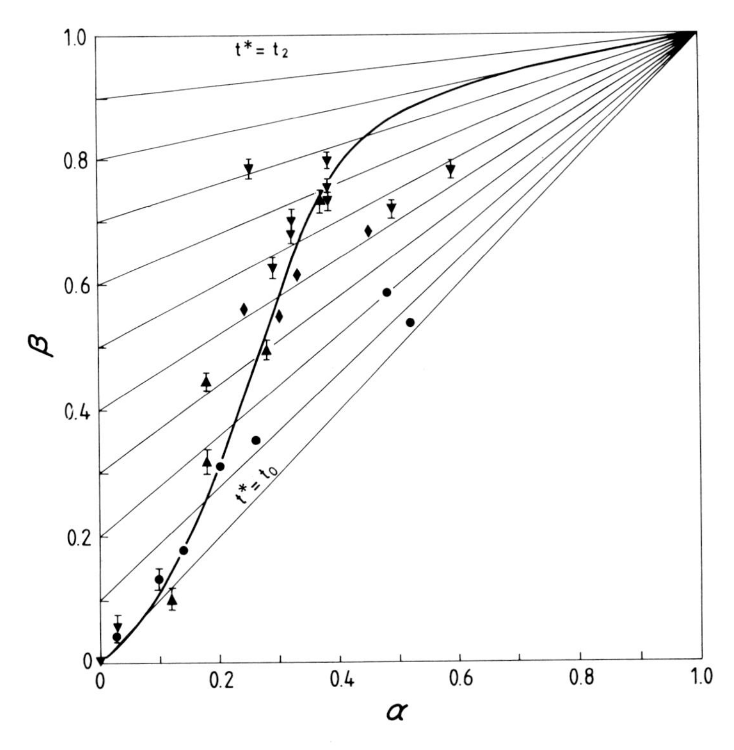

Fig. 2.1. Theoretical relationship between the fraction of ^{40}K lost, α, and the fraction of radiogenic ^{40}Ar lost, β, at t_2, the time of disturbance of the K-Ar system, t* being the measured conventional age. (Mitchell and Taka 1984)

producing similar losses of K and ^{40}Ar without any significant change in the K-Ar dates, (2) at an intermediate stage, the K was thought to be stabilized in the lattice relative to ^{40}Ar, resulting in a decrease in the K-Ar dates, and (3) at an advanced stage, the introduction of H_2O molecules into the interlayer positions induces again selective ^{40}Ar loss with concomitant decrease in the K-Ar dates. According to this model, most of the loss of radiogenic ^{40}Ar would occur when the K loss is between 20 and 40%. This model prediction fails to explain the results of the study by Kulp and Engels (1963) presented above.

Several studies focused also on the trends of changes in the Rb-Sr ages of biotites arising from various degrees of weathering under natural conditions. Zartman (1964) found that weathering lowered the Rb-Sr dates by about 6 to 8%, but produced negligible change in the K-Ar ages of biotites from a granite in central Texas. Marvin et al. (1965) observed that weathering lowered both the Rb-Sr and K-Ar ages of biotites in a bentonite deposit in Utah. Goldich and Gast (1966) noted that weathering severely affected the Rb-Sr age, while causing only moderate decrease in the K-Ar age, of a biotite from a Precambrian gneiss in Manitoba. Fullagar and Bottino (1969) also found that weathering causes variable degrees of reduction in the Rb-Sr ages of biotite. Marvin et al. noted that adsorption of Rb and concomitant loss of radiogenic ^{87}Sr and gain of common Sr are the primary causes of reduction in the Rb-Sr ages of biotites due to weathering. Goldich and Gast found no change in the K/Rb ratio even though the degree of weathering of biotite varied, leading them to conclude that the loss of radiogenic ^{87}Sr was the main cause for the decreases in the Rb-Sr ages.

The above studies clearly suggest that the responses of the Rb-Sr and K-Ar systems of biotites to conditions of weathering are complex. To shed light on this subject, Clauer (1978; 1981a) and Clauer et al. (1982a) made detailed chemical and isotopic studies of weathered biotites in a soil profile developed over a migmatite in Chad Republic (Fig. 2.2). They examined both the K-Ar and the Rb-Sr systems of a series of progressively more weathered biotites. The fresh or unweathered biotite contained about 9.4% K_2O, 579 μg/g Rb, and 16.6 μg/g Sr with the $^{87}Sr/^{86}Sr$ ratio of 1.589. The fresh biotite had a Rb-Sr model age of 610 ± 26 Ma and an apparent K-Ar age of 583 ± 13 Ma. The weathering of this biotite typically produced oxidation of Fe^{2+}, removal of K, Rb, Ti, and Mn, and an increase in H_2O^+. The K/Rb ratio decreased from 140 to 80 with increased weathering. The decrease in the K/Rb ratio is perhaps attribuable to stronger fixation of Rb, whose hydration energy is less than that of K (Kittrick 1966, 1969). The authors noted that the chemical changes appear to have occurred even at the very incipient stage of weathering of the biotite, before the development of any morphologically distinct authigenic mineral phase could be detected or identified by X-ray diffraction.

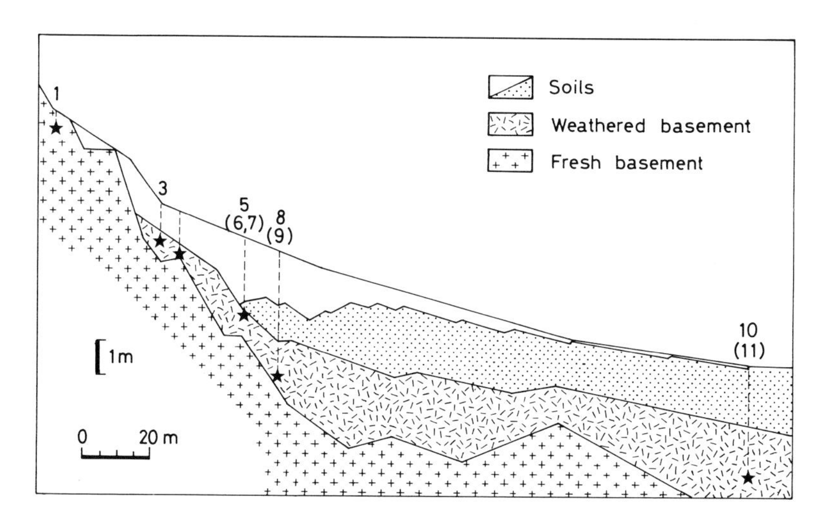

Fig. 2.2. Schematic representation of the soil profile of the Chad Republic. The *numbers* refer to the samples and the *stars* to their location in the profile. (Clauer et al. 1982a)

Study of the biotites in the soil profile from the Chad Republic has shown that the Ar content of the biotites increased with an increase in the degree of weathering, the Ar being possibly accommodated by either adsorption onto or incorporation into the particles. Also, the K-Ar dates varied from about 583 Ma for the fresh biotite to about 210 Ma for the moderately weathered one and to about 346 Ma for the most deeply weathered one. These changes in the K-Ar dates with increased weathering reveal that as little as a 10% loss of K produces a reduction of about 15% in the K-Ar date and that a loss of about 30% of K causes a 50% decrease in the K-Ar date. The relationship between $^{40}Ar/^{36}Ar$ and $^{40}K/^{36}Ar$ values for the different weathered samples, shown in a K-Ar isochron diagram (Fig. 2.3), indicate a curvilinear trend, which begins with the point representing the fresh sample located on the reference line, and ends with those of the apparently most weathered samples being near the initial value of the reference line as defined by the atmospheric $^{40}Ar/^{36}Ar$ ratio. These results are in contradiction with those of the ion exchange experimental study by Kulp and Engels (1963), and with the Ar-loss model proposed by Mitchell and Taka (1984). Evidently, natural weathering of biotite is governed by variables, some of which have not been considered in laboratory experiments or in the computer model.

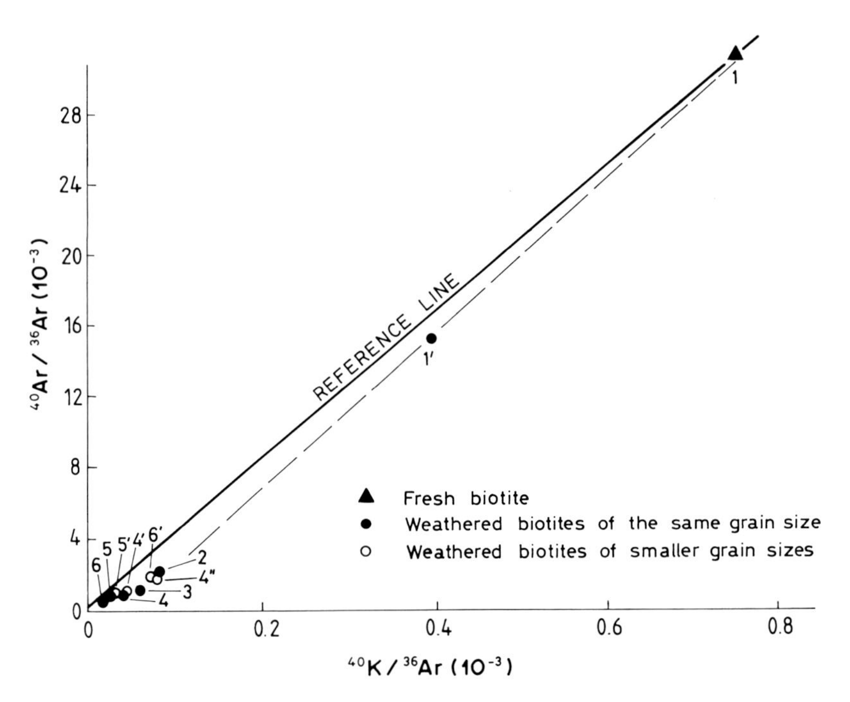

Fig. 2.3. Effect of progressive weathering on the K-Ar system of the Chad biotites. For *numbering*, see Fig. 2.2. (Clauer 1981a)

The effect of weathering on the Rb-Sr system of biotite, as revealed by the study of the same biotites in the same soil profile from the Chad Republic, is seen in a decrease in the Rb/Sr ratio with increased weathering. The Sr contents increase from about 17 μg/g for the unweathered biotite to about 92 μg/g for the most weathered one. Also, the $^{87}Sr/^{86}Sr$ ratios decrease with increases in Sr contents. As with the K-Ar dates, the Rb-Sr dates decrease due to weathering. The trend in the reduction of the Rb-Sr dates is similar to that of the K-Ar dates, as the greatest reduction in the Rb-Sr date was in the moderately weathered biotite, which yielded a date of about 130 Ma, whereas the most weathered one yielded a date of about 240 Ma. The reduction in the Rb-Sr date in the early stage of weathering can be attributed to the combined effects of both the loss of radiogenic ^{87}Sr and the adsorption of common Sr, but the date reduction at the advanced stage of weathering may be ascribed more to adsorption of common Sr and loss of Rb than to loss of radiogenic ^{87}Sr. This is supported by evidence that the Sr content increased markedly, while the $^{87}Sr/^{86}Sr$ ratio and the Rb content decreased only moderately at the advanced stage of weathering. The decreases in the Rb-Sr isotopic values with increased degree of weathering occur along a curved trajectory below the reference isochron line for the unweathered biotite samples (Fig. 2.4). This trend in the changes of the Rb-Sr dates is broadly similar to the direction of change in K-Ar values during the weathering.

A comparison between the Rb-Sr and the K-Ar systems of the variously weathered biotites in the soil profile from the Chad Republic is helpful in gaining some understanding of the path of chemical reorganization during natural weathering. In the upper part of the soil profile, at the onset of weathering where removal of K from biotite is less than 3%, the decrease in

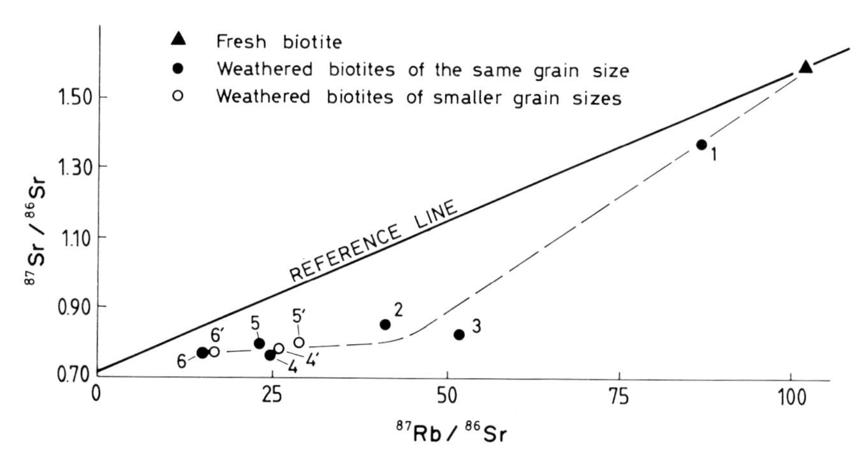

Fig. 2.4. Effect of progressive weathering on the Rb-Sr system of the Chad biotites. For *numbering*, see Fig. 2.2. (Clauer 1981a)

^{87}Sr relative to total Sr is greater than the decrease in ^{40}Ar relative to total Ar. In the intermediate part of the soil profile marked by a greater loss of K from biotite, the magnitudes of decrease in both the ^{87}Sr and the ^{40}Ar are nearly the same. In the lower part of the weathering profile where more than 32% of K is removed from biotite, the magnitude of the decrease in ^{87}Sr relative to total Sr now appears to be slightly higher than that of the decrease in ^{40}Ar relative to total Ar. The reduction in the K-Ar and Rb-Sr dates began from the initial stage of weathering, where the magnitude of reduction in the Rb-Sr date is slightly higher than that in the K-Ar date. Both the K-Ar and the Rb-Sr dates were dramatically reduced at advanced stages of weathering, the Rb-Sr dates always remaining significantly much less than the K-Ar dates and the relative reductions in the dates by each system closely mimicking the other (Fig. 2.5). The reductions in both the K-Ar and the Rb-Sr dates are less at the more advanced stage than at the intermediate stage.

The most common observation of all studies on natural weathered biotites is that the Rb-Sr dates are distinctly lower than the K-Ar dates when the K removal is about 5% or more. Goldich and Gast (1966) suggested that this effect arises from two factors, one of which could be the exponential effect in the ^{40}Ar/^{40}K age relation, and the other could be the preferential

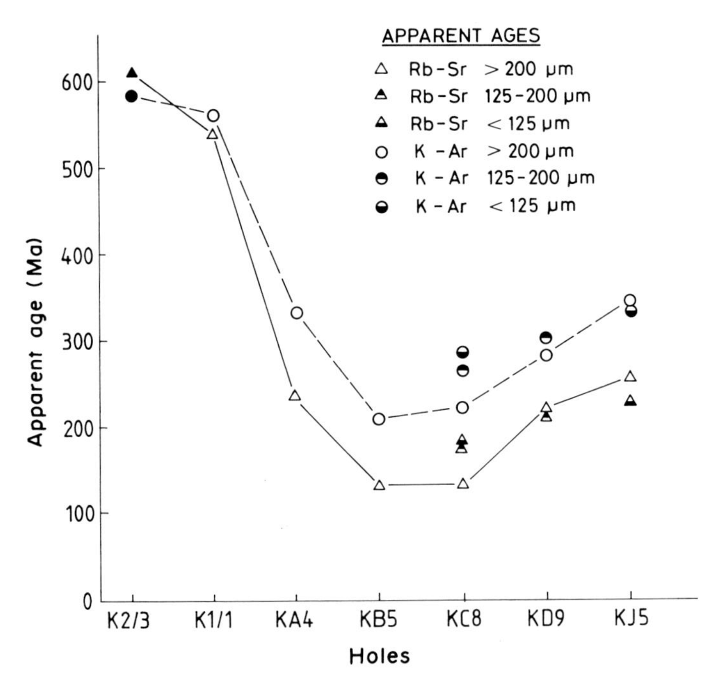

Fig. 2.5. Comparison of the Rb-Sr and K-Ar dates of the Chad biotites relative to their location in the soil profile. For the legends of the *holes*, refer to Fig. 2.2. (Clauer et al. 1982a)

adsorption of Rb over K. However, this suggestion fails to explain all the details of changes in K, Ar, Rb and Sr content and $^{87}Sr/^{86}Sr$ ratios of biotites when variously weathered, as suggested by the studies of Clauer (1981a) and Clauer et al. (1982a). Based on the combined chemical and isotopic evidence presented in these studies, we tend to think that the weathering-related crystallites partially entrap Sr and Ar from the ambient fluids in micro-environments within the soil profile. Such Sr and Ar are incorporated in lattice sites previously occupied by K, but are less radiogenic in composition than the intra-crystallite Sr and Ar. The incorporation presumably occurs in more or less closed-system microenvironments, where the fluid isotopic compositions varied due to differential dissolution of minerals of varying $^{87}Sr/^{86}Sr$ ratios and ^{40}Ar contents.

The analyses of oxygen and hydrogen isotope compositions of structural OH in biotite under weathering conditions provide information concerning the structural rearrangement which might occur during the alteration; but data for such information are presently scarce. The current state of knowledge on this matter is primarily from studies by Lawrence and Taylor (1972) and Clauer et al. (1982a). Lawrence and Taylor (1971) analyzed the $\delta^{18}O$ of biotites from a soil profile developed on a granite in Elberton County, Georgia. They found no change in the oxygen isotopic value for the biotites between the fresh rock and the most intensely weathered zone, clearly indicating that no isotopic exchange occurred between the mineral and the local meteoric water. These results are at variance with those obtained by Clauer et al. (1982a) who noted changes in both the $\delta^{18}O$ and the δD values of biotites from the soil profile in the Chad Republic. The $\delta^{18}O$ of these biotites increased from +5.7 per mill for the unweathered biotite to about +7.3 per mill for the biotite with incipient weathering, whereas at the same time the δD values decreased from −63 per mill to about −84 per mill. Any additional increase in the degree of weathering produced only a maximum additional increase of 1 per mill in the $\delta^{18}O$ value and a maximum additional decrease of about 6 per mill in the δD value. Thus, much of the change in the ^{18}O and D values for the biotites occurred at the very early stage of weathering, and subsequent changes related to increased weathering were of much smaller magnitude. This happens despite increases in the H_2O content, increases in the oxidation of Fe, and decreases in K contents attendant with the weathering of biotites.

As most of the changes in the stable isotope compositions appear to have occurred in the biotite samples before the crystallization of any detectable amount of authigenic mineral phases, Clauer et al. suggested that the formation of some "protominerals" at the initial stage of weathering may explain the changes in the ^{18}O and D values of biotites. Simple mixing of fresh biotite and the postulated "protomineral" is inadequate to explain the δD values, which remained nearly unchanged in the more weathered zone. The cause of this very rapid decrease in the δD values of biotites at the very initial stage of weathering, where distinct development of clay minerals has yet to take place, remains unknown. The decrease demonstrates that isotopic exchange occurred between the structural OH of the biotites and their ambient fluids during the weathering process. Laboratory studies of O'Neil and Kharaka (1976) demonstrated that, at surface temperatures, isotope exchange between micaceous minerals and water results in a decrease of the δD value of the mineral without any significant change in the $\delta^{18}O$ value. The experimental study of Komarneni et al. (1985) showed that Fe-oxidation of biotite leads to a decrease in the $\delta^{18}O$ by 1.6 to about 4.6 per mill. Contrary to the results of these experiments, the data on the Chad biotites was marked by an increase in the $\delta^{18}O$ values by about 1.5 to 3.0 per mill. Additional studies are needed to resolve the clear disagreement between the experimental and the natural data.

The concept of "protomineral" formation at the incipient stage of weathering is difficult to substantiate because the material most likely represents a poorly crystallized metastable phase formed at sites of structural defects in the biotite particles, but Wollast (1967), Paces (1973), Petrovic (1976), Soubies and Gout (1987), and Banfield and Eggleton (1988) have suggested precipitation of amorphous products during early stages of biotite weathering. Clauer et al.'s (1982a) suggestion of the presence of a "protomineral" is based on changes in the oxygen and the Sr isotopic compositions of biotites. An interesting aspect of the stable isotope changes for the biotites from the Chad Republic is that the δD values remained nearly constant at about -90 per mill, or changed very little with increased degree of weathering beyond the stage of "protomineral" formation, where the advanced stage of weathering is marked by increased amounts of newly formed kaolinite. The constancy of the δD value suggests isotopic equilibrium between the hydrous minerals and the groundwater. The near constancy in the Sr isotopic values also suggests Sr isotopic equilibrium between the hydrous layer silicate minerals under a closed system with respect to Sr. By taking the isotope fractionation factor of hydrogen between kaolinite and water at 0.970, as suggested by Savin and Epstein (1970a) and Lawrence and Taylor (1972), the alteration fluid in equilibrium with the weathered biotite of the Chad profile probably had a δD value of -60 per mill. This corresponds to a $\delta^{18}O$ value of -8.8 per mill for the fluid, calculated from the D-^{18}O relationship of global meteoric waters. The present-day local meteoric water at the location of the soil profile in Chad Republic has a δD value of -20 per mill and a $\delta^{18}O$ value of -3.2 per mill. Hence in comparison with the local meteoric waters, the calculated isotopic $\delta^{18}O$ value of the water during the weathering of biotite appears to be extremely low. Although one may offer as an explanation for this difference that the alteration of biotite took place in the past when the climate was relatively humid, such an explanation should be carefully weighed against the possibility that a similar isotopic anomaly could also arise from weathering of the mineral in micro-environments under low water-to-mineral ratios, in which case the isotopic composition of the fluid may not at all correspond to that of meteoric waters.

Rare-earth-element (REE) distributions have proved useful in tracing the origin of minerals and fluids in low-temperature environments (Duddy 1980). Cullers et al. (1975) observed significant differences in the relative distributions of REE among several kaolinite, smectite, and chlorite minerals, which they ascribed to differences in the weathering processes. From a study of REE in a soil developed on altered granodiorite, Nesbitt (1979) noted that in comparison to unaltered parent rocks, residual illite-kaolinite rich clay deposits were depleted in total REE and enriched in light REE (LREE) relative to heavy REE (HREE), whereas the partially altered granodiorite was enriched in total REE, but depleted in LREE relative to HREE. Nesbitt suggested that mineralogical controls are an important factor controlling the REE fractionation of these deposits.

The study of Clauer et al. (1982a) furnished some information as to mobilization and fractionation of REE during variable degrees of weathering of biotite minerals. Unlike the study of Nesbitt (1979), which showed that residual kaolinite-illite products become LREE-enriched, depleted in total REE and low in Sm/Nd ratio in comparison to the parent rock, that of Clauer et al. showed that biotites containing variable concentrations of kaolinite due to weathering, tended to have pronounced enrichments in HREE relative to LREE, increases in total REE, and high Sm/Nd ratios, as compared to unweathered biotite (Fig. 2.6). The results of these two studies

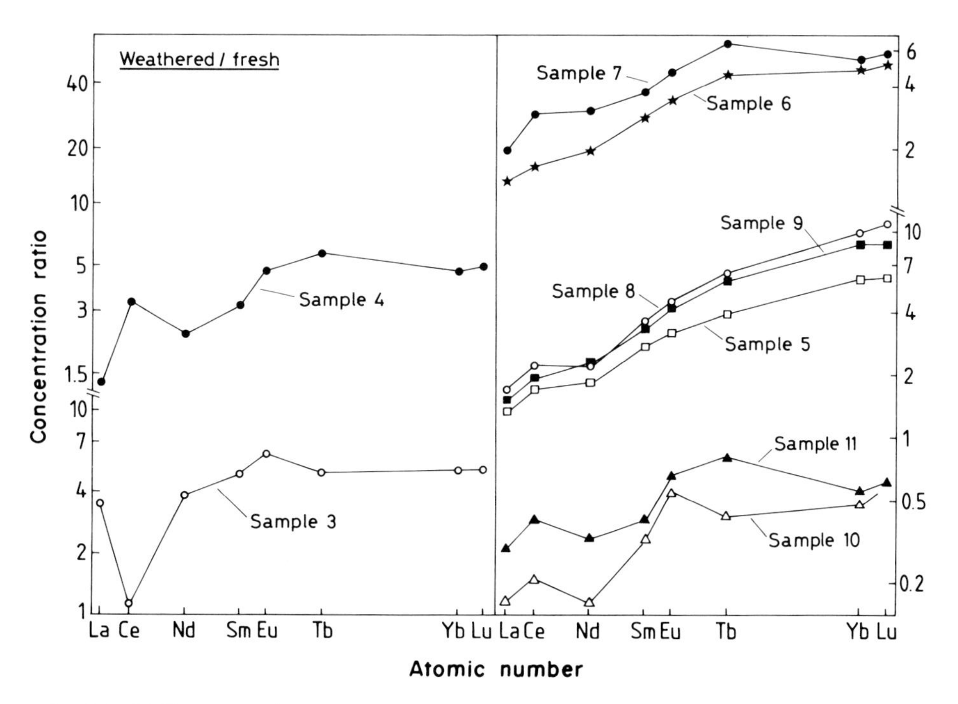

Fig. 2.6. REE patterns of the weathered Chad biotites relative to the unweathered biotite. For sample *numbering*, see Fig. 2.2. (Clauer et al. 1982a)

clearly indicate that the behavior of REE in weathering environments is extremely varied and that additional studies are essential to understand the causes of these variations. However, the results so far demonstrate that the Sm/Nd ratios of the same mineral in a soil profile can vary significantly.

In summary, weathering of biotite proceeds via oxidation of Fe^{2+}, depletion of elements such as K, Ti, Mn, and increase of the H_2O^+ content. The Rb contents may or may not increase significantly, but the K/Rb ratios decrease markedly. The REE contents and the Sm/Nd ratios apparently increase. The $^{87}Sr/^{86}Sr$ and $^{40}Ar/^{36}Ar$ ratios are also dramatically reduced, while the Sr and Ar contents tend to increase. The incipient stage of weathering of biotite seems to proceed with the formation of some poorly crystalline or amorphous "protominerals" which have yet to be morphologically described. Changes in the chemical and isotopic compositions of the biotites predict the development of these "protominerals" (Fig. 2.7). Microenvironments in soils seem to play an important role in the chemical and isotopic compositions of clay minerals formed from weathering of biotite.

1.1.2 The Weathering of Feldspars

The weathering of feldspars has long been known to occur with incongruent dissolution producing secondary solid hydrous aluminum-silicate phases. La Iglesia et al. (1976), for instance, described the formation of kaolinite from a feldspar as a homogeneous precipitation after feldspar hydrolysis. Recent morphological studies of feldspar grains isolated from soils have provided additional insight into the mechanism of weathering of these minerals in natural environments (Wilson 1975; Berner and Holdren 1977, 1979; Eswaran and Wong Chan Bin 1978; Keller 1978; Rodgers and Holland 1979; Calvert et al. 1980; Anand et al. 1985). These studies provided crucial evidence against the once-popular hypothesis that weathering induces a tightly held protective layer of alumino-silicate composition at the surface of the feldspar grain, the development of which controls the diffusion of ions and the rate of dissolution of the feldspar mineral (Berner and Holdren 1979; Lei Chou and Wollast 1984). Examining the products of weathering, which developed in microcracks formed in oligoclase and orthoclase cobbles in a soil zone, Rodgers and Holland (1979) found that the microcracks of orthoclase contain only kaolinite, but those of oligoclase contain a central zone of kaolinite surrounded by a zone of smectite. Noticing that the boundary between the edge of the microcracks and the clay particles is very sharp and that the composition of the clay minerals in the microcracks differs greatly between orthoclase and oligoclase, the authors suggested that transport of solutes is sufficiently slow to permit the establishment of concentration gradients within the microcracks.

Products of weathering of feldspars can be compositionally and morphologically varied depending on a number of factors, such as activities of

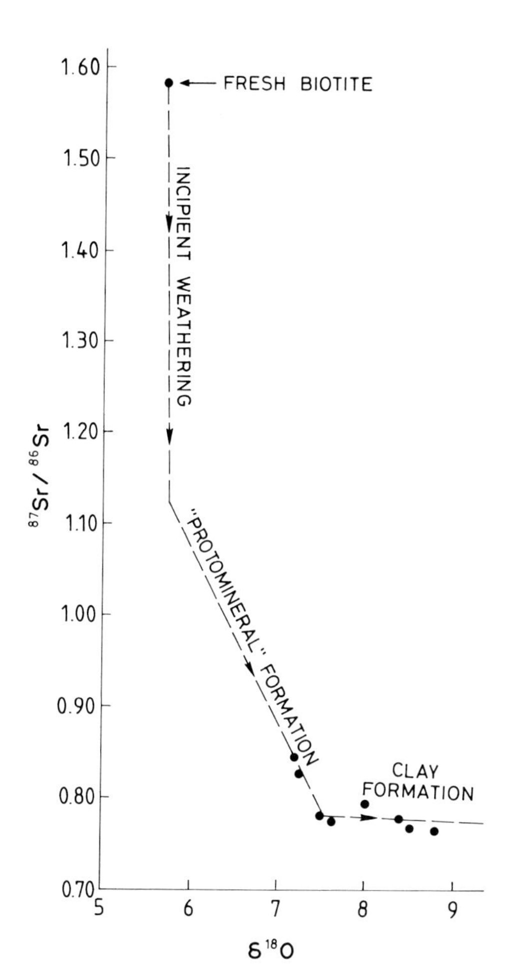

Fig. 2.7. The course of change in the relationship between the $^{87}Sr/^{86}Sr$ ratio and the $\delta^{18}O$ value from unweathered to weathered biotites in a soil profile in the Republic of Chad

solutes, the availability of water as controlled by the physical environment, pH, and organic activity (Helgeson et al. 1969; Droubi et al. 1976; Busenberg 1978). Halloysite and kaolinite are the common end products when the reactions occur in the presence of very large amounts of water over a long time period. The initial solid product of weathering of feldspar often consists of a poorly crystalline metastable phase (Delvigne and Martin 1970; Eggleton and Buseck 1980; Tazaki 1986), the metastable phase controlling the initial composition of the fluid. Keller (1978) described two types of kaolinite, a platy and an elongated form, which appear to form from feldspar dissolution. Presumably, the platy kaolinite precipitates during a long, slow process, whereas the elongated form is generated during a more rapid process. These different morphologies could also reflect different generations of authigenic minerals as a result of compositional differences of the solutions.

The discussion above suggests that weathering of feldspar grains commences with incongruent dissolution of minute patchy areas on the surface of the grains and that a sharp chemical boundary exists between the unaltered surface and the authigenic clay minerals developed in these patchy spots. Isotopic data are a potentially useful means of characterizing the reaction mechanisms during weathering of mineral grains but, to date, studies are very limited on the behavior of isotopic systems in weathering feldspars. Zartman (1964), investigating partly weathered granites from Texas, compared the Rb-Sr dates on microcline in a weathered profile to that from the underlying fresh granite. He found that the Rb and the radiogenic ^{87}Sr contents of the two populations of microcline were not very different, the small differences between the two probably relating to original variations in mineral chemistry. The two Rb-Sr dates of 1005 ± 25 Ma and 1000 ± 25 Ma were identical, suggesting that weathering had little or no effect on the Rb-Sr dates of microcline. The lack of alteration of the Rb-Sr isotopic systematics of microcline when subjected to weathering was in sharp contrast to the distinct alteration of the Rb-Sr isotopic systematics of biotite in the same weathering zone, as evident from the biotite in the fresh granite giving a date of 1005 ± 15 Ma, as compared to the biotite in the weathered zone yielding a date of 955 ± 20 Ma.

Clauer (1981a) examined the Rb-Sr dates of plagioclase minerals in a soil zone developed on migmatite in the Chad Republic. He observed that plagioclase extracts from two separate subzones in the soil profile gave dates that were identical to that of the unweathered plagioclase minerals at about 610 Ma, but two other plagioclase extracts from the same soil profile gave discordant dates, one having a slightly lower date and the other having a distinctly higher date than that of the fresh rock (Fig. 2.8). On the basis of the common nature of weathering of feldspar, the two discordant dates probably reflect the presence of clay impurities in the plagioclase extracts. Alternatively, the high or the low date may reflect selective leaching of Sr with either low or high $^{87}Sr/^{86}Sr$ ratios during weathering. In contrast to the Rb-Sr dates of the plagioclase separates, that of the coexisting biotite fraction was considerably lowered, amounting to as much as 130 Ma by weathering. A similar pattern of age disturbance between the two minerals was previously observed by Zartman (1964), knowing that plagioclase feldspars are considered to be as sensitive to weathering as biotites (Goldich 1938). The lack of modification of the isotopic dates of some feldspars, but not in others, could be related to laboratory artifacts in the removal of clay particles from the feldspar grains by means of ultrasonic cleaning treatment.

Unlike the Rb-Sr dates, which varied considerably, the K-Ar dates of the feldspar extracts from soils formed in the Chad locality were in good agreement (Fig. 2.9). This difference between the Rb-Sr and K-Ar systematics of feldspars grains in a soil zone could be related to either a preferential weathering-induced disturbance of Ca-rich domains in the minerals, or the presence of some impurities in the plagioclase extracts

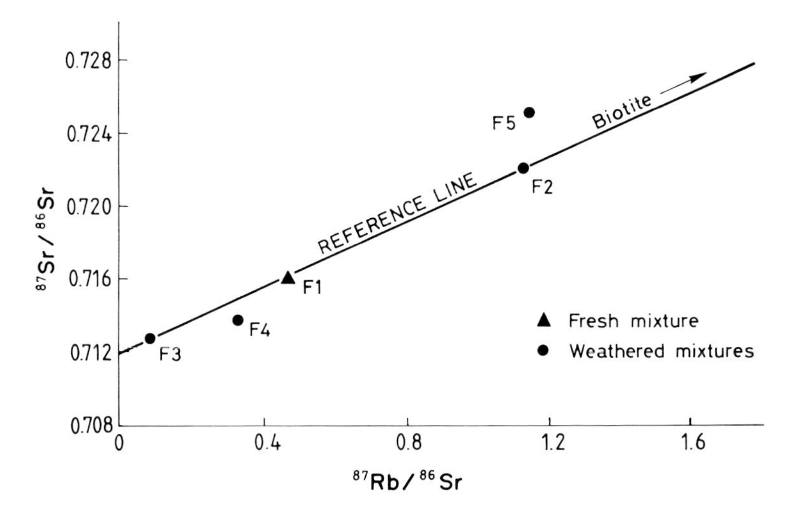

Fig. 2.8. The effect of weathering on the Rb-Sr system of feldspars in a soil profile in the Republic of Chad. For *numbering*, see Fig. 2.2. (Clauer 1981a)

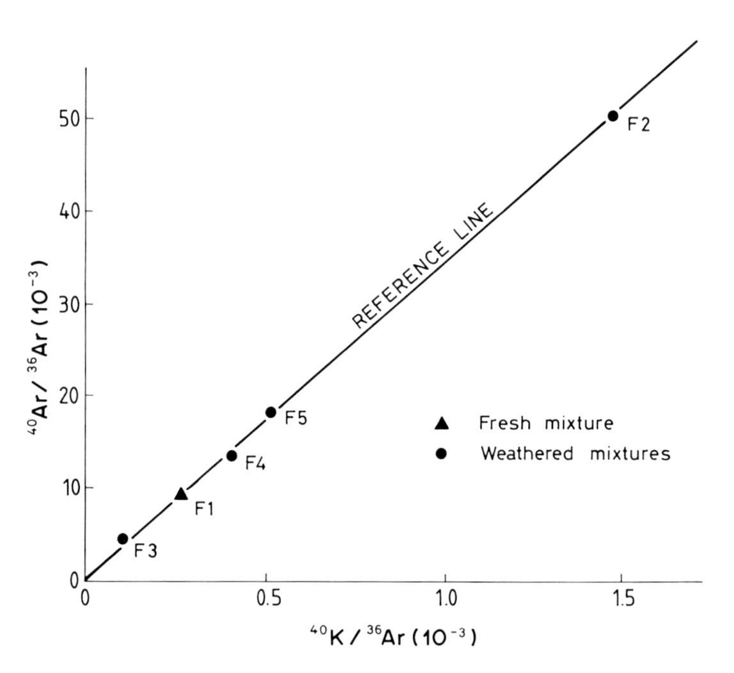

Fig. 2.9. The effect of weathering on the K-Ar system of feldspars in a soil profile in the Republic of Chad. For *numbering*, see Fig. 2.2. (Clauer 1981a)

which influenced the Rb-Sr dates and not the K-Ar dates. Owing to a lack of detailed microstructural or micromorphological evidence about the feldspars, no convincing explanation can be given at this time to account for the disagreement between the two sets of isotopic values for the same plagioclase mineral. However, it can be assumed that the presence of clay impurities in feldspar grains can be an important factor in discrepancy. Boger et al. (1987) followed a careful process of removing adhering clay particles from surface of feldspar grains for the purpose of dating such feldspar grains. Their study has shown that the isotopic dates of carefully selected feldspar grains free of any adhering clays can be used to trace the provenance of sediments containing the feldspar minerals.

1.1.3 The Weathering of Muscovite

The dioctahedral sheet silicates are much more resistant to weathering than the trioctahedral ones (Millot 1970; Wilson 1975), muscovite being long known to be highly resistant to chemical weathering in Earth-surface environments. However, muscovite minerals are known to have experienced reduction in size in soil zones. Park and Pilkey (1981) examined muscovite grains separated from two weathered granites under SEM and found no abrasion or scratches on the mica flakes, but they noticed that microorganisms are abundant on the mica flakes. They suggested that the microorganisms affect the grains physically by splitting and boring.

The orientation of the proton in octahedral OH groups is believed to be a major factor explaining the resistance of muscovite to weathering (Bassett 1960). The distance between the interlayer K^+ and the hydrogen of the hydroxyl groups is greater in dioctahedral than in trioctahedral minerals, and hence the repulsive force between K^+ and hydrogen is less in the dioctahedral form than in the trioctahedral one. Rotation of Si-O tetrahedra around vacant octahedral sites in the dioctahedral structures is believed to cause distortion of the tetrahedral sheet atoms with respect to the adjacent octahedral sheet, also giving a greater stability to dioctahedral structures.

The resistance to changes in the Rb-Sr and K-Ar isotopic composition of muscovite subjected to weathering has been documented by Clauer (1981a). He observed no detectable changes in the model ages between a muscovite sample collected from a fresh pegmatite vein in the Congo Republic, and that from very deeply weathered veins. The isotopic systems seem to remain unaffected as long as no crystallographic modifications occur. Komarneni et al. (1985) did not find any change in the oxygen isotope composition of muscovite following experimental K-depletion. This resistance to oxygen isotope exchange with ambient fluids is additional evidence of the stability of muscovite.

1.2 Sr Isotopic Compositions of Clay Minerals Derived from Biotite and Feldspars

Weathering of biotite can follow a number of paths, one of which is the formation of kaolinite. Clauer (1979b) made Sr isotopic investigations on kaolinite formed from weathering of biotite in a migmatite located in the Republic of the Ivory Coast. The unweathered biotite had an $^{87}Sr/^{86}Sr$ ratio of 2.670 and Rb and Sr contents of 853 and 35 µg/g, respectively, whereas the pure kaolinite fraction derived from this biotite after intense weathering yielded an $^{87}Sr/^{86}Sr$ ratio of about 0.7125 and Rb and Sr contents of 3.2 and 11.6 µg/g, respectively (Table 2.1). The investigation of weathered biotites from the soil profile in the Chad Republic also showed that kaolinite formed in the biotite particles as a result of weathering. The analyses of unweathered biotite yielded an $^{87}Sr/^{86}Sr$ ratio of 1.589 with Rb and Sr contents of 579 and 16.6 µg/g, respectively. In contrast, weathered biotite with identified traces of kaolinite had an $^{87}Sr/^{86}Sr$ ratio of 0.7648 with Rb and Sr contents of 515 and 61 µg/g, respectively. The chemical analyses by EDX coupled with SEM observations revealed that kaolinite developed with spotty distributions in biotite particles, suggesting that the biotite particles are irregularly leached during the weathering process (N. Clauer, unpubl. data). The isotopic data need to be explained in terms of the physical distribution of the kaolinite in biotite. The data on the Ivory Coast kaolinite show that, despite deriving from biotite minerals with very high $^{87}Sr/^{86}Sr$ ratios, this authigenic mineral attains low $^{87}Sr/^{86}Sr$ ratios. Kaolinite with similarly low $^{87}Sr/^{86}Sr$ ratios appears also to have formed in the soil profile of the Chad Republic. Regardless of how the isotopic values are attained by kaolinites, what is important here is that kaolinites derived from biotites are most likely to have low $^{87}Sr/^{86}Sr$ ratios.

Kaolinites developed from feldspars in some soil profiles in the Congo Republic were found to have very high $^{87}Sr/^{86}Sr$ ratios, between 0.7604 and 2.313, Rb contents between about 23 and 60 µg/g, and Sr contents between about 1 and 49 µg/g (Clauer 1979b; Table 2.1). Kaolinite particles of the

Table 2.1. Rb-Sr isotope data of kaolinites formed from different precursors. (Clauer 1979b)

Location	Precursor	Rb (µg/g)	Sr (µg/g)	Rb/Sr	$^{87}Sr/^{86}Sr$ ($\pm 2\sigma/\sqrt{N}$)
1 Congo	Feldspar	46.9	49.2	0.95	0.82689 ± 39
2 Congo	Feldspar	60.2	122.4	0.49	0.76029 ± 27
3 Congo	Feldspar	47.6	7.9	6.03	0.9804 ± 10
4 Congo	Feldspar	23.3	0.9	25.89	2.3129 ± 10
5 Ivory Coast	Biotite	3.2	11.6	0.28	0.7125 ± 10
6 Ivory Coast	Feldspar	46.6	7.2	6.47	1.09424 ± 56

upper part of the profile above the water table had higher $^{87}Sr/^{86}Sr$ ratios but lower Rb and Sr contents than that below the water table. The Sr isotopic variations in the entire soil profile could be related to a mixing of Sr which is trapped in the mineral during its formation from weathering of feldspar, with Sr of groundwater which is adsorbed on the surface of the mineral. Figure 2.10 is a plot of the $^{87}Sr/^{86}Sr$ ratios relative to reciprocal of the Sr contents of the kaolinites, describing a linear relationship between the two parameters. The linear trend suggests that the kaolinites contained Sr primarily from two sources, each having different isotopic compositions. Based on this linear trend, the Sr isotopic ratio of the groundwater Sr adsorbed on the particles may be estimated at about 0.783. The concept of two types of Sr held in kaolinite minerals has also been advocated by Mosser and Gense (1979), but much of the Sr is thought to be adsorbed at the edges of the particles (Fordham 1973). The increase in the Rb contents of kaolinites in the lower zone of the soil profile, below the water table, can be similarly explained as due to a fraction being adsorbed, while the remainder is trapped.

Heterogeneities in the Sr isotopic composition of kaolinites through a soil profile have also been recognized in a profile developed on a gneissic basement in eastern Cameroon (N. Clauer and J.P. Muller, unpubl. data).

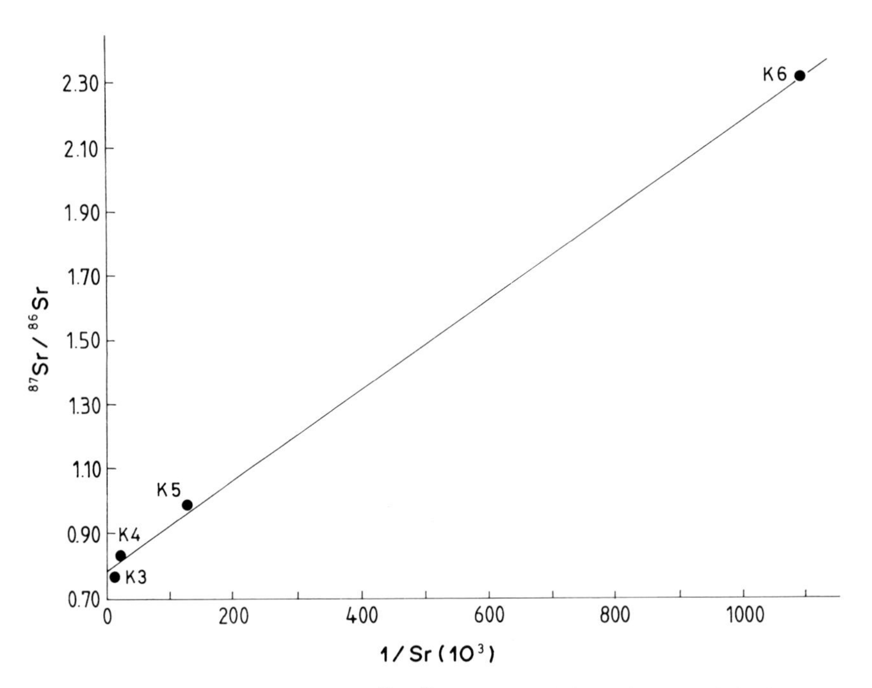

Fig. 2.10. Relationship between the $^{87}Sr/^{86}Sr$ ratios and the reciprocal of the Sr contents of kaolinites in a soil in the Republic of Congo. (Clauer 1979b)

Pure kaolinite samples from a lower loose saprolite zone at a depth of 700 cm, which are the youngest in the profile (Muller 1987), had a mean $^{87}Sr/^{86}Sr$ ratio of 0.72185 ± 0.00006, and that from a depth of 610 cm had a mean $^{87}Sr/^{86}Sr$ ratio of 0.71959 ± 0.00005. A kaolinite fraction collected from a nodule of the intermediate indurated zone at a depth of 435 cm, which still displayed the rock structure, had the highest $^{87}Sr/^{86}Sr$ ratio of 0.72236 ± 0.00003. The lowest $^{87}Sr/^{86}Sr$ ratio of 0.71420 ± 0.00006 was found for the oldest kaolinite sample of the topsoil zone, at a depth of 85 cm. These ranges of isotopic data suggest that authigenic clay minerals in this soil profile formed in isolated microenvironments.

The isotopic data from weathered biotites and feldspars and from authigenic clay minerals derived from these precursors suggest that the same type of clay minerals could have different Sr isotope compositions depending on their location in the soil profile and the type of the precursor. The isotopic data also lead to the idea that microenvironments created within soil profiles are at least as important as the climate or the type of parent rock. Varied Sr isotopic compositions for clay minerals formed in soil profiles also suggest that these minerals are more reflective of microenvironments in the progressively weathered parent materials than of macroenvironments in the continental weathering domain. The importance of microenvironments was also considered by Velde and Meunier (1987), who postulated that several different clay parageneses might coexist in a weathered rock.

1.3 Sr Isotopic Characteristics of Fluids in Weathering Profiles

The preceding discussion has emphasized that authigenic clay minerals generated in the same soil profile may have different $^{87}Sr/^{86}Sr$ ratios depending on the precursor and the physicochemical conditions of the environment during the mineral-fluid reactions. For thorough understanding of the mechanisms governing clay genesis during continental weathering processes, knowledge of chemical and isotopic compositions of ambient fluids is helpful. Fluids which may be identified with the clays in a soil zone are difficult to collect in the field. Consequently, information about the isotopic composition of soil fluids in equilibrium with clay minerals is altogether lacking. Nevertheless, isotopic characteristics of such fluids may be indirectly assessed from analyses of soluble salts which can be leached from the weathered materials.

Clauer (1976, 1981a) leached with distilled water the soluble materials of the saprolite whole rocks from upper and intermediate zones of the soil profile of the Chad Republic and examined the $^{87}Sr/^{86}Sr$ ratio of the solutions. The water-soluble component of the soil in the upper part of the profile gave an $^{87}Sr/^{86}Sr$ ratio of about 0.709, whereas that in the intermediate part of the profile yielded a value of about 0.715. Although no water leachate was made on materials from the very low part of the profile, two different size fractions from extensively weathered biotites at this level gave very similar $^{87}Sr/^{86}Sr$ ratios of about 0.765. The Sr content of one of the two fractions was twice that of the other. Similarity in the Sr isotope compositions between these two fractions of a weathered biotite sample, despite their very different Sr contents, suggests that the $^{87}Sr/^{86}Sr$ ratio of Sr trapped in the structure was identical to that adsorbed on the mineral particles. The high Sr in one of the fractions can be due to assimilation of Sr from the ambient pore fluid with an $^{87}Sr/^{86}Sr$ ratio of about 0.765. In summary, the $^{87}Sr/^{86}Sr$ ratio of the pore fluid of the soil profile in the Republic of Chad varied from 0.709 in the upper part of the profile, through about 0.715 in the intermediate part, to about 0.765 in the lower part.

Additional information about the $^{87}Sr/^{86}Sr$ ratio of ambient fluids in soil profiles comes indirectly from the isotopic study by Clauer (1979b) on authigenic kaolinites collected both below and above the water table from a deeply weathered soil profile in western Africa. The kaolinite below the water table contained about 60 µg/g Sr with an $^{87}Sr/^{86}Sr$ ratio of 0.7603, as compared to the kaolinite sample located above the water table, with about 49 µg/g Sr and an $^{87}Sr/^{86}Sr$ ratio of 0.8269 (Table 2.1). Assuming that the difference of 11 µg/g of Sr between the two kaolinite samples is primarily due to adsorption by the mineral, a mass balance calculation shows that the external Sr adsorbed onto the mineral particles in contact with the water had an $^{87}Sr/^{86}Sr$ ratio of about 0.7155.

The Sr isotopic composition of ambient fluids in soil profiles can also be deduced from analyses of changes in the Rb-Sr systematics of the parent materials being weathered. The Sr of each bulk parent material is essentially partitioned into the weathered product and the ambient solution, which is nearly Rb-free. Thus the Sr isotopic composition of the solution may be computed from the intercept on the ordinate of a Rb-Sr isochron diagram by the line connecting the Rb-Sr isotopic values of the parent material and the weathered residue. Taking this approach, we have reanalyzed the data given in the publication of Dasch (1969) to demonstrate that the $^{87}Sr/^{86}Sr$ isotopic compositions of soil fluids vary significantly through a soil profile. The data of Dasch (1969) on samples of red saprolite developed on the Elberton granite in Georgia are recast in a Rb-Sr diagram (Fig. 2.11). The data points for the least weathered samples plot close to the theoretical isochron for the fresh granodiorite, but that of the two most weathered samples plot below, to the right of the isochron. The relatively small deviation of the data points for the least weathered samples away from the isochron probably reflects the influence of feldspar weathering in causing initially preferential leaching from Sr-enriched (or Ca-enriched) domains in the feldspar grains. This preferential leaching would lead to increases in both the Rb/Sr and the $^{87}Sr/^{86}Sr$ ratios for the residual materials. The increased deviation of the Rb-Sr isotopic values for the most highly weathered products away from the isochron probably reflects additional complexity as the effect of weathering

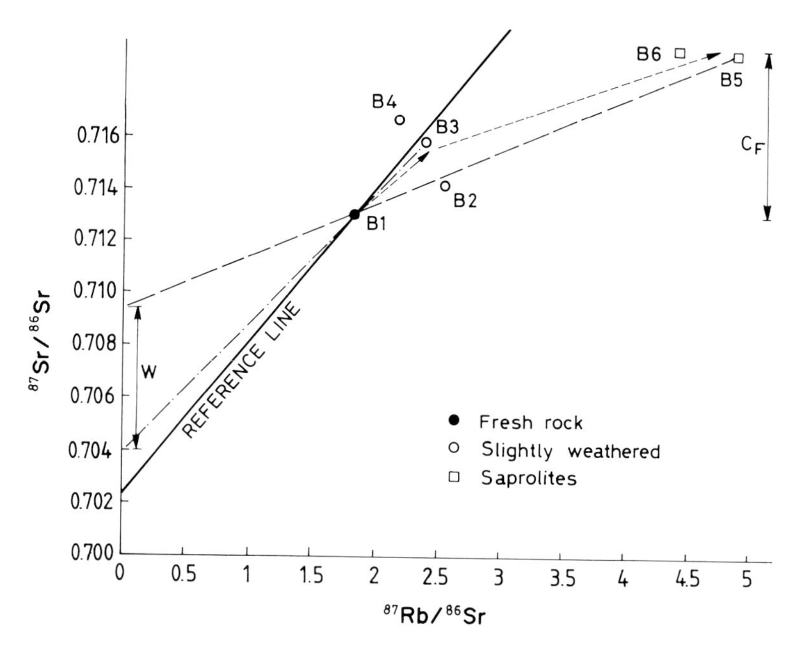

Fig. 2.11. Rb-Sr data of an unweathered and several weathered rock samples of Elberton granite, Georgia. The *arrows* marked with the signs W and C_F indicate the range of Sr isotopic compositions of waters and clay minerals, respectively, due to weathering of principally feldspars in the rock. (After Dasch 1969)

of biotite is now superimposed on the effect of weathering of feldspar. The Sr isotopic composition of the fluid may be deduced from the intercept of the line connecting the data points for the fresh material and its weathered products, on the ordinate of a Rb-Sr isochron. As shown in Fig. 2.11, the $^{87}Sr/^{86}Sr$ ratios of soil fluids related to the soil development in Georgia varied from about 0.7040 for the least weathered samples to about 0.7095 for the most weathered samples.

Dasch (1969) also presented the Rb-Sr isotopic data of both a soil profile and its underlying fresh rock belonging to the 1700-Ma-old Boulder Creek granite in Colorado. The entire soil zone is about 30 m thick and consists of a lower bleached facies and an upper red facies. The soil zone was developed prior to the time of deposition of the Pennsylvanian Fountain Arkose which overlies the profile. The data point of the fresh rock and the data points of saprolites from both the bleached and the red facies are recast in a Rb-Sr isochron diagram (Fig. 2.12). The intercept on the ordinate for each line between the fresh rock and individual soil samples then represents the $^{87}Sr/^{86}Sr$ ratio of the soil fluid in association with the respective saprolite sample. The "bleached rock" facies and the "red rock" facies belong to two different modes of soil development. The upper parts of the former zone contain anomalously high amounts of Sr relative to the lower parts. The

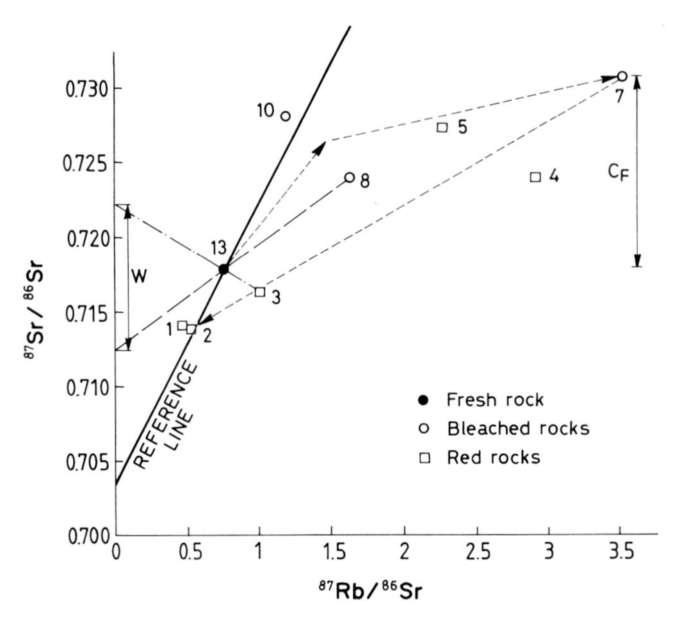

Fig. 2.12. Rb-Sr data of an unweathered and several weathered rocks of Boulder Creek granite, Colorado. For *symbols*, see Fig. 2.11. (After Dasch 1969)

bleached samples appear to be genetically linked to the underlying fresh granodiorite. The data points of the two most weathered samples (7 and 8) from the bleached facies are the farthest away to the right of the isochron, whereas that of the least weathered sample (10) is slightly above but closer to the isochron. The weathered samples from the bleached facies have higher $^{87}Sr/^{86}Sr$ and Rb/Sr ratios than the fresh rock. The Rb-Sr isotopic data for the weathered samples may be explained in relation to feldspar-dominated weathering systems. Based on this reasoning, the $^{87}Sr/^{86}Sr$ ratios of soil fluids progressively increased from 0.7125 to 0.7145, as calculated from the intercept on the ordinate of each line connecting the fresh rock and a weathered sample. The evolutionary history of the Rb-Sr isotopic systems of the upper red facies remains somewhat unclear, because of the uncertainty of its origin. Nevertheless, the data make evident that the more extensively weathered samples with relatively low Rb/Sr and $^{87}Sr/^{86}Sr$ ratios lie closer to the theoretical isochron than the less weathered samples, suggesting that the ambient soil fluids experienced Sr isotopic changes with increase in the degree of weathering.

From the discussions above, it becomes clear that the ambient fluids related to the authigenesis of minerals in a soil profile developed over any crystalline rock are heterogeneous in Sr isotopic compositions. The heterogeneity results from varying degrees of weathering of mainly feldspar and biotite minerals in the parent rock and reflects fluid-mineral interactions in numerous nearly isolated, closed systems within the soil profile. A major significance of the heterogeneity in the Sr isotopic compositions of ambient fluids is that the solid products of weathering through a soil profile would also be heterogeneous in Sr isotopic composition. The evolution of the isotopic systems of the weathered products and the isotopic relations of different secondary products through a soil profile are discussed below.

1.4 A Sr Isotopic Model for Clay Authigenesis in Soil Profiles

Plagioclase feldspars generally weather more readily than biotites (Goldich 1938). This differential behavior between the two types of minerals should be reflected in the Sr isotopic data of their weathered products. The studies discussed above have emphasized that the weathering of plagioclase minerals tends to increase the Rb/Sr ratios of the whole rocks faster than their $^{87}Sr/^{86}Sr$ ratios, whereas the weathering of biotite minerals produces a reduction of both the Rb/Sr and the $^{87}Sr/^{86}Sr$ ratios of the whole rocks. The trend of isotopic changes in the whole rock with continued weathering of feldspar and biotite is depicted in a Rb-Sr isochron diagram (Fig. 2.13). The representation shows that as long as hydrolysis of the feldspar minerals

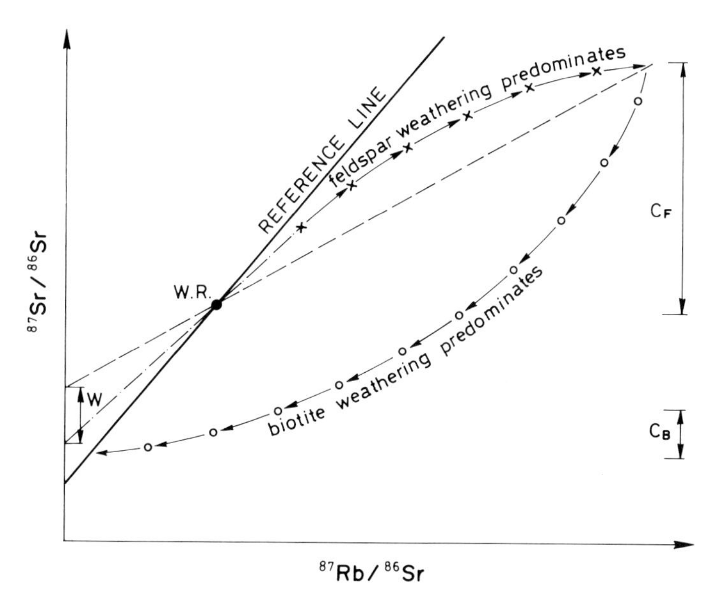

Fig. 2.13. Theoretical scheme for a modeled evolution of the isotopic Rb-Sr signatures of waters and clays in a soil profile. The *arrow* marked with the sign C_B indicates the range of Sr isotopic compositions of clay minerals due to weathering of principally biotites. For the other *symbols*, see Fig. 2.11 and text; *W.R.* whole rock

prevails, the changes in the Rb-Sr systematics of the unweathered rock together with the solid residue may be described by a curvilinear displacement of the $^{87}Sr/^{86}Sr$ and the Rb/Sr ratios below the theoretical isochron of the fresh whole rock. The very first Sr released to the water from the whole rock is probably selectively leached from the most Ca-rich domain in the plagioclase crystals, causing the $^{87}Sr/^{86}Sr$ ratio of the leached Sr to be lower than that of the bulk parent material. The clay minerals formed during this early phase of weathering will have higher $^{87}Sr/^{86}Sr$ and Rb/Sr ratios than those of the bulk parent material. As the weathering continues, the $^{87}Sr/^{86}Sr$ ratios of the leachates will be increased and the residual products will continue to be higher in both the Rb/Sr and the $^{87}Sr/^{86}Sr$ ratios. The Rb/Sr and the $^{87}Sr/^{86}Sr$ ratios of the bulk material continue to move upward below the reference isochron in response to selective leaching of Sr from Ca-rich domains in the plagioclase crystals. When the weathering of the plagioclase grains has progressed to the stage that no significant chemical heterogeneity exists in the unweathered part of the plagioclase crystal, any subsequent weathering then would lead to the leachate and the solid residue having the same Sr isotopic composition.

According to the model of weathering of feldspar that we have just described, authigenic clay materials in soils should be widely varied in their $^{87}Sr/^{86}Sr$ and Rb/Sr ratios. This suggestion of a spread in the isotopic composition of clay minerals is supported by the Rb-Sr isotopic evidence from weathered products from the feldspar-rich Elberton granite (Dasch 1969). In this case, the $^{87}Sr/^{86}Sr$ values for the authigenic clay material are suggested by the arrow C_F in Fig. 2.11. They most likely ranged from 0.713, which is the $^{87}Sr/^{86}Sr$ ratio of the fresh rock, to at least 0.720, which is the highest $^{87}Sr/^{86}Sr$ ratio obtained in the study. Similar variations in the $^{87}Sr/^{86}Sr$ ratios of the clay products presumably existed in the bleached and the red facies of the soil profile developed on the Boulder Creek granite (Dasch 1969; arrow C_F in Fig. 2.12). An important feature to note here is that the successive generations of clay products formed from the weathering of the same parent feldspar could yield an apparent isochron with a positive slope, as the data points of the weathered products lie along a linear array in a Rb-Sr isochron diagram (Fig. 2.11).

When weathering of biotite is dominant in a rock, the trend of changes in the Rb-Sr isotopic system will be roughly along a concave downward path below the theoretical isochron line for the fresh material prior to its weathering. The initial changes are due to selective loss of radiogenic ^{87}Sr from biotite flakes accompanied by incorporation of somewhat large amounts of normal Sr from ambient fluids. At the very advanced stage of weathering of biotite, marked by intense leaching of K and other alkali ions, selective loss of ^{87}Sr from biotite is less important than the increased adsorption of normal Sr from ambient fluids. At this stage the $^{87}Sr/^{86}Sr$ ratios of the end products are mainly dominated by Sr from fluids whose $^{87}Sr/^{86}Sr$ ratios are controlled by the dissolution of feldspar. The end products formed can yield varied Rb/Sr ratios, due to changes in the amounts of Sr being adsorbed, but only limited variations in the $^{87}Sr/^{86}Sr$ ratios. Consequently, the end products from this very advanced stage of weathering may fit a Rb-Sr isochron with a nearly zero slope. Alternatively, the isochron may be viewed as a mixing line between essentially two end products, a clay attaining a high Rb/Sr ratio by evolving from a weathered biotite (arrows C_B in Figs. 2.13 and 2.15, and curve B(1)–B(2) in Fig. 2.14), and a clay acquiring a relatively low Rb/Sr ratio through derivation from highly weathered feldspar (arrows C_F in Figs. 2.11, 2.12, 2.13 and 2.15, and curve A in Fig. 2.14).

The results of the study by Blaxland (1974) on naturally weathered biotite and feldspar minerals and whole rocks of Butler Hill granite in Missouri may now be processed through the model described above to predict the Sr isotopic compositions of clays formed from weathering of plutonic rocks. The $^{87}Sr/^{86}Sr$ and Rb/Sr ratios of the unweathered granite and variously weathered fractions of the rock are shown in a Rb-Sr isochron-diagram (Fig. 2.15); these ratios reach maximum values at an intermediate stage and then decrease to much lower values than that of the unweathered sample, almost conforming to the theoretical isochron trend. Following our model, the clay minerals deriving from weathering of feldspars

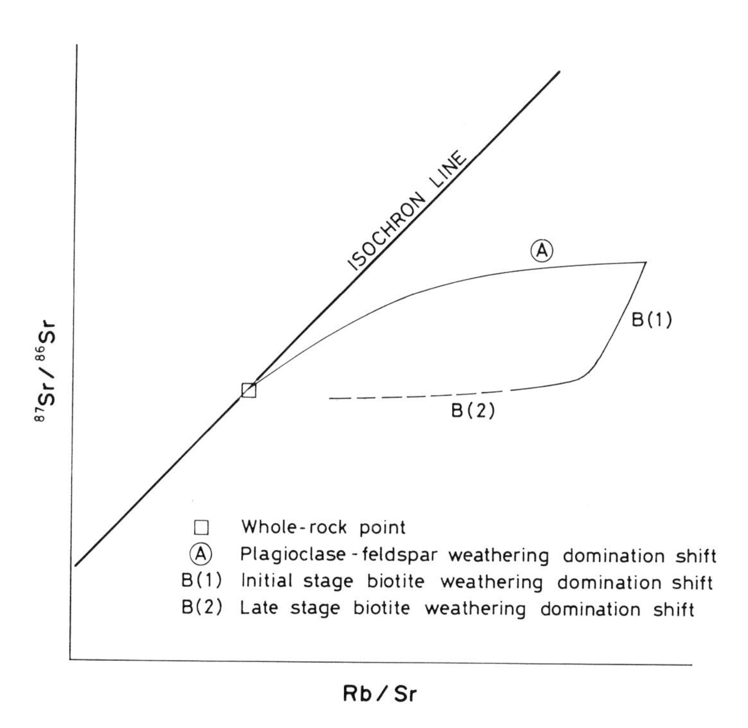

Fig. 2.14. Theoretical scheme for the effect of weathering on the Rb-Sr of whole rocks in a soil profile

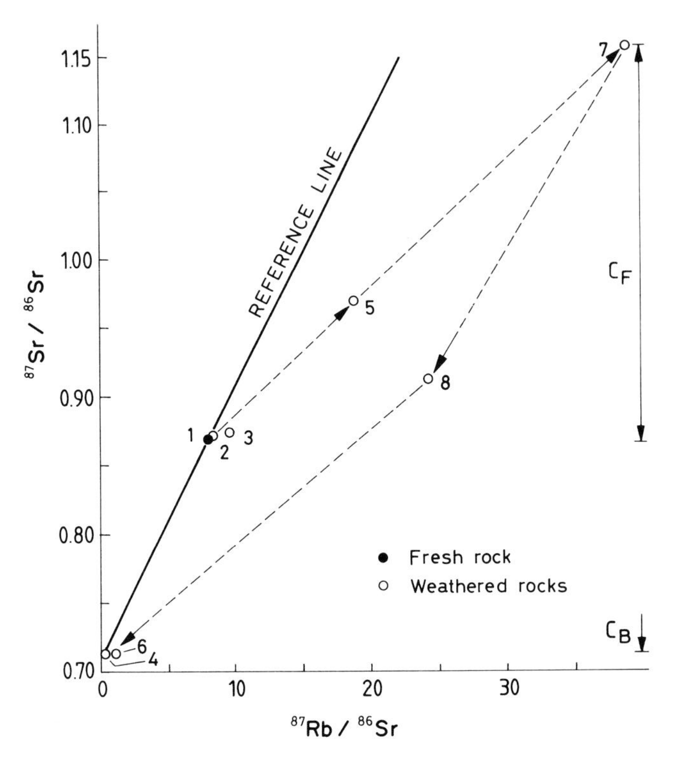

Fig. 2.15. Rb-Sr isotopic data of an unweathered and several weathered samples of Butler Hill granite, Missouri. For *symbols*, see Fig. 2.11. (Blaxland 1974)

of Butler Hill granite would have $^{87}Sr/^{86}Sr$ ratios ranging between 0.870 and 1.150 (arrow C_F in Fig. 2.15). Such a range of values is not unusual, as Clauer (1979b) reported that kaolinites formed from weathering of feldspars in different soil profiles in Western Africa had $^{87}Sr/^{86}Sr$ ratios between 0.827 and 2.313. The model also dictates that clays formed from the weathered biotites, in the highly weathered samples, should have quite low $^{87}Sr/^{86}Sr$ ratios of about 0.710 (arrow C_B in Fig. 2.15), which is also close to the $^{87}Sr/^{86}Sr$ ratios found elsewhere for such minerals.

The results of a study by Brass (1975), who examined the Rb-Sr system of soil profiles developed on arkoses in New Zealand, can also be explained in terms of the model described above. Brass observed several different Rb-Sr alignments for these soil samples, which systematically had lower slopes than that of the fresh, unweathered rocks (Fig. 2.16). In one extreme example, the analytical data of the soil samples gave a nearly zero slope for the line, indicating a total loss of radiogenic ^{87}Sr in recent time and suggesting an isotopic equilibrium with environmental Sr (Fig. 2.17). Brass suggested, merely by speculation, that the initial $^{87}Sr/^{86}Sr$ values defined by the lines corresponded to the $^{87}Sr/^{86}Sr$ ratios of the ambient fluids. According

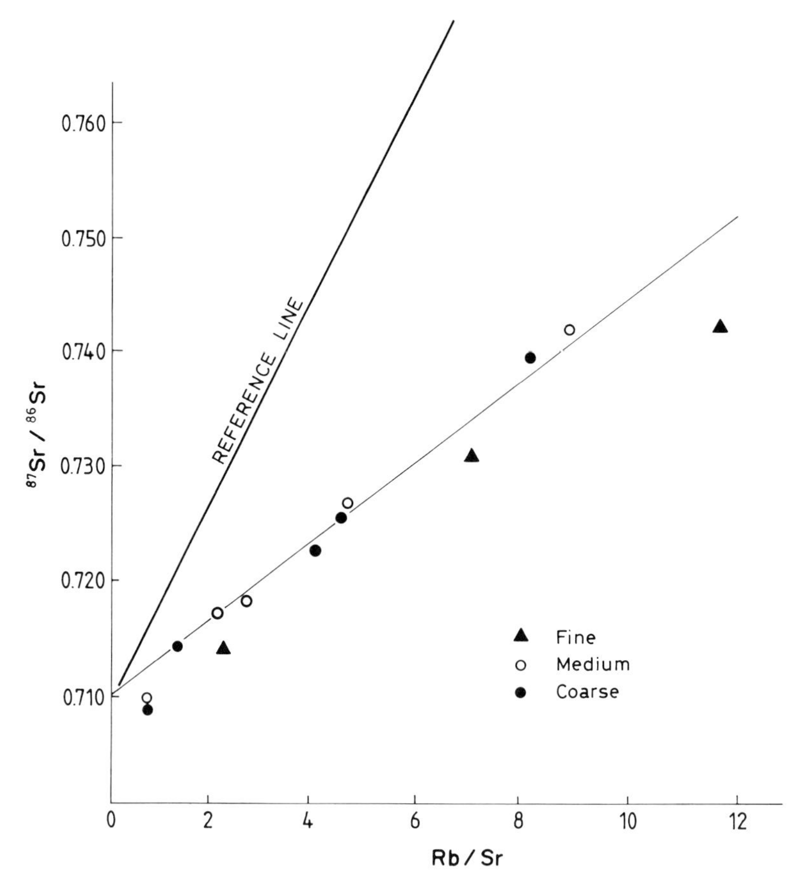

Fig. 2.16. The Rb-Sr isotopic systems of different fractions of soil samples in a soil profile developed on an arkose in New Zealand. (Brass 1975)

to our model depicting the changes in the $^{87}Sr/^{86}Sr$ ratios of the weathered products and associated fluids, each line given by Brass (1975) reflects an artifact due to the mixing of clay minerals which evolved with different $^{87}Sr/^{86}Sr$ ratios during different stages of the weathering process.

It seems that according to the model, the most severely weathered samples should plot very close to the theoretical isochron. This may explain why Fullagar and Raglan (1975) obtained ages on highly weathered rocks that were reasonably close to those of the fresh rocks. The model discussed above may also be helpful in analyzing the changes that occurred to the Rb-Sr isotopic composition of an Archean plutonic rock body studied by Worden and Compston (1973). The authors found that the data points of variously weathered whole rock samples of the pluton were scattered along the isochron defined by the data points of the fresh whole-rock samples. The authors remained uncertain about the interpretation of the colinearity of the

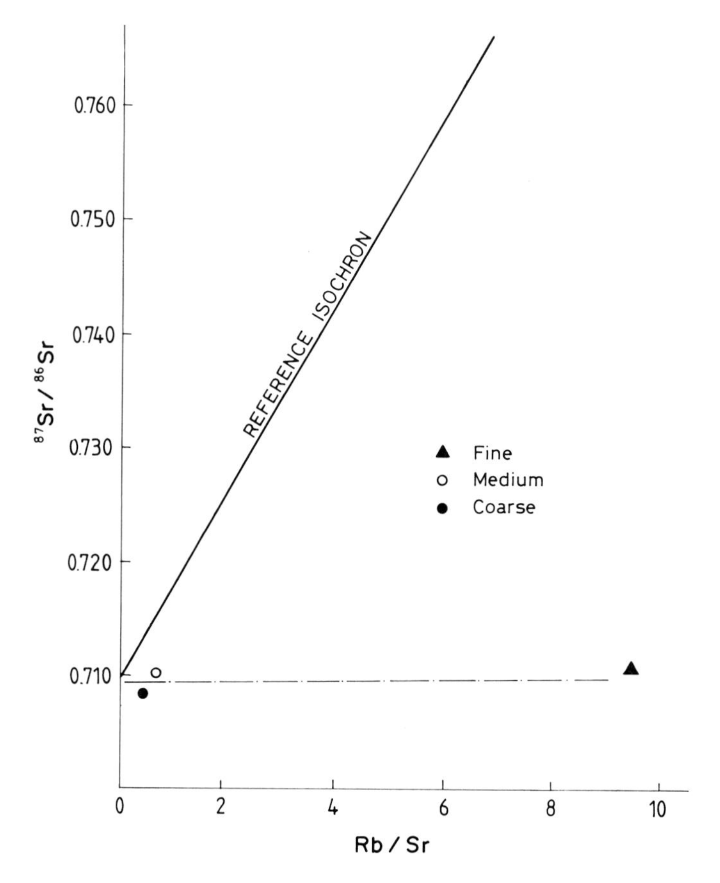

Fig. 2.17. The Rb-Sr isotopic systems of different fractions of soil samples in a soil profile developed on an arkose in New Zealand. (Brass 1975)

data points in the isochron diagram, whether resulting from hydrothermal activity or from a weathering process shortly after the formation of the pluton. Following our model of isotopic redistribution in weathering environments, we suggest that the data are amenable to an interpretation highlighting the effects of weathering on the plutonic rocks before any significant growth of radiogenic Sr occurred in the various mineral components. By averaging the data points for each step in the weathering process, Clauer (1976) showed, as displayed in Fig. 2.18, that with increases in the degree of weathering of the rock samples, the data points moved progressively to the right of the diagram and then, for the most weathered samples, shifted far to the left of that of the fresh samples. We earlier discussed the loop-shaped trend in the shift of the Rb-Sr isotopic data for weathered whole-rock samples relative to that for the corresponding fresh samples. In the case of the pluton studied by Worden and Compston, the shift without the

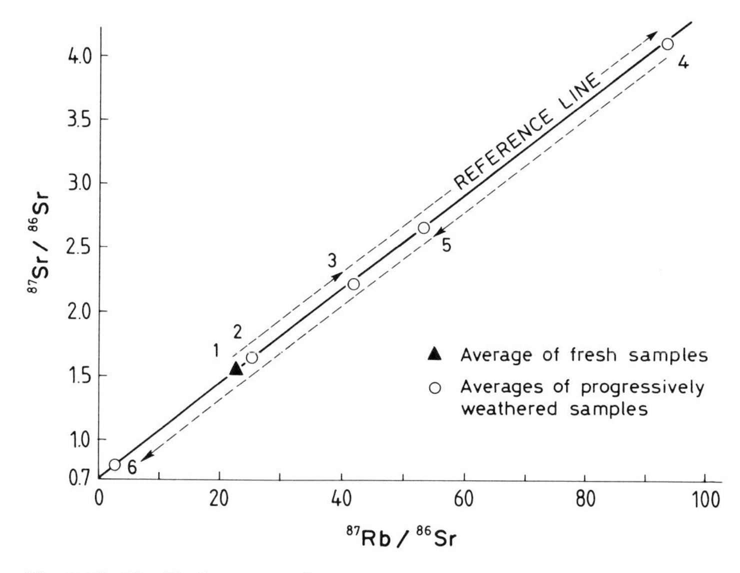

Fig. 2.18. The Rb-Sr system of progressively weathered Archean plutonic-rock samples shortly after formation. The *numbers* represent the weathering sequence from fresh rock samples (*1*) to increasingly weathered materials (*2* to *6*). (After Worden and Compston 1973)

prominent loop could reflect weathering of the rock samples in the very early phase of their formation history, before any significant radiogenic ^{87}Sr was formed by decay.

1.5 Stable Isotope Geochemistry of Clay Minerals from Soil Profiles

The oxygen and hydrogen isotope composition of a particular clay mineral formed in a soil environment is controlled by: (1) the isotopic composition of the ambient water, (2) the isotopic composition of the parent materials, (3) the ratio of the amount of water to parent material, and (4) the temperature-dependent equilibrium isotope fractionation factors between the newly formed clay mineral and the ambient water. Savin and Lee (1988) summarized the values of equilibrium oxygen and hydrogen-isotope fractionation factors for different clay mineral-water systems at surface weathering temperatures. Some degrees of uncertainty are attached to these values for each different clay mineral-water system, because the fractionation factors were determined either from analysis of naturally occurring minerals whose conditions of formation, composition, or status of physical purity are questionable, or established by using empirical bond-type calculations which require several assumptions, or extrapolated from high temperature values. As the isotopic exchange rates have been found to be extremely slow at weathering temperatures, the equilibrium isotope fractionation factors could not be experimentally determined.

The small range of surface temperatures between 0 and about 35 °C produces only small changes in the equilibrium isotope fractionation factors. Also, at these temperatures, no isotopic exchange in the clay minerals is believed to occur in the soil environment (Lawrence and Taylor 1971; James and Baker 1976). Studies on the stable isotope composition of clay minerals in soil environments are very few. Most of these studies have assumed that clay minerals formed in soil profiles under conditions of very high ratios of water to parent material. One advantage of such an assumption is that the stable isotope compositions of clay minerals may then be used to interpret the isotopic composition of local waters or to estimate the temperature of formation using the values of isotopic composition of local meteoric water. The importance of local meteoric water was claimed some 20 years ago by Savin and Epstein (1970a), who found that naturally occurring, low-temperature, nonmarine kaolinite samples from widely different localities, are enriched in ^{18}O and depleted in D relative to local meteoric waters, but that the isotopic composition of the kaolinites can be correlated with the isotopic values of the local meteoric waters. The linear relationship indicated that a range of kaolinites were formed in isotopic equilibrium with their respective local meteoric waters. Savin and Epstein also found that, unlike that of kaolinite, the oxygen and hydrogen isotope composition of smectites of low temperature origin varies widely, and lacks good colinearity relative to the oxygen-hydrogen isotopic trend of meteoric waters. Lawrence and Taylor (1971), analyzing several Quaternary soils from different regions of the USA, found that the isotope composition of kaolinite samples in general correlated with the isotope composition of corresponding present-day meteoric waters. The $\delta^{18}O$ ranged from 0 to +29 per mill, and the δD varied from +25 to −170 per mill; but they also observed that the isotope compositions of kaolinite samples within the same locality were varied (Fig. 2.19). Furthermore, they observed that the isotope compositions of kaolinite and smectite samples from the same soil zone were sometimes nearly identical and that in general the isotope compositions of smectite samples agreed more closely to the "kaolinite line" than to the "smectite line" defined previously by Savin and Epstein (1970a). Lawrence and Taylor, therefore, suggested that either the fractionation factors for the two minerals are similar, or the compositions of the smectites or the formation temperatures of the smectites in the soil they studied were different from those of the smectites investigated by Savin and Epstein. Lawrence and Taylor did not consider the potential effects of other factors such as evaporation of water in the soil environments and crystallization in microenvironments with low water-to-parent material ratios.

The effect of evaporative process on the stable isotope compositions of soil waters was recognized by Allison and Hughes (1982), who observed differences in ^{18}O of as much as +6 per mill. Gouvea Da Silva (1980) also

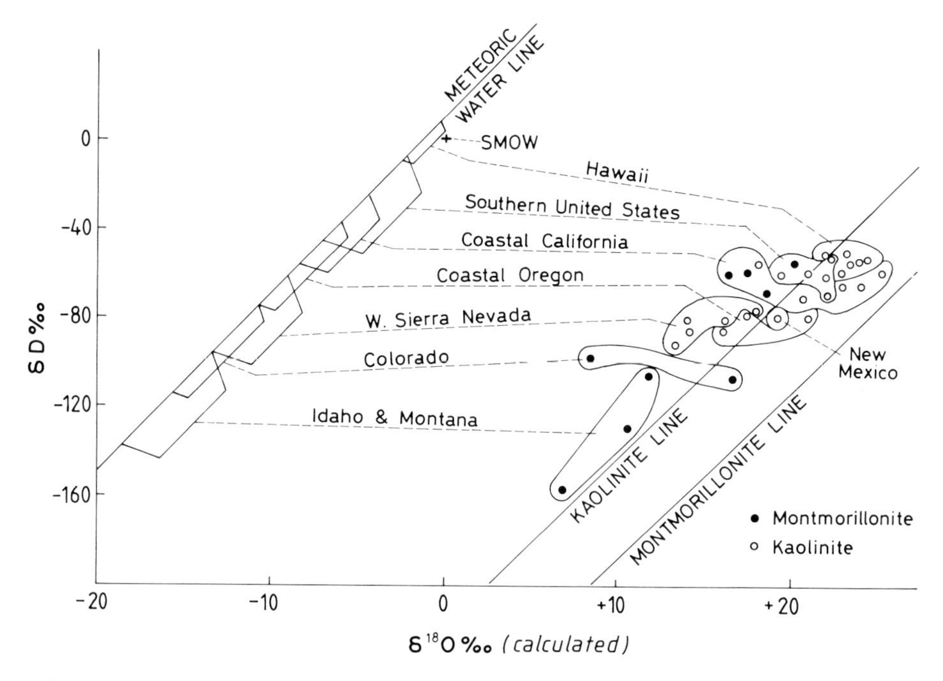

Fig. 2.19. The stable isotope compositions of kaolinite samples in soil profiles in North America and Hawaiian Islands in relation to corresponding local present-day meteoric waters. (Lawrence and Taylor 1971)

mentioned that considerable modification of the chemical and isotopic composition of soil solutions occurs due to the evaporative process. Spiers et al. (1985) found from analyses of clay separates from sediments in two different soil profiles in northeastern Alberta that the $\delta^{18}O$ value of untreated clay separates, which contained both detrital and authigenic poorly crystalline materials, decreased with increasing profile depth, the net decrease being about 2 per mill. They ascribed these changes in $\delta^{18}O$ to the formation of poorly crystalline authigenic clay materials in soil solutions whose ^{18}O contents varied across the profile as a result of evaporation of infiltrated meteoric waters.

Illustrations of the formation of clay minerals in soil environments with very low water-to-parent material ratios are very few. Clauer et al. (1982a) noted a change of about 2 per mill in the ^{18}O content of biotites weathered to various degrees to kaolinite. As Clauer et al. noted, if these neoformed clays are conceived as being the result of local dissolution and reprecipitation reactions in a microenvironment with a small water-to-parent mineral ratio, the kaolinite could not have been in isotope equilibrium with local meteoric water. If the mineralization occurred in the presence of large amounts of local meteoric water, the solutions would have been depleted in

^{18}O and D contents relative to present-day local waters, indicating that these minerals had formed during a period of cold climate.

Bird and Chivas (1988) analyzed the stable isotope composition of some Permian kaolinites and suggested that these minerals formed under conditions of extensive leaching by waters which were at polar to sub-polar temperatures. Bird and Chivas also maintained that oxygen isotope exchange was negligible between authigenic kaolinite and waters in the environment since crystallization. Lawrence and Taylor (1972) found that smectite in the Big Sur soil profile in California consists of variable mixed phases – a gray variety with 3.5 to 4.5% Fe_2O_3, and a brown variety with 6.5 to 8.5% Fe_2O_3. Based on a mass balance calculation, they estimated that the average $\delta^{18}O$ value of the gray smectite is about +21.5 per mill and that of the brown smectite is about +18.7 per mill. The authors did not say how this wide range of isotopic values could be explained in terms of crystallization of the smectites in an environment of high water-to-mineral ratio.

Clauer et al. (1989) investigated the oxygen isotope compositions of kaolinite minerals in a soil profile developed over a gneissic basement rock in eastern Republic of Cameroon. The soil profile consisted of three main superposed zones of variable extent: a loose saprolite zone at the base, a nodular zone with indurated ferruginous nodules embedded in soft and mainly red clay material at the intermediate level, and a yellowish clay-rich topsoil zone (Muller and Bocquier 1986). The $\delta^{18}O$ values of the kaolinite fractions from the basal saprolite zone, and also from nodules in the intermediate zone averaged to $+17.3 \pm 0.4$ per mill (Fig. 2.20). Kaolinite samples from the intermediate red clay zone yielded an average $\delta^{18}O$ value of $+18.2 \pm 0.2$ per mill, while those from the soft red material gave an average value of $+19.1 \pm 0.2$ per mill. The $\delta^{18}O$ values of kaolinites from the topsoil zone ranged from +17.4 to +18.7 per mill. These isotopic variations for pure kaolinite fractions are not compatible with the assumption of a complete or a progressive isotopic equilibrium with the environmental or meteoric waters across the soil profile. The differences in the $\delta^{18}O$ values of the kaolinites across the soil profile must derive from crystallization of several generations of kaolinites forming either at different temperatures, or, more probably, under different microenvironmental chemical conditions in the soil zone.

In summary, it can be seen that the $\delta^{18}O$ values of kaolinite minerals can be expected to be measurably variable across the same soil profile, which strongly suggests that the ambient fluids also varied in their oxygen isotope compositions. Such variations in the oxygen isotope composition of ambient fluids make it unlikely that high water-to-mineral ratios existed during crystallization of the authigenic clay minerals; however, they are possible if the formation of clay minerals in soils is controlled by microenvironments.

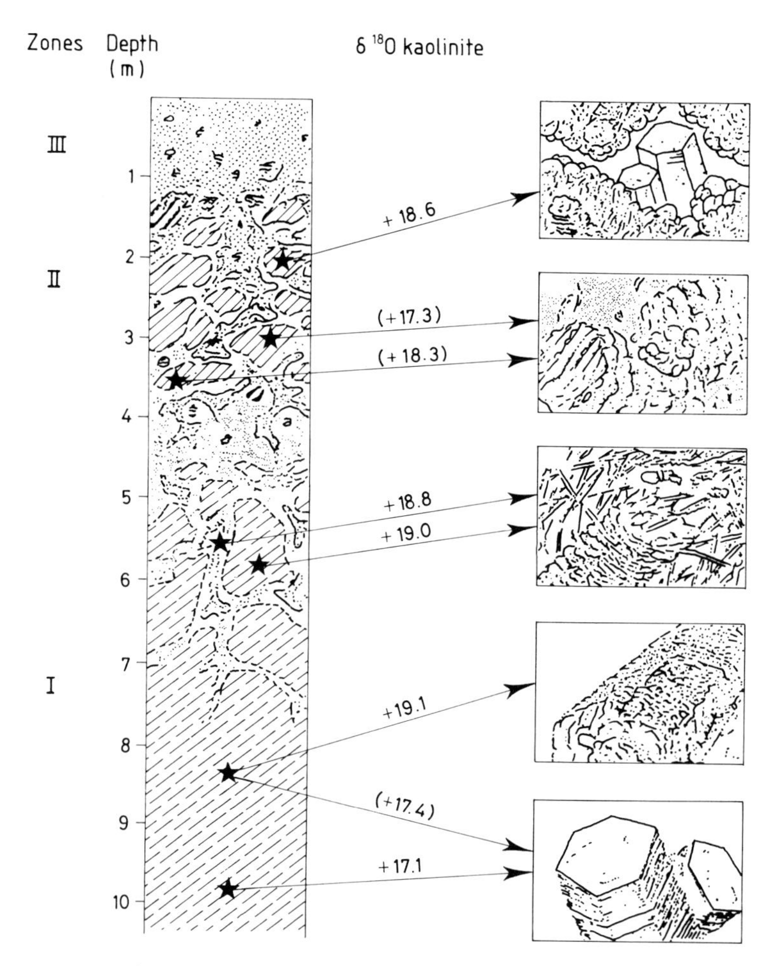

Fig. 2.20. $\delta^{18}O$ values of pure kaolinite samples of a soil profile in the Republic of Cameroon. A description of the soil profile is given on the *left* with the sample locations, the $\delta^{18}O$ values are given in the *middle* and enlargements of the kaolinite habitus are given to the *right*. (After Clauer et al. 1989)

1.6 Experimental Clay Authigenesis and Evaluation of Mass Transfers in Soil Profiles

Thellier and Clauer (1989) analyzed the Rb-Sr isotopic compositions of clay minerals and solutions from an outdoor experiment carried out by Thellier (1984) under semiarid weather conditions. The experiment focused on the effects of continued movements of capillary moisture, under evaporative conditions, on the chemical and mineralogical compositions of soils (Thellier et al. 1988). The soils were put into several columns of different lengths ranging from 11 to 51 cm. Each column was screened at the bottom and open at the top. The bottom 2 cm of each column was kept submerged in a water solution contained in a tank. To prevent evaporation of the water solution from the surface in the tank, and thus to allow evaporation only through the surface of the soil column, a thin layer of paraffin oil was added to the surface of the solution. The original soil consisted of 20% clay, 22% silt, and 58% sand, which is a typical composition for soils of equatorial Africa. The clay fraction consisted of 55% smectite, 40% kaolinite, and 5% illite. The silt and sand fractions consisted essentially of feldspar, muscovite, and quartz. Continued movement of capillary moisture resulted in the crystallization of some authigenic clay-type silicate minerals in the upper sections of the columns. Observations by TEM revealed that the detrital smectite flakes were surrounded by tiny smectite-type fibers and associated with long rigid palygorskite-like laths (Fig. 2.21). Based on the chemical composition of the solution used for the experiment, thermodynamic calculations attested to the formation of these Mg-rich crystals from solutions in the soil (Thellier et al. 1988).

Thellier and Clauer (1989) noted that the $^{87}Sr/^{86}Sr$ and the Rb/Sr ratios of the clays through the different columns broadly defined a mixing line in a

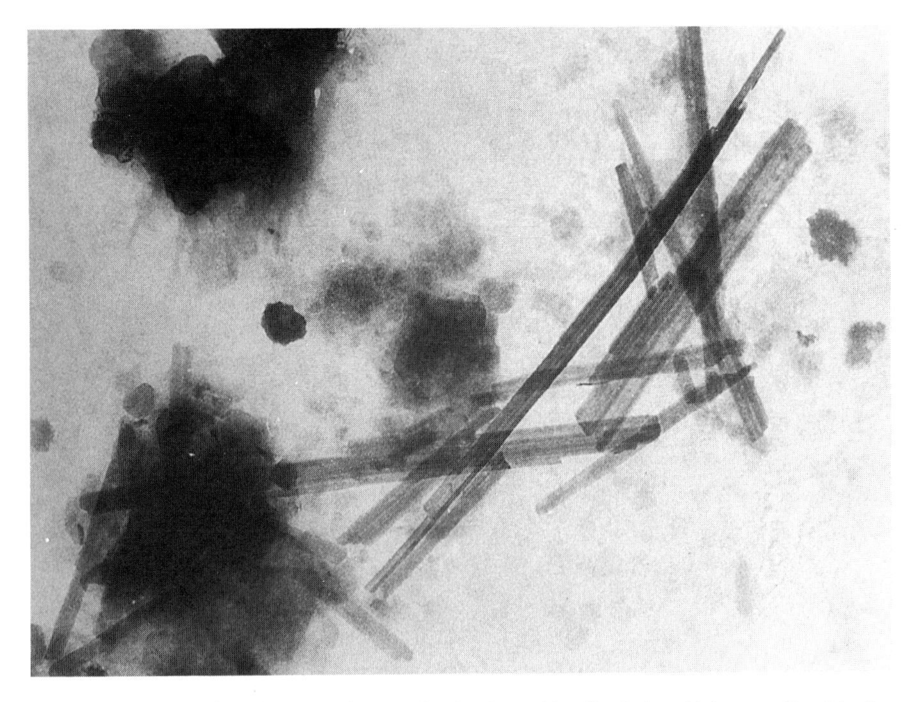

Fig. 2.21. Transmission electron micrograph of palygorskite-like laths which crystallized in the uppermost part of experimental columns. (Thellier et al. 1988)

Rb-Sr isochron diagram attesting to the occurrence of two end-members, the original clay material with an $^{87}Sr/^{86}Sr$ ratio of about 0.734 and the solution with an $^{87}Sr/^{86}Sr$ ratio of about 0.717 (Fig. 2.22). However, the mineralogical details revealed that considerable repartitioning of Rb and Sr must have occurred as a result of authigenesis, and that the mixing line does not result from a mere addition or substraction of either element from the fluid to the solid phase or vice versa. Thellier and Clauer also noted that the authigenic palygorskite-like minerals appeared to have the same Sr concentration and $^{87}Sr/^{86}Sr$ ratio as, but less Rb concentration than, the smectite precursor, suggesting that the authigenesis resulting in unchanged isotopic compositions had to occur in a closed system. The consequence of the decrease in the Rb/Sr ratio of the clay assemblage resulting from replacement of smectite by palygorskite is an increase in the Rb-Sr date. Thellier and Clauer, therefore, cautioned against uncritical use of Rb-Sr dates for paleogeographic reconstructions of sediments, because pedogenic clays themselves, within a single soil, can have widely varied Rb-Sr isotopic compositions due to varied moisture conditions in varied microenvironments.

N. Clauer (unpubl. data) separated plagioclase grains from the soil samples following the above evaporation experiment and analyzed their Rb and Sr contents and their $^{87}Sr/^{86}Sr$ ratios to determine whether these minerals supplied the elements for the authigenesis of the clays during the experiment. Relative to the feldspar sample from the lower part of the column, that from upper part lost 7.7% Rb and 0.6% Sr, while the $^{87}Sr/^{86}Sr$ ratio

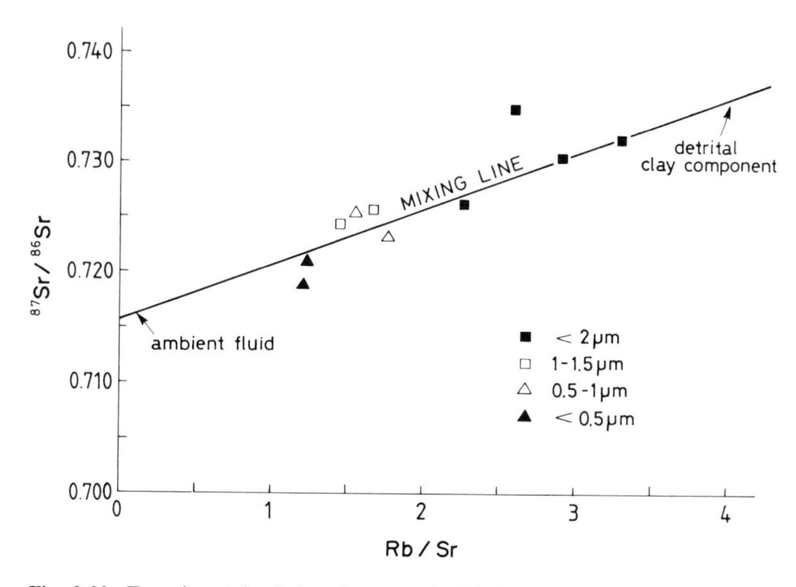

Fig. 2.22. Experimental relations between the Rb-Sr isotopes of the migrating waters and the clay fractions of different sizes. (Thellier and Clauer 1989)

only slightly decreased from 0.74867 ± 0.00017 to 0.74342 ± 0.00016. The very small loss of Sr and the small decrease in the Sr isotopic composition of the plagioclase grains upward in the columns suggest that this mineral was not extensively leached during the experiment and that it did not act as a precursor for the authigenic palygorskite-like clay formed at the upper part of the columns. These results thus indirectly confirm that the precursors of the Mg-rich clay minerals formed during the experiment were the flakes of smectite.

1.7 Isotopic Dating of Clay Authigenesis in Soil Profiles

Knowledge of the ages of paleosoils has been of much interest to both geologists and soil scientists because of the potential for such materials to yield information about climatic conditions in the past. Clays in soil profiles are residual products and are not as intimately associated with detrital components as clays in sedimentary rocks. Presumably because of a lack of understanding of the isotopic systems of authigenic clay minerals in soils and the lack of any reliable method for separation of pure clay fractions, there are very few isotopic studies of authigenic clays in paleosoils (Long and Agee 1985). Macfarlane and Holland (1991) determined the Rb-Sr isotopic compositions of bulk samples of paleosols developed on Precambrian plutonic rocks of Australia and South Africa. They found that the Rb-Sr system of these rocks was reset during a period of post-weathering metasomatism related to local and/or regional thermal disturbances, leaving an uncertainty about the exact period of weathering. The critical importance of understanding the behavior of isotope systems for dating clay formation in soil profiles is illustrated by the few studies discussed below. In the presentation of these studies, we have ignored the problem of possible eolian clay contamination which can occur, as has been suggested by studies of Hurley et al. (1961) and Dymond et al. (1974), because the current lack of understanding of the behavior of the isotopic systems in such environments precludes a rational judgment on eolian contamination.

1.7.1 The K-Ar Isotopic System of Clay Minerals in Soil Profiles

K-Ar isotopic analyses of clay products formed in soil profiles can furnish valuable information about the time of formation of these minerals and also about the openess of the parent material to water and air during the formation of the new minerals. In a situation of high infiltration rate or high water-to-rock ratio, any Ar incorporated by the newly formed clay component within the soil profile at the time of formation can be expected to have the same isotopic composition as that of the open atmosphere. If the initial $^{40}Ar/^{36}Ar$ ratios of the clays are found to be higher than that of the atmosphere, the result is a proof that clay crystallization occurred in micro-environments with Ar contents directly influenced by the alteration of the parent materials.

In the Hoggar Mountains in Algeria, a Miocene basalt with a K-Ar age of 17.8 ± 1.3 Ma (Bonhomme in Leprun 1979) was deeply weathered, producing a 1-m-thick soil profile covered by a massive stratified ferruginous caprock. The clay fraction of the caprock consisted mainly of kaolinite mixed with hematite. After removal of the hematite, the kaolinite minerals yielded a K-Ar date of 132 ± 6 Ma (Table 2.2). As compared to the fresh rock, the kaolinite had 80% less K_2O content, but 50% more radiogenic ^{40}Ar (N. Clauer and J.C. Leprun, unpubl. data).

In one of the Tertiary continental basins of the French Massif Central, a basanite flow with a K-Ar age of 7.4 ± 0.1 Ma was weathered before the deposition of the next 6.5 ± 0.1-Ma-old flow (Brousse et al. 1975). The soil profile developed on the 7.4 Ma-old flow consists of three facies: a smectite-rich (75%) zone at the bottom, a reddish zone with a mixture of kaolinite (70%) and smectite (30%) in the middle, and a ferruginous caprock containing a mixture of kaolinite (75%) and smectite (25%) at the top. The K-Ar dates of the <2 µm clay separates ranged from 35 ± 5 Ma for the bottom zone, to 107 ± 7 Ma for the middle zone, to 117 ± 5 Ma for the top zone (Table 2.3; N. Clauer and J.C. Leprun, unpubl. data). These values are much higher than the time of development of the soil profile. The K_2O

Table 2.2. K-Ar isotope data of kaolinite formed from weathering of a Miocene basalt in Hoggar, Algeria. (N. Clauer and J.C. Leprun, unpubl. data)

Samples	K_2O[a]	Ar*[a]	^{40}Ar*[b]	Age (±2σ)
kaolinite	0.34	53.36	1.50	132 ± 6
whole rock	1.73	30.73	1.00	17.8 ± 1.3

[a] Amounts in percent.
[b] Amounts in 10^{-6} cm^3/g STP.
* stands for radiogenic; ages in Ma.

Table 2.3. K-Ar isotope data of weathering products from a soil profile in the Massif Central, France. (N. Clauer and J.C. Leprun, unpubl. data)

Samples[a]	Mineralogy[a]	K_2O[b]	Ar*[b]	^{40}Ar*[c]	Age (±2σ)
Caprock	75%K + 25%S	0.29	58.91	1.13	117 ± 5
Int. clay	70%K + 30%S	0.15	47.91	0.55	107 ± 7
wea. bas.	25%K + 75%S	0.15	14.75	0.17	35 ± 5

[a] int. clay, Intermediate clay; wea. bas., weathered basalt; K, kaolinite; S, smectite; ages in Ma.
[b] amounts in percent.
[c] amounts in 10^{-6} cm^3/g STP; * stands for radiogenic.

contents of the clay fractions varied little through the soil profile. In all cases the K_2O contents of the clays in the soil were much lower than that of the parent rock. The absolute amounts of radiogenic ^{40}Ar in the clay fractions, on the other hand, increased upward across the profile. Removal of hematite coating from one of the separated clay fractions diminished the amount of atmospheric Ar, but did not significantly modify the K-Ar date. The high K-Ar dates of the authigenic clay fractions could not be due to the presence of any residual unaltered parental materials which are about 7.4 Ma old and would have induced a decrease of the values. We, therefore, argue that the anomalously high K-Ar dates of authigenic clay fractions in soil profiles relate to preferential entrapment of radiogenic ^{40}Ar relative to K by the authigenic clays at the time of crystallization, presumably in micro-environments which had little or no communication with the surface environment.

1.7.2 The Rb-Sr Dating of Authigenic Clays from Soil Profiles

A detailed Rb-Sr isotopic study was made by N. Clauer and J.C. Leprun (unpubl. data) on clay fractions extracted from a 18-m-thick soil profile developed on a granodiorite near Dori in northeastern Republic of Burkina Faso (Leprun 1979). The fresh parent rock consisted of large grains of quartz and plagioclase with minor amounts of microcline, green hornblende, and biotite, and trace amounts of ilmenite. The <1 μm clay fraction in the deepest zone of the soil consisted of about 55% kaolinite, 40% smectite, and 5% illite. Above this zone, between −16 and −6.8 m, is a red zone containing deeply weathered microcline, amphibole, and plagioclase. Tiny hexagonal kaolinite particles were found to coat weathered feldspar and amphibole relicts. Kaolinite was the dominant clay mineral in an intermediate zone containing ferruginous nodules, and in the ferruginous caprock near the top of the profile (Table 2.4).

The Rb-Sr whole-rock isotopic age of the fresh granodiorite was found to be about 1.6 Ga, with an initial $^{87}Sr/^{86}Sr$ ratio of about 0.703 (Table 2.5 and Fig. 2.23). This isotopic date is geologically reasonable for this region of Africa (Bonhomme 1962). The Rb and Sr contents of all clay fractions, except those from the bottom part of the soil profile, were lower than those of the unweathered granodiorite. The clay fractions from the bottom part of the soil profile contained less Sr but more Rb than the unweathered whole rock. Clay fractions of <1 μm size consisted of varied amounts of kaolinite, smectite, and illite. The data points of four out of five of these <1 μm fractions plotted along a line in an isochron diagram, giving a date of about 300 Ma with an initial $^{87}Sr/^{86}Sr$ ratio of about 0.7120. Isotopic determinations were also made on subfractions of two pure kaolinite samples which belong to the reddish intermediate zone.

Also analyzed were acid leachate and residue of one of the size fractions of each sample. The untreated fractions, the acid leachate, and the acid-

Table 2.4. Mineralogical data of clays in a soil profile developed on a granodiorite in the Republic of Burkina Faso. (N. Clauer and J.C. Leprun, unpubl. data)

Samples	Depths[a]	Facies[b]	Mineralogy[b]
G <1 μm	0.4	Caprock	100% K
F <1 μm	3.8	Vers. clay	100% K + traces I
E <1 μm	6.8	Pink clay	100% K + traces S
<0.2 μm (1)			100% K
0.2–0.6 μm (2)			100% K
0.6–1 μm (3)			100% K
D <1 μm	12.0	Pink clay	100% K + traces S
<0.2 μm (1)			100% K
0.2–0.6 μm (2)			100% K
0.6–1 μm (3)			100% K
C <1 μm	16.0	Pink clay	100% K + traces S
B <1 μm	17.0	Pist. clay	55% K + 35% S + 10% I
A <1 μm	18.0	Pist. clay	55% K + 40% S + 15% I

[a] Depths in the profile in meters.
[b] vers. clay, Versicolor clay; pist. clay, pistacchio clay; K, kaolinite; S, smectite; I, illite.

Table 2.5. Rb-Sr isotope data of clays in a soil profile developed on a granodiorite in the Republic of Burkina Faso. (N. Clauer and J.C. Leprun, unpubl. data)

Samples[a]		Rb (μg/g)	Sr (μg/g)	$^{87}Rb/^{86}Sr$	$^{87}Sr/^{86}Sr$ ($\pm 2\sigma/\sqrt{N}$)
E	<0.2 μm (1)	24.8	64.3	1.120	0.71610 ± 22
	0.2–0.6 μm (2)	15.3	86.1	0.514	0.71453 ± 11
	0.6–1 μm (3)	15.7	163.3	0.279	0.71356 ± 15
	<1 μm (L)	69.1	2635	0.076	0.71332 ± 6
	<1 μm (R)	12.8	104.7	0.353	0.71385 ± 7
D	<0.2 μm (1)	15.8	78.3	0.586	0.71226 ± 16
	0.2–0.6 μm (2)	13.5	142.8	0.275	0.71199 ± 9
	<1 μm	7.84	75.4	0.301	0.71165 ± 9
	<1 μm (R)	7.25	73.4	0.286	0.71194 ± 6
C	<1 μm	80.6	188.7	1.238	0.71798 ± 13
B	<1 μm	103.6	146.9	2.044	0.72174 ± 6
A	<1 μm	125.4	193.3	1.880	0.72057 ± 24
Whole rock 1		81.8	772	0.307	0.71081 ± 20
Whole rock 2		155.1	666	0.675	0.71991 ± 9
Whole rock 3		129.7	732	0.513	0.71654 ± 10

[a] (1) to (3) stand for the three and two size fractions of the samples E and D, respectively; whole rocks 1 to 3 stand for three different samples; L and R, leachate and residue, respectively.

insoluble residue of each kaolinite sample defined separate isochrons. The isochrons for the two kaolinite separates were subparallel, giving approximately the same age of about 150 Ma with distinctly different initial $^{87}Sr/^{86}Sr$ ratios of about 0.711 and 0.712. These two values for the initial Sr isotopic

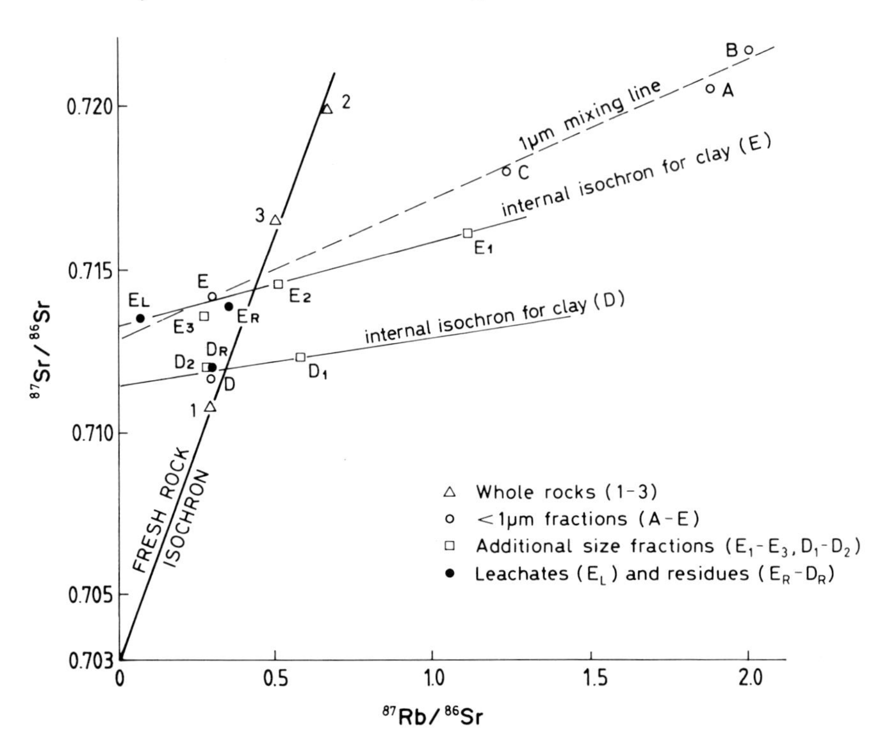

Fig. 2.23. Rb-Sr isochron diagram for the fresh granodiorite whole-rock samples and the different clay fractions of the Dori profile in the Republic of Burkina Faso. For *symbols*, see Tables 2.4 and 2.5

composition of the clays suggest that these minerals formed in microenvironments with limited communications among them.

The two parallel Rb-Sr isochrons which give nearly the same age of 150 Ma but different initial $^{87}Sr/^{86}Sr$ ratios, most likely reflect the time of formation of the soils, whereas the apparent isochron with the 300 Ma date is considered to be a mixing line. The two 150 Ma isochrons were defined by two separate samples of pure kaolinite which occurred in close proximity to each other, yet had different Sr isotopic ratios at the time of formation. The differences in the initial Sr isotopic ratios reflect the chemical differences of their respective formation environments, which had to be isolated from each other as they are very proximate. The 300-Ma line has been considered to be a mixing line because the data points refer to mineralogically heterogeneous fractions of diverse origin, and they display a high degree of scattter in the alignment. Such a kind of alignment was obtained by Dasch (1969) on weathered products of Elberton granite, as discussed in a previous section. The claim about the 150-Ma isochron has yet to be tested by other independent means.

2 Clay Weathering and Alteration in Soils

Hurley et al. (1961) reported two K-Ar dates (325 and 336 Ma) on <1 μm clay fractions from soils developed on sedimentary rocks of New York State and consisting of mixtures of illite and chlorite. These values being not very different from stratigraphic age of the parent material, the authors considered that the overall influence of weathering in these rocks was negligible. However, based on the fact, discussed above, that radiogenic isotopes are possibly concentrated by clay minerals generated in soil profiles, it is clear that the dates obtained by Hurley et al. are artifacts integrating losses and concentrations of elements including the radiogenic daughter products. Clauer (1970), in the frame of a Rb-Sr dating study on Paleozoic and Precambrian shales and slates of the Vosges mountains, made a few Rb-Sr determinations on clay fractions extracted from obviously weathered shales, relative to unweathered samples. He observed that the data points of the weathered clay fractions systematically plotted below the isochrons obtained for the unweathered fractions, and concluded that continental weathering altered the Rb-Sr system of clay minerals.

Parron and Nahon (1980) found that, depending on the intensity of the weathering, glauconite weathers to a glauconite/smectite (=glaucony) mixed-layer phase or to kaolinite and Fe-oxide phases. Courbe et al. (1981)

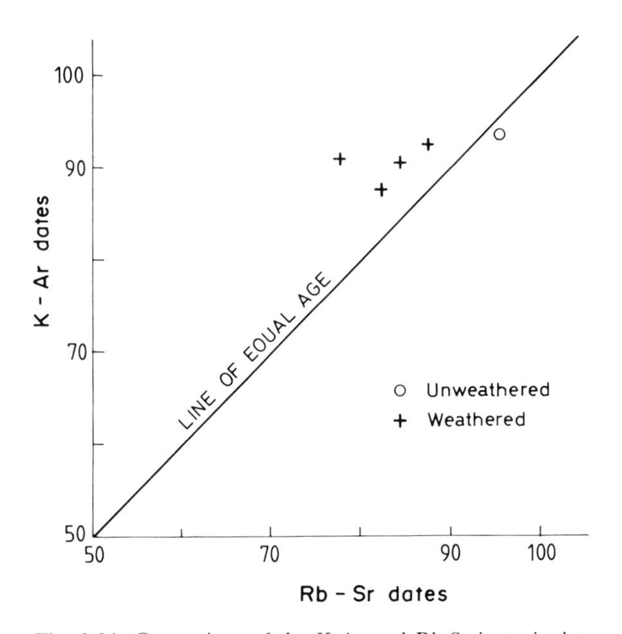

Fig. 2.24. Comparison of the K-Ar and Rb-Sr isotopic dates of an unweathered and several weathered glauconites

and Loveland (1981) reported that partial weathering of glauconite produced new K-depleted mixed-layer minerals covering the rims of the glauconitic precursors. Odin and Hunziker (1974) studied the effects of weathering on the K-Ar system of well-crystallized glauconites and observed that the weathering caused a loss of as much as 15% K by the mineral. At the same time, a nearly identical amount of radiogenic ^{40}Ar was lost, so that the K-Ar dates of the minerals remained unaffected. Clauer (1976) analyzed the Rb-Sr isotopic compositions of aliquots of the same glauconites and found that, whereas the Rb contents remained nearly the same, the Sr contents increased due to weathering, changing from about 13.5 μg/g in the unweathered glauconites to as much as 28.9 μg/g in the weathered ones. The $^{87}Sr/^{86}Sr$ ratio decreased from 0.7794 in the unweathered glauconites to 0.7381 in the weathered ones. The simultaneous effects on the Rb-Sr and K-Ar dates, as a result of weathering, are presented in Fig. 2.24 which indicates that the Rb-Sr dates decreased by as much as 17.8%, while the K-Ar dates remained relatively unaffected. Why the Rb-Sr system of glauconites seems to be variously affected by continental weathering, but not the K-Ar system, remains to be explained.

3 Summary

As continental weathering of minerals by waters often results in the formation of products with varied amounts of newly formed solid products and dissolved chemical constituents that are released to the waters, paths of formation of these minerals from their parent precursors can be well delineated from analyses of varied kinds of isotopic records in the newly formed clays. While the evidence is useful in defining the environmental conditions in the formation of some specific clay minerals from the effects of weathering, knowledge of the configuration of any isotopic evidence can greatly advance understanding of the degree of chemical modification that can normally occur for the clay minerals as a result of erosion, transportation, and deposition subsequent to their pedological formation.

Authigenic clay minerals found in soil environments have widely varied $^{87}Sr/^{86}Sr$ ratios, dependent essentially on the composition and the age of the parent mineral or minerals being weathered. The Sr isotopic ratios of any given clay mineral at the time of its formation from geologically old materials are generally high (>0.725) when the clay mineral is derived from weathering of rocks, such as many sialic rocks that are rich in alkali feldspar minerals. The ratios are moderate to low (0.725–0.710) for the same clay formed from biotite and hornblende minerals, and very low (<0.709) when the clay formed from rocks enriched in calcic or soda-calcic plagioclase minerals, such as many mafic rocks. The history of newly formed clays begins with the formation of "protominerals" whose oxygen and Sr isotopic compositions are governed principally by inheritance of these isotopes from the primary minerals that are being weathered in many individual microenvironments. The newly formed protominerals tend to have some degree of preferential incorporation of radiogenic ^{87}Sr and ^{40}Ar relative to the parent minerals. Thus, a specific type of clay minerals across a soil profile can have varied initial Sr and Ar isotopic compositions. These minerals may also be able to retain such diversities in the isotopic compositions for an indefinite period of time in the soil profiles. This means in turn that these minerals are potentially datable by isotopic methods if the varied initial isotopic compositions can be determined accurately. A remarkable feature in the formation of clays from the weathering of feldspars across a soil profile is that the Rb-Sr isotopic data of clay products generated from the same family of parent feldspar minerals may yield apparent Rb-Sr isochrons with positive slope. However, at advanced stages of weathering of biotite minerals, the Sr isotopic compositions of the clays may be influenced not only by the isotopic compositions of parent biotite, but also by that of leachates derived from weathering of feldspar minerals. The resulting clay products derived from such mixed influences can have Rb-Sr isotopic compositions that can be described by apparent isochrons whose slopes can be widely varied from near zero to some positive values.

The importance of microenvironments in the formation of authigenic clay minerals in soils is also evident from oxygen isotope data. Although a popular notion suggests that the stable isotope data of clay minerals in soils reflect the isotopic composition of local meteoric waters, variations in the oxygen isotopic composition have been found in several instances for pedogenic clays formed in the same soil profile. Such variations in the oxygen isotope compositions of the clays and, therefore, the isotopic variations in soil pore fluids can be explained in terms of crystallization of the minerals in soil environments with limited water-to-mineral ratios.

Pedogenic clay minerals, even of kaolinitic composition, being variously enriched in radiogenic isotope contents, deposition of such clays in oceans without any significant modifications during the transport to the oceans will render the oceanic clays to have model isotopic dates that are invariably higher than the age of formation. An implication of the presence of such isotopically enriched clays in the ocean sediments is that all of the high isotopic dates for many modern ocean clays are not necessarily due to the presence of only recycled illitic materials from ancient sedimentary rocks.

Isotopic study of weathering processes in continental soil profiles has also shown that an important source of dissolved Sr carried from the continents to the oceans is released from weathering of feldspars and not only from weathering of carbonates, as commonly assumed in the modeling of continental input of Sr to the oceans. Weathering of feldspars, which occurs at the onset of weathering of plutonic and metamorphic rocks, produces clay minerals with high $^{87}Sr/^{86}Sr$ ratios and releases Sr with low $^{87}Sr/^{86}Sr$ ratios into the run-off waters. This release buffers the run-off Sr and explains why Albarède and Michard (1987) observed significant changes in the isotopic composition of Sr released in the run-off waters of a drainage basin only several tens of million years after major orogenic activities.

Chapter 3

Isotope Geochemistry of Clay Minerals in Young Continental and Oceanic Sediments

Continental erosion is affected mainly by riverine processes and, to a smaller degree, by wind and glacial activity. The debris eroded by these agents are deposited in fluvial, glacial, lacustrine and other continental depositional centers and in various parts of the oceanic basins, the latter constituing almost the bulk of the sediments deposited during the entire history of the Earth. A recent estimate given by Milliman and Meade (1983) suggested that about 13.5×10^{12} kg of sediment are annually delivered to the oceans by the rivers and, according to Gibbs (1967), 4×10^{12} kg of the material consist of $<2\,\mu m$ size particles. Assuming that nearly 60% of the $<2\,\mu m$ fraction consists of clay minerals, which is the same amount present in an average shale, about 2.4×10^{12} kg of clay minerals are delivered each year to the present-day oceans by the rivers.

Clays in modern ocean sediments have complex origin. Some are detrital, having been derived from varied terrigenous sources, whereas others are formed in ocean basins as syngenetic or diagenetic minerals resulting from a variety of processes within the ocean basin. Some of the terrigenous clays delivered to the ocean may experience varied degrees of ion exchange with sea water before burial beneath the sea floor. Additional modifications of these clays are possible in the early burial history because of ion exchange with pore waters in the shallow subsurface. Marine authigenic clay minerals may be traced largely to precipitation from ions dissolved in sea water, dissolution-precipitation reaction involving terrigenous silicate minerals, and alteration of oceanic volcanic rocks under widely varied temperatures ranging from that of cold ocean bottom environment to that of high temperature hydrothermal conditions at a ridge environment. These syngenetic clay minerals will approach chemical and isotopic equilibrium with macro- and microenvironments in the ocean. Varying degrees of post-depositional modification of these newly formed clays beneath the ocean floor cannot be ruled out because of the varied chemical and mineralogical composition of sediment pore waters and temperature increase attendant with burial. In addition, new clay minerals may be formed under shallow burial conditions beneath the sea floor by reaction between pore water and unstable minerals, amorphous phases, or rock fragments. This chapter will focus on the radiogenic and stable isotopic composition of continental debris delivered to the ocean, the extent of clay modification when entering the ocean environment, the formation of synsedimentary clays by various processes in the ocean environment, and diagenetic clay mineralization and modification in sediments beneath the ocean floor.

1 Recent Continental Erosional Debris and Clay Sediments

Isotopic studies of clays in recent continental sediments can be valuable in understanding the climates, as well as the erosional pattern in a drainage basin in the recent history of the Earth; but to date, little effort has been made in this respect. Even the scarce isotopic data that exist often lack critical information about the mineralogy of the clays and thereby fail to achieve their full potential. Isotopic studies have been made on continental bulk detritus in lakes or rivers to gain information about the mean crustal residence age of the source rocks (McCulloch and Wasserburg 1978; O'Nions et al. 1983; Miller and O'Nions 1984; Davies et al. 1985; Michard et al. 1985), but the analyses of the bulk sediments, and not the clays, are not of much use for the delineation of the physicochemical processes that regulate the conditions at or near the surface of the Earth.

1.1 Sr Isotope Geochemistry

Studies on Sr isotopic composition of modern river sediments have provided a reasonably good perspective on the range of variations in isotopic signatures of suspended loads of rivers draining widely different terrains and have contributed to increased understanding of the isotopic compositions of clays which are now being delivered to the oceans by the major rivers. Goldstein and Jacobsen (1988) reported that the suspended loads of a large number of rivers ranged in $^{87}Sr/^{86}Sr$ ratios from 0.704 to 0.800, and in Rb/Sr ratios between less than 0.01 and 1.60. The two ratios in the suspended loads are poorly correlated (Fig. 3.1), reflecting the differential dissolution of crustal rocks of varied isotopic composition. The very high $^{87}Sr/^{86}Sr$ ratios were found for suspended sediments from three rivers draining across very old Precambrian crystalline rocks in west Greenland. Those suspended loads with $^{87}Sr/^{86}Sr$ ratios which are less than that of marine Sr (0.707–0.709) belong to rivers with predominantly young volcanic rocks in the drainage basin, as illustrated by many rivers in Japan and the Philippines. The Sr isotopic compositions of suspended sediments of most major rivers are higher than that of marine Sr. Although Goldstein and Jacobsen presented the Sr isotopic data for a large number of river loads, these data are of limited use for reconstructing the chemical nature of present-day continental

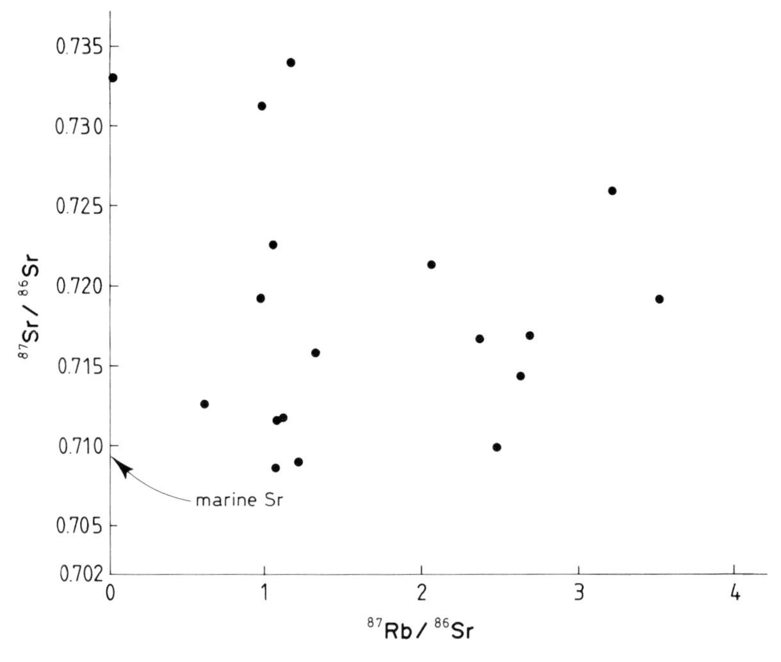

Fig. 3.1. Rb-Sr isotopic compositions of suspended loads of rivers. (Goldstein and Jacobsen 1988)

weathering and erosion because they lack information about the mineralogic composition of the suspended materials.

Patterns of weathering of rocks and minerals in a drainage basin may be understood from the relationship of the $^{87}Sr/^{86}Sr$ ratios between suspended loads and corresponding dissolved loads of rivers. Goldstein and Jacobsen (1988) compared the $^{87}Sr/^{86}Sr$ ratios of dissolved and suspended loads of some major rivers in the world (Fig. 3.2). As shown in the figure, the $^{87}Sr/^{86}Sr$ ratios of both the dissolved and the suspended loads are often nearly identical when the $^{87}Sr/^{86}Sr$ ratios of the suspended loads are less than 0.709. This relationship is evident for rivers which drain across mainly young volcanic rocks. The close isotopic similarity between the dissolved and the suspended loads is essentially due to a lack of isotopic differences among minerals of the young volcanic rocks in the basin (Chaudhuri and Clauer 1992a). The same cannot be said about the Sr isotopic compositions of suspended and dissolved loads for rivers draining across old immature siliclastic sedimentary rocks and crystalline silicate rocks. The differential weathering of minerals in rocks within a drainage basin would make the Sr in the dissolved phase isotopically different from that in the residual solids transported as suspended loads in rivers. Chemical weathering of silicate rocks would often begin with preferential dissolution of phases having low Rb/Sr and $^{87}Sr/^{86}Sr$ ratios, such as plagioclase feldspars, making the

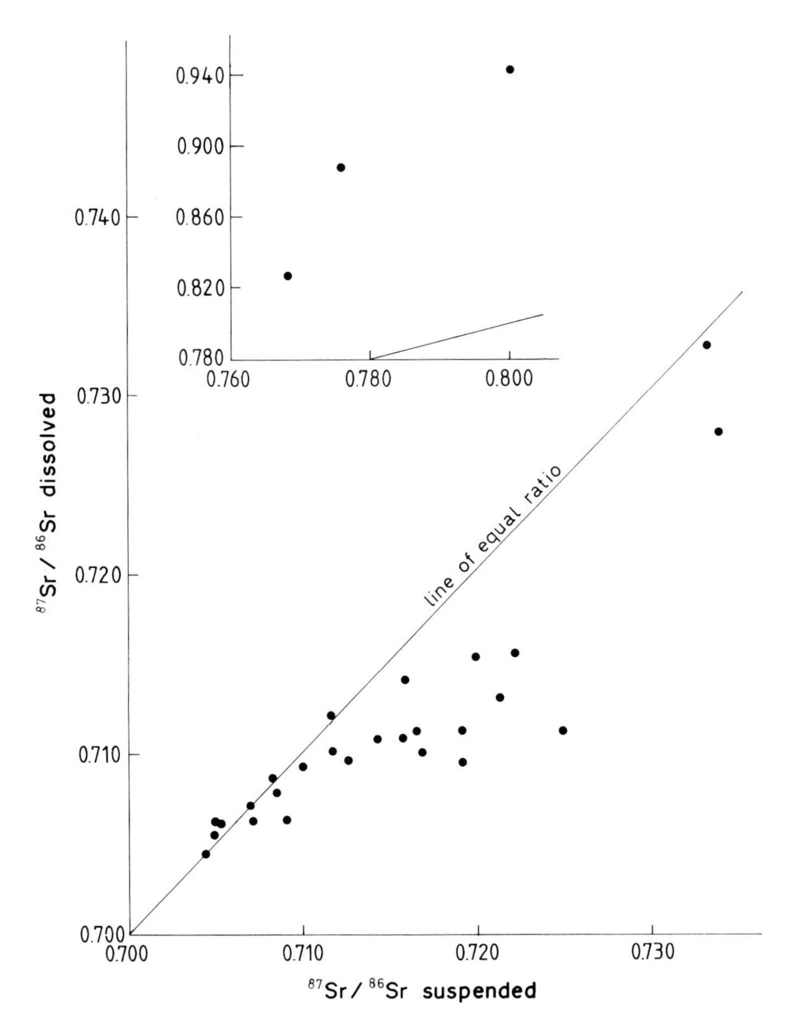

Fig. 3.2. Relations in Sr isotopic compositions between suspended and dissolved loads of rivers. The *upper diagram* belongs to data of Greenland rivers. Many rivers draining marine carbonate rocks have lower $^{87}Sr/^{86}Sr$ ratios in the dissolved loads than in the suspended loads. (Goldstein and Jacobsen 1988)

dissolved loads lower in $^{87}Sr/^{86}Sr$ ratio than the corresponding suspended loads. Dissolution of carbonate minerals which are generally low in $^{87}Sr/^{86}Sr$ could further increase the Sr isotopic difference between the dissolved and the suspended loads. Hence, the combined dissolution of feldspar and carbonate minerals could potentially explain why for the vast majority of rivers the suspended loads have higher $^{87}Sr/^{86}Sr$ ratios than the dissolved loads. A few rivers have been found whose suspended loads were lower in $^{87}Sr/^{86}Sr$ than the dissolved ones, for example the rivers in Greenland (Fig.

3.2 upper part). In such cases, selective dissolution of K-bearing silicate phases with high $^{87}Sr/^{86}Sr$ ratios, such as biotites, coupled with complete absence of carbonate minerals or other readily soluble low Rb/Sr-bearing mineral phases in the drainage basin, are the primary factors.

Suggestions have been made that model ages may be used to characterize the age of source materials for river sediments (Biscaye and Dasch 1971). As Rb-Sr dates of suspended sediments represent an integrated value of different isotopic dates for weathered products and eroded lithic fragments derived from continents, calculated model ages should not be used without rigid constraints. A model construction must be based on detailed understanding of the isotopic evolution of weathered and eroded products in a generally complex drainage basin. In the absence of this understanding, model-age calculations provide information of little value.

Isotopic data for clay sediments in lakes are very rare, although such data are useful to understand the isotopic partitioning between waters and rocks in a drainage basin (Faure et al. 1963). Jones and Faure (1969) found that sediments and waters in Wanda Lake, Antarctica, are in Sr isotopic equilibrium with each other. The bottom sediments yielded $^{87}Sr/^{86}Sr$ ratios of 0.7148 to 0.7149, and waters of the lake and the Onyx river which supplies the lake had ratios of 0.7149 ± 0.0001 and 0.7146 ± 0.0002, respectively. A later study by Jones and Faure (1978) showed that salts from soils around the lake had the same Sr isotope composition. Hart and Tilton (1966) observed that the Pb isotopic compositions of sediments and waters from Lake Superior are very similar, suggesting that the Pb isotopes are not fractionated during continental weathering. The isotopic similarities between sediment and lake water in a drainage basin underlain by geologically very ancient rocks, as noted by Jones and Faure and also by Hart and Tilton, are surprising when the isotopic fractionations between sediment and water are very common for most rivers. A possible explanation could be that in the studies of the two lakes, the isotopic compositions of both sediment and water are controlled essentially by a highly soluble mineral phase, but no mineralogical information for the sediment is given in these studies for any support of this conjecture.

Deltaic and estuarine sediments have yet to be studied in detail, so that we know very little about isotopic differentiation among clay sediments in these transitional environments. Rama Murthy and Beiser (1968) observed that the $^{87}Sr/^{86}Sr$ ratios of Amazon deltaic sediments ranged between 0.711 and 0.715 with Rb/Sr ratios of about 0.59. Clauer (1976) analyzed $<2\,\mu m$ clay fractions of two samples of deltaic sediments of the Rhone River in southeastern France. The clay minerals consisted of 65% illite, 20% smectite, and 15% chlorite. The average $^{87}Sr/^{86}Sr$ ratio of the samples was about 0.720 and the Rb and the Sr contents were about 210 and $150\,\mu g/g$, respectively. Biscaye (1972) measured the Sr isotopic compositions of carbonate-free sediments of the Rio de la Plata estuary. The $^{87}Sr/^{86}Sr$ and the Rb/Sr ratios of the upper and the middle Plata estuary samples were

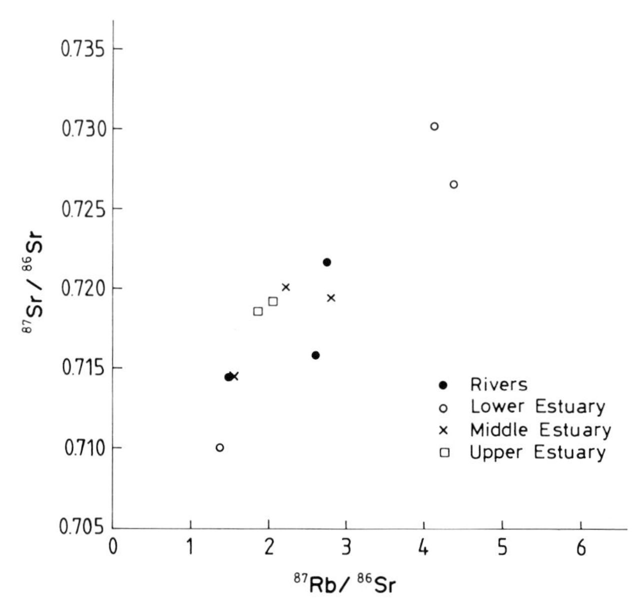

Fig. 3.3. Relationship between the $^{87}Sr/^{86}Sr$ and $^{87}Rb/^{86}Sr$ ratios of carbonate-free sediments of Rio de la Plata river and estuary. (Biscaye 1972)

found to be within the range of values found for the river sediments (Fig. 3.3). The sediments in the lower estuary, on the other hand, have been found to be isotopically different from those in the upper and the middle estuaries. The cause for the difference remains unknown although the authors hinted that the difference could be related to variation in the source. The estuarine sediments collectively define a clear trend in an isochron diagram, and the trend is different from the one exhibited by sediments from the adjacent shelf (see Fig. 3 in Biscaye and Dasch 1971). Differential transport and sedimentation at the lower estuary for finer sediments with higher Rb/Sr and $^{87}Sr/^{86}Sr$ ratios can be a factor for the isotopic trend in estuarine sediments.

1.2 Nd Isotope Geochemistry

Data from weathered soil profiles and river waters have shown that interactions between water and continental rocks and minerals at or near surface conditions produce considerable fractionation of the REE relative to crustal or shale abundance (Goldstein and Jacobsen 1988; Sholkovitz and Elderfield 1988; Elderfield et al. 1990). Braun et al. (1990) reported that weathered residual products in a lateritic soil profile developed on a syenitic rock were enriched in LREE, whereas pore waters were enriched in HREE relative to the syenitic host rock. While the dissolved fractions of many rivers are enriched in HREE relative to shale, suspended particles from the same rivers are enriched in LREE relative to shale. Considering that REE in rivers are transported in suspended particles, colloidal particles (10–0.01 μm in size) and solution phase, Sholkovitz (1992) found that a large fractionation of REE occurs between colloidal and solution phases and that colloidal particles carry much of the total REE present in the dissolved pool consisting of colloidal and solution phases. The colloidal phase is LREE enriched, while the solution is HREE-enriched relative to shale abundance. The fractionation of the REE between the two phases is related to differences in the relative affinities for surface adsorption and for complexation to ligands in solution. The relative solubility among REE in a complexed form, with carbonates, organics, phosphates, etc., increases with increase in the atomic number. The Sm/Nd ratio of the solution phase is about 5 to 14% higher than that of the colloidal phase.

Sm-Nd isotopic studies of clays in river sediments are limited (Goldstein et al. 1984; Stordal and Wasserburg 1986; Goldstein and Jacobsen 1988). Goldstein et al. (1984), analyzing Sm/Nd isotopic compositions of suspended sediments of 20 major rivers and also of sediments in banks and bottoms of a few rivers, found that the Sm/Nd ratios of the sediments ranged between 0.176 and 0.242 and the $\varepsilon_{Nd}(0)$ values between −3.3 and −16.1. Analyses of different size fractions of a bulk sample from the Amazon River revealed that the change from coarse to fine particle size produced a slight increase in the Sm/Nd ratio with attendant increases in both the Sm and the Nd contents by a factor of about 3, but no change in the Nd isotopic composition. The small increase in the Sm/Nd ratio is not clearly understood for lack of information on the mineralogy of the different fractions. We noted above that the Sm/Nd ratios of clay sediments tend to increase with increased degree of weathering of parent materials, and this may at least partly explain why the fine fraction tends to have a higher Sm/Nd ratio than the coarse fraction of the sediment from the Amazon River.

Goldstein and Jacobsen (1988) also reported Sm-Nd isotopic data for suspended loads from 25 major rivers in the world. The Sm/Nd ratios ranged from 0.160 to 0.284 and the $\varepsilon_{Nd}(0)$ values from +7.1 to −44. The studies by Goldstein et al. (1984) and Goldstein and Jacobsen (1988) have shown that the sediments transported by present-day rivers are lower in their Sm/Nd ratio than ocean basalts (0.23–0.39), but are similar to many Andean-type volcanics. In contrast, the river sediments are lower in $^{143}Nd/^{144}Nd$ ratios than either the ocean basalts or the Andesitic volcanics. The Sm-Nd isotopic characteristics of most river sediments are indeed intermediate between that of andesites and Proterozoic-Archean continental metasediments and gneisses, suggesting a large input of old continental rocks in the modern-day river sediments.

The study of Goldstein and Jacobsen (1988) revealed that the $^{87}Sr/^{86}Sr$ and $\varepsilon_{Nd}(0)$ values of suspended loads in rivers bear a nonlinear inverse relationship (Fig. 3.4). These authors also noted a similar nonlinear relationship between the two isotopic parameters for the dissolved loads of rivers (Goldstein and Jacobsen 1987). They concluded that this relationship may be explained in terms of mixing of two end-member components, one of which consists of weathering products of young island arc materials with average $^{87}Sr/^{86}Sr = 0.7045$, $\varepsilon_{Nd}(0) = +7$, and $Sm/Nd = 0.06$ and the other of weathering products of old (Archean) crustal materials with average $^{87}Sr/^{86}Sr = 0.800$, $\varepsilon_{Nd}(0) = -42$, and $Sm/Nd = 0.32$. Goldstein and Jacobsen (1988) suggested that 25–30% of suspended materials of North American rivers is derived from young orogenic areas. They estimated the following average values for the suspended loads: 0.716 for the $^{87}Sr/^{86}Sr$ ratio, 0.49 for the Rb/Sr ratio, −10.6 for the $\varepsilon_{Nd}(0)$ value, and 0.19 for the Sm/Nd ratio. By comparison, the estimated average values for the dissolved loads are: 0.710 for the $^{87}Sr/^{86}Sr$ ratio, 0.03 for the Rb/Sr ratio, −8.4 for $\varepsilon_{Nd}(0)$ value, and 0.20 for the Sm/Nd ratio.

Nd model ages for suspended loads of rivers can be calculated to develop models of contemporaneous upper continental crustal evolution, assuming that the erosional mass represents an average sampling of the upper continental crust of the time considered. Goldstein et al. (1984) found a near constant Nd model age (relative to depleted mantle) of about 1.7 Ga. Since this value is close to the Nd model age of about 2 Ga for many post-

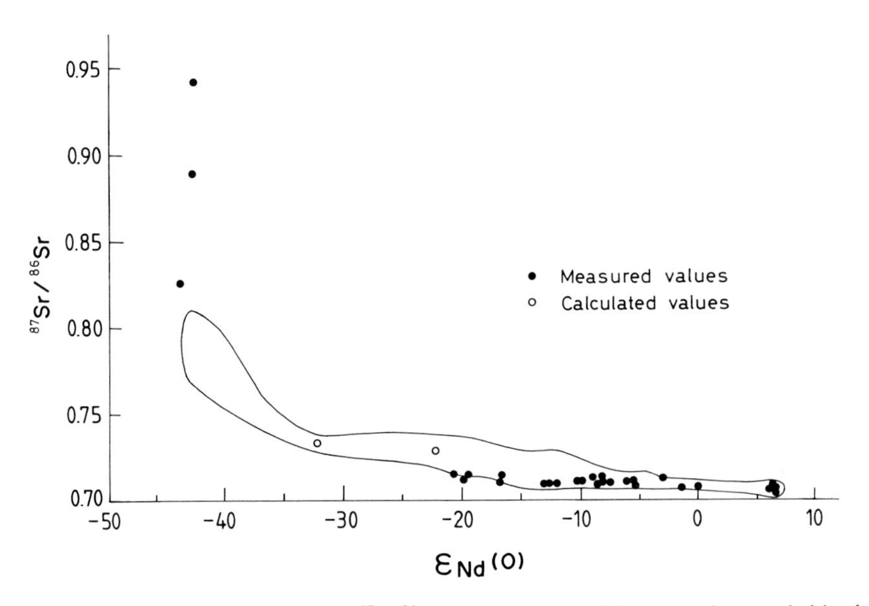

Fig. 3.4. Relationship between the $^{87}Sr/^{86}Sr$ ratio and the $\varepsilon_{Nd}(0)$ values of suspended loads of rivers. The *encircled area* includes the trend between the two parameters. The linear relationship between the Sr isotope ratio and the $\varepsilon_{Nd}(0)$ deviates strongly for waters with high Sr values and low $\varepsilon_{Nd}(0)$ values. (Goldstein and Jacobsen 1988)

Archean sedimentary rocks, many have suggested that new crustal additions from the mantle have been quite small during the post-Archean period. The model age of modern continental erosional debris should be interpreted with some caution, because the data may be biased by suspended sediments being sampled from large rivers which appear to form at favorable tectonic settings such as trailing edges and continental collision zones, in contrast to those from small rivers which develop at active tectonic settings, as suggested by Milliman and Meade (1983). Goldstein and Jacobsen (1988) noted in their study of suspended sediments of rivers in a wide variety of drainage basins, that sediments from rivers with drainage basin developed on passive margin tectonic settings have higher Nd model ages than those from rivers with drainage basins developed on active tectonic settings. McLennan et al. (1990) have also shown that tectonic settings have strong influence on the Sm-Nd isotopic data of deep-ocean sediments. Additional bias may be created in the model-age data for sediments by the effects of sediment sorting (McLennan et al. 1990). Furthermore, the history of evolution of the upper crust from Sm-Nd model ages may not be strictly valid for lack of strict representation of the contemporary upper crustal material because, as Veizer and Jansen (1979, 1985) and McLennan (1988) noted, the sedimentary mass at a given time being cannibalistically recycled is incorporating variable proportions of older crust.

1.3 Ar Isotope Geochemistry

Information about K-Ar isotopic composition for modern river sediments is very sketchy. The study of Hurley et al. (1961) is practically the only reference material on this subject. The authors found that the K-Ar dates of the clay-sized fraction could be either nearly similar to, or considerably higher than, that of the bulk sediment. For example, clay-sized fractions and bulk material for the Mississippi River deltaic sediments had extensive overlap in the K-Ar dates, ranging from 214 to 390 Ma; by contrast, the clay-sized <2 and $<4\,\mu m$ fractions with K-Ar dates between 795 and 810 Ma were considerably older than the bulk material with K-Ar dates of 280 to 357 Ma for the Red River deltaic sediments. The results from the Mississippi River sediment indicate an absence of heterogeneity in the K-Ar dates for the various components that make up the sediment. The Mississippi River clay sediments contain a large amount of smectite with smaller amounts of illite, kaolinite, and chlorite, whereas the Red River clay sediments have much smaller amounts of smectite.

In view of the fact that the Mississippi River drainage basin is underlain by rocks of varied compositions and ages, the similarity in the K-Ar dates between the bulk material and the $<0.4\,\mu m$ clay fractions is probably related to either the overwhelming influence of the clay-size fractions in the size-frequency distribution for the sediment or the formation of the coarse fractions by a process of coagulation of the finer fractions. The study of Hurley et al. (1961), therefore, suggests that the correlation between K-Ar dates and particle size should be very carefully evaluated in terms of mineralogy and texture of the materials.

1.4 Stable Isotope Geochemistry

Yeh and Eslinger (1986) reported the oxygen isotope composition of the $<0.1\,\mu m$ fractions of sediments from the Missouri and the Mississippi rivers. The fine clay fraction consisted mainly of illite/smectite mixed-layers with 10 to 45% illite layers. The $\delta^{18}O$ values of the clays ranged from +15.8 to +18.2 per mill in the northern part of the drainage basin and averaged +20.5 per mill in the southern part of the drainage basin. Assuming that the illite/smectite mixed layers contained 30% illite on the average and that the formation temperature of these minerals was between 15 and 25 °C, the authors calculated that the waters in isotopic equilibrium with the minerals had $\delta^{18}O$ values between −4 to −6 per mill. These data tend to suggest that the illite/smectite mixed-layer mineral precipitated from the present-day waters, but this interpretation cannot be supported by the K-Ar results of Hurley et al. (1961) which indicate that the clay particles are old and derived largely from Paleozoic rocks, without invoking that the meteoric water isotopic composition in the region remained largely unchanged since the Paleozoic.

Salomons et al. (1975) found a progressive increase from about +15 to +20 per mill in the $\delta^{18}O$ values of the clay fractions from Atlantic Ocean as the sediments were taken progressively toward the estuary of the Rhine River. The authors attributed this change to an admixture of materials transported landward by currents and tides; but further north, in the vicinity of the Elms River, they found no variation in the $\delta^{18}O$ of the clay fraction. The $\delta^{18}O$ values of the clays in the ocean are well within the range of the values for clays of continental environments, signifying that the original oxygen frameworks of the clay minerals mostly remain unaffected after entering into the oceans.

2 Terrigenous Clays in Young Ocean Basins

Many investigators have reported about the modifications of terrigenous clays in the form of K and Mg uptake from sea water (Weaver 1958; Carroll and Starkey 1960; Russell 1970; Drever 1971; Roberson 1974). The most convincing evidence presented thus far about the ion exchange between terrigenous clays and sea water is by Sayles and Mangelsdorf (1977), who carried out an ion exchange experiment which avoided potential biases in the composition of the solution that could arise from washing of the clays during the experiment. These authors have shown that montmorillonite clays with varied charges and compositions, and kaolinite clays have cation exchange capacities remaining essentially unchanged by coming into contact with sea water. However, the terrigenous clays experience prominent changes in the composition of the exchangeable cations, as they appear to gain K, Na, and Mg, while losing Ca by entering into oceans, so that by inference we may assume that the clays may gain Rb and lose Sr upon coming into contact with sea water. According to these authors, the amount of Ca that is released by the terrigenous clays in oceans each year is approximately 10% the amount of dissolved Ca that is delivered to the oceans by rivers annually. A later work by Sayles and Mangelsdorf (1979) has confirmed this general pattern in the change of the composition of the exchangeable cations. The studies by Sayles and Mangelsdorf clearly suggest that at least some degree of modification in the Rb-Sr and K-Ar isotope systems for terrigenous clay minerals is possible soon after their entry into oceanic environments.

Besides ion exchange with sea water, terrigenous clay minerals deposited at the sediment-water interface may undergo a dissolution-precipitation reaction in a marine environment to produce layered silicate minerals. Glauconite is a typical example of this type of authigenic mineral formation in a marine environment. Several have suggested that authigenic clay minerals may form by reaction between terrigenous clays and sea waters soon after detrital sediments come into contact with sea water (Mackenzie and Garrels 1966; Mackin 1986).

Sediments in deep oceans have been the focus of a wide variety of isotopic studies. The basement basaltic rocks may be bare having no sedimentary cover or they may be variously covered by sediments, especially in the topographic lows where the basement rocks may be covered by few hundred meters of sediment. Biscaye (1965) noted that the depositional age for deep-ocean silicate sediments at the floor of the Atlantic Ocean is generally less than 1 million years and that they are largely detrital in origin. Much of north Atlantic deep sediments at the surface of the ocean floor appears to be terrigenous in origin, as suggested by the study of Behairy et al. (1975). The authors noted that the average concentrations of illite, chlorite, smectite, and kaolinite clays in deep sea sediments in the north Atlantic Ocean are about 55, 10, 15, and 20%, respectively. These abundances in deep sea sediments in the north Atlantic Ocean are broadly similar to that in suspended particulate materials from the same province. By contrast, the deep sea sediments in the south Atlantic Ocean average 47% illite, 26% smectite, 10% chlorite, and 17% kaolinite, while the particulate materials consist of 48% illite, 11% smectite, 20% chlorite, and 21% kaolinite. The difference in the clay mineral composition between particulate materials and deep sea sediments in the south Atlantic Ocean lies in the smectite content. The high smectite content for deep sea sediments in the south Atlantic Ocean suggests that some fraction of smectite in these sediments derived from submarine alteration of volcanic rocks and debris. The smectite contents among deep-sea sediments in major ocean basins are varied, averaging about 15% for the north Atlantic Ocean, 26% for the south Atlantic Ocean, and 41% for the Indian Ocean.

Isotopic analyses of clays in deep-ocean sediments have proved extremely useful in constraining the processes of detrital clay modification and syngenetic clay formation in ocean environments. The isotopic data of clays in these sediments also provide insight into the nature of dispersal of sediments in an ocean basin. A significant body of information now exists about Rb-Sr, K-Ar, and oxygen isotopic compositions of clays in deep ocean sediments, but Sm-Nd and U-Th-Pb isotopic data are limited (Dymond et al. 1973; Goldstein et al. 1984; Grousset et al. 1988).

2.1 Sr Isotope Geochemistry

Rama Murthy and Beiser (1968) reported that the $^{87}Sr/^{86}Sr$ ratios of the $<2\,\mu m$ fraction of some ocean sediments collected from the equatorial region of the Atlantic Ocean ranged from 0.7098 to 0.7234. This range of values is within that for clay sediments of major rivers in the adjacent continental areas and the deltas. Three clay samples had $^{87}Sr/^{86}Sr$ ratios between 0.7067 and 0.7082, which are lower than that for the $^{87}Sr/^{86}Sr$ ratio of the present-day sea waters. These values suggested that the sediments contained clay minerals that were derived from volcanic materials of detrital origin, or were generated in situ in response to submarine volcanic activities. Rama Murthy and Beiser noted that the $^{87}Sr/^{86}Sr$ ratios of the clays were positively related to the sum of illite and kaolinite contents, but very poorly related to the illite content alone. The authors suggested that kaolinite could be a carrier of radiogenic Sr to the deep-sea sediments. This appears reasonable from the evidence provided by kaolinites in weathered profiles which we discussed in the previous chapter, illustrating that kaolinites formed in weathered profiles can have Sr with high $^{87}Sr/^{86}Sr$ ratios. The evidence produced by Rama Murthy and Beiser also suggests that these minerals keep their Sr isotopic signature in the marine environment.

A great deal of knowledge about the Sr isotope compositions of silicate fractions of recent oceanic sediments comes from the work of Dasch (1969). This pioneer work on Sr isotope geochemistry has been used to conclude that terrigenous clay minerals preserve their isotopic identity after coming into contact with sea water, implying that hardly any Sr exchange occurs between clays and sea water. The bulk samples that Dasch analyzed were treated in dilute acid to remove acid soluble phases. The XRD analyses of the silicate residue samples showed that they also contained nonclay minerals like detrital micas and feldspars which are capable of rendering high $^{87}Sr/^{86}Sr$ and Rb/Sr ratios. Hence, the data of Dasch gathered from analyses of bulk silicate samples are not true measures of the isotopic composition of "bottom clay minerals" in modern ocean sediments. Detrital clay fractions can carry high $^{87}Sr/^{86}Sr$ and $^{40}Ar/^{36}Ar$ ratios giving unusually high model ages, but these ages do not necessarily imply derivation of the clays essentially from very old source rocks. Neoformed kaolinite and smectite clays in soil profiles can have high isotopic dates, because these authigenic minerals inherit excess ^{87}Sr or ^{40}Ar from the precursors.

A detailed examination of Dasch's data reveals some intriguing facts. The $^{87}Sr/^{86}Sr$ ratios of the sediments in the Atlantic Ocean bottom are limited to less than 0.743. About 69% of the samples have Sr isotope ratios that are above 0.709, which is the value of the present-day sea water, and one-third of these have values that are above 0.720. The remaining 31% have $^{87}Sr/^{86}Sr$ ratios which are below the sea water value. These low values suggest that ocean sediments contain varied amounts of volcanogenic clay components of either terrigenous or oceanic origin. We have recast the Sr isotopic data of Dasch (1969) by relating them to the clay mineralogy of the bulk silicate fractions. The $^{87}Sr/^{86}Sr$ ratios are very broadly correlated positively with the illite content, but negatively with the smectite content. The trend in the relationship between the smectite content and the Sr isotopic composition suggests that the average $^{87}Sr/^{86}Sr$ ratio for pure smectite samples is close to about 0.704 (Fig. 3.5). This average value is closely similar to that of many oceanic basalts, indicating that at least some of the smectites in the ocean sediments derived primarily from submarine alteration of oceanic volcanic rock fragments. Our analysis also shows that the $^{87}Sr/^{86}Sr$ and the Rb/Sr ratios of the ocean sediments are broadly positively correlated, and that the spread in the relationship between the $^{87}Sr/^{86}Sr$ and the Rb/Sr ratios of the ocean sediments is about the same as that of the river loads (Fig. 3.6). This linearity reflects mixing between two end-member components, one of which is illitic and the other is smectitic in composition.

Ikpeama et al. (1974) noted a negative correlation between the Sr content and the $^{87}Sr/^{86}Sr$ ratio for different types of sediments in the Red Sea. They concluded that the sediment was a mixture of two detrital components: a sialic one having a low amount of Sr with a high $^{87}Sr/^{86}Sr$ ratio and a simatic one having a high amount of Sr with a low $^{87}Sr/^{86}Sr$ ratio. Boger and Faure (1974, 1976) found that the sediments in the Red Sea described a hyperbolic relationship between the $^{87}Sr/^{86}Sr$ ratio and the Sr content (Fig. 3.7A) or a linear relationship between the Sr isotopic composition and the reciprocal of the Sr content (Fig. 3.7B). A two-component mixing was also observed for sediments of the Ross Sea in Antarctica (Shaffer and Faure 1976; Kovach and Faure 1977a, b, 1978). The simatic component of the ocean sediments could be a smectitic clay that formed from alteration of volcanogenic products in ocean sediments.

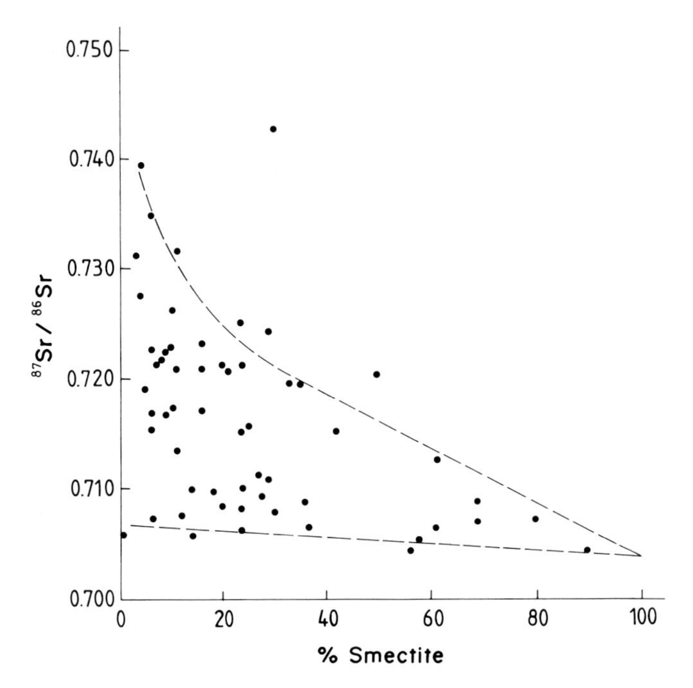

Fig. 3.5. Relationship between the $^{87}Sr/^{86}Sr$ ratios and the smectite contents of acid-leached recent sediments from the North Atlantic Ocean bottom. (After Dasch 1969)

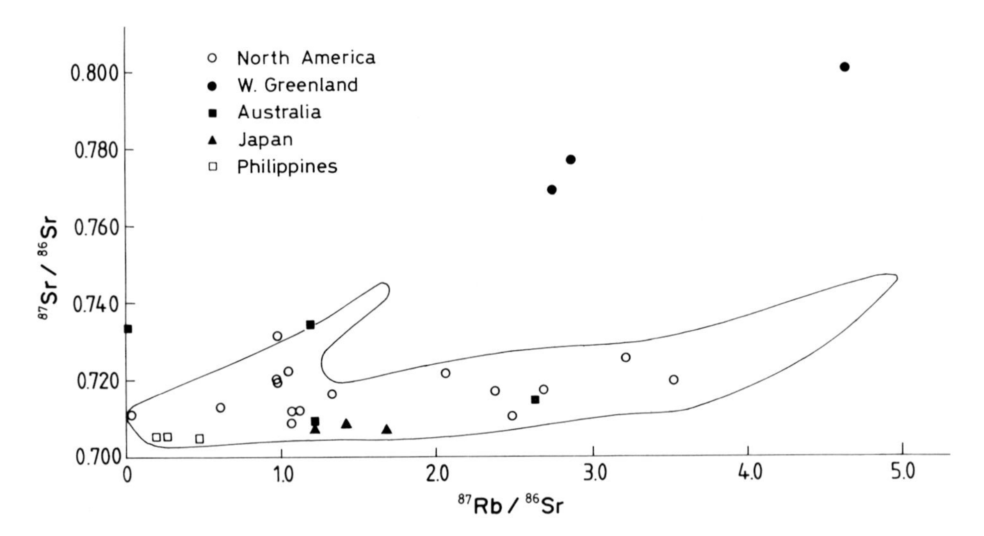

Fig. 3.6. Relationship between the $^{87}Sr/^{86}Sr$ and $^{87}Rb/^{86}Sr$ ratios of suspended loads of rivers. The *encircled area* represents the spread of the same ratios for modern carbonate-free sediments of the Atlantic Ocean. (Goldstein and Jacobsen 1988)

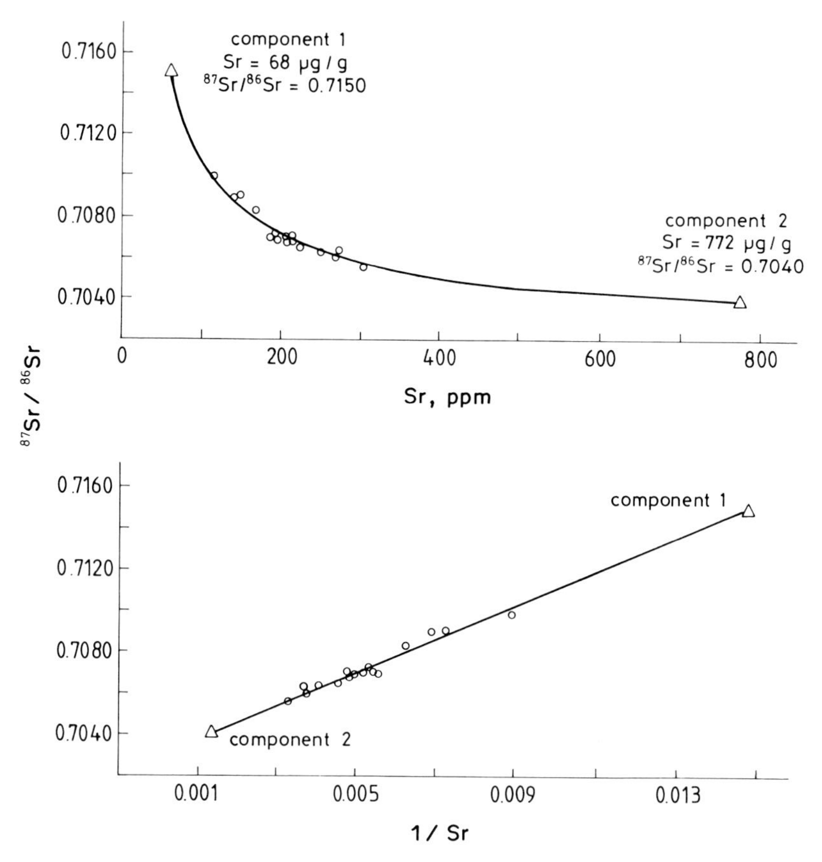

Fig. 3.7. Hyperbolic and linear relationship between the $^{87}Sr/^{86}Sr$ ratio and the Sr content or its reciprocal in the mixing of two detrital components analyzed in Red Sea sediments. (Boger and Faure 1974, 1976)

Biscaye and Dasch (1971) reported on the Sr isotopic composition of carbonate-free sediments from the Argentine basin in the Atlantic Ocean. As much as 40% of the samples had the $^{87}Sr/^{86}Sr$ ratios between 0.7040 and 0.7090, which is lower than the Sr isotopic value of the sea water (Fig. 3.8). The spread in the $^{87}Sr/^{86}Sr$ ratios of the sediments is similar to that published by Dasch (1969). Analyses by Biscaye (1972) on sediments from the Plata estuary and related rivers in the adjacent continental margin area have shown that the $^{87}Sr/^{86}Sr$ and the Rb/Sr ratios of the sediments are higher than that of the sediments in the adjacent oceanic area. This difference in the Sr isotopic composition between the estuarine and the Argentine basin sediments, as explained by Biscaye and Dasch, is due to the transport of large amounts of volcanogenic materials from the south into the Argentine basin. The oxygen isotope data of Lawrence (1979) lend support to Biscaye and Dasch's conclusion.

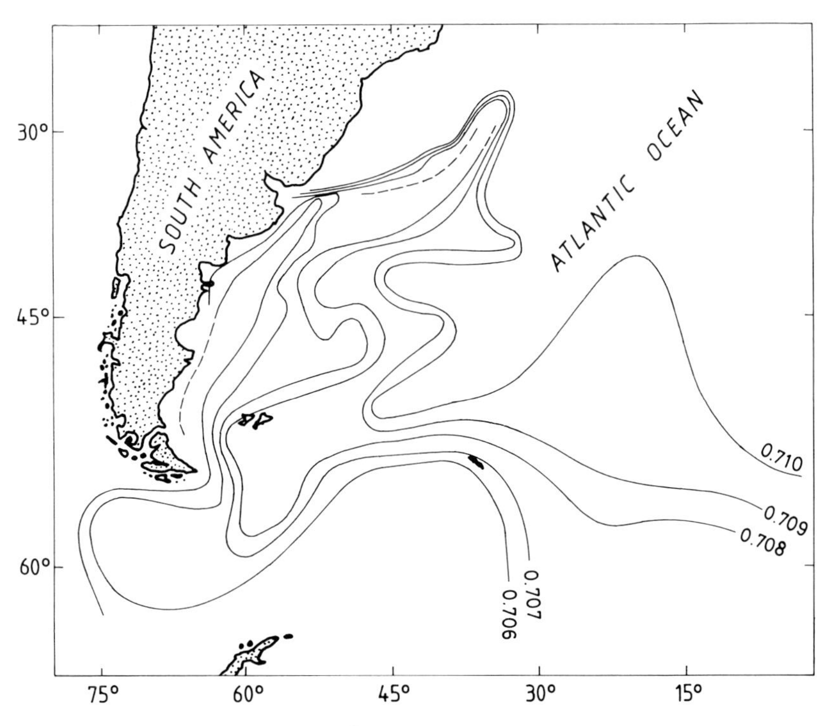

Fig. 3.8. Distribution of the $^{87}Sr/^{86}Sr$ ratios of carbonate-free Holocene sediments of the Argentine Basin and the contiguous areas. (Biscaye and Dasch 1971)

Venkatarathnam et al. (1972) published some Rb-Sr data on $<2\,\mu m$ sediment fractions in the eastern Mediterranean Sea. Mineralogic data have been given for two of these samples and although both samples contained smectite, kaolinite, chlorite, and illite, one was enriched in kaolinite and the other in illite. The kaolinite-rich sample had an $^{87}Sr/^{86}Sr$ ratio of 0.7185 and a Rb/Sr ratio of 0.84, whereas the illite-rich sample had a slightly lower $^{87}Sr/^{86}Sr$ ratio at 0.7155 but a higher Rb/Sr ratio at 1.50. The relatively high $^{87}Sr/^{86}Sr$ ratio for the kaolinite-rich sample is not unusual, as we have already mentioned that kaolinites formed in weathering profiles can have very high $^{87}Sr/^{86}Sr$ ratios.

Clauer (1976) reported Rb-Sr isotopic analyses of an Fe-rich smectite sample collected from the continental shelf off Abidjan, Ivory Coast. Leaching of the clay sample by 1 N HCl acid removed about 1.6% and 56.5% of the original amounts of Rb and Sr, respectively. The Rb and Sr contents of the untreated sample were $89\,\mu g/g$ and $51.2\,\mu g/g$, respectively. The acid treatment clearly caused increases in both the Rb/Sr ratio by a factor of about 2 from 1.74 to 4.04 and the $^{87}Sr/^{86}Sr$ ratio from 0.7212 to 0.7415. The leachate had an $^{87}Sr/^{86}Sr$ ratio of 0.7097 which is slightly higher than the present-day sea water value. The $^{87}Sr/^{86}Sr$ ratio of 0.7415 for the leached smectite agreed well with the value of about 0.7400 reported by Dasch (1969) for leached sediments from the nearby Walvis Bay. The $^{87}Sr/^{86}Sr$ ratio of the leached smectite is also similar to the values reported for clay fractions extracted from soils in central Ivory Coast (Clauer 1979b). Thus, the clay components in these ocean sediments appear to be isotopically similar to those formed in soil profiles in adjacent continental land mass, suggesting that these terrigenous clays largely retain their inherited Sr isotopic signatures through the processes of transport to and deposition in ocean environments.

Several isotopic studies have recognized the influence of wind-borne particles on the isotope compositions of deep-ocean sediments (Dasch 1969; Biscaye et al. 1974; Grousset et al. 1983, 1988). Since Garrels and Mackenzie (1971) have suggested that about 0.5×10^{11} kg/a of sediment are transported into the oceans by wind and storms, this flux of modern atmospheric sediment to the ocean is far from negligible. Although this flux amounts to about 5% of the river sediment flux, the abundance of eolian components is probably much more than 5% in the fine deep sea sediments (Rex and Goldberg 1958; Clayton et al. 1972). Biscaye et al. (1974) collected dust samples over the Atlantic Ocean between Dakar in Senegal and the Barbados Islands and found that these dust sediments had $^{87}Sr/^{86}Sr$ ratios between 0.7146 and 0.7213. Also, to the south of the Gulf of Guinea and the Walvis Bay, along the coast of Africa, dusts collected over the ocean had $^{87}Sr/^{86}Sr$ ratios between 0.7241 and 0.7471 with Rb/Sr ratios between 0.76 and 1.67. The $^{87}Sr/^{86}Sr$ ratios of deep-ocean sediments from these localities were found to be generally similar to those of the dusts (Dasch 1969). These isotopic values also agree reasonably well with the $^{87}Sr/^{86}Sr$ ratios of clay minerals formed from the weathering of rocks on the nearby continent (Clauer 1979b). The results of the different studies suggest that the eolian flux from Africa is a major source for the deep sea sediments east of the Mid-Atlantic ridge.

Grousset and Chesselet (1986) and Clauer et al. (in progress) analyzed the Sr isotopic compositions and the REE distributions of Holocene samples collected along the Mid-Atlantic Ridge between Iceland and the Azores. Nine $<2\,\mu m$ clay fractions consisting mainly of smectite (60–75%) with smaller amounts of chlorite (25–40%) and illite (0–15%) were leached with 1 N HCl to remove Sr from calcite and also from particle surfaces. The $^{87}Sr/^{86}Sr$ ratios of the acid-leached clay fraction were widely scattered between 0.70634 and 0.71994, but positively related to the distance of the samples from Mid Atlantic Ridge (Fig. 3.9). The clays with the very low $^{87}Sr/^{86}Sr$ ratios most likely formed from alteration of oceanic basalts. Authigenesis of this clay material from volcanogenic products was also suggested by the REE patterns with a slight positive Eu anomaly normalized to the REE contents of the North American Shale Composites (NASC). The high $^{87}Sr/^{86}Sr$ ratios for the samples away from the ridge were due to increased amounts of detritus derived probably from the Scandinavian shield

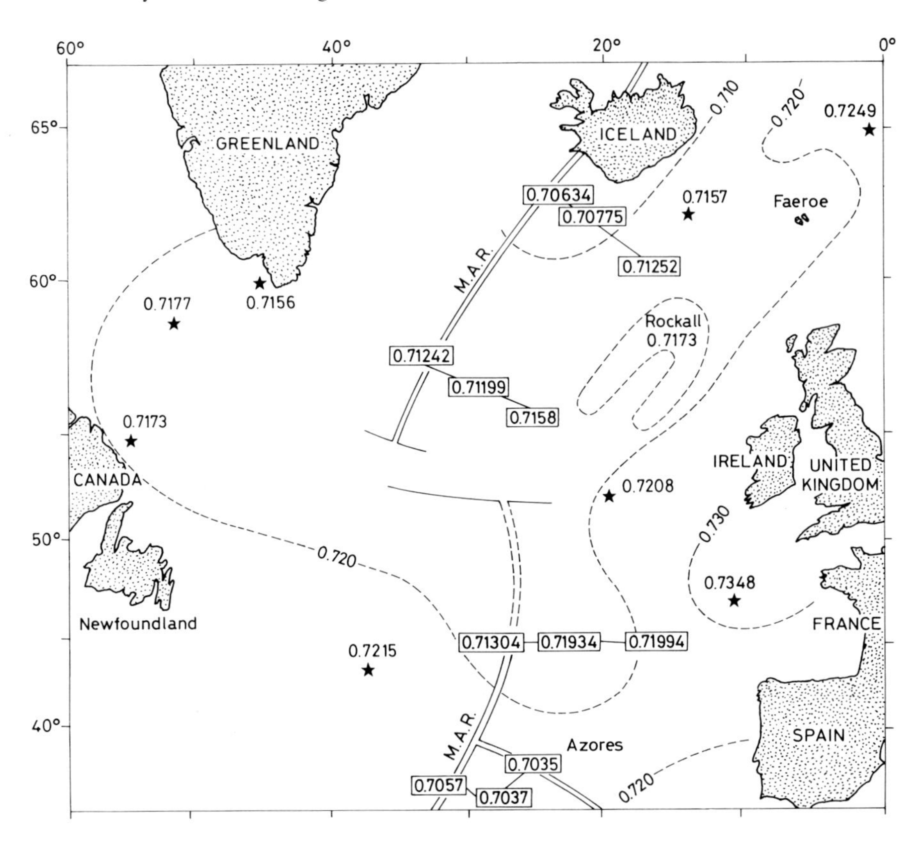

Fig. 3.9. Sr isotopic composition of Holocene samples collected in the eastern Atlantic basin along three transects perpendicular to the Mid-Atlantic Ridge (*M.A.R.*). (Clauer et al. in prep.). The isotopic contours shown by the *dashed curves* and the values located with the *stars* are from Dasch (1969).

in the northern part of the ocean. Mass balance calculations on the $^{87}Sr/^{86}Sr$ ratios of the samples collected from the vicinity of the Azores zone in the eastern part of the ocean suggest a transport to and across the ridge from the North American continent either by wind or surface currents (Grousset et al. 1983). This suggestion of a continental source for the clay materials is supported by the evidence of very flat REE patterns normalized to the NASC.

2.2 Ar Isotope Geochemistry

Information about K-Ar data of modern deep-ocean sediments is extremely sketchy. The study of Hurley et al. (1963) is the one that has been used by many to build a case for the origin of K-bearing clay minerals in recent ocean sediments. Hurley et al. noted that the K-Ar dates of <2 µm clay samples of surface sediments from various parts of the Atlantic Ocean ranged from 163 to 464 Ma. These clay fractions contained disordered 10 Å illite as the major K-bearing phase and 1.9 to 2.2% K_2O. The K-Ar dates of the clay fractions differed widely from that of the bulk samples. For example, in the Caribbean Sea the <2 µm clay fractions had K_2O contents between 1.9 and 2.5% and K-Ar dates between 153 and 168 Ma, whereas the bulk samples had K_2O contents between 1.7 and 2.0% and K-Ar dates between 222 and 255 Ma.

In some parts of the Atlantic Ocean, the clay fraction had higher K-Ar dates than the corresponding coarse sediments. For example, the <2 µm clay fraction of a red pelagic sediment off the southeastern coast of North America had a K_2O content of 2.2% and a K-Ar date of 464 Ma, whereas the >2 µm fraction with a K_2O content of about 2.7% yielded a K-Ar date of 390 Ma. The high K-Ar date of 464 Ma for the red pelagic clay was similar to the K-Ar values for clay fractions of many Paleozoic shales on the east coast of North America. The study of Hurley et al. confirms that the clay fractions of recent sediments of the Atlantic Ocean are mainly terrigenous in origin, and strongly suggests that their K-Ar isotopic systems remain largely unmodified by transport to and deposition in the deep oceans.

As the sediments in the Atlantic Ocean basin have been found to yield mainly detrital isotopic signatures, Huon and coworkers (Huon et al. 1991; Huon and Ruch 1992; Jantschik and Huon 1992) considered the use of the K-Ar signature of these clays to construct the depositional history of the basin during the Pleistocene. Jantschik and Huon (1992) found that interglacial sediments, consisting of mica, chlorite, smectite and kaolinite, had K-Ar dates between 350 and 500 Ma, whereas those deposited during the periods of enhanced ice-rafting gave K-Ar dates between 800 and 1400 Ma (Fig. 3.10). The differences in the values between the interglacial and glacial sediments reflect differences in the provenances of these sediments. The authors interpreted that the interglacial sediments with "lower" K-Ar dates were derived from Caledonian and Precambrian terrains of North America, Greenland, and Scandinavia, whereas the glacial sediments were derived largely from Precambrian rocks.

2.3 Nd Isotope Geochemistry

Grousset et al. (1988) analyzed the Nd isotope composition of deep sea sediments in both the North Atlantic Ocean and the Mediterranean Sea. The $\varepsilon_{Nd}(0)$ values of the sediments ranged from -14.5 to $+4.9$. The relatively less radiogenic Nd values reflect sources of the sediments being from Western Africa and Canada (Fig. 3.11). The materials might have been transported by major wind systems: the westerly wind carrying dust from the

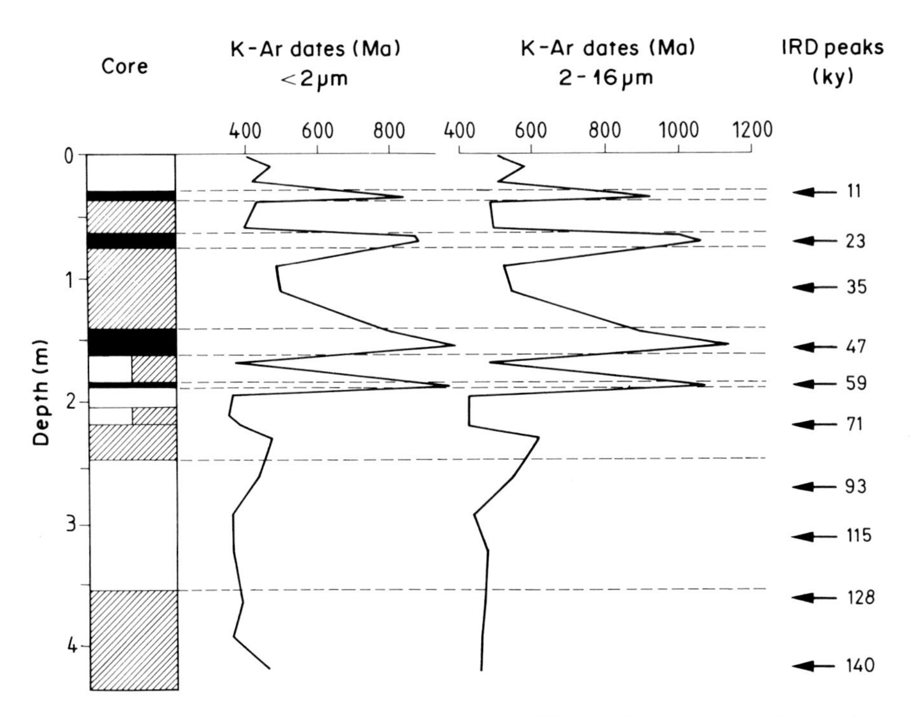

Fig. 3.10. K-Ar dates of carbonate-free fractions of different sizes in cored sediments from Atlantic Ocean relative to the periods of high ice-rafted detritus (IRD-peaks) at 11, 23, 47 and 59 ka. (Jantschik and Huon 1992)

Sahara desert in the African continent and the easterly wind carrying soil particles from the North American continent. The authors observed a zone of more radiogenic $\varepsilon_{Nd}(0)$ values, indicating influence of recent volcanism, along an axis from the Azores to Gibraltar which lies between the two eolian-dominated areas.

More recently, McLennan et al. (1990) reported Sm-Nd isotopic data on muds and sands in Recent deep sea turbidite deposits formed in a wide variety of tectonic settings. They observed that sands and muds from passive margin settings (trailing edge basins and continental collision basins) have Sm/Nd ratios between 0.15 and 0.22 and $\varepsilon_{Nd}(0)$ values between -4.4 and -25.7, whereas those from active margin settings (strike-slip margins, back-arc margins, continental-arc margins, fore-arc basins) have Sm/Nd ratios between 0.17 and 0.33 and $\varepsilon_{Nd}(0)$ values between -14.2 and $+8.3$. The authors observed a general relationship between tectonic settings and Nd model ages by having relatively higher model ages for turbidites from passive margin settings (1.11 to 2.58 Ga) compared to those from active margin settings (0 to 1.85 Ga). The model ages were thought to approximate average provenance ages. The authors also observed from analyses of several pairs

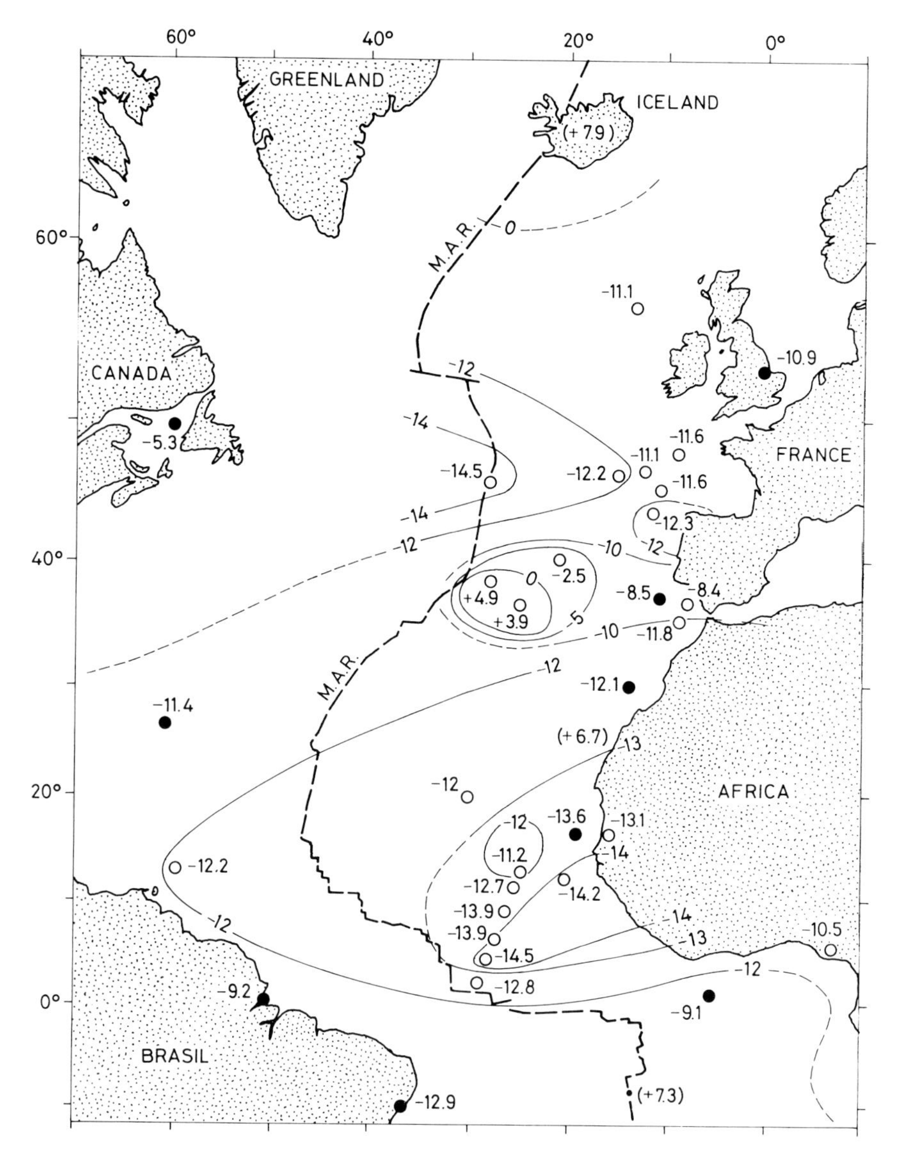

Fig. 3.11. Distribution of $^{143}Nd/^{144}Nd$ ratios of aerosols (*open circles*) and Holocene carbonate-free sediments of the Atlantic Ocean. (Grousset et al. 1988)

of sands and muds in the turbidite deposits that these rocks might be identical to, or differ by being either higher or lower from each other in their $\varepsilon_{Nd}(0)$ values, the difference between the two being as much as 7 units. They ascribed the difference to a factor of sorting during sedimentary transport and deposition.

2.4 Stable Isotope Geochemistry

Savin and Epstein (1970b) observed that the <325 mesh fractions of carbonate-free ocean-floor sediments ranged in $\delta^{18}O$ from +11.5 to +28.5 per mill and in δD from −55 to −87 per mill. The minerals present in these carbonate-free samples were illite, chlorite, smectite, kaolinite, quartz, feldspar, and oxides of Fe and Mn, and occasionally some phillipsite. Calculations on a quartz-free basis showed that the clay minerals of the ocean sediments ranged in $\delta^{18}O$ from +15.3 to +20.0 per mill and in δD from −57 to −93 per mill. The isotopic compositions of clay mineral suites of different ocean cores are shown in Fig. 3.12. Savin and Epstein described the clay-mineral suites by a four component smectite-illite-chlorite-kaolinite system with the respective $\delta^{18}O$ and δD compositions of the end members being given as: smectite (+17, −7), illite (+15, −6), chlorite (+15, −15), and kaolinite (+25, −3). The isotope compositions of the end-member illite, smectite, and chlorite are clearly different from those to be expected if the minerals were in equilibrium with sea water. The isotope data for the kaolinite were less conclusive about its origin. Savin and Epstein concluded that most of the clay minerals in recent marine environments are detrital in origin and that oxygen isotopic exchange between sea water and recent detrital clays is negligible.

The resistance of clay minerals to any alteration of their oxygen isotope compositions at surface or near-surface temperatures has been documented also by observations of clay minerals in soil profiles and by studies of mineral-water exchange under laboratory conditions, as discussed earlier (Lawrence and Taylor 1972; O'Neil and Kharaka 1976), but this view of clay

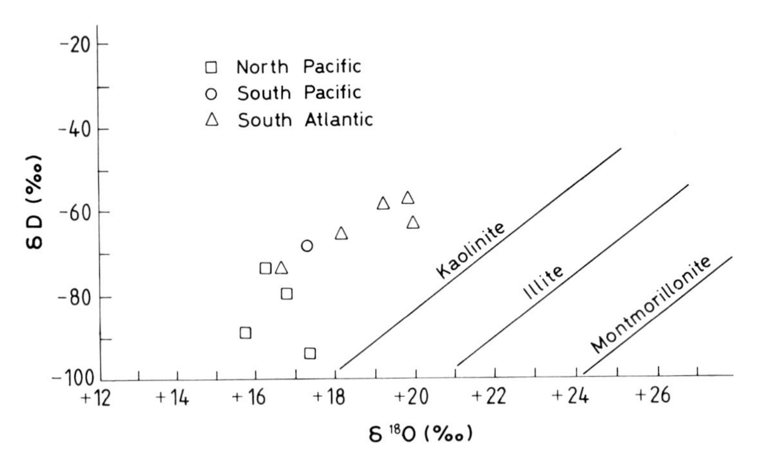

Fig. 3.12. Oxygen and hydrogen isotope compositions of a suite of oceanic clay sediments. The *lines* represent for each mineral the equilibrium with meteoric waters at a low temperature. (Savin and Epstein 1970b)

minerals remaining inert with respect to oxygen isotope exchange at surface conditions is not shared by all (James and Baker 1976). Yeh and Savin (1976) noted, based on studies of cores of 14000- and 31000-year-old sediments at subbottom depths of 10–15 and 33–39 cm in the north Pacific Ocean, no significant exchange for the >0.5 μm clay fractions. As the >0.5 μm clay fractions constitute nearly the bulk sediments, isotopic exchange between bulk sediments of terrigenous origin and sea water still appears to be insignificant.

Yeh and Epstein (1978) investigated hydrogen isotope chemistry of the same deep-sea sediments which were previously analyzed for oxygen isotope compositions by Yeh and Savin (1976). The fine clay fractions (<0.1 μm) of these sediments had δD values of −78 per mill, as compared to the coarse clay fractions (>0.1 μm) which had values between −68 and −77 per mill. They noted that the D/H exchanges for the <0.1 μm fractions ranged from 8 to 28%. As the <0.1 μm fraction did not constitute more than about 5% of the bulk sediment, Yeh and Epstein, observing δD values between −68 and −77 per mill for >0.1 μm clay fractions, concluded that no significant hydrogen isotope exchange occurred between the clay minerals and sea water at least during the past 2–3 Ma since the deposition of the clay sediments. Since the isotopic exchange was insignificant, Yeh and Epstein also suggested that the D/H ratios of clay samples in oceans may be used for determining the provenance of sediments if they are no older than 2 to 3 million years.

Tsirambides (1986) analyzed $\delta^{18}O$ values of different size fractions of sediments from the uppermost (7.5 to 15 cm) ocean subbottom at locations close to the Mascarene plateau, NE of Madagascar Island, and the Somalian basin between the African continent and the Carlsberg ridge. An illite/smectite mixed layer was the dominant clay mineral (30–60%) in the <0.1 μm fraction, whereas illite was predominant (45–75%) in the 0.5–0.1-μm fraction. The $\delta^{18}O$ values of the 0.5–1-μm fractions were mostly between +16.2 and +20.0 per mill, whereas those of the <0.1 μm fractions were between +15.7 and +26.3 per mill. The relatively wider scatter in the $\delta^{18}O$ values of the fine fractions seems to be related to variations in the amount of an illite/smectite component enriched in ^{18}O. Such an illite/smectite mineral phase could have been formed from weathering of volcanic ash produced during the Late Tertiary and Early Quaternary.

Lawrence (1979) analyzed the oxygen isotope composition of the <20 μm size fraction of the surface sediment from the Brazil basin in the south Atlantic Ocean. He concluded from the oxygen isotope data that these sediments originated from two sources, one of which was a kaolinite-rich fluvial load derived from erosion of the Brazilian Shield and the other an illite-smectite-chlorite rich sediment which was transported by the Antarctic bottom current flowing northward into the Brazilian basin. The $\delta^{18}O$ values of the kaolinite-rich sediments were between +21 and +26 per mill, whereas that of the illite-smectite-chlorite-rich sediments carried by the

Antarctic bottom current ranged from +12 to +15 per mill. However, these isotopic values cannot be ascribed solely to the clays, because significant amounts of quartz and other nonclay silicate minerals are expected to occur in the $<20\,\mu m$ fraction analyzed. It might also be noted that the isotope data of sediments collected along major continents might be biased by the possible occurrence of dust transported from nearby continent.

3 Authigenic Clays in Young Deep-Ocean Basins

Evidence for the formation of authigenic clay minerals (saponite, nontronite, celadonite, etc.) is abundant in hydrothermally altered oceanic basalts. Alteration of volcanogenic particles in ocean environments has also resulted in considerable amounts of authigenic clay minerals, such as smectites and zeolites. Formation of low-temperature K-bearing clay mineral phases is best illustrated by the occurrence of glauconite in many off-shore sediments. The isotopic studies of these different authigenic minerals have proved useful to obtain some very critical information about the process of clay mineral formation in oceanic environments. A question which is commonly raised about the origin of these authigenic minerals is whether or not their isotopic evolutions were largely governed by the isotopic compositions of the average ocean water. For this reason we will first give a summary of the ranges of the different isotopic compositions in modern ocean waters before discussing the isotopic aspects of clay authigenesis.

3.1 Isotopic Characteristics of the Present-Day Ocean Waters

3.1.1 Sr Isotope Data

Sea water is a medium which integrates the responses of various global processes. Strontium is a minor cation in the sea water with a concentration of about 7 to 8 μg/g. The residence time of Sr in the ocean is about 5 million years and this is much longer than one-cycle mixing time of ocean waters, which is about 1000 years or so (Goldberg 1965). Because of the very long residence time relative to the length of a mixing cycle, the ocean waters appear to have an almost uniform Sr isotopic composition. The average $^{87}Sr/^{86}Sr$ ratio for present-day sea water Sr is 0.70906 ± 0.00003 (2σ; Faure 1982). However, the uniformity in the isotope composition of Sr in sea water should be viewed with some caution, because local heterogeneities in the isotopic compositions have existed, as they are evident from the results of several studies on some chemical sediments in modern oceans, reflecting that the precipitation rate was faster than the rate by which the isotopic difference of the water may be erased through mixing of ocean water.

Examples of significant differences have been found for the Sr isotopic compositions of water masses next to hydrothermal vents in the oceanic ridge systems (Albarède et al. 1981; Fig. 3.13). The difference in the $^{87}Sr/^{86}Sr$ ratios records the mixing between hot hydrothermal fluids cycled through oceanic crust and cooler sea water. Hot waters (40 °C) of a small geothermal spring located 45 km SSW of Ft. Myers on the west Florida continental shelf contained about 12.9 μg/g Sr with an $^{87}Sr/^{86}Sr$ ratio of 0.70858 ± 0.00004 (Stille et al. 1992). This relatively low value suggests that the ocean waters from this area might have interacted with the Eocene carbonate rocks of the Floridan platform. The Sr isotopic values of waters from restricted bay areas can also be different from average value of sea water (Aberg and Wickman 1987).

The $^{87}Sr/^{86}Sr$ ratio of sea water is controlled by two major inputs: continental Sr primarily supplied by river discharge and smaller amounts by groundwater runout (Chaudhuri and Clauer 1986), and Sr from oceanic crust supplied primarily by hydrothermal systems, especially along ocean ridge systems (Albarède et al. 1981; Clauer and Olafsson 1981). Submarine limestone recrystallization is an additional source of Sr which controls the balance of the Sr isotopic composition of sea water (Elderfield and Greaves 1981). We will see in Chapter 4 that this marine $^{87}Sr/^{86}Sr$ ratio changed during geological time.

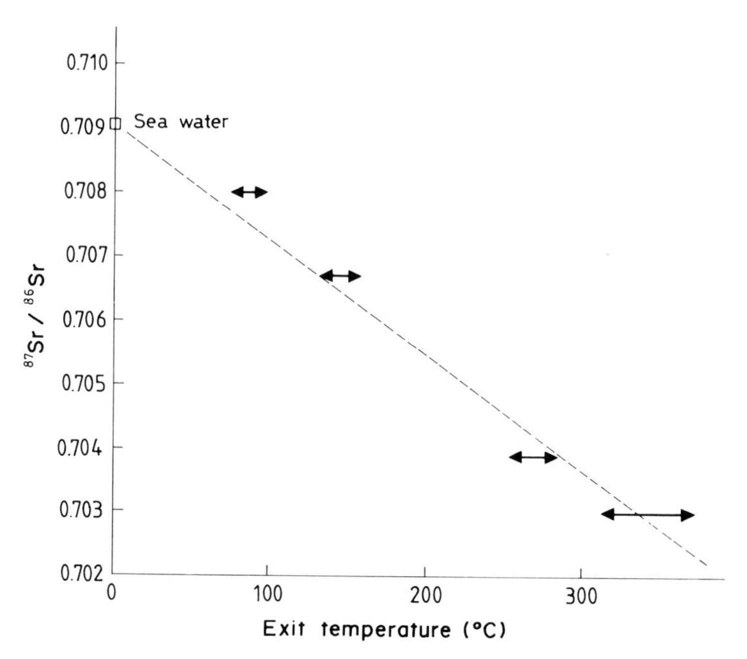

Fig. 3.13. Relationship between the $^{87}Sr/^{86}Sr$ ratio and the exit temperature of hydrothermal fluids of the East Pacific Rise. (Albarède et al. 1981)

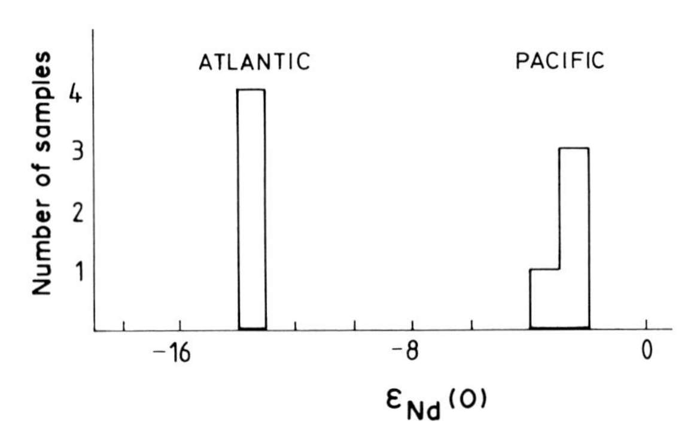

Fig. 3.14. $\varepsilon_{Nd}(0)$ values of the Atlantic and Pacific Ocean waters. (Piepgras et al. 1979; Goldstein and O'Nions 1981)

3.1.2 Nd Isotope Data

The concentrations of Nd are varied among major ocean waters and range from 0.41 to 4.21×10^{-6} µg/ml (Piepgras et al. 1979). Concentration variations also exist with depth in a single ocean. Like that of Nd, the concentrations of Sm vary between 0.35 and 0.80×10^{-6} µg/ml, the average Sm/Nd ratio of sea water being about 0.24.

The ocean waters are significantly varied in $^{143}Nd/^{144}Nd$ ratios, ranging from an average value of about 0.51195 for the Atlantic Ocean to 0.51250 for the Pacific Ocean (Fig. 3.14). The lack of uniformity in the Nd isotopic compositions among ocean waters is reflected in the short residence time of Nd which has been estimated to be between 300 and 7150 years (Faure 1986; Goldstein and Jacobsen 1988). The relatively high $^{143}Nd/^{144}Nd$ ratio for the Pacific Ocean has been attributed to derivation of Nd from young source rocks, either from young continental crust around the Pacific Ocean or from active circum-Pacific magmatism (Stille 1992). Continents are believed to be the major source of Nd in the ocean waters, because oceanic Mn nodules and metalliferous sediments carry an overwhelming Nd isotopic signature of continental derived Nd (O'Nions et al. 1978).

3.1.3 Stable Isotope Data

Waters in deep oceans have $\delta^{18}O$ values which are within a few tenths of 0 per mill. Deep ocean waters average about +0.12 per mill for the north Atlantic Ocean, about +0.02 per mill for the Pacific Ocean, and about −0.45 per mill for the Antarctic Ocean (Craig and Gordon 1965). Surface ocean waters are more variable in their $\delta^{18}O$ values, but those with average salinity have $\delta^{18}O$ between +1 and −1 per mill. The ^{18}O concentrations also vary with depth. The variations in the oxygen isotopic compositions of ocean waters are related to evaporation, precipitation, input of river water, freezing and melting of ice, and mixing of water masses.

The oxygen and hydrogen isotope compositions of ocean surface waters are approximately linearly related, due primarily to the combined effect of evaporation of surface waters and precipitation of atmospheric moisture. Craig (1961) reported that the D contents are related to the ^{18}O contents in surface waters of some major oceans by:

$$\delta D = m\, \delta^{18}O,$$

where m is 7.5 for the north Pacific Ocean, 6.5 for the north Atlantic Ocean, and 6.0 for the Red Sea.

The oxygen and hydrogen isotope compositions of waters of restricted seas can be significantly different from that of large open oceans. This is illustrated by waters of the Gulf of Aqaba which connects to the Red Sea, with the inflow of surface water and the outflow, beneath the surface water, into the Red Sea. The $\delta^{18}O$ values of the surface waters of the gulf range from +2.2 to +2.5 per mill, as compared to +1.8 for the $\delta^{18}O$ for the Red Sea surface waters (Anati and Gat 1989). The $\delta^{18}O$ values of the waters in the Gulf of Aqaba increase to +2.95 per mill at a depth of about 150 m. Anati and Gat explained that the high values of the waters are due to evaporation exceeding precipitation and runoff in this arid climate.

3.2 Clay Authigenesis and Modification at Ocean Floor Conditions

Although terrigenous clays are highly abundant for the large part of the modern ocean deposits, authigenic and syngenetic clays can represent significant amounts for some of these deposits. Glauconites in several offshore sediments and smectite-celadonite associated with alteration of ocean-floor basalts and volcanic debris testify to the existence of wide varieties of presently existing conditions in which authigenic clay minerals can form. Illite is the most represented clay component in Recent ocean sediments, but of detrital origin. However, formation of illite in oceanic environments has also been sought from both thermodynamic computations and experimental data. Garrels (1984) showed, on the basis of the activities of K, H and Si, that micaceous clay minerals could be in equilibrium with the sea water. Harder (1974) was able to synthezise illite at low temperature (3–20 °C) by precipitation of Al-hydroxides from Si-Mg-K-bearing solutions. Thus, kinetic barriers appear to be a major cause for the lack of formation of illitic minerals in present-day ocean environments. Eberl and Hower (1976) have shown that the requirement of an activation energy of 19.6 ± 3.5 kcal/mol for the conversion of smectite to illite is too high for the replacement to occur at Recent ocean bottom temperatures in a few tens of million years. The data of Eberl and Hower also suggest that the increase in temperature by one order of magnitude shortens the time for conversion of smectite to illite by nearly two orders of magnitude. For instance, an increase in temperature from 2 to 50 °C reduces the time of formation of illite from smectite from 100 million years to about 1 million years. Although in many places, the present deep-ocean bottom temperatures are about 2–3 °C, in the Precambrian the ocean bottom temperature could have been significantly higher (Perry 1967; Knauth 1992), increasing the possibility of the formation of illite from replacement of smectite in a short time after the deposition.

3.2.1 Glauconites from Continental Off-Shores

Studies on isotope geochemistry of recent glauconies are few, and much of the present state of knowledge about these minerals in recent ocean bottom sediments is built upon the results of investigations on glauconies from the Gulf of Guinea (Giresse and Odin 1973; Odin et al. 1979; Keppens and O'Neil 1984; Clauer et al. 1992a, b; Stille and Clauer 1994). Varied Recent glaucony mineral phases from smectite-rich to glauconite-rich end members were found in biodigested fecal pellets associated with muds which contain kaolinite and smaller amounts of smectite in the Gulf of Guinea (Giresse 1965). The clay minerals in the muds were generated as products of weathering of rocks in the African continent, especially in the intertropical area (Paquet 1970). The clays, having inherited radiogenic ^{87}Sr and ^{40}Ar from their precursors, were subsequently transported to the ocean by the Congo River.

Odin (1975) suggested, on the basis of studies on glaucony pellets of the Gulf of Guinea, that glauconitization occurs in a specific microenvironment where Fe-smectite first crystallizes and then evolves to glauconite by progressive incorporation of K. In support of this formation model, Birch et al. (1976) suggested that Fe enters the octahedral positions of the clay structures very early, probably by a mechanism that is independent of K fixation. Odin and Dodson (1982) claimed that Recent marine glauconitic pellets in the Gulf of Guinea probably formed by crystal growth from ions of the interstitial fluids. Studies on the Sr, Nd and oxygen isotopic compositions by Clauer et al. (1992a, b) and Stille and Clauer (1994) provided additional constraints and further refinement of the above proposed model of formation of glauconites. The $^{87}Sr/^{86}Sr$ ratios of HAc and HCl leachates from these samples ranged between 0.70875 and 0.70987 and demonstrated that leaching with dilute HCl caused no preferential extraction of ^{87}Sr from the minerals relative to leaching with HAc, and that the Sr external to the minerals is of marine origin. Clauer et al. found, as Odin et al. (1979) observed in their analyses of the K-Ar dates, that the Rb-Sr dates decreased as the K_2O content increased (Fig. 3.15). The K-Ar dates decreased from 473 to 50 Ma, and the Rb-Sr dates from 415 to 64 Ma. Odin (1975) claimed

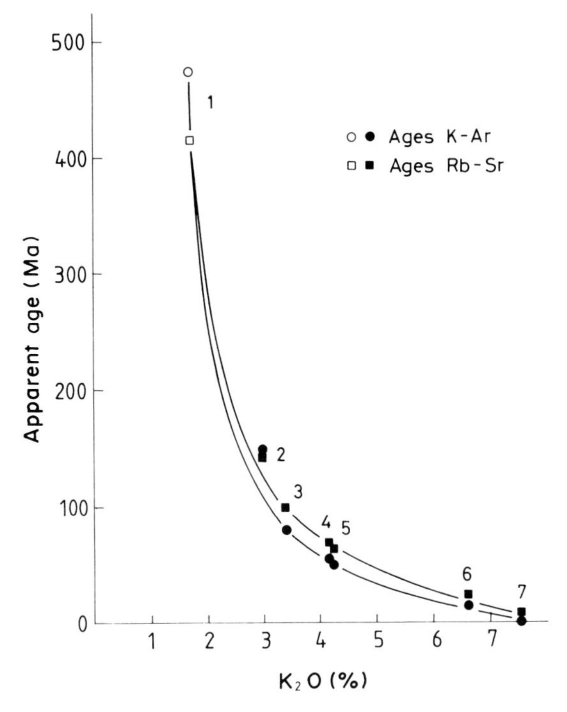

Fig. 3.15. Trends in the K-Ar and the Rb-Sr dates with respect to changes in K_2O contents of recent glauconitic pellets (*filled symbols*) evolving from clays in the associated mud (*open symbols*) in the Gulf of Guinea. (Clauer et al. 1992b)

that the decrease in the K-Ar dates is more due to loss of radiogenic ^{40}Ar than to gain of K_2O. This claim was based on the evidence that in the Gulf of Guinea, those glauconitic pellets with as much as 6.6 wt.% K contained about 10% inherited radiogenic ^{40}Ar, which is in contradiction with the hypothesis that the pellets at this advanced stage of their evolution should be free of the isotopic memory of precursors by complete dissolution (Odin et al. 1979; Odin and Matter 1981). Clauer et al. (1992a, b) interpreted that the decrease of the Rb-Sr dates with increase in the K_2O content was primarily due to addition of Rb and not loss of radiogenic ^{87}Sr. This interpretation is based on the relationship between the $^{87}Sr/^{86}Sr$ ratios and the K_2O contents which showed that those glauconite pellets whose K_2O contents are less than 4.5% had the same $^{87}Sr/^{86}Sr$ ratio as the clay fraction of the mud hosting the pellets (Fig. 3.16). The Nd isotope geochemistry analyses of the same Recent glauconies showed that the $^{143}Nd/^{144}Nd$ ratios of the glauconies and the associated clays ranged between 0.51168 and 0.51176 (Clauer et al. 1992b; Stille and Clauer 1994). The authors recognized two trends in the Nd isotopic compositions with respect to the K_2O contents of the glauconies. Unlike the $^{87}Sr/^{86}Sr$ ratios which remained invariant

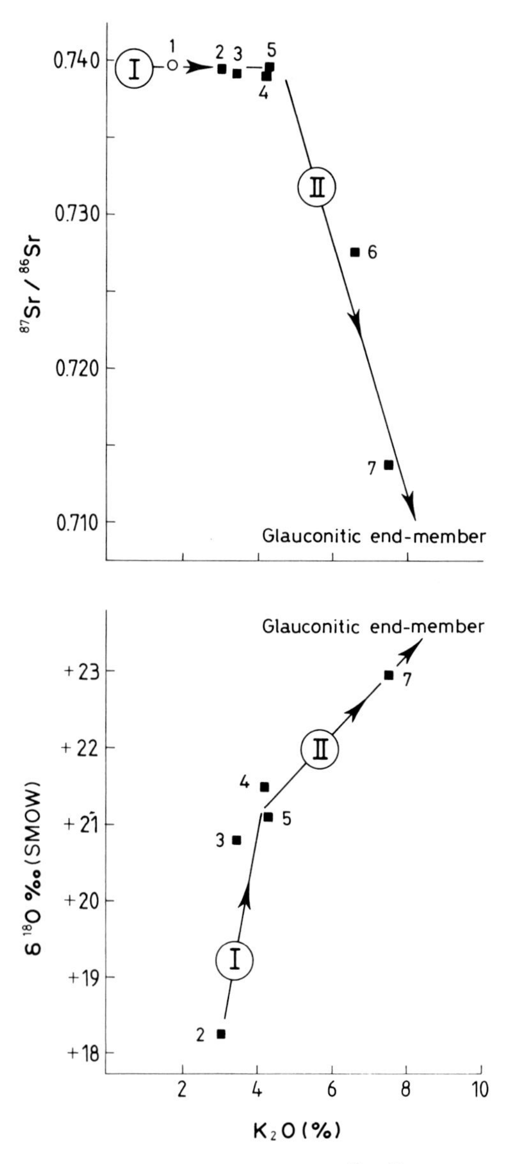

Fig. 3.16. Paths of evolution of the $^{87}Sr/^{86}Sr$ ratio and the $\delta^{18}O$ value with respect to change in the K content of recent glauconitic pellets (*filled symbols*) evolving from clays in the associated mud (*open symbol*) in the Gulf of Guinea. (Clauer et al. 1992b)

between the detrital clay fraction and those glauconies with less than 4.5% K_2O, the $^{143}Nd/^{14}Nd$ ratio appeared to have changed significantly. This change was not due to sea water influence (arrow I on Fig. 3.17), as the isotopic shift was not towards the sea water value. The influence of the

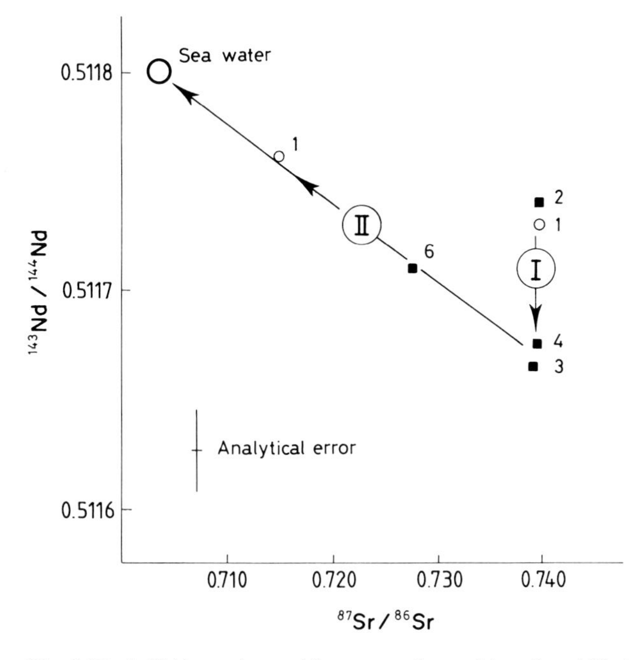

Fig. 3.17. Sr-Nd isotopic trend for recent glauconitic pellets (*filled symbols*) evolving from clays in the associated mud (*open symbols*) in the Gulf of Guinea. (Clauer et al. 1992b)

marine Nd on the Nd isotope signatures of the glauconies is detectable only when the K_2O content is more than 4.5% (arrow II on Fig. 3.17).

Odin and Dodson (1982) believed that granular glauconies are the result of a marine authigenesis in a substrate consisting of various components which are susceptible to alteration. Previously, Odin and Matter (1981) assumed that the authigenesis of glauconite occurred in an open system. However, this open system condition is not in accord with the K-Ar results, as glauconitic pellets without visible relicts of precursors and containing at least 6.6% K_2O had 10% inherited radiogenic ^{40}Ar. The glauconitization process is more complicated than the model proposed by Odin (1975), and this complexity is underscored by the study of Amouric and Parron (1985), who observed, from HRTEM analyses, a structural discontinuity between smectite and neighboring glauconite layers and concluded that the glauconite layers formed by a process involving dissolution-precipitation.

The close similarity in the Sr isotopic compositions between the evolving glauconies and the associated detrital clay of the mud in the Gulf of Guinea strongly supports the existence of a genetic link between the glauconies and the clay minerals of the mud (Clauer et al. 1992a, b). This genetic link has been further corroborated by oxygen-isotope evidence presented by Keppens and O'Neil (1984). The collective Sr and oxygen isotope evidence suggests that glauconitization resulting in 4.5% K_2O of the evolved phase must have occurred in a closed system. The progressive evolution of glauconies was accompanied by increases in K_2O contents to more than 4.5%, as well as decreases in the $^{87}Sr/^{86}Sr$ ratios approaching that of the sea water. Clauer et al. (1992a) therefore suggested a two-step glauconitization process, beginning with a process of dissolution-precipitation of the detrital clay material in the pellets in a confined environment without participation of sea water (arrow I on Fig. 3.16) and continuing later in its evolution with precipitation-crystallization in an open system condition allowing involvement of sea water (arrow II on Fig. 3.16). The second step might correspond to crystal growth by direct precipitation from the solution, a suggestion which was previously made by Clauer and Millot (1978). According to the Clauer et al.'s (1992a, b) two-stage model for the growth of glauconies, the isotopic records of any natural glaucony are only reflecting a given state in the evolution that is presumably a continuum between the least evolved and theoretically the most evolved stages.

The Nd-isotope evidence independently corroborates the model based on the Sr data. In addition, the Nd data suggest the occurrence of a supplementary component in the pellets that is not of marine origin. This component has to be very low in Sr, explaining why its presence has not been detected in the Sr isotope signatures. The Nd-isotope composition of this supplementary component was also different from that of the detrital silicate phase. The Nd derived from dissolution of the supplementary component was reused by the glauconite material formed during the early phase of the glauconitization. Previous leaching experiments of clay minerals with acids have shown that Fe oxy-hydroxides associated with clay particles often contain high amounts of REE but low amounts of Sr (Clauer et al. 1993), and this is also evident from sequential leaching experiments on the Recent Guinean glauconies (Stille and Clauer 1994). The supply of Fe which is needed for the formation of the glauconitic minerals, partly comes from this oxy-hydroxide phases (Odin 1975). It can, therefore, be suggested that dissolution-precipitation of clay minerals in a closed system mainly uses up the elements intimately associated with the clay particles, explaining why the oxygen isotope composition in evolving glauconies is apparently not in isotope equilibrium with sea water at the end of the dissolution-crystallization phase of the process (arrow I on Fig. 3.16, lower diagram). The slight change observed in the oxygen isotope composition can then be related to the using up of the elements of the Fe oxy-hydroxides. This confirms the entering of Fe into the mineral structures before that of K, which occurs mainly during the second evolutionary stage, as suggested by Birch et al. (1976).

Odin and Dodson (1982) have also discussed the duration of the evolution from the alteration of precursors to highly evolved glauconites free of any excess ^{40}Ar. They estimated that the period of glauconitization commonly covers between 1000 and 10000 years, and maybe even 100000 to 1000000 years for highly evolved ones. The duration of the early diagenetic evolution of common clay minerals in Recent ocean sediments is still to be accurately determined, so that we may only assume that the duration of syndepositional evolution of these minerals is about the same time as that of glauconies.

3.2.2 Authigenic Clay Minerals from Ocean Basalt Alteration

Alteration of volcanic materials in the marine environment frequently results in the formation of varied amounts of palagonite and associated clay minerals as common secondary products. The alteration environments vary from metamorphic conditions with temperatures reaching a few hundred degrees centigrade to submarine weathering conditions with low temperatures. The discharge of acidic hydrothermal brines into the ocean is another means of formation of different clay minerals. Hydrothermal alteration of basalt results in a wide variety of clay minerals. The study by Donnelly et al. (1980) identified a wide variety of clay minerals as replacement products occurring in glasses and palagonites and also as crystallized minerals in cracks and vugs. They concluded that variations in clay mineralogy reflect both the extent of the alteration of the basaltic rocks and the aging of these minerals. According to these authors, the alteration of young oceanic crust produces limited quantities of Al-poor smectite phases whose compositions range from saponite to celadonite, whereas the alteration of older crusts commonly results in either Al-rich saponite or celadonite. Bass et al. (1973), Kempe (1974), Kastner (1976), Humphris et al. (1980), Mevel (1980), Pritchard (1980) among many others, have provided information about formations of celadonite and saponite as products of submarine alteration of basalts.

Palagonitization is the replacement of basaltic glass by poorly crystallized or amorphous colloidal, clay-like material (Honnorez 1972). The replacement product or palagonite is red, brown, yellow rinds on a variety of exposed basalt or basaltic glass surfaces such as margins of pillow basalt, margins, vesicle walls, cracks, or crevice walls of hyaloclastite grains. Because of the colloidal, clay-like nature of the palagonite, it often adsorbs trace alkali, alkaline earth, and rare-earth elements in large concentrations. Palagonitization appears to be an isovolumetric replacement, and the clay mineralization occurs during or after palagonitization, as these minerals have been found to replace pre-existing magmatic minerals, glass, palagonite and also to crystallize in cracks and vugs, together with other secondary minerals (zeolites, calcite, etc.).

3.2.2.1 Oxygen Isotope Geochemistry

Basalt alteration at conditions of greenschist metamorphism (200–300 °C) have often resulted in formation of chlorite minerals (Muehlenbachs and

Clayton 1972); but at lower temperatures, hydrothermal alteration of basalts has produced palagonite and various associated clay minerals such as smectite, celadonite, smectite/celadonite or smectite/illite mixed-layers, and chlorite (Banks 1972; Melson and Thompson 1973). Kristmannsdottir (1977) reported an occurrence of trioctahedral Fe and Mg smectites in Reykjanes (Iceland) basalts hydrothermally altered at a temperature of about 200 °C. Lawrence and Drever (1981) observed abundant saponite in the vicinity of a doleritic sill intrusion into basaltic basement rocks at DSDP Site 395A. One of the saponite samples from this DSDP Site yielded a $\delta^{18}O$ value of +9.8 per mill, giving an estimated formation temperature of about 157 °C. The authors maintained that saponite formed at this elevated temperature by basalt-sea water interaction due to intrusion of the sill. Stakes and O'Neil (1982) reported saponite-rich pillow breccias from a dredge haul at a site about 350 km to the east of the East Pacific Rise. These saponite samples are Fe- and Mg-rich varieties and replace the outer portions of the pillows, and also occur as vein-filling deposits within the relatively unaltered interior. The authors reported that the $\delta^{18}O$ values of the Fe- and Mg-rich saponites ranged from +8.7 to +11.4 per mill and estimated that the crystallization temperatures of these minerals ranged between 135 and 170 °C in sea water-dominated systems with water/rock ratios of 50 or higher. The studies mentioned above suggest that various Fe-rich saponite minerals are products of moderately high temperature reactions and not of sea floor weathering or diagenetic processes at normal bottom temperatures.

Hydrothermal acidic brines, sometimes known as "black smoker" fluids, with temperatures of more than 350 °C were first observed from the East Pacific Rise (EPR). These fluids have undoubtedly caused alteration of rocks and minerals of the upper oceanic crust before being discharged into the ocean. Haymon and Kastner (1986) reported that very high temperature alteration of tholeiitic basaltic glass and calcic plagioclase during the venting of hydrothermal fluids on the crest of the EPR produced layered crusts composed of aluminous beidellitic smectite, randomly interstratified Al-rich chlorite/smectite, chlorite and X-ray amorphous aluminosilicate material. The clay assemblages with various other secondary minerals presumably formed from replacement of fresh basaltic glass and calcic plagioclase within 1 mm of vesicle walls and surfaces of cracks and fractures exposed to hydrothermal fluids. A sample of the beidellitic smectite produced from such a high temperature hydrothermally altered basalt, yielded a $\delta^{18}O$ value of +4.1 per mill, whereas an associated sample of chlorite/smectite mixed-layer mineral had a $\delta^{18}O$ value of +3.5 per mill. Assuming that the two minerals equilibrated with sea water or with the "black smoker" fluids whose $\delta^{18}O$ value is about +1.6 per mill, Haymon and Kastner estimated that the formation temperature for the minerals ranged between 290 and 360 °C.

Formation of authigenic clay minerals by precipitation from sea water as a result of expulsion of hydrothermal brines into oceans at centers of submarine ridges and mounds has been noted by Bischoff (1972). The author reported that a hydrothermal brine at a temperature of about 250 °C is discharged into the Red Sea, resulting in precipitations of Fe-rich dioctahedral smectite (nontronite) and Al-rich montmorillonite/beidellite. Cole (1983) estimated from oxygen isotope data that the nontronites with $\delta^{18}O$ values between +12.1 and +18.1 per mill, formed at temperatures between 78 and 130 °C and that the montmorillonite/beidellite minerals with $\delta^{18}O$ values between +8.0 and +10.5 per mill, formed at temperatures between 206 and 162 °C.

Basaltic glasses altering to smectitic and celadonitic clay minerals at moderate to low temperatures, that are still significantly higher than that of cool normal ocean bottom, have been reported in several different studies. These smectites commonly range in composition from nontronite to saponite and the celadonites can have as much as 9.9 K_2O (Melson and Thompson 1973; Juteau et al. 1979; Robinson et al. 1979; Donnelly et al. 1980; Richardson et al. 1980; Lawrence and Drever 1981; Mehegan and Robinson 1982; Stakes and O'Neil 1982; Staudigel and Hart 1983). Estimated temperatures of formation for many of these smectites and celadonites are often in the range between 20 to 75 °C. Lawrence and Drever (1981) reported that the $\delta^{18}O$ values of some vein-filling saponites in a basaltic basement rock at DSDP Site 395 ranged between +19.9 and +24.3 per mill, suggesting that these minerals formed at temperatures between 23 and 58 °C. Stakes and O'Neil (1982) noted that both a sample of vein-filling nontronite with the $\delta^{18}O$ value of +22.5 per mill and a sample of vein-filling celadonite with the $\delta^{18}O$ value of +19.5 per mill formed at about 35 °C. McMurtry and Yeh (1981) analyzed some authigenic, poorly crystallized smectites from the EPR and the Bauer basin. The $\delta^{18}O$ values of these minerals ranged between +24.2 and +30.9 per mill, suggesting that they formed at temperatures between 30 and 50 °C.

The origin of smectitic clay minerals in pelagic oozes near sites of hydrothermal brine discharges is often not very clear, as illustrated by the occurrence of clay deposits in the area of the Galapagos hydrothermal mounds formed about 130 000 years ago. McMurtry et al. (1983) reported that dark green authigenic Al-poor nontronite is predominant in the mound sediment, whereas Fe-montmorillonite is the most abundant clay mineral in the surrounding pelagic ooze. The nontronites from the mound have K_2O contents of about 2.7%, whereas the Fe-rich montmorillonites have K_2O contents of about 0.9%. The authors have shown by the use of Al_2O_3-MgO-Fe_2O_3 variation diagrams that the compositions of the Fe-rich montmorillonites lie between that of the authigenic nontronite found on the mound and that of terrigenous Al-montmorillonite/beidellite mixed-layer minerals, suggesting that the Fe-rich montmorillonites are mixtures between the two end-members. The $\delta^{18}O$ values of pure fine ($<2\,\mu m$) nontronites from the mound ranged between +21.5 and +25.0 per mill, indicating that the minerals formed at temperatures between 25 and 47 °C. The present in situ temperature of the mound is no more than 15 °C. The difference between the present temperature of the mound and the temperature of the clay mineralization suggests cooling of the hydrothermal systems at the mounds over the past 130 000 years, or a change in the patterns of circulation of fluid and precipitation of nontronite within the mounds. By comparison with the nontronite, the Fe-montmorillonite in the pelagic ooze next to the hydrothermal mounds had $\delta^{18}O$ values between +23 and +25 per mill. Assuming a marine authigenic origin, temperatures of their formation were estimated to be between 27 and 39 °C, which were about 4 to 7 °C higher than the present-day in situ temperature. The difference in the temperature may be variously explained, including transport of the clays to the present location by bottom currents since their formation at hydrothermal conditions at spreading centers, terrigenous origin and varied mixtures of authigenic Fe-nontronite and detrital Al-montmorillonite.

Clay minerals of hydrothermal mounds can have a complex history of formation, as suggested by Buatier et al. (1988), who investigated Fe-rich clay minerals from hydrothermal mounds located about 22 km south of the Galapagos rift axis (DSDP Legs 54 and 70). The clays were of two types differing in morphology, one was lath-shaped and the other veil-like (Fig. 3.18). The lath-shaped particles were glauconites with regular 10 Å spacing, whereas the veil-like particles were nontronites and Fe-montmorillonites with irregular 10 and 13 Å spacings. The $\delta^{18}O$ values of most clay fractions averaged at about +24.1 per mill, independent of the particle type (Buatier

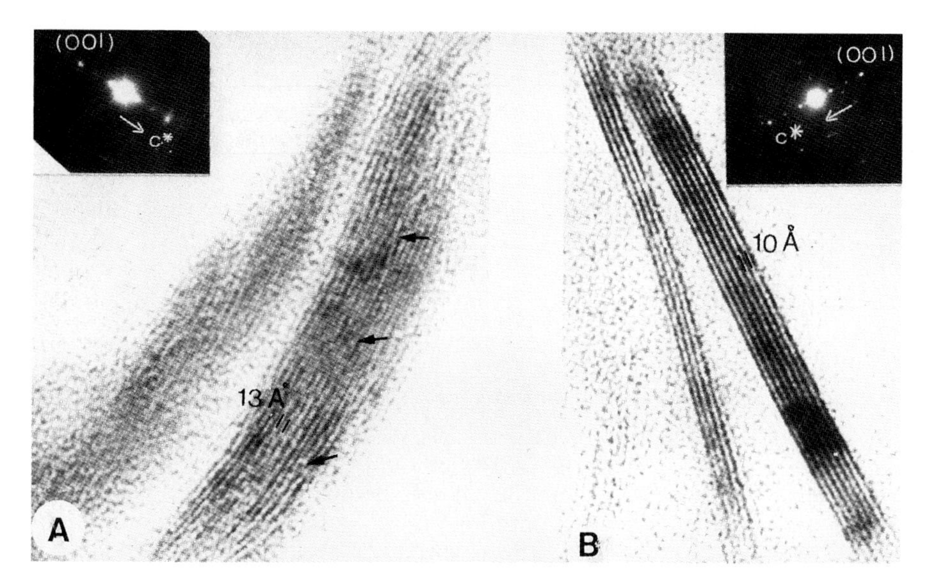

Fig. 3.18. TEM micrographs with electron microdiffraction diagrams. **A** Veil-shaped celadonite and **B** lath-shaped glauconite, from hydrothermal mounds near the Galapagos rift axis. (Buatier et al. 1988)

1989). Observations by TEM revealed transition from veil to lath morphology. This morphologic change appeared unrelated to any variation in temperature which was estimated to be about 40 °C, assuming that the $\delta^{18}O$ value of the fluid was about +2 per mill. Two veil-rich fractions had $\delta^{18}O$ values of +21 per mill, suggesting occasional influxes of water from sites below the mounds at temperatures as high as 60 °C. The veil-like smectites from the shallow part of the core had $\delta^{18}O$ values of +26.3 per mill, suggesting a formation temperature as low as 20–25 °C from fluids primarily of sea-water origin.

3.2.2.2 Sr Isotope Geochemistry

Hydrothermal deposits of smectitic and celadonitic clays occurring as vein-filling minerals in oceanic basalts have been the subjects of several studies on Rb-Sr isotopic dating (Hart and Staudigel 1978, 1980, 1986; Richardson et al. 1980; Staudigel and Hart 1985). The results of these studies have provided information not only about the time of formation of these vein-filling minerals relative to the time of formation of the host basaltic rock, but also about the sources of Sr in these clays. The study by Hart and Staudigel (1986) gives a good insight into the origin of these vein-filling minerals and reports Rb-Sr isochrons for such clay minerals from three geographically different localities (Figs. 3.19–3.21), the dates ranging from 70 to 121 Ma. More importantly, where the oceanic basement basalt is about 135–155 Ma old, the age of celadonite-smectite formation is about 121 Ma, making the vein clay mineralization to be younger by about 15 to 34 Ma. Similarly, at DSDP Sites 417/418, the basaltic crust formed about 119–122 Ma ago, but the hydrothermal clay minerals in it had a date of about 108 Ma, indicating that the clay mineralization occurred about 10–14 Ma after the formation of the crust. The significance of the dates for the vein clay minerals is far from clear. Assuming that the dates are isotopic "closure" ages, the hydrothermal system could have persisted for a few tens of million years, or the clay mineralization at each site could have been the result of renewed hydrothermal activity after the oceanic crust formation. The dates of the vein clay mineral formation do not necessarily imply complete cessation of all circulation, because, as Hart and Staudigel noted, relatively low temperature fluids could continue to circulate through the basaltic rocks without having any significant precipitation of clay minerals in these veins.

The initial Sr isotopic ratios defined by the isochrons of the smectite and celadonite clay minerals at the time of the "closure" of the Rb-Sr system, shown in the study of Hart and Staudigel (1986), ranged between 0.7055 and 0.7070 (Figs. 3.19–3.21). These Sr isotopic values are significantly higher than the Sr isotopic ratios of unaltered oceanic basalts, but lower than those of contemporaneous sea water, suggesting that the clay minerals at different sites incorporated varied amounts of Sr from both basaltic crust and sea

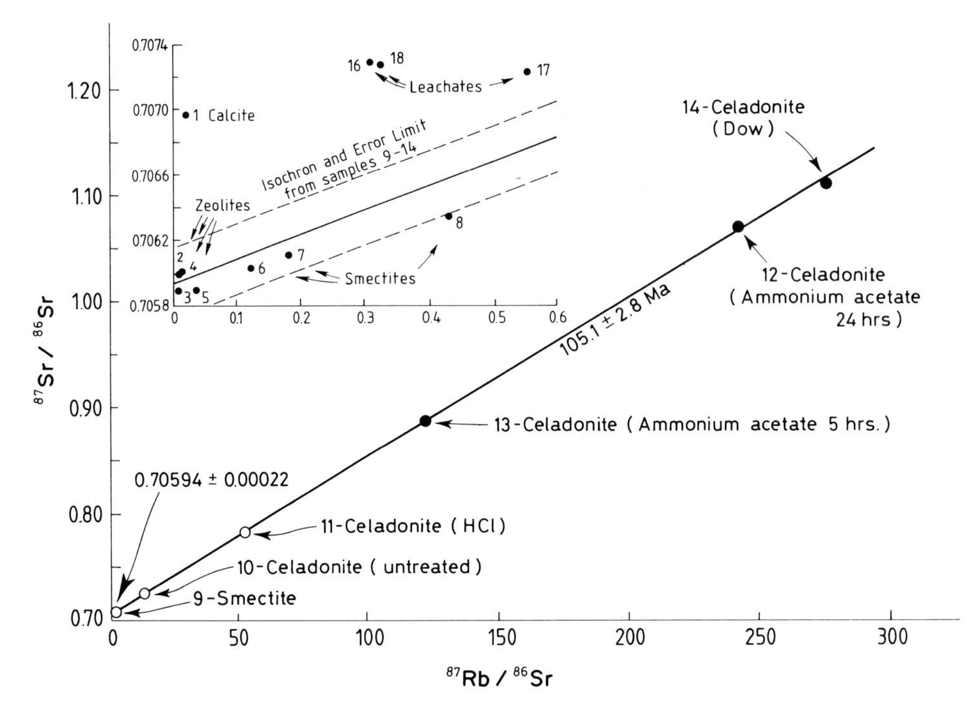

Fig. 3.19. Rb-Sr isochron diagram for DSDP Hole 462A vein minerals. *Insert* shows the data points for the samples with low Rb/Sr ratios and leachates from celadonite sample 10. The Sr isotopic ratios of the leachates are identical to that of sea water (about 0.7074) about 105 Ma ago. The initial Sr isotopic ratio of the unaltered basalt is 0.7036–0.7038. (Hart and Staudigel 1986)

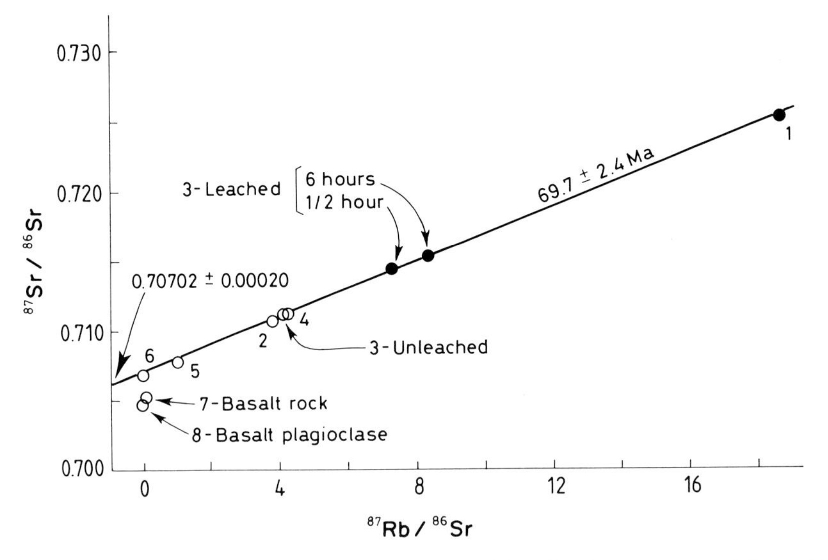

Fig. 3.20. Rb-Sr isochron diagram for DSDP Hole 261 vein minerals (sample 2 is a calcite, the others are celadonite and smectite ones). (Hart and Staudigel 1986)

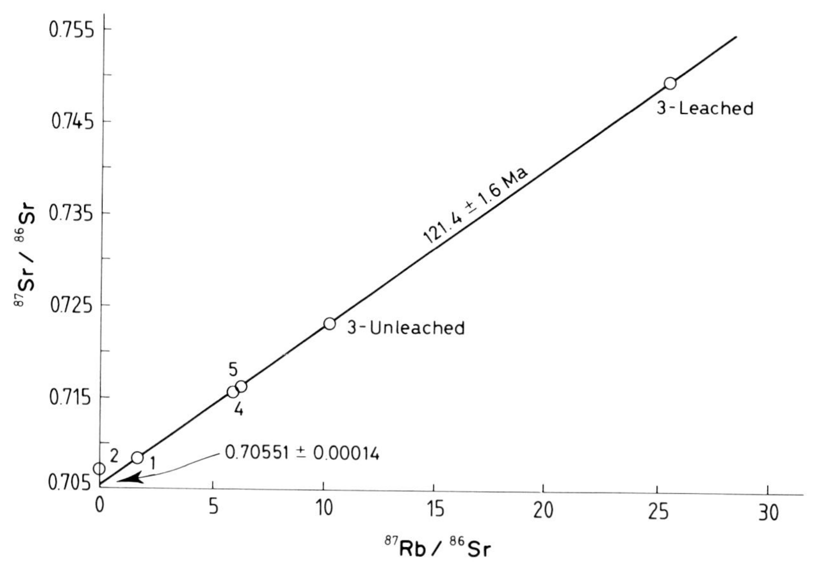

Fig. 3.21. Rb-Sr isochron diagram for DSDP Hole 516F vein minerals and basalt host. (Hart and Staudigel 1986)

water. The isotopic data have also shown that secondary calcites from the rock containing the smectite-celadonite clays, were slightly higher in Sr isotopic ratio than the clays at the time of their formation. As the Sr isotopic ratios of the calcites were very close to or slightly lower than that of contemporaneous sea water, the solution at the time of vein-filling calcite formation was more enriched in contemporaneous sea water than the solution at the time of formation of the clay minerals.

Hart and Staudigel (1986) also observed some degree of internal Sr isotopic heterogeneity together with chemical heterogeneity for some of the celadonite minerals. Chemical and Sr isotopic data of unleached and corresponding NH_4Ac leached celadonite samples from three different sites (DSDP Sites 261, 462A, and 516F) have shown that the leaching generally increased the Rb/Sr and $^{87}Sr/^{86}Sr$ ratios of the celadonite sample without any essential change in the Rb-Sr dates. The increase in the Rb/Sr ratio of the celadonite following leaching was more due to the loss of as much as 80% of the Sr than to the loss of the Rb ($<7\%$). Although the ages remained essentially unchanged following the leaching, the Sr isotopic ratios of leachates examined for a celadonite sample from DSDP Site 462A were not in agreement with the Rb-Sr isochron initial value. The present-day $^{87}Sr/^{86}Sr$ ratios of the leachates were closely similar at about 0.7072, while the Rb/Sr ratios ranged between 0.108 and 0.192. The disagreement between the leachates and the leached samples in regard to the isochron may suggest, unless the leaching experiment caused some selective ion exchange, that another altered phase was associated with the celadonite.

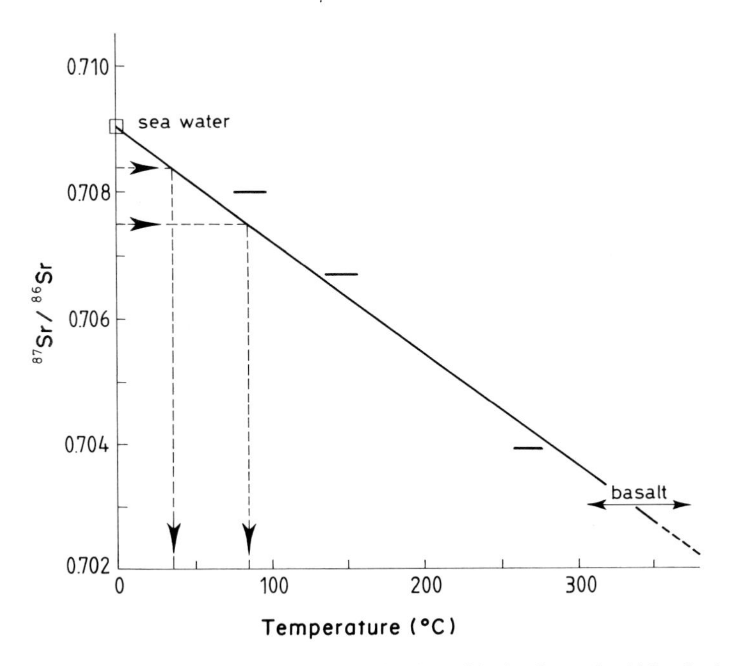

Fig. 3.22. The temperature of crystallization of hydrothermal acid-leached clay minerals from DSDP Leg 54 using the scale of the temperature-dependent variation in the $^{87}Sr/^{86}Sr$ ratios of the EPR hydrothermal fluids

Buatier (1989) reported that the $^{87}Sr/^{86}Sr$ ratios of some acid-leached clay minerals from hydrothermal mounds near Galapagos rift axis (DSDP Legs 54 and 70) ranged from 0.70768 to 0.70849. Considering these minerals to be entirely authigenic and to have precipitated directly from hydrothermal fluids, the variations in the Sr isotopic ratios could be temperature-dependent, in a way in which the $^{87}Sr/^{86}Sr$ ratios of hydrothermal fluids expelled at 21 °N on the EPR were temperature-dependent (Albarède et al. 1981). Based on isotopic parameters of the EPR fluids, the temperatures for precipitation of the clay minerals from the Galapagos rift could have ranged from 40 to 70 °C, which is reasonably close to the temperatures previously calculated from stable isotope data (Fig. 3.22).

3.2.2.3 K and Rb Geochemistry

The secondary vein-filling clay minerals in basalts are considerably varied in their K and Rb contents. In general, the Rb content increases with the K content, a trend which occurs also for palagonite and volcanic glass. High-temperature hydrothermal alteration of basalt releases K to sea water, whereas low-temperature alteration of basalt causes uptake of K from sea water. The crossover between the K uptake by the rock and the K release from the rock probably occurs at temperatures between 70 and 150 °C (Seyfried 1977). As the K and Rb data available for many of the vein-filling clay minerals come from those which have generally formed at moderate hydrothermal temperatures of less than 100 °C, K/Rb ratios of these clay minerals were influenced by the palagonitic or glassy precursor of the clay minerals and the sea water. Several palagonites from DSDP Site 418A apparently formed at moderate temperatures and had K/Rb ratios between 400 and 800, whereas smectitic clays associated with these palagonites had K/Rb ratios between 200 and 500 (Hart and Staudigel 1980, 1986; Staudigel and Hart 1983). By comparison, the average K/Rb ratio of sea water is about 3500 and that of unaltered oceanic basalt is about 1060. Thus, despite the potential uptake of both K and Rb from sea water by palagonite and smectite or celadonite clays, a highly preferential uptake of Rb over K occurs for palagonites and associated clay minerals, the latter having higher preference than the former probably in response to the crystallinity difference between the two phases (Fig. 3.23). The study by Hart and Staudigel (1986) has shown that internal heterogeneity exists in regard to the K/Rb ratio for a sample of vein-filling celadonite. The celadonite, leached with NH_4Ac for about 4 and 20 h, yielded K/Rb ratios of 499 and 193, respectively, in contrast to the K/Rb ratio of about 253 for the same unleached celadonite and, as discussed previously, the Sr isotopic compositions of the two leachates of the celadonite were also varied.

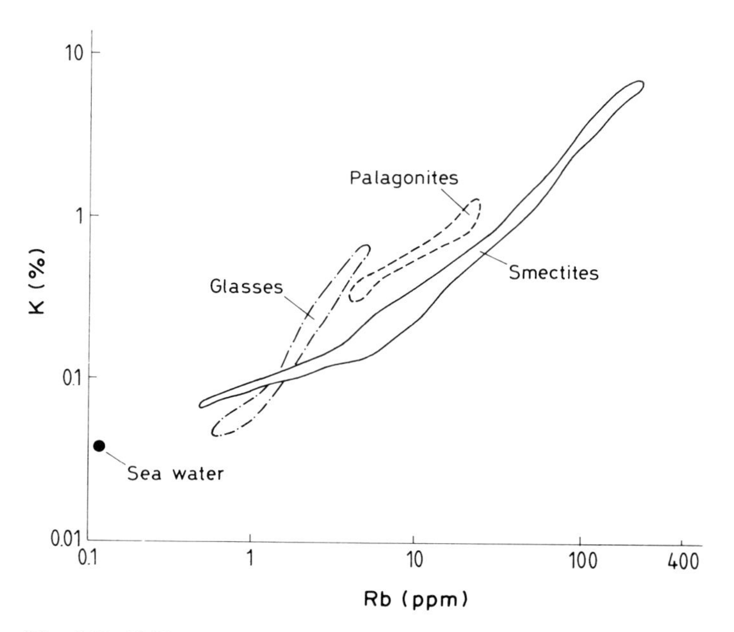

Fig. 3.23. K-Rb concentrations in basalt glass, palagonite, and smectite from DSDP Hole 418A. (Hart and Staudigel 1980)

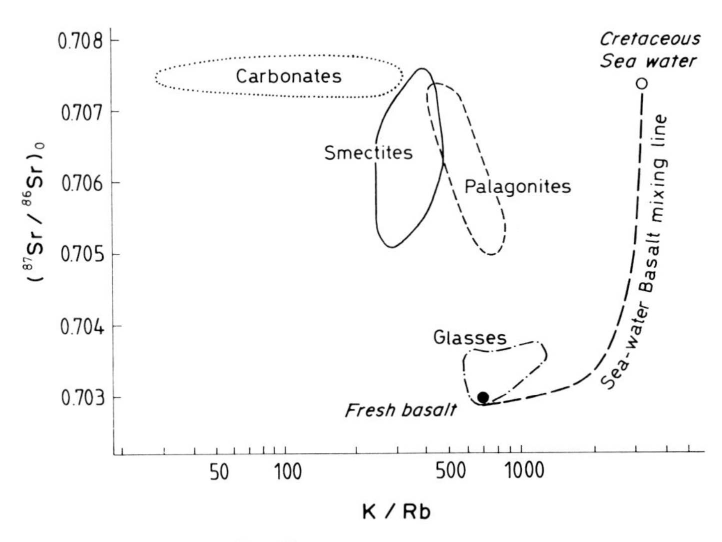

Fig. 3.24. K/Rb versus $^{87}Sr/^{86}Sr$ for glasses, palagonites, smectites, and carbonates from DSDP Holes 417 and 418. (Hart and Staudigel 1982)

Staudigel and Hart (1983) noted from a study of glass alterations at DSDP Sites 417A and 418A that, while K and Rb were taken up by glasses altering to palagonites and secondary clay minerals, about 85% of the Sr was lost to and also 40 to 60% Sr was gained from the sea water. The data presented in Fig. 3.24 show that the alteration of glass to palagonite and secondary clay minerals leads to decreases in K/Rb ratios and increases in the $^{87}Sr/^{86}Sr$ ratios, indicating that incorporation of sea water Sr is accompanied by partitioning of the alkali elements gained from the sea water.

3.2.2.4 REE Geochemistry

The relative abundances of the REE and their concentrations in vein-filling clay minerals in the altered ocean crust are generally unknown. Data are available for a few altered glass and palagonite materials which are precursors to smectite and celadonite minerals. The information from these materials may be used to make some reasonable conclusions about the relative abundances of REE in the clay minerals. Because of the very low abundance of the REE in sea water relative to altered basalts from which the clay minerals have formed, the water/rock ratio in the reactions does not appear to be high enough to cause any significant departure from the overwhelming influence of the REE from the rock. Analyses of altered glasses and palagonites have produced mixed results. Both Donnelly et al. (1980) and Rice et al. (1980) reported that alteration of the basaltic crust has caused increase in the REE for the altered materials without any significant change in the REE pattern from those of unaltered basalt. Staudigel et al. (1980) have found that palagonite is depleted in REE content relative to glass by as much as 40% without influencing the relative abundance, and that the depletion in the amount of REE correlated positively with K content. Unlike these investigators, Ludden and Thompson (1978) found relative enrichment in LREE in palagonitized rinds of dredged basalts. They found from comparison between crystalline interiors and palagonite rinds of individual basalt pillows that the selective enrichment in the LREE is characterized by a change from a LREE-depleted profile to a flat or slightly enriched profile.

The different results on the REE contents of weathered basalts clearly show that palagonitization or alteration of glass can have a significant effect on the REE budget for the altered materials. The secondary clay minerals whose progenitors are commonly the palagonites and the altered glasses, can also be expected to have varied REE profiles. The variations in the profiles of the REE for the secondary phases relative to that for their precursors or unaltered basalts may be largely controlled by water chemistry (pH, alkalinity, etc.) and water-to-altered phase or rock ratio. Michard (1989) found that REE concentrations of hydrothermal solutions, generally with LREE enrichment, increase as the pH decreases, independent of rock type and temperature. Sm/Nd isotopic data on vein-filling clay minerals are needed to provide further insight into their origin.

3.2.2.5 Pb Isotope Geochemistry

As the low-temperature alteration of basalt results in enrichment of U along with alkali elements, Hart and Staudigel (1986) were attracted to the possible dating of vein-filling clay minerals in altered basalts by the U-Th-Pb isotopic method. Their data have shown that the potential for dating vein-filling clay minerals by this isotopic method is quite low. Both U and Pb concentrations with values of 0.075 and 0.643 μg/g, respectively, for celadonite clays from DSDP Sites 261 and 516F were similar to that of N-type mid-ocean ridge basalts (MORB). By comparison, clays with concentrations of 0.06–0.7 μg/g were enriched in Th relative to their host basalt. Hart and Staudigel also reported that Rb/U and Th/U ratios of the clays were about 750 and 850, respectively. These values are extremely high in comparison to the values of 30 and 20 for oceanic basalts. Such high values for the celadonites remain unexplained.

P. Vidal and N. Clauer (unpubl. data) determined the Pb isotopic compositions of some of the <2 μm clay samples from DSDP Leg 54 near the Galapagos spreading center that were previously analyzed for the oxygen and strontium isotopic compositions by Buatier (1989). The range of the $^{208}Pb/^{204}Pb$ and $^{206}Pb/^{204}Pb$ ratios of the clay fractions was similar to that of the Galapagos basalts reported by White (1979; Fig. 3.25). However, the $^{207}Pb/^{204}Pb$ ratio of these clay fractions is slightly above that of the basaltic rocks, suggesting a slight contamination by a continental-derived detrital component which was too small to be detected in the analyses of the Sr

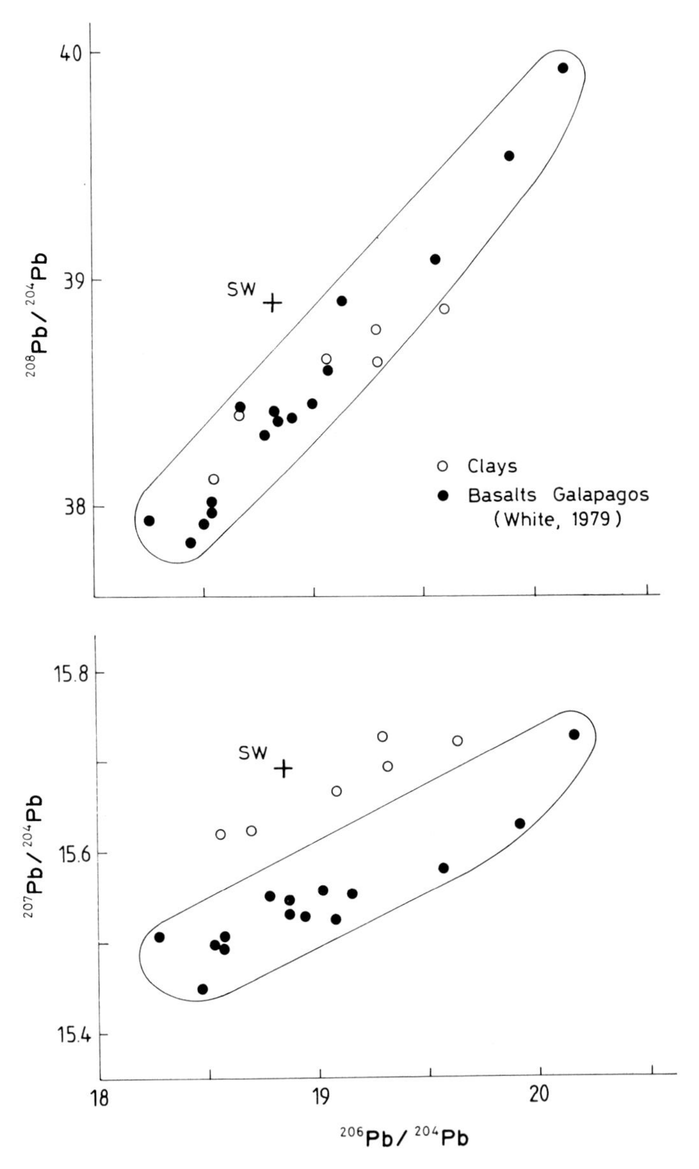

Fig. 3.25. Distribution of Pb isotopic compositions of hydrothermal clay minerals from DSDP Leg 54. The *bounded area* represents the distribution of the corresponding isotopic values of the Galapagos basalts. (P. Vidal and N. Clauer, unpubl. data)

isotopic compositions or whose Sr isotopic composition was probably very close to that of the authigenic component.

3.2.3 Nontronites from Submarine Alteration of Intraplate Volcanics

Alteration of volcaniclastic sediments away from the ocean ridges can also result in formation of secondary clays. An indurated mass of volcanic origin was found covered by a thin layer of deep-sea red clays at a locality between the Tuamotu Archipelago and the Marquisas Islands (Hoffert et al. 1978). The process of induration of the mass also included clay mineral authigenesis and hence chemical and isotopic analyses of these sediments proved useful in understanding the formation of clay minerals in ocean bottom and sub-bottom environments. The authors analyzed Sr isotopic compositions of clay minerals separated from a 20-cm-long core of sediments containing a middle Eocene indurated mass at the bottom 10 cm and Plio-Quaternary brown mud at the top 10 cm. The brown mud contained palagonite, phillipsite, Mn-micronodules, smectite, and some Ca-carbonate minerals. The indurated mass contained volcanic glass, phillipsite, and a small amount of smectite. The clay minerals coated the surface of the glass grains and acted as cement for the indurated mass. Clauer's (1982a) data on the Rb-Sr isotopic compositions of acid-leached phillipsite samples separated from the Eocene indurated sediment gave a Rb-Sr isochron-age of 14.7 ± 3.3 Ma with an initial $^{87}Sr/^{86}Sr$ ratio of about 0.7044 (Fig. 3.26). This isochron suggests that the induration leading to the closure of the Rb-Sr system of the phillipsite samples occurred about 25 million years after the deposition of the sediments which were clearly volcaniclastic in origin. The initial $^{87}Sr/^{86}Sr$ ratio of the isochron at 0.7044 is very close to the $^{87}Sr/^{86}Sr$ ratios of the basaltic rocks of

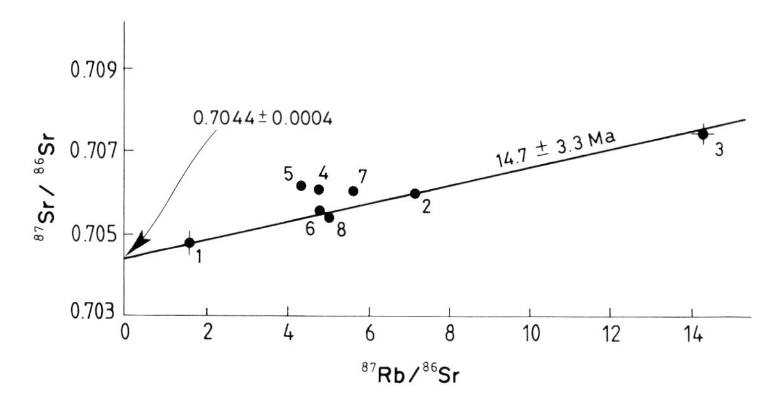

Fig. 3.26. Rb-Sr isochron diagram for phillipsite samples in an indurated mass in the southeastern Pacific Ocean bottom. (Clauer 1982a)

the region (Peterman and Hedge 1971; Hedge 1978), indicating that the zeolites were derived from volcanic precursors. This result is similar to that for clays from several oceanic ridge materials as reported by Hart and Staudigel (1986).

The Sr isotopic results obtained on the clay minerals and the zeolites of the indurated mass and the overlying brown mud may be explained in terms of a three-stage genetic model for the smectite and the phillipsite. In the first stage, both smectite and phillipsite formed from volcanic glass. Phillipsite crystallized in an environment isolated from sea water and the Sr trapped in its structure had an $^{87}Sr/^{86}Sr$ of about 0.703–0.704 (Fig. 3.27A). The Rb/Sr ratio of this phillipsite was higher than that of the glass, because of the release of Sr to the water during a dissolution-crystallization process. The Rb/Sr ratios progressively increased by continuous release of Sr to the immediate intergranular environment. The $^{87}Sr/^{86}Sr$ ratio of the smectite (0.70550 ± 0.00012) was higher than that of the phillipsite, but lower than that of the sea water. Such behavior was also recorded by Hart and Staudigel (1986). The smectite seems to have trapped some sea water Sr, indicating that the crystallization environment of the mineral was different from that of the phillipsite, and that the crystallization of these two minerals could have occurred at different times. The Rb/Sr ratio of the smectite was higher than that of the volcanic glass. During the second stage, after crystallization and induration, phillipsite and smectite minerals evolved under closed system conditions enabling the minerals to accumulate radiogenic ^{87}Sr (Fig. 3.27B: the subparallel horizontal lines integrating the analytical data rotate around their intercepts). During the third stage, following removal of the uppermost zone of the indurated mass, probably due to erosion by bottom currents, both mineral phases adsorbed variable amounts of marine Sr at their surfaces or easily exchangeable sites, as the mixing lines in the Rb-Sr isochron diagram intersect the ordinate axis at a common point with a Sr isotopic value closely similar to that of the marine Sr (Fig. 3.27C).

The REE and Sr isotopic data suggest that the induration process did not proceed in chemical equilibrium with sea water (Figs. 3.28–3.29). This occurred under compaction conditions attendant with progressive isolation from sea water. The contact with sea water was later renewed, after erosion of the top zone by bottom currents, as evident by the progressive variations in the texture, mineralogy and chemistry of sediments from bottom to top of the indurated mass in the core. The $^{87}Sr/^{86}Sr$ ratio of the smectite in the core of a polymetallic nodule recovered from a nearby locality suggests that this smectite mineral phase might have been transported by bottom currents from another location to its present place after sub-marine weathering of the indurated mass. The clay particles appear very similar to flaky particles of some deep-sea red clays collected from other places in the Pacific Ocean (Clauer et al. 1982b; Clauer and Hoffert 1985).

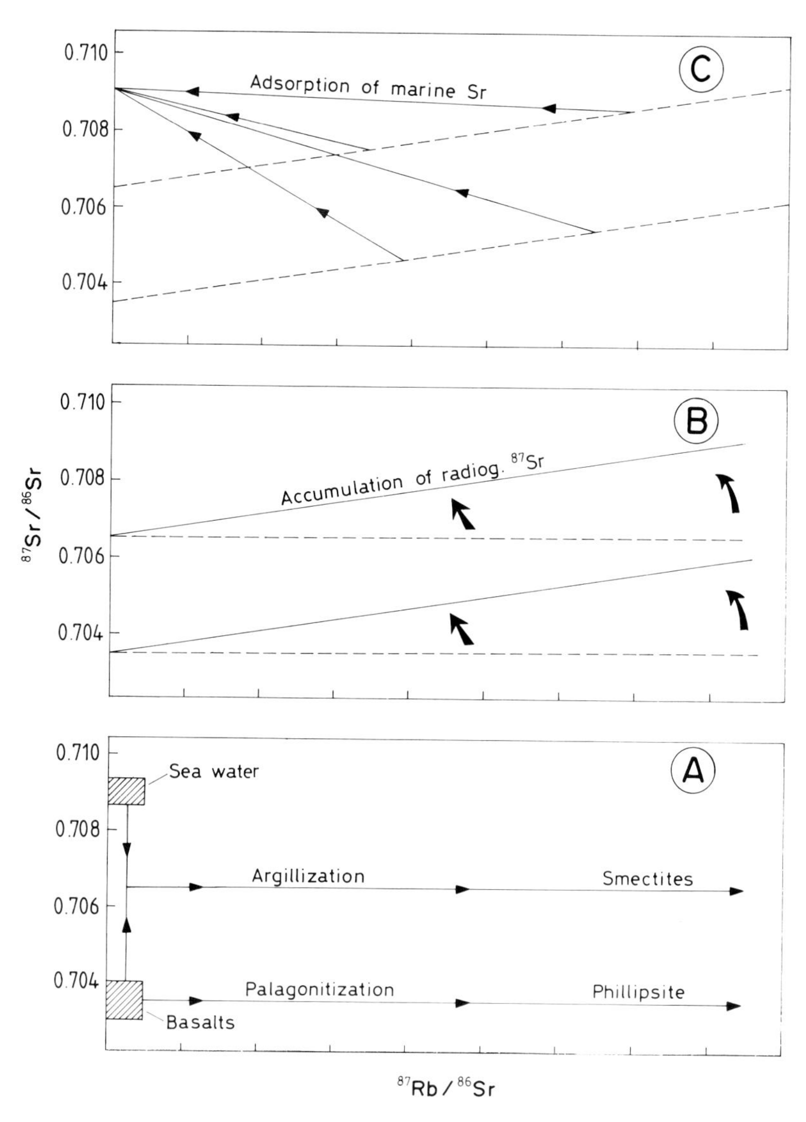

Fig. 3.27. Three-stage evolutionary model for the Rb-Sr system of smectite and phillipsite in an indurated mass in the southeastern Pacific Ocean bottom. For explanation, see text. (Hoffert et al. 1978)

3.2.4 Smectites from Metalliferous Sediments

Dymond et al. (1973) analyzed the Sr isotopic compositions of metalliferous sediments occurring within the Mendocino and Murray fracture zones in the north Pacific Ocean. These sediments contained 63 to 85% Fe-rich smectite with traces of quartz, plagioclase, illite, and kaolinite/chlorite. The samples

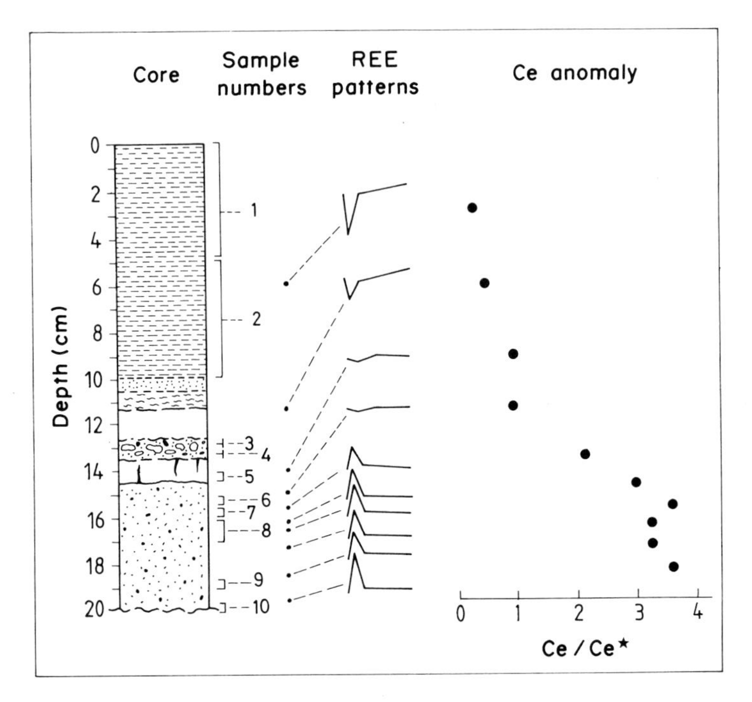

Fig. 3.28. Description of a 20-cm-long core of sediments in the southeastern Pacific Ocean bottom. Distribution of the REE and magnitude of the Ce anomaly. (Hoffert et al. 1978)

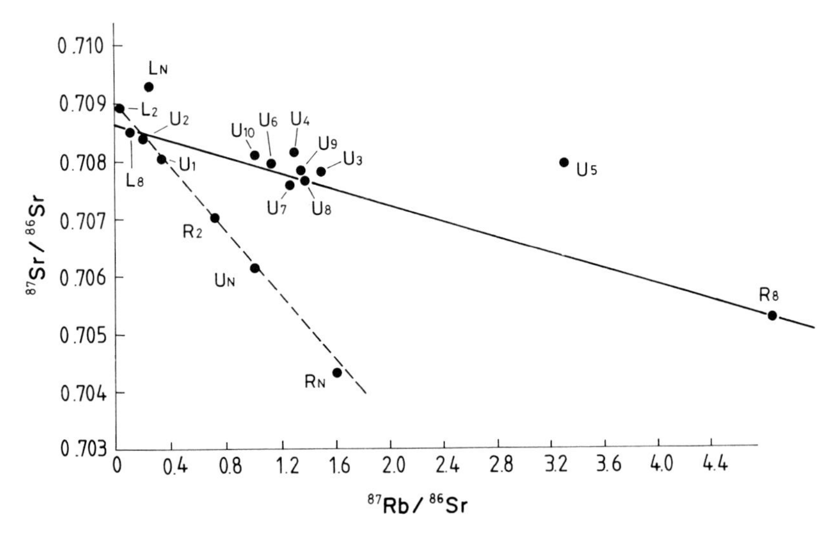

Fig. 3.29. Rb-Sr data of separated fractions of the 20-cm-long core from southeastern Pacific Ocean. The letters *L*, *U*, and *R* stand for leachate, untreated, and residue, respectively, of the samples numbered along the core in Fig. 3.28. (Hoffert et al. 1978)

washed with distilled water were high in Sr contents (269 to 364 μg/g), with $^{87}Sr/^{86}Sr$ ratios nearly identical to that of the sea water. Two populations of clays were recognized from acid-leached samples, one with Sr contents between 52 and 77 μg/g and $^{87}Sr/^{86}Sr$ ratios between 0.7052 and 0.7063, and the other with Sr contents between 27 and 58 μg/g and $^{87}Sr/^{86}Sr$ ratios between 0.7175 and 0.7327. The smectite-rich clay fraction with the low Sr isotopic ratio certainly originated from alteration of fragments of oceanic basalts, following which the clay adsorbed large amounts of sea water Sr. The clay materials with relatively high $^{87}Sr/^{86}Sr$ ratios were clearly terrigenous in origin.

Thomson et al. (1984) reported results of Sr isotopic analyses on clay fractions of pelagic sediments collected from Nares Abyssal Plain, east of the tip of Florida in the north Atlantic Ocean. The sediments in this area are well laminated and the sequences comprise two types of clay material: a red-brown type of pelagic origin, and a relatively coarse, gray variety of turbiditic origin. The mineral compositions (40–55% illite, 15–30% kaolinite, 0–10% chlorite, and 0–10% smectite) of these two types of clay being similar, implies that they had similar origin but were deposited under different conditions. The clay particles in the pelagic red-brown sediments had an average $^{87}Sr/^{86}Sr$ ratio of 0.72283 and a mean Sr content of about 151 μg/g, whereas that in the turbidites yielded an average $^{87}Sr/^{86}Sr$ ratio of 0.72840 and a mean Sr content of 151 μg/g. The $^{87}Sr/^{86}Sr$ ratios of the clays in the turbidites are approximately close to the two values of 0.7262 and 0.7317 which Dasch (1969) previously found for siliciclastic sediments in the same vicinity. The Sr isotopic data suggest a terrigenous source for the gray turbidite clays. Thomson et al. observed that the $^{87}Sr/^{86}Sr$ ratios of HCl leached red clay samples ranged from 0.72800 to 0.72890. Having found that the isotopic values of the acid-leached red clays are similar to the $^{87}Sr/^{86}Sr$ ratios of unleached gray turbidite clays, Thomson et al. claimed that the two clays had the same origin. This claim may not be justified, because the laboratory methods differed, the gray turbidite not being acid-leached.

Barrett and Friedrichsen (1982) found that the Sr isotopic compositions of some metalliferous sediments from Galapagos mounds (DSDP, Leg 70) ranged from 0.70909 which is equivalent to that of the sea water value, to 0.70891 which is slightly lower. The Pb isotopic composition of the sediments reflected basement exhalations, as the values ranged from 17.8 to 18.9 for the $^{206}Pb/^{204}Pb$ ratio, from 15.53 to 15.63 for the $^{207}Pb/^{204}Pb$ ratio, and 37.6 to 38.7 for the $^{208}Pb/^{204}Pb$ ratio. The analyses of ^{18}O and D values of the sediments indicated that the sediments formed at a temperature between 4 and 16 °C. As for Buatier's (1989) Sr data for Fe-rich clays, the range of temperatures at which these clay minerals precipitated from exhalation fluids can be estimated by comparing the Sr isotopic data of the clays to the empirical scale of the variation in the Sr isotopic composition of the EPR hydrothermal fluids as a function of temperature. The estimated temperature from the Sr isotopic data of the metalliferous sediments (Barrett and

Friedrichsen 1982) is in accord with that calculated from stable isotope compositions.

3.3 Evolution and Paleogeography of Deep-Sea Red Clays

Since the discovery of a "Red Clay" deposit in the south Pacific Ocean by Thompson (1874), this sediment, consisting mainly of Fe-smectite, has been found in many other localities in deep-ocean bottoms. Following the work of Murray and Renard (1884), it was considered authigenic, because it occurred in a region where detrital components are generally lacking. But since then, the meaning of deep-sea red clays has changed as they were also considered to be of detrital origin. Here, we consider these clays to be of authigenic origin at a regional scale in an ocean basin, based on studies by Clauer et al. (1982b) and Clauer and Hoffert (1985), who elaborated on Sr isotopic evidence in some Pacific deep-sea clays. Several ocean-bottom deep-sea red clays from north of the Marquisas fracture zone and west of the Marquisas Islands, which these authors and Hoffert et al. (1978) analyzed, were enriched in Fe-smectite clays. Acid-leached fractions of these clays had $^{87}Sr/^{86}Sr$ ratios ranging from 0.70694 to 0.70858, the low $^{87}Sr/^{86}Sr$ ratios of some of these clay fractions suggesting that they inherited Sr primarily from marine volcanic sources. For lack of evidence of any recent volcanic activity in this region of the Pacific Ocean, the clay material was believed to have been transported to its present location following its formation elsewhere in an oceanic environment.

Some understanding of the chemical environment for the diagenetic growth of authigenic clays in red clay sediments comes from the study of sediments from the Central American platform in the Pacific Ocean. The sediment consists of smectite and was abnormally enriched in Fe, Mn, Ba, Zn, Ni, and Pb (Dasch et al. 1971). The bulk sediment that was washed by deionized water, contained 774 μg/g Sr with an $^{87}Sr/^{86}Sr$ ratio of 0.70907, which is very close to the Sr isotopic ratio of Recent sea water. Following HCl leaching, the <20 μm size fraction yielded a Sr content of 34 μg/g with an $^{87}Sr/^{86}Sr$ ratio of 0.70652, which clearly indicates that the Sr in the acid-resistant silicate fraction was mainly of volcanic origin.

Church and Velde (1979) analyzed the oxygen and Sr isotopic compositions of Pliocene-Miocene sediments at shallow sub-bottom depths (43–58 and 99–100 cm) in the central southwestern Pacific Ocean. The sediments contained palygorskite and smectite with unusually high $\delta^{18}O$ values of about +56 per mill, that remains unexplained. The $^{87}Sr/^{86}Sr$ ratios of the palygorskite and the smectite were about 0.7192 and 0.7154, respectively. The $^{87}Sr/^{86}Sr$ ratio of the palygorskite is far too high to be an authigenic phase, and hence the authors reasoned that the mineral had to form from some detrital precursors. An alternative interpretation for these results is that authigenic minerals may grow in modern ocean environments under the condition of a closed system characterized by a high mineral-to-water ratio.

Clauer et al. (1982b) investigated the mineralogy, chemistry and Rb-Sr isotope chemistry of the clay fractions of a 380-cm-long core of deep-sea red clays from a southern Pacific ocean bottom location about 4.7 km deep, which is below the carbonate compensation limit. The location of the core was in proximity to volcanic seamounts, and hence the influence of volcanogenic debris was overwhelming in the sediments, as compared to a terrigenous influence. The core represented the uppermost part of a 60 m-thick, homogeneous, fine, dark brown muddy sequence (Hoffert 1980). Palagonite was the main component of the coarse fraction at the upper part of the core, whereas phillipsite was abundant in the section down to 160 cm. The lowermost 120 cm of the core consisted of brown aggregates of nontronitic smectite with subordinate amounts of phillipsite, quartz, and Fe-oxides. Morphological observations based on TEM indicated that flakes of nontronite were surrounded by fibers of the same chemical composition, suggesting a transport of the flakes from another place to the present-day location prior to the in situ crystallization of the fibers (Fig. 3.30). The physical modifications of nontronite have been accompanied by increases in the Fe, K, and Rb contents and concomitant decreases in the Mg, Ca, Ti, Na, and Sr contents. The comparison of Sr isotopic compositions of the clays with that of the interstitial waters suggests migration of radiogenic ^{87}Sr from the solid to the fluid (Fig. 3.31). The isotopic composition of Sr derived from leaching of the mineral particles also supports this migration scheme. The combined chemical and isotopic evidence points to a syngenetic evolution of the clay materials.

Clauer and Hoffert (1985) examined the problem of determining the sedimentation rates for the deep-sea red clays by comparing two cores of sediment 100 km apart from each other in the region of the Marquisas Islands in the southern Pacific Ocean. The sediments of both drill cores were similar in their clay mineral composition. The only difference was that the uppermost part of the sediment core which is located afar from the Marquisas fracture zone, consisted mainly in palagonite with an abundance of 80% in the coarse fraction (>63 μm). The acid-leached residue of Fe-smectite in the sediments near the water-sediment interface in the core away from the Marquisas Islands had an $^{87}Sr/^{86}Sr$ ratio of 0.70718 which is significantly lower than that of the sea water, indicating the presence of some amount of volcanic Sr in these smectites. The $^{87}Sr/^{86}Sr$ ratios of the Fe-smectite below the water-sediment interface in the same core remained nearly constant between 0.70864 and 0.70878. By contrast, the Sr isotopic ratios of the same smectites in the sediments of the core drilled closer to the Marquisas Islands increased with depth (Fig. 3.32). The downward uniformity in the $^{87}Sr/^{86}Sr$ ratio for the smectites in the core away from the Marquisas Islands strongly suggests that the minerals in the entire core originated elsewhere from the same precursor. The predominance of

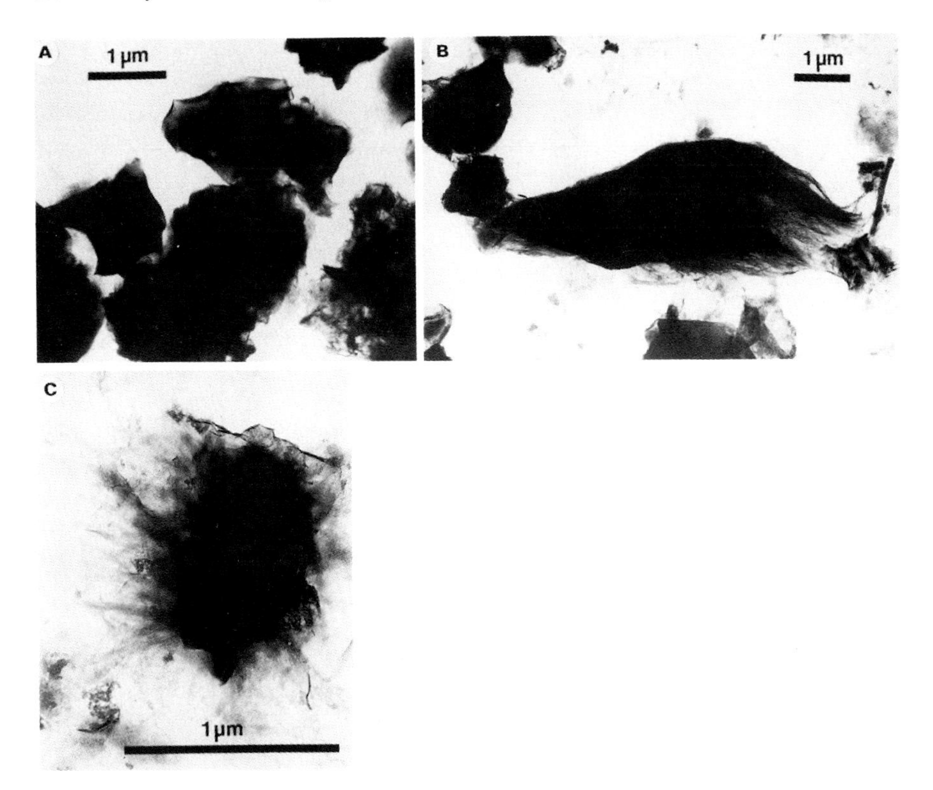

Fig. 3.30. TEM micrographs of nontronitic smectite particles of the deep-sea red clays in the southern Pacific Ocean bottom. **A** Veils surround flakes. **B** Fibers are growing on flakes. **C** Flakes are surrounded by fibers. (Clauer et al. 1982b)

palagonite in the coarser fraction of the sediment suggests that these sediments were deposited recently by turbidite-type currents and subsequently the sediments were altered to phillipsite.

Sedimentary red-clay facies occurs extensively across several hundreds of km^2 area in the Pacific Ocean, originating in part from submarine weathering or hydrothermally related alteration of volcanogenic debris. The sediments exhibit considerable similarities in mineralogical, morphological, and chemical makeups. The evidence from such sediments suggests that authigenic sediment formed at the ocean bottom might be transported over long distances to be redeposited elsewhere, making the use of the traditional method of determining the sedimentation rate and also applying it uniformly to an entire basin questionable. The history of deep-sea red clays at a given site seems to correspond to a complex set of submarine sedimentary processes which include weathering and dissolution of parent rocks, authigenesis of minerals, erosion, transport and deposition of the newly formed minerals, and finally epigenesis to diagenesis.

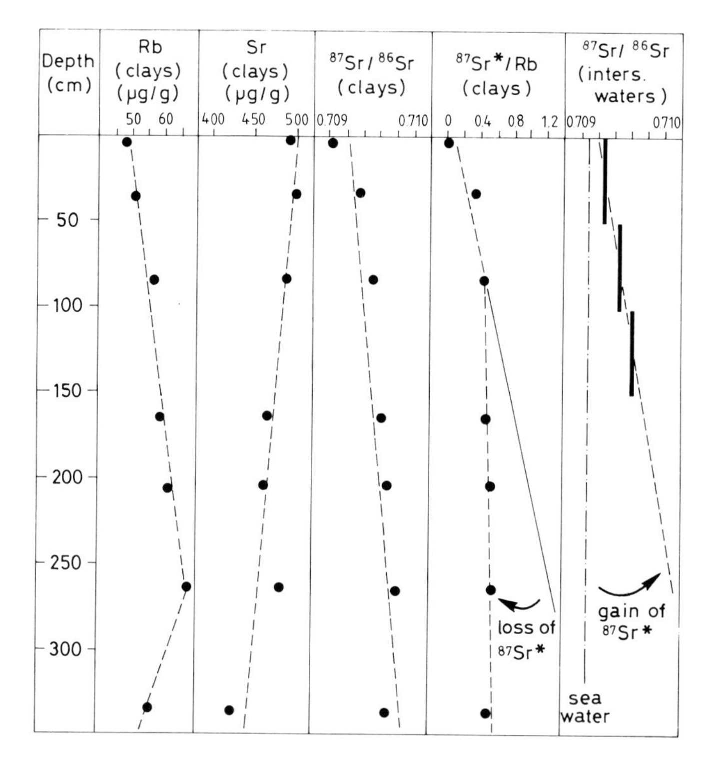

Fig. 3.31. Rb and Sr contents and $^{87}Sr/^{86}Sr$ ratios of acid-leached nontronites of deep-sea red clays in the Pacific Ocean bottom. The comparison between the production of radiogenic ^{87}Sr ($^{87}Sr^*$) in the solid material and the evolution of the $^{87}Sr/^{86}Sr$ ratio of the interstital waters relative to depth is also displayed. (Clauer et al. 1982b)

4 Isotopic Evolution of Clays in Buried Deep-Ocean Sediments

The present-day deep ocean floor is covered by varied amounts of sediments, the thickness of which may be more than 600 m at given locations. At many DSDP sites, volcanic ash beds were found, and varying amounts of glassy and volcanogenic mineral and rock fragments were described within the pelagic sediments. The volcanic ash beds are often variously altered to form smectite-rich clay minerals, and the volcaniclastic materials dispersed in the pelagic sediment, primarily of terrigenous origin, can amount to as high as 10%. These volcanic materials, therefore, serve as major precursors to clay minerals that form at intermediate subbottom depths below the ocean floor.

As discussed above, terrigenous clay minerals should also theoretically approach a chemical equilibrium in their depositional environment with the ocean water. Although their modifications appear to be minor under ocean bottom conditions, diagenetic modifications of clay minerals have indeed

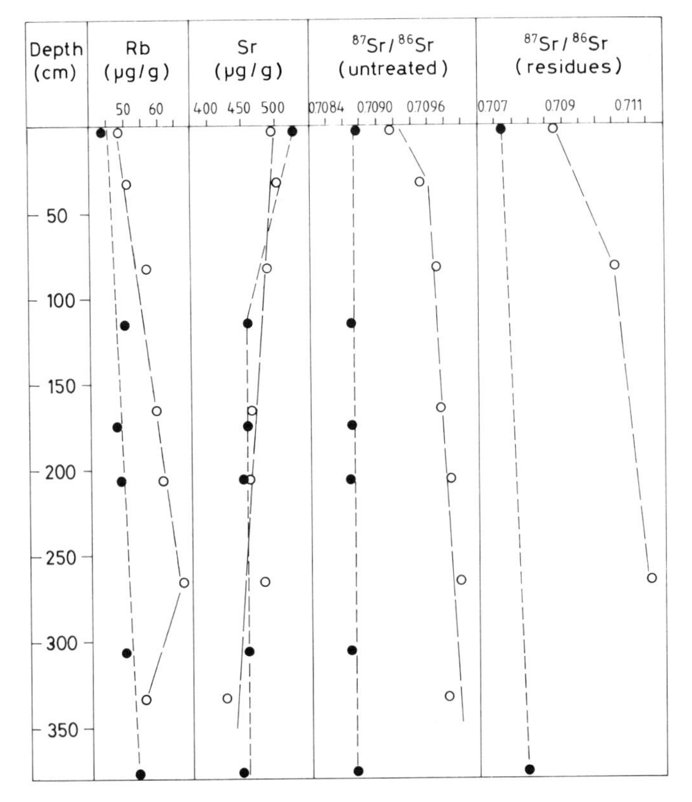

Fig. 3.32. Comparison between the Rb-Sr data of the nontronite clay fractions of the two south Pacific cores, one close to (*open circles*) and the other (*filled circles*) away from Marquisas Islands. (Clauer and Hoffert 1985)

been recognized in connection with shallow burial, within a few million years after deposition. These modifications potentially should modify the pore water chemical and isotopic compositions, and knowledge of the isotopic modifications of both the continental- and volcanic-derived clay minerals and associated fluids provides a firm basis for understanding of the chemical modification at the interface between the major mineral phases of the sediments.

4.1 Isotopic Characterization of Sediment Pore Waters

Chemical and isotopic data from pore waters in deep-sea sediments give some clues as to mineralogical and chemical changes that occur in sediments resting above the basaltic crust. The study of Lawrence et al. (1979) gives some broad understanding of the trends in chemical and isotopic changes of pore waters with depth in a clay-rich sediment. Gieskes and Lawrence (1981) and, more recently, Lawrence (1989) have reviewed the scope and significance of the pore water chemistry on potential mineral or rock and water interactions in a sedimentary column beneath the ocean floor.

Lawrence et al. (1979) reported the chemical, Sr isotopic, and stable isotope data of pore waters in a sedimentary column (DSDP Site 323), which consists of about 640 m of terrigenous sediment with a few thin interbedded ash layers and volcanic debris occasionally dispersed in the sediment that overlies a 60-m-thick altered volcanic ash layer at the top of the ocean basaltic crust. As observed in many studies, the pore waters in this sedimentary column are generally characterized by decreases in Mg and K and increases in Ca and Sr with increases in depth. These chemical trends have been explained by diffusive transfer of the elements between altering basaltic layers underneath and the ocean water above the thick sedimentary column (Kastner and Gieskes 1976; Lawrence et al. 1979). Any significant perturbation in these concentration-depth profiles may be attributed to alteration of volcanic ashes or debris within the sedimentary column. The decreases in Mg and K contents of pore waters with depths suggest sinks for these elements in the altering basaltic rocks. Smectite and celadonite minerals occurring as products of alteration of basaltic rocks and volcanic debris underlying the thick sedimentary column are likely sinks for these elements. Many different studies of submarine basalt alteration have shown that Ca is lost by the altering rocks which may explain the positive gradient for the Ca profile with depth. The profile for Sr can be similarly explained, but it can also be influenced significantly by the increased recrystallization of calcite with increase in depth, releasing Sr to the pore waters. The alteration of layers of volcaniclastic debris and basalts underlying the sediment is a process that might have begun before any significant amount of sediment accumulated on the top and continued through time to the present period, although most likely at a much slower rate than before. As alteration of the volcanogenic materials appears to dominate the chemical compositions of the pore waters and chemical exchanges with the overlying sea water are minimal or not extensive, the alteration of the volcanic layers is often isochemical (Gieskes and Lawrence 1981).

Although pore waters in many deep sea sediments have unidirectional vertical concentration gradients for different elements, whose concentrations either decrease or increase with increase in depth, some instances are known, as presented in Fig. 3.33, where the concentration gradients of different elements are reversed for pore waters in sediments near the altering basaltic crust or the volcanic layers (Baker et al. 1991). The reversed trends in the concentration gradients for pore waters terminate in the basement basaltic rock at which point the concentrations are similar to that of the modern ocean water. Both the Sr and the oxygen isotopic compositions of the pore waters at this terminating point are also similar to that of modern ocean water. This chemical and isotopic evidence suggests the existence of a large

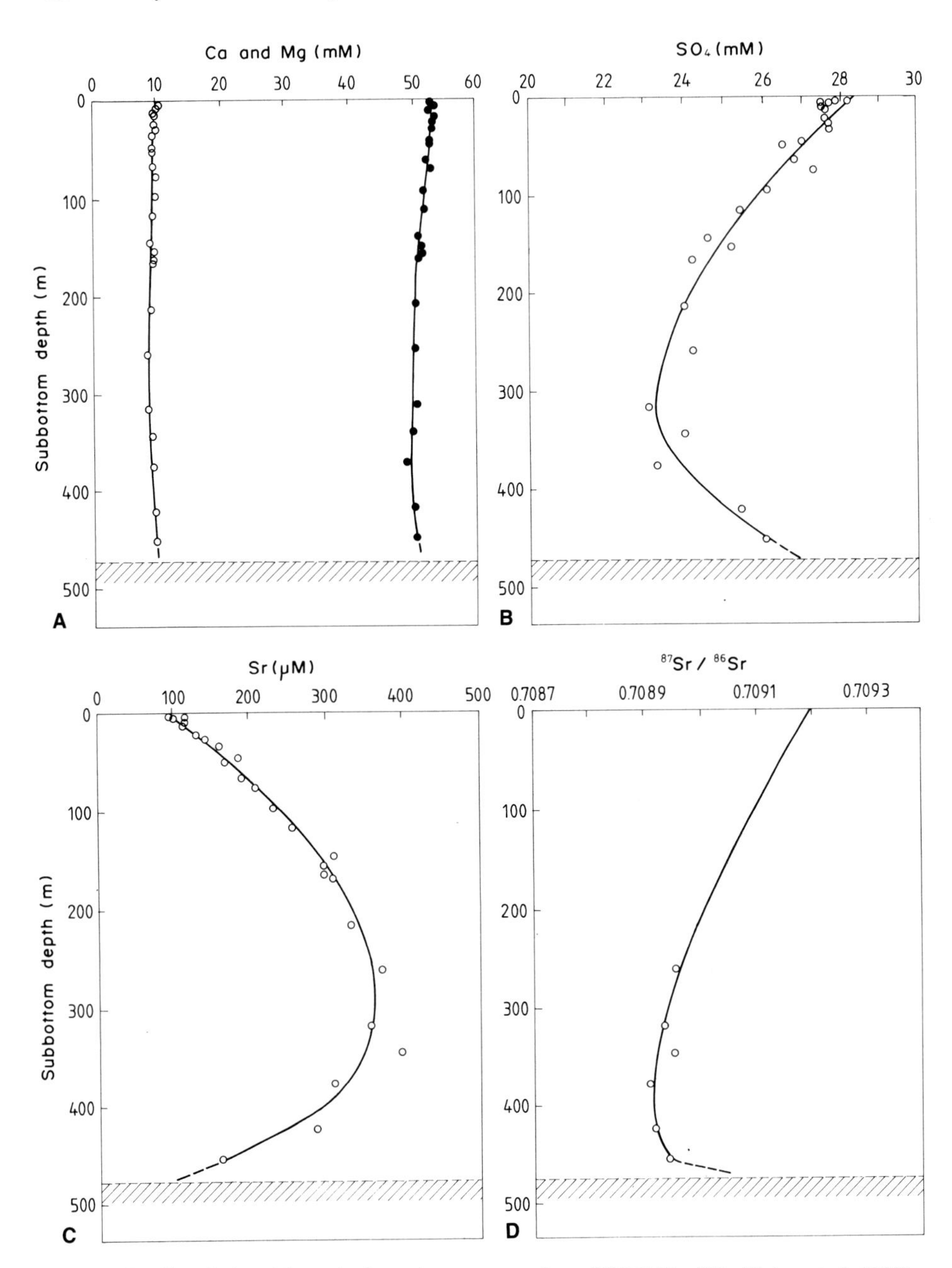

Fig. 3.33. Chemical and isotopic data of pore waters from DSDP Site 572. (Baker et al. 1991)

scale lateral advection of modern sea water through basaltic crust. Baker et al. concluded that this discharge of water out of the basaltic crust is rapid, having an estimated residence time of about 20000 years, and that it must be removing large quantities of heat from off-ridge regions of the basaltic crust. The lateral circulation, apparently resulting from subcrustal heating, is maintained through high permeable upper basaltic basement while being prevented from vertical migration above the basement rocks due to the cover by low permeability sediment. Such large-scale lateral advection of water with a composition nearly that of sea water suggests that interaction with the basaltic crust under the cold ocean bottom temperature is minimal at least within time spans of several tens of thousand years.

The $^{87}Sr/^{86}Sr$ ratio of interstitial waters in oceanic sediments is sensitive to epigenetic or diagenetic alteration of both silicate and carbonate sediments (Hofmann et al. 1972; Clauer et al. 1975; Elderfield et al. 1982). Although studies of pore waters in ocean sediments are few, with increasing depth below the sediment-water interface, the $^{87}Sr/^{86}Sr$ ratios of interstitial waters decrease relative to the value of recent sea water, suggesting diffusive transfer of volcanic Sr from basaltic rocks at depth (Hawkesworth and Elderfield 1978; Lawrence et al. 1979; Elderfield et al. 1982). In Ca-carbonate-rich sediments, recrystallization of older carbonate minerals at depths can be a cause for the decrease in $^{87}Sr/^{86}Sr$ of shallow interstitial waters (Gieskes et al. 1986). Indeed, data reported by DePaolo (1986) and Hess et al. (1986) suggest that the interstitial waters may not be in isotopic equilibrium with the carbonate fossils, as the isotopic values of the pore waters have been found to be both higher and lower than that of the foraminifera sediments (Fig. 3.34). The degree to which the Sr isotopic composition of pore waters is influenced by clay mineral modifications remains unknown.

As the total amounts of most cations during sediment-water reactions is essentially in sediments, the identification of any minor reaction between waters and sediments based on changes in the cations is frequently difficult. By contrast, the amounts of oxygen in pore waters and sediments are often comparable, and hence the extent and the nature of interactions between pore water and sediment beneath the ocean floor may be detected in changes in the isotope compositions of the pore waters. An example of variation of oxygen isotope compositions of pore waters in shallow buried sediments is given in the study by Lawrence et al. (1979). Figure 3.35 indicates the variation in the oxygen isotope compositions of the pore waters through a 700 m section of sediment at DSDP Site 323. The lower 60 m of this section, referred to as the basal sediment, consisted largely of authigenic smectite with very little illitic interlayering and contains a few percent of material $>10\,\mu m$ in size and also a minor amount of detrital illite. This smectite-rich section is believed to be of Cretaceous to Paleocene age. The upper 640 m of the section, Oligocene to Recent in age, consisted essentially of continental detrital clays composed of illite/smectite mixed-layer minerals, illite, chlorite,

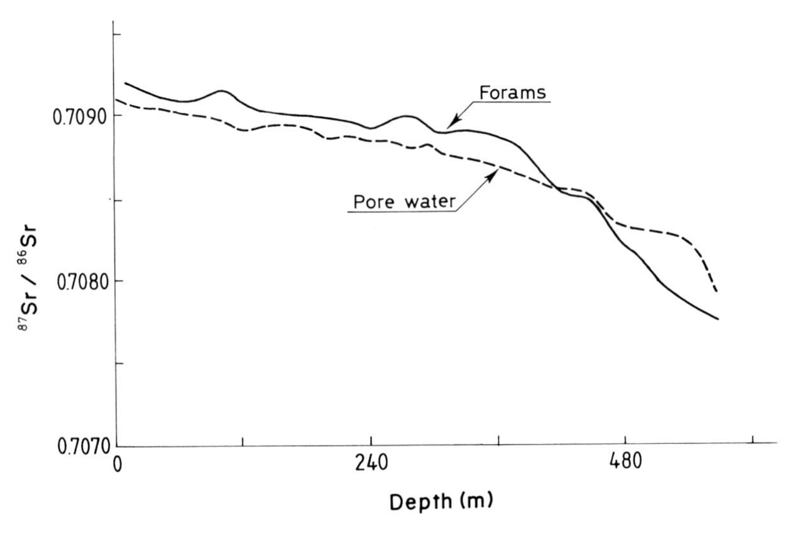

Fig. 3.34. The $^{87}Sr/^{86}Sr$ ratio of pore waters relative to that of foraminifera fossils of the associated sediments at varied depths below the ocean bottom. (Hess et al. 1986)

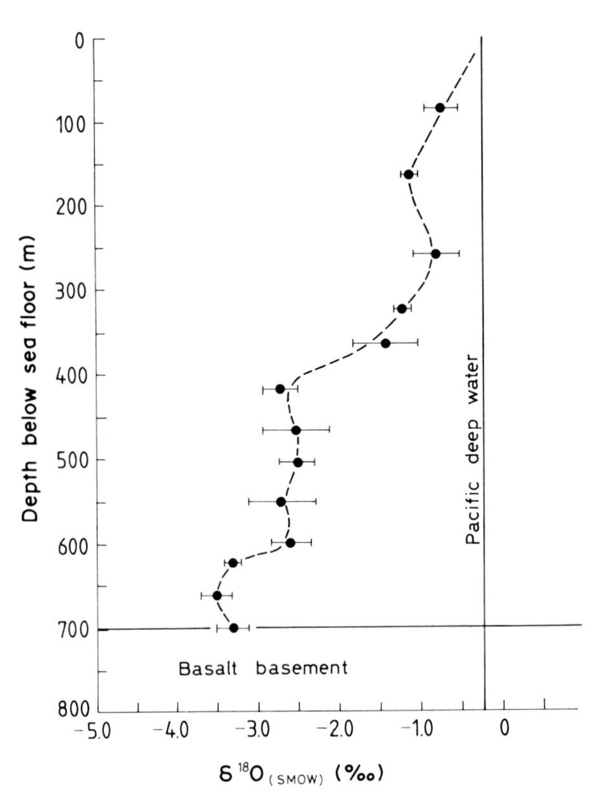

Fig. 3.35. The $\delta^{18}O$ of pore waters in sediments from the Bellinghausen Abyssal Plain (DSDP Site 323) relative to burial depth. (Lawrence et al. 1979)

and kaolinite, and a small amount of silt which contained quartz, feldspar, and mica. Small amounts of authigenic smectite and K-feldspar have been identified at the 400–450-m level in this upper sediment unit. The basal sediment, consisting essentially of authigenic smectite, was originally composed of volcanic materials, whereas the upper sediment consisted essentially of unaltered terrigenous clay sediments.

The $\delta^{18}O$ of the pore waters at DSDP Site 323 decreased from 0 near the surface to −3.5 per mill at a 660-m subsea depth. However, this decrease in the ^{18}O content of the water with depth is not uniform, as the oxygen isotope profile is marked by two distinct changes in the gradient, one at about the 600–700 m depth interval and the other at about the 400–500 m depth interval, and a less distinct change in the gradient at about the 150–200 m depth interval. The distinct changes in the gradient reflect low-temperature alteration of volcanic components in the sediments. Using a material balance calculation, Lawrence et al. (1979) showed that low temperature alteration of approximately 25 m of basalt, or 50–70 m of volcanic ash could explain the observed $\delta^{18}O$ in the pore waters. These authors found no mineralogical and geochemical evidence for the suggestion of recrystallization of clay minerals in the terrigenous upper sediments. However, Yeh and Eslinger (1986) reported authigenesis of clay minerals in the $<0.1\,\mu m$ clay fraction of the same sediments previously investigated by Lawrence et al. (1979). Any significant influence of recrystallization of such small clays on the oxygen isotope composition of pore water is doubtful, because these clays constitute only a small fraction of the total clay fraction in the sediment. What is not known, however, is the importance of the potential isotopic exchange of the surface layers of clay minerals. Whatever the exact cause of the decrease in the ^{18}O content of the pore water with depth, the study by Lawrence et al. clearly indicates that any model of authigenic clay mineral formation after the burial of the sediments must take into account the decrease in the $\delta^{18}O$ of the pore waters with burial.

4.2 Isotope Compositions of Clays in Shallow Buried Ocean Sediments

Data are absent regarding isotopic equilibrium between sea water and clay minerals other than glauconitic ones in recent oceanic sediments, but evidence exists suggesting an approach toward a chemical equilibrium. Clauer et al. (1984), for instance, observed morphological and chemical changes in detrital smectitic clay fractions of Paleocene and Cenomanian sedimentary rocks from the Atlantic Ocean, but the authors were not able to determine accurately when these modifications occurred. The changes were mainly visible in very small size ($<0.2\,\mu m$) particles with lath-type overgrowths (Table 3.1). Interaction of the particles with interstitial water resulted in decreases in both the Rb-Sr and the K-Ar dates, arising from

Table 3.1. Rb-Sr and K-Ar data of Atlantic Ocean clay fractions. (Clauer et al. 1984)

Samples	Rb (μg/g)	Sr (μg/g)	$^{87}Sr/^{86}Sr$ ($\pm 2\sigma/\sqrt{N}$)	Rb-Sr age (Ma ± 2σ)	K_2O (%)	^{40}Ar rad ($10^{-6} cm^3/g$)	K-Ar age (Ma ± 2σ)
386-34-3-100 (<0.2 μm)							
Untreated	156.9	86.8	0.72118 ± 19		2.87	16.22	167 ± 5
Leachate	453	3288	0.70998 ± 8	166 ± 7			
Residue	145.5	31.8	0.74062 ± 19				
386-34-3-100 (0.2–0.4 μm)							
Untreated	126.0	63.3	0.72380 ± 13		2.28	15.56	200 ± 6
Leachate	274	2014	0.70967 ± 16	187 ± 8			
Residue	125.6	28.8	0.74232 ± 12				
415A-11-1-12 (<0.2 μm)							
Untreated	51.9	73.7	0.71117 ± 13		0.62	2.97	142 ± 4
Leachate	96.9	1219	0.70844 ± 13	121 ± 14			
Residue	50.2	41.8	0.71412 ± 13				

increased loss of radiogenic isotopes relative to corresponding parent radioactive isotopes. Notwithstanding this loss, the isotopic dates of the particles still remained significantly above their stratigraphic age of 60 Ma. The Rb-Sr date decreased from about 187 to about 121 Ma and the K-Ar date from about 200 to about 142 Ma.

A glauconite-bearing unit of probable Pliocene age separates pelagic clays from overlying glacial-marine sediments at a subbottom depth of about 330 m on the Voring Plateau in the Norwegian Sea. Smalley et al. (1986) made Rb-Sr determinations on very dark to light green glauconite splits and found that the Rb contents ranged narrowly between 209 and 227 μg/g without any visible relationship to the color of the pellets. On the contrary, while the Sr contents of the very dark to dark green pellets ranged between 4.4 and 6.4 μg/g, that of the sample consisting of light green pellets was significantly higher at 22.4 μg/g. The data points of the light green and dark green glauconite splits from the deepest sample fitted a line in a Rb-Sr isochron diagram with that of a sample of foraminifera tests giving the $^{87}Sr/^{86}Sr$ ratio of the contemporaneous sea water. The Rb-Sr date of the line was 11.6 ± 0.2 Ma with an intercept $^{87}Sr/^{86}Sr$ ratio of about 0.70905. This date was interpreted by the authors to be the time of sedimentation of the glauconites, signifying that the 350 m of burial did not affect the Rb-Sr system of the glauconite minerals; but three data points plotted above the isochron and gave Rb-Sr model ages between 13.9 and 16.3 Ma, assuming the value for the isotopic composition of the initial Sr is identical to that of the foraminifera. Smalley et al. considered that the pellets with Rb-Sr date above 11.6 Ma had to be in chemical disequilibrium with the sea water at the time of their formation. The authors offered two explanations for this: either the pellets formed at an earlier time and were resedimented, but in this case even if they had been of different ages a rough trend toward an average sea water value should have been observed, or the pellets contained relics of Sr from a substrate material. No evidence for reworking of sediment could be found, but the authors mentioned that the glauconitic pellets with the older model ages could contain relics of micaceous substrate material. In that case, the Sr of the samples would not have been biased by an incomplete resetting of the precursor minerals but by contamination from a substrate mineral on which the glauconite pellets were fixed. No SEM observations were made to check this possibility and no K_2O values are available to test the degree of evolution of the glauconites. The Rb contents were in general about 200 μg/g, whatever the color of the glauconite splits.

Clauer et al. (1990) investigated the Rb-Sr and oxygen isotope compositions, in conjunction with chemical, electron microscopic, X-ray diffraction, and microprobe analyses, of smectite-rich clay size fractions separated from samples of Albian-Aptian and Paleocene shales at subbottom depths between 300 and 900 m, in the north Atlantic Ocean. Observations by TEM revealed three types of smectite particles: (1) flake-like particles with diffuse and irregular edges often rolled up, (2) lath-type particles regularly shaped, and (3) mixtures of both types (Fig. 3.36). The lath-type particles were highly abundant in the smaller size fractions (<0.2 μm) and the flake-type in the coarser fractions (1–2 μm). The authors suggested that the laths are of authigenic origin and the flakes of detrital origin, as did Bouquillon and Chamley (1986) for similar clay particles in surface sediments of the Indian Ocean. The REE distribution patterns of the smectite-rich clay fractions were similar to that of the NASC (Fig. 3.37; Haskin et al. 1968; Piper 1974). No Ce negative anomaly, which is typical for authigenic materials acquiring REE primarily from sea water (Goldberg et al. 1963; Hogdhal et al. 1968), was found in any size fraction whether enriched in laths or flakes. The laths apparently derived from preexisting flakes without significant input of REE from interstitial waters, reflecting a dissolution-precipitation process in a closed system.

The δD values of the smectite minerals were scattered between -79 and -108 per mill. This contrasts with the δD values between -60 and -80 per mill for many authigenic clay minerals in ocean sediments (Savin and Epstein 1970b). The retention of these unusually low δD values during growth of authigenic laths supports the idea that the pore waters were not in rapid communication with the open sea waters during diagenetic recrystallization of the clay particles. The $\delta^{18}O$ of the smectite minerals in the Albian-Aptian and Paleocene sediments in the Atlantic Ocean ranged from $+17$ per mill to about $+23.6$ per mill, which is less than that of many authigenic minerals, but much closer to the values of detrital clay particles (Savin and Epstein 1970b; Yeh and Savin 1976). The evidence further supports the evolution of clay minerals in an environment with low water-to-rock ratio.

Different size fractions of several samples were leached with dilute (1 N) HCl to differentiate the Sr that was adsorbed onto the mineral surfaces from

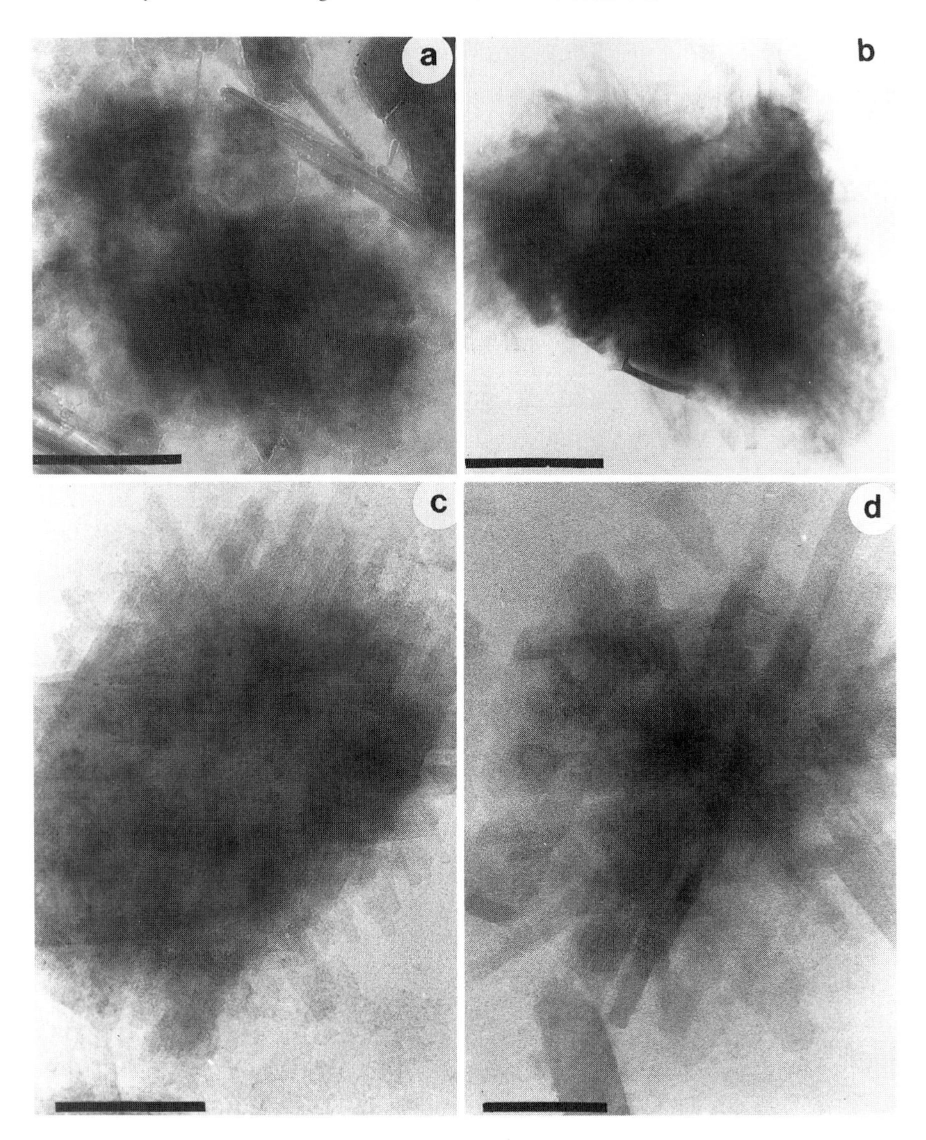

Fig. 3.36. TEM micrographs of smectite particles from Albian-Aptian and Paleocene black shales of the north Atlantic Ocean. Three types of features can be recognized: **a, b** flakes with diffuse edges. **c** Flakes surrounded by laths. **d** Laths. (Clauer et al. 1990)

that trapped in the mineral structures during crystallization. The Sr isotopic data of both the leachates and the acid-leached clay fractions of some samples from the Albian-Aptian and Paleocene shales are shown in Fig. 3.38. The lines for different size fractions have different slopes; the smaller the size, the lower the apparent ages. These lines are mixing lines, as the data points belong to samples containing varied proportions of detrital flake-

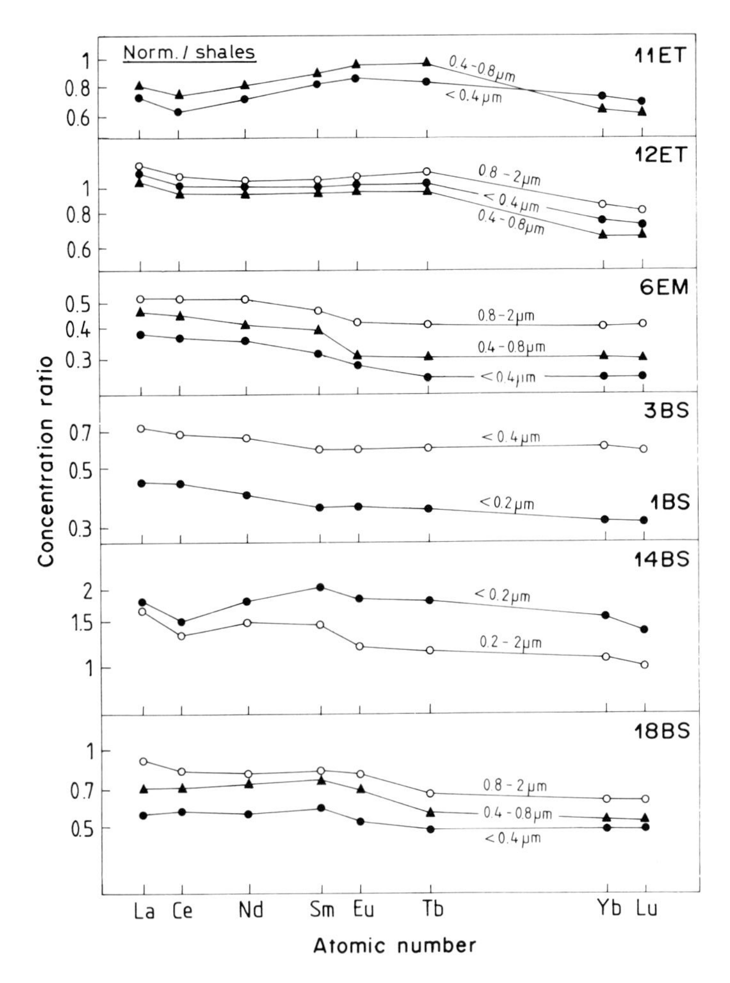

Fig. 3.37. REE distribution patterns of different size fractions of smectite clays from Albian-Aptian and Paleocene black shales of the north Atlantic Ocean. (Clauer et al. 1990)

type and authigenic lath-type materials. The $^{87}Sr/^{86}Sr$ values of the leachates were systematically above that of the contemporaneous sea water and they correlated to the amount of lath-shaped particles present in the fractions analyzed. The decreases in the slope of the mixing lines suggest that the transfer of radiogenic ^{87}Sr from clay particles to the interstitial waters was dependent not only on their difference in $^{87}Sr/^{86}Sr$ ratios relative to sea water value, but also on their Rb/Sr ratios.

The time when recrystallization of the smectites occurred was calculated from changes in the isotopic dates as a function of the abundances of the authigenic lath-type material (Fig. 3.38). The extrapolated ages at 100% authigenic laths corresponded to about 100 Ma for the Albian-Aptian material and 60 Ma for the Paleocene samples. These values are close to the stratigraphic ages of the sequences and suggest that the recrystallization process happened soon after deposition.

Briqueu and Lancelot (1983) analyzed Rb-Sr isotopic compositions and K contents of an abyssal clay sample from an active subducting plate in the

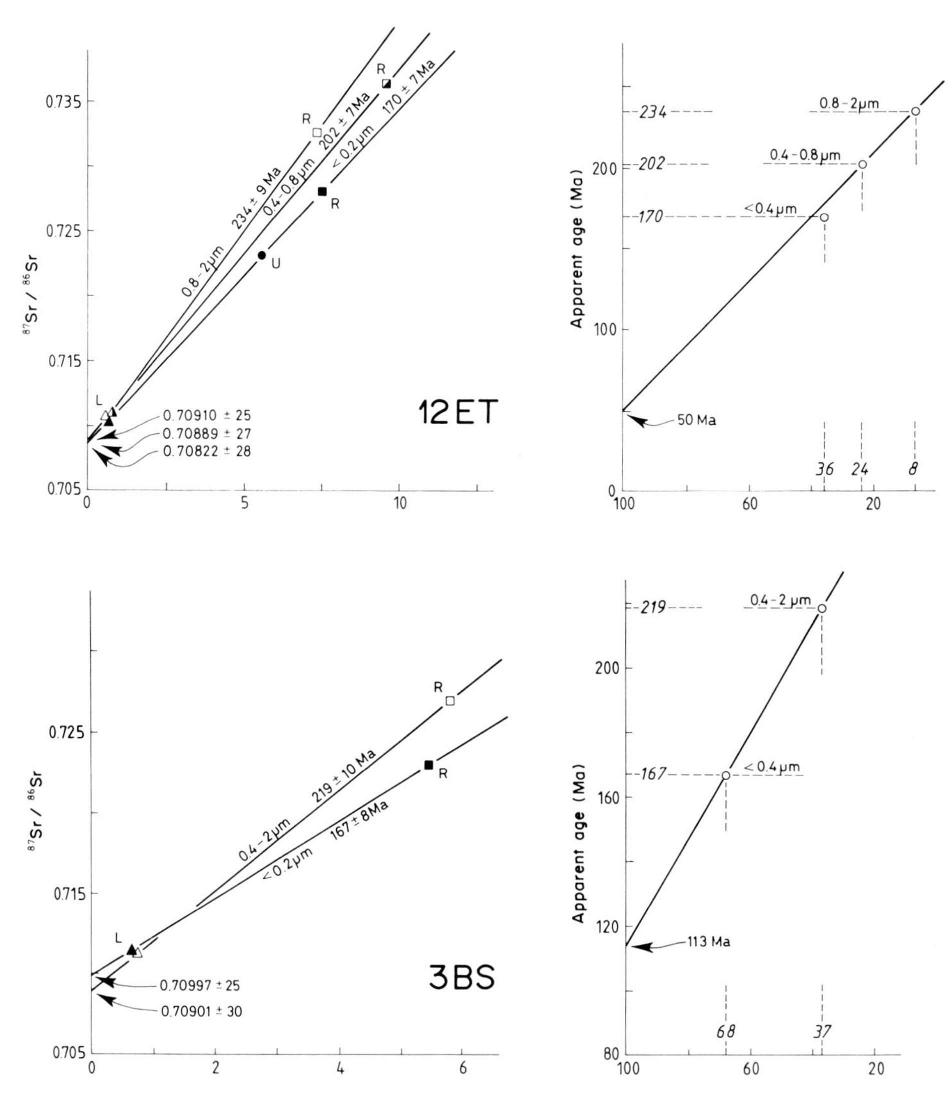

Fig. 3.38. Rb-Sr isotopic data of size fractions of several smectite clay fractions from Albian-Aptian and Paleocene black shales of the north Atlantic Ocean. The Rb-Sr isochron diagrams display mixing lines obtained after acid-leaching for each size fraction: the dates decrease and the initial $^{87}Sr/^{86}Sr$ ratio increase while the size decreases. The dates of each size fraction were also related to the amount of lath-type particles in the fractions and estimation of the period of crystallization was obtained by extrapolating to 100% lath-type particles. (Clauer et al. 1990)

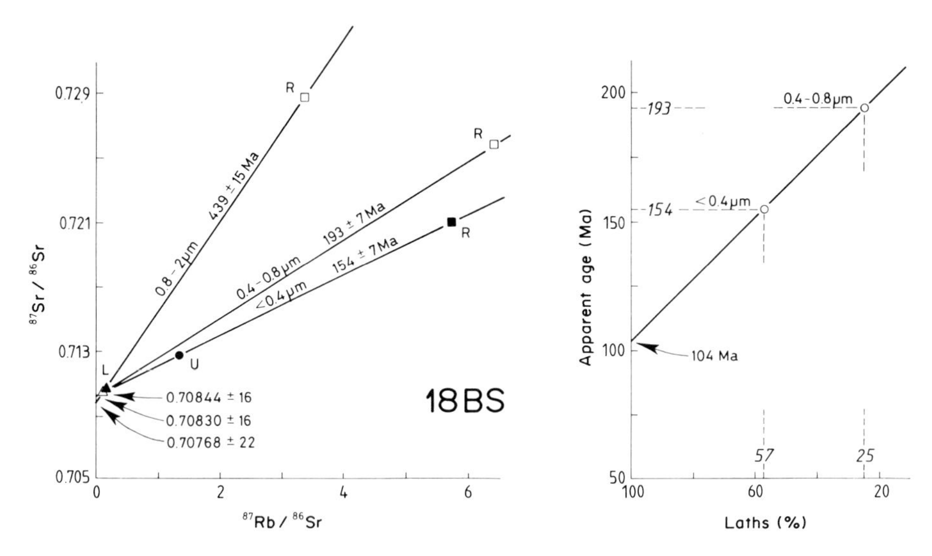

Fig. 3.38. *Continued*

region of the Vanuatu island arc in the western Pacific Ocean. The abyssal clays came from a unit about 19 m thick within a 649-m-thick sedimentary section resting on altered tholeiitic basalt. The time of deposition for the entire sediment spanned from Middle Miocene to Pleistocene. The unit with abyssal clays, presumably of Miocene age, occurs about 64 m below the sea floor and consists of glass shard ash, zeolites, and micronodule-rich clays. The initial $^{87}Sr/^{86}Sr$ ratio, after making correction for the radiogenic growth of ^{87}Sr, decreased from 0.70741 for the untreated clay sample to 0.70632 for the acid-leached sample. The leachate yielded an isotopic value which is about the same as that of contemporaneous sea water. The Rb/Sr ratios increased from 0.24 for the untreated sample to 0.48 for the acid-leached aliquot, the increase being primarily due to the decrease in the Sr concentration from 175 μg/g to 83 μg/g. The Rb concentration remained constant between 40 and 42 μg/g and the K content decreased from 1.74% for the untreated sample to 1.20% for the acid-leached sample, resulting in a decrease in the K/Rb ratio from 414 for the untreated sample to about 300 for the acid-leached clay primarily due to loss of K by acid leaching. The Sr isotopic data make it evident that the Sr isotopic composition of the acid-leached clay is intermediate between that of sea water and tholeiitic or calc-alkaline volcanic rocks of the area, suggesting that the Sr in the abyssal clays consists of both sea-water Sr and oceanic crustal Sr. The adsorbed labile Sr, the amount of which is about the same as that of the "fixed" nonexchangeable Sr in the clay, had an $^{87}Sr/^{86}Sr$ ratio which is nearly the same as that of contemporaneous sea water, suggesting that the labile Sr was derived essentially from oceanic environment. Furthermore, the decrease in the K/Rb ratio for the acid-leached clay relative to the untreated one also suggests adsorption of some alkali elements from sea water. These results agree with those of Clauer et al. (1982b) and Clauer and Hoffert (1985) from red clays in the eastern and central Pacific ocean bottoms.

The oxygen isotopic exchange between clay sediments and ambient waters most likely involves chemical reaction. Any apparent oxygen isotopic exchange by the terrigenous clays in oceans may suggest either the actual amount of isotopic exchange between the detrital clay and sea water, or formation of authigenic clay minerals, or recrystallization of clays. Hence, knowledge of the apparent amount of post-formational isotopic exchange between ocean sediments and ocean waters is useful to determine the degree to which sea-water chemistry is buffered by the formation of authigenic minerals. Yeh and Savin (1976) estimated the extent of oxygen isotope exchange between ocean sediments and sea water by analyzing the oxygen isotope compositions of different size fractions of sediments from three separate cores in the north Pacific Ocean. Estimated depositional ages of the sediments ranged from 14000 to 2.7 million years. The clay minerals in the sediments from two of the cores contained about 70 to 80% illite, 15 to 20% chlorite-kaolinite, and less than 5% smectite, whereas that from the third core consisted of 40 to 70% illite, 10 to 60% smectite, and less than 13% chlorite-kaolinite. The clay mineralogy of two illite-rich, smectite-poor clay samples did not vary much with size fraction, but that of the smectite-rich sample varied markedly with size fraction, characterized by increase in the smectite content and decrease in the illite content with decrease in size. The $\delta^{18}O$ values of illite-rich clay samples ranged from about +13.7 to +16.8 per mill for the coarse fractions (>0.5 µm) and from +15.5 to +17.6 per mill for the fine fractions, whereas that of smectite-rich clay fractions ranged from +14.8 to +15.6 per mill for the coarse fractions and from +18.1 to +23.7 per mill for the fine fractions. After making reasonable estimates of the initial $\delta^{18}O$ and the equilibrium $\delta^{18}O$ values for the continental clays deposited in deep-sea environment, the authors calculated from the $\delta^{18}O$ values that the isotopic exchange that could have occurred between detrital clays and sea water varied between 12 and 16% for clay fractions in the size range between 0.1 and 0.5 µm and amounted to as much as 42% for clay fractions of <0.1 µm size. The accurracy of such estimation is dependent on the assumed values for the initial $\delta^{18}O$ and the equilibrium $\delta^{18}O$. As the <0.1 µm fraction constituted about 5% of the total clay sediment and the fraction with a size ranging between 0.1 and 0.5 µm constituted about 35% of the total clay sediment, the authors estimated that about 4 to 9% oxygen isotope exchange occurred between the entire detrital clay sediment and pore water in a period of about 10000 to less than 3 million years. The estimated total amount of oxygen isotope exchange between terrigenous clay sediment and sea water was considered by the authors to be sufficiently high to suggest that clay reconstitution may be an important factor in regulating the chemistry of sea water.

Yeh and Epstein (1978) estimated the amount of hydrogen isotope exchange between clays and ambient water from clays from two separate depths of about 0.30 and 110 m below the sea floor in the north-Pacific sediment core studied by Yeh and Savin (1976) for changes in the oxygen isotope composition. The deposition of the deep clay occurred about 3 Ma before the deposition of the shallow clay. By comparing the hydrogen isotope compositions of the shallow clays with that of the deep clays of the same size, the authors found difference in the δD values only for the $<0.1\,\mu$m clay size fraction. The δD values of three 0.1–2 μm clay fractions for the deep sample varied between -72 and -77 per mill, which was the same for the three clay fractions of identical size for the shallow clay sample. By contrast, the δD value of the $<0.1\,\mu$m clay fraction of the shallow sample was -78 per mill and that of the deep sample was -73 per mill. The data suggest that at least in the last 2–3 Ma period no significant hydrogen isotope exchange occurred for clays $>0.1\,\mu$m in size. The fraction $<0.1\,\mu$m made up only 5% of the total clay sediment. The authors calculated that the amount of hydrogen isotope exchange between such fine clay particles and ambient sea waters ranged from 8 to 28%, which is about the same magnitude for oxygen isotope exchange previously reported by Yeh and Savin. Because the extents of hydrogen and oxygen isotope exchanges were compatible, the apparent isotopic exchanges have been regarded by Yeh and Epstein (1978) as due to formation of authigenic minerals or recrystallization of the sediments, and not to actual isotope exchange between the terrigenous clay and ocean waters. These authors also concluded from lack of any significant isotopic exchange for clay fractions $>0.1\,\mu$m that the D/H ratios of ocean sediments may be used to trace their provenance, provided the sediments were deposited within the last 2–3 million years and possibly no later.

Yeh and Epstein (1978) also reported that a sample of very fine montmorillonitic clays of $<0.1\,\mu$m size in a red clay sediment of Upper Eocene age collected from DSDP Site 40 in the northeast Pacific Ocean, had a δD value of -15 per mill and a δ^{18}O value of $+34.9$ per mill. The authors suggested, based on knowledge of the range of isotopic values of sea water since the Upper Eocene and on reasonable estimates for the hydrogen and the oxygen isotope fractionation factors between montmorillonite and water at different equilibrium temperatures, that the montmorillonite most likely formed at sea floor conditions.

Lawrence et al. (1979) analyzed the oxygen isotope compositions of clays in a 700-m-thick sedimentary sequence drilled at DSDP Site 323 in the Bellinghausen Abyssal Plain in the southeastern Pacific Ocean. The upper 640 m of the sequence consisted of terrigenous sediments of Oligocene to Recent age. The basal 60 m consisted of claystone of Cretaceous to Paleocene age. Underlying the basal unit was Cretaceous basalt with numerous clay and carbonate veins. A sample of vein-filling celadonite clay from basement basalt, having a K_2O content of 9.3% and a charge deficiency of 1.75 per unit cell, yielded a $\delta^{18}O$ value of +21.4 per mill, which corresponded to an estimated temperature of 30 ± 5 °C for its formation (Kastner and Gieskes 1976). The sediment of the upper sequence consisted of illite/smectite mixed layer with approximatively 40% illitic interlayers, discrete illite, chlorite, and kaolinite in the clay fraction, but was dominated by quartz, feldspar, and mica in the silt fraction. The clay fraction of the basal sedimentary sequence was composed primarily of an authigenic smectite mineral suite. The silt size fraction of this basal sediment was dominated by K-feldspar, plagioclase, quartz, zeolite, and smectite. The basal sediments mostly were of volcanic origin, and were highly altered primarily to smectite. In the sediment of the upper sequence, the $<2\,\mu m$ clay fractions had an average $\delta^{18}O$ value of about +19.1 per mill, whereas coarse fractions had an average value of +13.5 per mill. No systematic change in either the $\delta^{18}O$ value or the mineralogy occurred for the $<2\,\mu m$ clay fraction in the upper sediment relative to depth, suggesting that no significant amount of recrystallization occurred for the detrital $<2\,\mu m$ clay fraction in the upper sediment. The relatively low value for the silt fraction relates to its high amounts of detrital quartz, feldspar, and mafic silicate minerals. The average $\delta^{18}O$ value of +19.1 per mill for the clay fraction in the sediments of the upper sequence at the DSDP Site 323 is similar to the mean value reported by Savin and Epstein (1970b) for the silicate fraction from recent ocean sediments.

In contrast to those in the sediments of the upper sequence, the silt and the clay fractions of the basal sedimentary sequence differed little in the $\delta^{18}O$ values, suggesting comparably high abundances of diagenetic minerals in both fractions. The average $\delta^{18}O$ value of the clay fractions of the basal sedimentary sequence was about +22.0 per mill, which is slightly higher than that of the upper sediments. The $\delta^{18}O$ values of clays in the basal sedimentary sequence, unlike that in the upper sedimentary sequence, correlated well with the Mg/Al, Si/Al and K/Al ratios (Figs. 3.39–3.41). The $\delta^{18}O$ values of pore waters decreased from 0 per mill at the top of the upper sediment to about −3.5 per mill at the bottom of the basal sediment (Fig. 3.35). The $^{87}Sr/^{86}Sr$ ratio of a sample of pore fluid in the basal sediment was about 0.7067, which is a mean value between that of Cretaceous or younger sea water and that of the bulk sediment sample of the basal sequence. The isotopic and chemical evidence from sediments and pore fluids and a mass balance calculation led Lawrence et al. (1979) to conclude that a discrete alteration or recrystallization of a very small amount of volcanogenic components in the terrigenous upper sediments can account for the observed trends of the pore fluids and that alteration of the detrital components was negligible in the upper 600-m-thick sediments.

Yeh and Eslinger (1986) analyzed the $\delta^{18}O$ of the $<0.1\,\mu m$ clay fractions of sediments at the DSDP Site 323 which was previously investigated by Lawrence (1979). The $\delta^{18}O$ values shifted randomly downhole between +16.8 and +21.8 per mill. The trend in the depth-^{18}O relationship probably reflects either varied amounts of terrigenous input from low latitude areas,

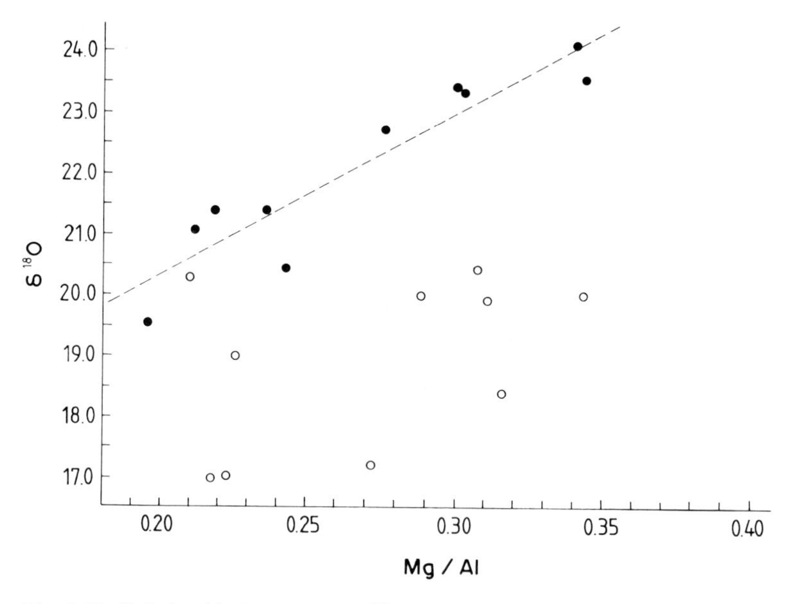

Fig. 3.39. Relationship between the $\delta^{18}O$ values and the Mg/Al ratios of clay-size fractions of sediments from Bellinghausen Abyssal Plain (DSDP Site 323). The *filled symbols* belong to clays in the terrigenous upper sediment and the *open symbols* belong to authigenic smectite-rich clays of the basal sediments. (Lawrence et al. 1979)

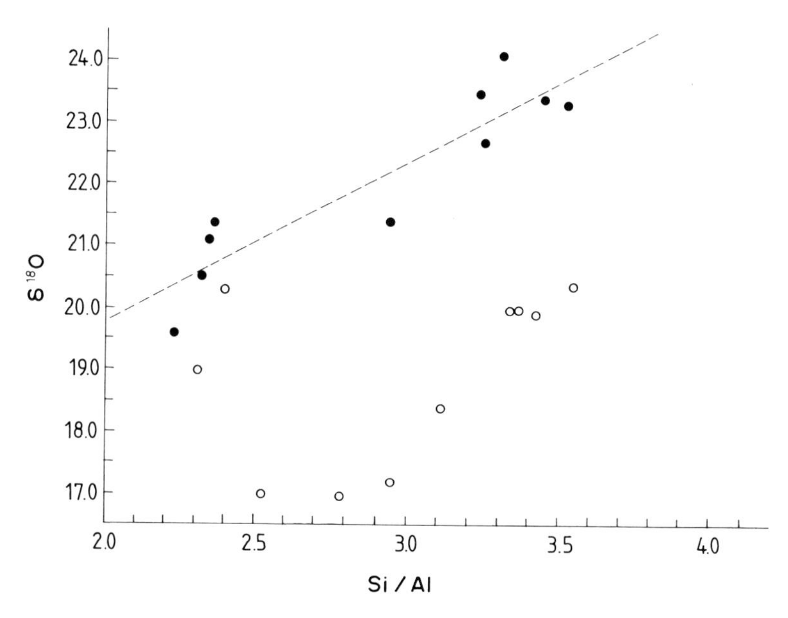

Fig. 3.40. Relationship between the $\delta^{18}O$ values and the Si/Al ratios of clay-size fractions of sediments from the Bellinghausen Abyssal Plain (DSDP Site 323). Explanations of signs in Fig. 3.39. (Lawrence et al. 1979)

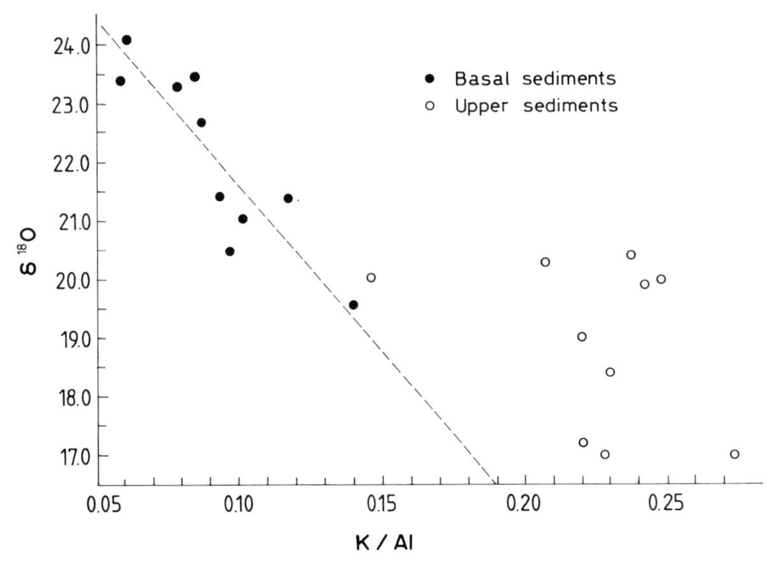

Fig. 3.41. Relationship between the $\delta^{18}O$ values and the K/Al ratios of clay-size fractions of sediments from the Bellinghausen Abyssal Plain (DSDP Site 323). Explanations of signs in Fig. 3.39. (Lawrence et al. 1979)

or presence of variable amounts of marine authigenic minerals in the very fine clay fractions. Authigenesis of clay minerals may be documented in very fine clay sediments, but the amount of such fine clay in the sediment is so small that extreme care is needed in separating this very fine authigenic clay in a pure form from the rest of the sediment.

Eslinger and Yeh (1981) analyzed oxygen and hydrogen isotope data of different size fractions of sediments covering 500 m of subbottom depth in the Aleutian Trench. The sediments, recording a depositional history of the last 300 000 years, presumably derived from mixed sources in the continent and were deposited mainly by ice-rafting and turbidity currents. The authors found that coarse clay fractions ($>0.1\,\mu m$) had $\delta^{18}O$ values between +9.7 and +12.0 per mill, and δD values between −46 and −74 per mill and they concluded that no post-depositional mineralogical or isotopic changes occurred for the coarse clay fractions. By contrast, the fine fractions of $<0.1\,\mu m$ size which consisted of smectite-rich clay mineral assemblages, had $\delta^{18}O$ values between +12.1 to +16.3 per mill and δD values between −55 and −67.5 per mill. Neither the $\delta^{18}O$ nor the δD values changed with increase in depth (Fig. 3.42). Hence, the authors contended that diagenesis was unimportant for the changes in the isotopic compositions of the clay fractions, and that the changes in the oxygen isotope composition were only due to changes in the source of the sediments. Eslinger and Yeh also claimed that the changes in δD with depth for the fine fraction ($<0.1\,\mu m$)

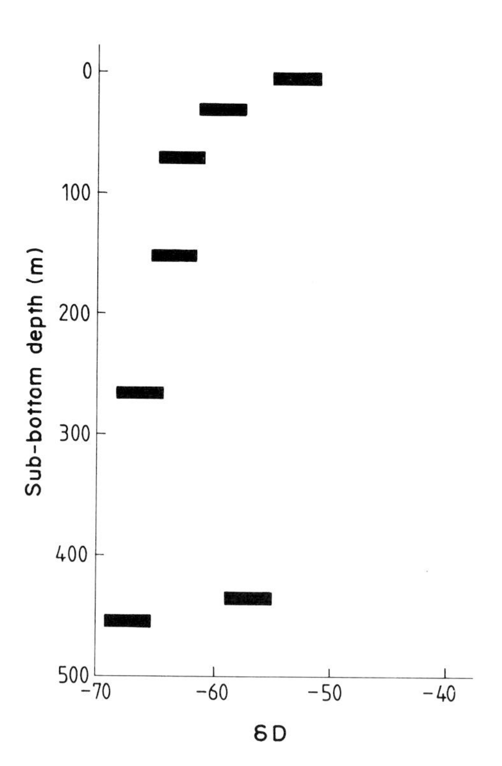

Fig. 3.42. The δD values of sediments from the Bellinghausen Abyssal Plain (DSDP Site 323) relative to burial depth. (Eslinger and Yeh 1981)

were either due to post-depositional isotopic exchange, or to climatic changes in the terrestrial weathering environment where the clays originated.

5 Summary

A major problem in studies of recent ocean bottom sediments is the difficulty in making the distinction between detrital terrigenous and authigenic oceanic clays. Isotopic data have high potentials in making such distinctions, but more isotopic studies on clays in both deep and shallow bottoms are needed to develop fully both the genetic and dispersal history of clay minerals in modern ocean sediments.

Isotope knowledge about clay minerals in sediments eroded from the continents has come largely from river suspended loads. Although the $^{87}Sr/^{86}Sr$ ratios have been found to range between 0.704 and about 0.800 for suspended loads from a large number of rivers, most of these values are lower than 0.735 but higher than that of recent sea water (0.709). The Rb/Sr ratios are between 0.01 and 1.60. The $^{87}Rb/^{86}Sr$ and the $^{87}Sr/^{86}Sr$ ratios are poorly correlated, reflecting varied effects of Rb-Sr isotopic compositions and apparent ages of rocks and their differential weathering. As compared to the Sr isotopic ratios of suspended loads, those of dissolved loads are low for a vast majority of the present-day rivers.

Isotopic dates based on the K-Ar method have served as a major source of information about ages of clay minerals in suspended sediments of rivers. The K-Ar dates have been found widely varied, as one study has shown the range of variation to be between 200 and 800 Ma. Although such high values suggest derivation of clays largely from very ancient sedimentary rocks, this interpretation could be somewhat misleading in view of the fact that clay minerals newly formed from weathering of crystalline rocks in geologically very young time periods are known to yield anomalously high isotopic dates as a result of incorporation of a significant amount of radiogenic isotopes at the time of their formation.

Oxygen isotope data of clays in modern rivers are few, but such data suggest that the clays could have formed more or less in equilibrium with meteoric waters in the present drainage basins at temperatures lower than 30 °C or so. The suggestion of oxygen isotope compatibility for the clays with recent meteoric waters does not necessarily mean that the clays formed in recent times. Detailed mineralogic data are needed to be integrated with the various different isotopic data to identify the multiple sources of clay minerals in rivers.

Clays in suspended loads among many rivers in the world have $\varepsilon_{Nd}(0)$ values between +7.1 and −44 and Sm/Nd ratios between 0.16 and 0.29. The Sm-Nd isotopic characteristics of the river clays are intermediate between those of andesites and Proterozoic-Archean continental metasediments and gneisses, suggesting a large component of old continental rock debris in the modern-day river sediments. The Nd isotopic values of the clays bear a nonlinear inverse relationship with the Sr isotopic values, indicating mixing in varied proportions between young island arc materials and very old crustal materials. The Nd model ages (relative to depleted mantle) are nearly constant at about 1.7 Ga for the suspended loads, which is roughly close to the model age of about 2.0 Ga for many post-Archean sedimentary rocks. Assuming that the erosional debris represent an average sampling of the upper continental crust, the average model age of 1.7 Ga for the suspended clays of rivers may then be interpreted as having very little addition of new mantle materials to the crust during the post-Archean period. However, such an interpretation should be viewed with caution, because suggestions have been made that sedimentary mass at a given time is cannibalistically recycled, resulting in the incorporation of varied amounts of older crust.

What do the different lines of isotopic evidence suggest about the origin of clays in ocean floor sediments? The Sr isotopic ratios for ocean floor clays, containing varied amounts of smectitic and illitic clay minerals, appear to be generally lower than 0.743. About two-thirds of these isotopic values are higher than 0.709, which is about the value for present-day sea water and the remainder lower than the value of the sea-water Sr. The Sr isotopic ratios for the clays correlate positively with the illite/smectite ratios. By extrapolation, when the trend in the relationship is extended to a pure smectite, the corresponding $^{87}Sr/^{86}Sr$ ratio is about 0.703–0.704, which is approximately similar to the isotopic values for Sr in ocean-floor basalts, suggesting that the smectitic component could have derived from alteration of oceanic basalts and that the ocean-floor sediments contain varied amounts of marine authigenic smectite clays. Although some amounts of these smectites could have formed as products of weathering in continental environments and later were transported by various agencies to the ocean basins, Nd isotopic evidence suggests that not all smectites can belong to the debris eroded from the continents. Sediments from deep-ocean floors in the north Atlantic Ocean and the Mediterranean Sea have been found to have $\varepsilon_{Nd}(0)$ values as high as +4.9. Also, muds associated with deep-sea turbidites from active margin settings are known with high positive $\varepsilon_{Nd}(0)$ values.

Information is very sketchy about K-Ar dates for deep ocean floor clay materials. The limited data indicate dates between about 150 and 460 Ma for some fine clays. These dates are similar to many dates that have been found for river clays, suggesting that the clays are primarily terrigenous in origin and may not have been greatly modified by entering into the ocean environments. More studies may reveal the true nature of the modifications, as terrigenous clay particles enter the ocean environments.

The ocean-floor clay minerals have $\delta^{18}O$ values commonly between +15 and +20 per mill. A study concluded that the $\delta^{18}O$ values of various components of the ocean floor clays may be given as: +15 per mill for chlorite, +25 per mill for kaolinite, +15 per mill for illite, and +17 per mill for smectite. These oxygen isotope values for the oceanic clays are similar to those of many terrestrial clays, suggesting not only that the ocean floor clays are essentially detrital in origin, but also that very little oxygen-isotope exchange occurs between the detrital clays and sea water under most of the ocean floor environmental conditions. Experimental observations also indicate that such oxygen-isotope exchange is insignificant, unless the clay particle size is very small, over a time period of 2 years for both kaolinite and well-crystallized illite minerals. Empirical isotopic data for some fine ($<0.5\,\mu m$) ocean-floor clays of 14 000–30 000 years old suggest that as much as 12% of the clays can be traced to marine evolution either by oxygen-isotope exchange of detrital clays with ocean waters, or by authigenic growth. The δD values of ocean-floor clays are commonly varied between −50 and −95 per mill; but, unlike oxygen-isotope exchange, hydrogen-isotope exchange is significant for clays in waters, suggesting that the hydrogen-isotope data may not be a reliable guide about environmental conditions that existed at the time of formation of the clay minerals.

Although opinions may vary about the amount of authigenic clays that may be present in ocean floor sediments, no disagreement exists about the formation of authigenic clay minerals in ocean floor environments. This is evident from many reports about glauconitic minerals in recent ocean sediments and also about smectitic and celadonitic minerals occurring as vein- or fracture-filling and replacement minerals in ocean floor basalts. The Sr isotope data of Recent glauconites indicate that the formation of glauconite is a two-step process, the initial step being replacement of smectitic precursor in a closed system with the near exclusion of marine Sr and the latter step being a crystal growth in an open system with chemical inputs from sea water. Duration of clay mineral authigenesis or modification is important for the evaluation of stratigraphic dating. Genesis of glauconitic materials seems to begin soon after deposition and their formation precedes burial, but this syndepositional evolution is more difficult to estimate for most other clay minerals. Strontium isotopic values of vein-filling or replacement clay minerals in basalts are internally varied and appear to range between those of ocean floor basalts (0.703–0.704) and ocean waters (0.709), indicating that these mineralizations occurred in environments with inputs of varied amounts of Sr from both the basaltic rocks and the ocean waters. The Rb-Sr isotopic data of the vein-filling clay minerals have shown that the mineralization could have occurred about 10–35 Ma after the formation of the basaltic crust. Any relation of these data to ocean water circulation or hydrothermal circulation through the basaltic crust is far from clear.

The $\delta^{18}O$ values of vein-filling and replacement celadonite and smectite minerals in ocean-floor basalts range widely from about +3.5 to +31 per mill. Assuming that the crystallizations occurred in high water-to-rock ratio environments, the isotopic variations relate to formation temperatures in the hydrothermal range from as low as about 25 °C to as high as 360 °C. The estimates for the temperatures should be viewed with caution, because the Sr isotopic data for the clays suggest that the crystallization of the minerals did not essentially occur in systems that were very much open to the circulation of sea water.

Deep sea sediments containing significant amounts of smectite and phillipsite minerals are found to occur in many places of the ocean floor considerably away from oceanic ridges or centers of other kinds of volcanic activities. The Sr isotopic ratios of these minerals, like those of vein-filling or replacement clay minerals in basalts, are lower than that of sea water, but higher than that of ocean floor basalts. Thus, the clay minerals in these ocean floor sediments could have originated elsewhere near areas of submarine volcanic activities that are far from the site of the present-day location.

The present-day basaltic ocean floor is variably covered by sediment, the thickness of which may reach as much as 600 m in some location. Such sedimentary covers may contain thin volcaniclastic layers interbedded with predominantly terrigenous sediments containing also some scattered volcanic fragments. Although modifications of terrigenous clays appear to be minor under deep-ocean conditions, authigenic clay minerals have been recognized in connection with shallow burial of the sediments, some few million or tens of million years after the deposition on the ocean floor. Collective geochemical and mineralogic evidence has indicated growth of some amount of authigenic smectitic minerals either as replacements of volcanogenic fragments in both the volcaniclastic units and the layers of terrigenous sediments, or as an appendage to other smectites that are probably of detrital origin. The Sr isotopic ratios of these shallow buried diagenetic clays have been found to be intermediate between that of marine Sr and that of oceanic crustal Sr. The $\delta^{18}O$ values for some of these fine-sized clays ($<0.5\,\mu m$) have been found to be between +18 and +24 per mill. By comparing with the oxygen isotope values of detrital clays from adjacent continents, the observed oxygen isotope values for the ocean clays suggest that they could be diagenetic in origin and the amount of such diagenetic clays in shallow buried ocean sediments could be in the range between 4 and 9%, signifying that oxygen isotope modifications for the sediments beneath the ocean floor are quite small.

Chapter 4

Isotope Geochemistry of Clays and Clay Minerals from Sedimentary Rocks

Authigenic clay minerals are records of a long and complex history of the sediments since their deposition. These minerals can be broadly classified as either syn-depositional or post-depositional, but the distinction between them in an ancient sedimentary rock is not always easy. This distinction is important because only the isotopic records of very early authigenic minerals may be examined to determine the time of deposition of the sediments. As is often the case, the separation of early and late diagenetic clay minerals from a rock is often frustrated by the overwhelming abundance of detrital clay minerals in the rocks. The fundamental requirement for tracing the different events in the genesis of the clays is then a clear understanding of the evolutions of the analyzed clay materials. Equally important is a critical assessment of responses of the different isotope systems of clay minerals in question that might have been subjected to varied depositional and post-depositional forces causing modifications in physical and chemical aspects of a mass of sediment (Clauer et al. 1992c).

Interests in isotopic analyses of clay minerals in ancient sedimentary rocks go much beyond establishing the time of deposition. Progressive burial of sediments exposes them to environments of increased pressure and temperature, causing a great deal of chemical modifications with new evolutionary conditions. For example, a comparison in the Rb/Sr ratios between carbonate-free recent ocean sediments and shales shows that the former generally have values between 0.71 and 1.75 with an average at 0.71, whereas the latter between 1.8 and 2.6 (data from Clauer 1973, 1976; Cordani et al. 1985). Isotopic analyses of clay minerals, therefore, provide an opportunity for reconstruction of these evolutionary paths and gain increased understandings of hydrodynamics and mass transfers of elements in a sedimentary basin. Isotopic records of clay minerals are most sensitive to chemical interactions. Compaction and accompanying dewatering alone apparently produces no discernible isotope "fingerprints" in the clay minerals (Clauer et al. 1990).

This chapter focuses on isotopic compositions of clay minerals in varied ancient sedimentary rocks during their burial and uplift histories. One of the aims is to give a detailed perspective on general problems concerning the fundamentals of isotopic dating of shales and clay minerals for the purpose of establishing the time of deposition of ancient sedimentary rocks. Another objective is to discuss recent advances that have been made in the field of burial diagenetic evolution of clay minerals as a result of isotopic researches on these materials.

1 Syndepositional Evolution of Clay Minerals in Argillaceous Sedimentary Rocks

Studies of recent shallow and deep-sea terrigenous sediments have demonstrated that syndepositional authigenic clay minerals are a significant but small fraction of the bulk clay materials that are dominantly terrigenous in origin. Long history of lithification and post-lithification of ocean sediments is covered by several episodes of clay mineral formations under various burial conditions. The most important criterion for the success of dating the time of deposition of any sedimentary rock is the clean separation of the syndepositional clays from other clay minerals and also nonclay minerals for the intended isotopic analyses. Many early studies for stratigraphic dating apparently made inappropriate assumptions about resetting of isotopic memories of detrital clays upon entering into the basin of deposition. Subsequent studies have shown that these assumptions have been too oversimplified, as in many cases the detrital isotopic memories are largely retained long after the deposition of the sediments. Few recent studies have shown that, simply by careful selection of samples from some sedimentary rocks, isotopic analyses may yield meaningful dates for the stratigraphic purposes.

1.1 Analyses of Whole-Rock Samples

Many shale whole rocks essentially consist of clay or mica-type minerals and quartz. Since the K contents of these rocks are taken to be almost entirely related to the abundance of the common K-bearing sheetsilicate minerals, analyses of bulk samples of shales became a popular method for many early isotopic investigations on establishing the time of deposition of sedimentary rocks. The Rb-Sr isochron method of dating was applied to sedimentary rocks for the first time in 1962 by Compston and Pidgeon, who assumed that Rb in fine-grained sediments is essentially concentrated in clay minerals and that the Sr may be isotopically homogeneous soon or at some time after the deposition. They further assumed that the time interval between the period of deposition and that of the Sr isotopic homogeneization may be short enough, so that the time of homogeneization would closely approximate the time of deposition of the sediment. This approximation has been considered to be appropriate for defining the times of deposition of early Paleozoic and Precambrian rocks. Analyzing several whole-rock samples of shales from three stratigraphically very old rock units, Compston and Pidgeon (1962) found that the Rb-Sr isochron dates of two of these units agreed well with their presumed stratigraphic ages. The samples from the third unit gave an ill-defined isochron amounting to an isotopic date far in excess of the stratigraphic age. The authors attributed this excess date for the third unit to the presence of large amounts of inherited radiogenic Sr in the sediments at the time of deposition. The apparent encouraging results of the study by Compston and Pidgeon (1962) prompted Whitney and Hurley (1964) to pursue a Rb-Sr isotopic investigation of shales from a paleontologically well-dated Pennsylvanian unit. They obtained a well-defined Rb-Sr isochron date, but this was 15% in excess of the stratigraphic age (Fig. 4.1). Whitney and Hurley questioned the usefulness of the whole-rock Rb-Sr chronology in determining the time of deposition for Paleozoic and younger rocks, but they believed that a high potential exists for the whole-rock Rb-Sr isotope method in determining the time of deposition of those unfossiliferous Precambrian sedimentary rocks whose mineral components derived largely from contemporary source rocks.

Following the suggestion made by Whitney and Hurley about dating unfossiliferous Precambrian sediments, Chaudhuri and Faure (1967) analyzed Rb-Sr isotopes of whole-rock samples of the Precambrian Nonesuch shales in Michigan. The different components in the sediments were considered to have derived from underlying magmatic rocks and the sediments were

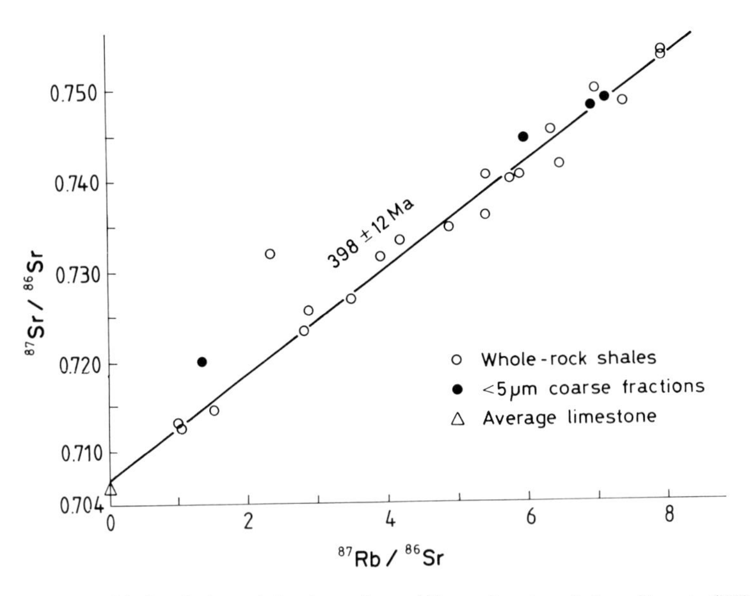

Fig. 4.1. Rb-Sr whole-rock isochron data of Pennsylvanian shale sediments (300–280 Ma). The slope of the line is in excess of the stratigraphic age of the sediments. (Whitney and Hurley 1964)

presumably deposited in a rift basin about 1100 Ma ago, shortly after the formation of the igneous rocks. The whole-rock samples of the Nonesuch shales yielded a Rb-Sr isochron date of about 1075 Ma. The authors suggested that an apparent Sr isotopic homogeneity probably resulted from diagenesis shortly after the deposition and that the isochron date was a good approximation of the time of deposition of the sediments. Several other studies carried out on Precambrian or Early Paleozoic sedimentary and metasedimentary rocks also gave well-defined Rb-Sr isochrons which were attributed to isotopic homogeneities by either diagenesis shortly after sedimentation or metamorphism (Moorbath et al. 1967; Obradovich and Peterman 1968; Faure and Kovach 1969).

In recent years, Sm-Nd isotope analyses of whole-rock samples have been made on Precambrian sedimentary rocks to determine the time of deposition. Stille and Clauer (1986), analyzing the Sm-Nd isotopic compositions of four whole-rock samples and one clay fraction of the Proterozoic Gunflint Formation, found that the Sm-Nd data points defined a line whose slope gave an isochron date of about 2.08 Ga (Fig. 4.2). This date was found to be much higher than either the Rb-Sr date previously measured by Faure and Kovach (1969) on similar rocks, or the Rb-Sr and K-Ar dates of about

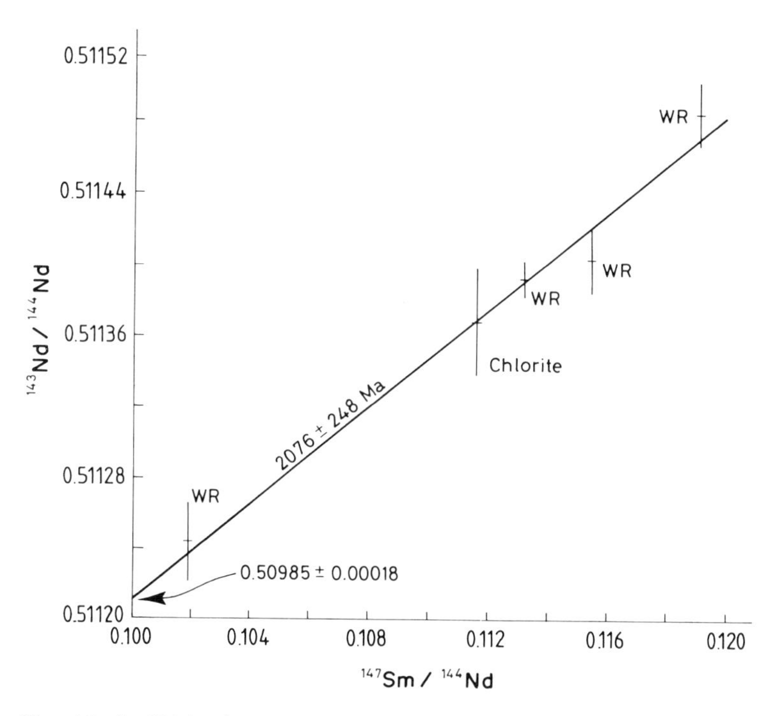

Fig. 4.2. Sm-Nd isochron data of four whole-rock samples and one clay fraction of the Proterozoic Gunflint Iron Formation. (Stille and Clauer 1986)

1.5 Ga measured by Stille and Clauer on the samples used in the Sm-Nd study. Stille and Clauer suggested that the whole-rock Sm-Nd date corresponded to the time of deposition because the date they obtained was close to the age of igneous intrusions within the analyzed sedimentary sequence, and that the low Rb-Sr and K-Ar isotopic dates represented a late diagenetic recrystallization event. Nägler et al. (1992) reported a Sm-Nd isotopic date of 1.52 Ga, based on the analyses of whole-rock samples, for early Ordovician pelites in the Iberian Massif in Spain. The date is clearly meaningless and the linear trend in the Sm-Nd isochron is definitely a mixing line.

In the late 1960s, the use of acid leached whole-rock samples of sediments for Rb-Sr analyses appeared to have considerable advantages, considering that leaching can remove the most mobile Rb and Sr fractions from the sediments leaving the clay fractions with only the most difficultly exchangeable components. Bofinger et al. (1968) compared a Rb-Sr isochron of untreated rocks to that of corresponding acid-leached rocks. The acid leaching removed Sr and little Rb, which increased the Rb/Sr and also the Sr isotopic ratios. The resultant effect was a reduction in the degree of scatter among the data points along the line shown by a smaller error of the calculated slope and a smaller MSWD factor (Fig. 4.3). The isotopic date extracted from slope of the line approximated the time of sedimentation. The authors claimed that the colinearity of the data points reflects an isotopic homogeneity of the sediments during early burial or low-grade metamorphism shortly after the deposition. Chaudhuri and Brookins (1969) also found that acid leaching improved the degree of colinearity of data points for a Rb-Sr isochron of whole-rock samples of Permian shales in Kansas, but the Rb-Sr date from acid-leached samples still exceeded the time of deposition of the sediments.

Cordani et al. (1978) presented several models to explain the colinearity of whole-rock data points in a Rb-Sr isochron diagram. In their isochron model, they assumed that a thorough mixing of the suspended particles after erosion and transportation results in a near Sr isotopic homogeneity at the time of deposition of the sediments. In fact, what Cordani et al. are claiming is that this mixing of particles corresponds to a bulk "Sr isotopic uniformity", and certainly not to a "Sr isotopic homogeneity", which strictly refers to both inter- and intragranular Sr uniformity. Clauer (1982b) examined in details the potentials of Rb-Sr dating of whole rocks and showed that the basic assumption of Cordani et al. rests on unsubstantiated ground. We maintain that the colinearities of whole-rock data points in isochron diagrams represent mixing lines, and not isochrons resulting from isotopic homogeneities. The agreement between a whole-rock isotopic date of a sedimentary unit and its stratigraphic age is in most cases fortuitous and can only be expected in dating sediments which derived directly from young underlying magmatic rocks, as Sr uniformities in these cases are probably realized. The data in the existing literature clearly suggest that very little

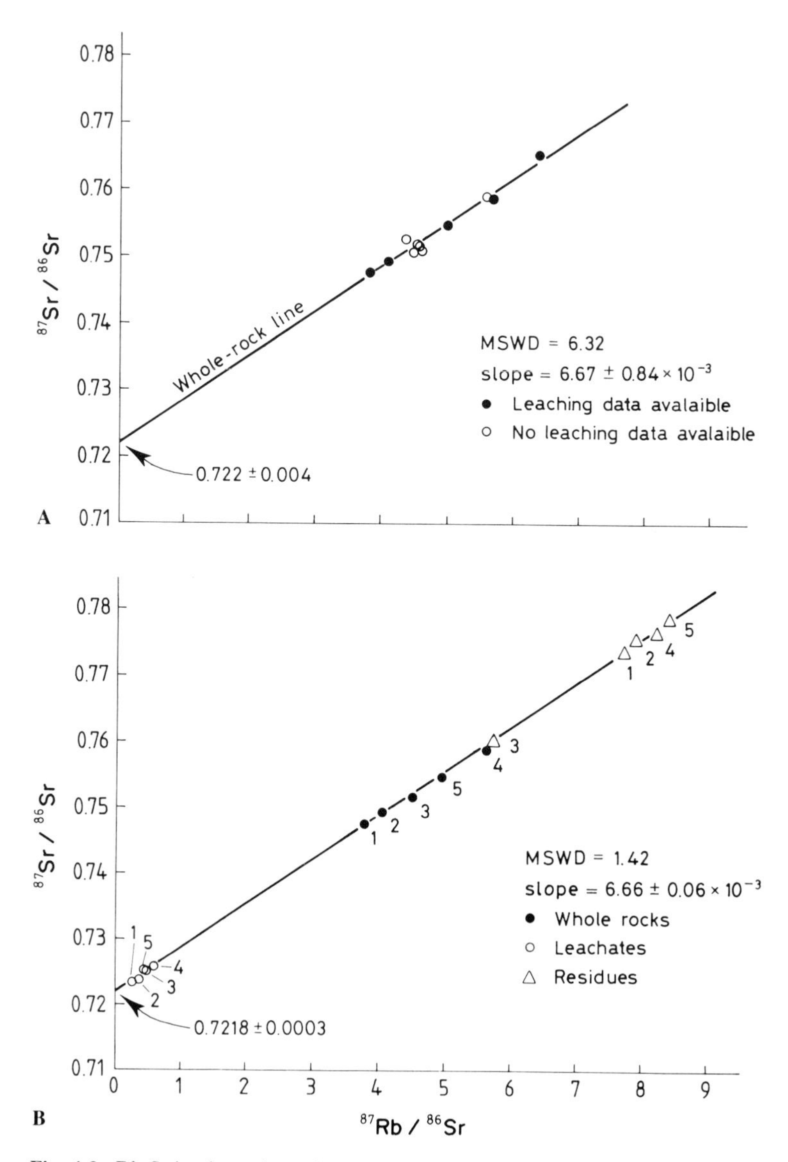

Fig. 4.3. Rb-Sr isochron data of untreated and HCl-leached whole-rock samples. **A** Data points of the untreated samples; **B** those of the leached, untreated whole-rock samples, and leachates. The *slope of the line* approximates the time of deposition. (Bofinger et al. 1968)

merit can be found in analyzing whole-rock samples for determining the time of sedimentation. Although the whole-rock analyses are almost useless for determining the time of sedimentation, such data could be potentially useful for some other purposes, such as reconstructions of crustal erosional histories and regional sedimentation patterns.

1.2 Dating Sedimentation Times by Analyses of Clay Fractions

Potassium-rich clay minerals in sedimentary rocks have been analyzed by many using different common isotopic methods to define the times of deposition of the rocks in question. Glauconite minerals, because of their commonly syndepositional origin in a sedimentary rock, are theoretically ideal minerals, when available, for determination of sedimentation times; but their isotopic records about the depositional time can be dubious, because, as other clay minerals, they may experience varied degrees of post-depositional isotopic exchanges with ambient fluids. Knowledge of those glauconites that have experienced no post-depositional chemical exchanges is necessary for any success in using the isotopic data from glauconite minerals to determine the time of deposition. In contrast to the isotopic records of glauconite minerals, those of common clay minerals such as illite and illite/smectite mixed-layer minerals in a sedimentary rock are often more complex, primarily due to varied origins.

Tracing any syndepositional record from an assemblage of clay minerals in a sedimentary rock hence requires that the interpretation of the isotopic data about the time of deposition of the sediment must be carefully constrained by detailed information obtained independently about the genetic history of the analyzed clay materials. Part of this information is provided by the knowledge of the secular isotopic variations of sea water. A brief outline of the secular variations of some isotopes in sea waters is given below, before presenting the isotopic aspects of clay formation in sedimentary environments.

1.2.1 Isotopic Secular Variation of Sea Water

In global sea water, the compositions of different isotopes that are commonly investigated in studies of clay minerals, are influenced essentially by fluxes from continental run-off and run-out, hydrothermal systems along plate boundaries in oceanic environments, and recrystallization of ocean floor sediments. Hence secular variations for many of these isotopic compositions for sea water result from relative variations in the different fluxes. Variations in continental flux induced by factors such as changes in climate, average elevation and orogenic activities, and to a smaller degree by changes in the total perimeter of the continental margins, as well as changes in the intensity of the submarine alteration of the oceanic crust during the different geologic epochs, have caused the time-dependent variations of isotopic compositions of some elements in sea water.

1.2.1.1 Sr Isotopic Variations

The trend of variation in the $^{87}Sr/^{86}Sr$ ratios with time for sea water is characterized by a major shift toward an increase in the value from 0.701 in the Archean to 0.703 at the beginning of the Proterozoic time at about 2.5 Ga, suggesting increasing influx of radiogenic ^{87}Sr from newly evolved continents and decreasing influx of mantle Sr to the oceans (Veizer 1989, 1992). The same ^{87}Sr/^{86}Sr ratio increased broadly from 0.703 during early Proterozoic to 0.709 at the end of the Proterozoic. Details about variations in the ^{87}Sr/^{86}Sr sea water ratio during the Proterozoic suffer from lack of data with good stratigraphic control. However, the ^{87}Sr/^{86}Sr data for Phanerozoic sea water are stratigraphically well constrained and indicate a progressive decrease from 0.7091 to 0.7073 during the Paleozoic and, following this, a progressive increase from 0.7073 to 0.7089 through the Mesozoic and the Cenozoic. Many prominent short-period fluctuations in the ^{87}Sr/^{86}Sr ratios during the Phanerozoic have been noted, reflecting largely increased influence of the continental Sr inputs or increased influence of the mantle Sr inputs from oceanic hydrothermal activities.

Peterman et al. (1970) related the shifts in the ^{87}Sr/^{86}Sr ratio to variations in the inputs between Sr from old sialic materials with high ^{87}Sr/^{86}Sr ratios and that from young volcanic materials with low ^{87}Sr/^{86}Sr ratios. Veizer and Compston (1974) also made a similar suggestion. Brass (1976) claimed that the varied input of continental Sr to the ocean is due to variations in the relative amounts between Sr derived from older rocks and that derived from younger rocks in connection with the processes of continental destruction and accretion. Clauer (1976) suggested that sea floor spreading and opening of ocean basins at different times influenced the well-defined low ^{87}Sr/^{86}Sr values for sea waters during the Phanerozoic and that the high Sr values resulted from weathering of older continental materials involved in orogenic activities. Spooner (1976) suggested that the Sr derived from hydrothermal activities at ocean ridges is the major cause for the changes in the isotopic composition of Sr in sea water through time. Although the idea of Spooner was seriously questioned by Brass and Turekian (1977), Sr isotope data of Albarède et al. (1981) and Clauer and Olafsson (1981) on hydrothermal fluids at ocean ridges and of Clauer (1981b) on radiolarians in proximity to ocean volcanic activities support Spooner's position. Readers are referred to Veizer (1989, 1992) for additional details on the secular variation of Sr isotopes.

Although the Sr isotopic curve published by Burke et al. (1982; Fig. 4.4) has been widely used to express the nature of the time-dependent variation in the isotopic composition of Sr in sea water, the precise nature of this variation is still a matter of debate. The flux and the isotopic composition of Sr from continent to the world ocean is still far from being entirely resolved. The suspended loads of rivers as they enter the oceans can be significant sources of Sr delivered into sea water (Chaudhuri and Clauer 1992a), the impact of such a continental flux on the Sr isotopic composition of global sea water being far from negligible (Nägler et al. 1993; Nägler et al. in progress) and having yet to be determined precisely.

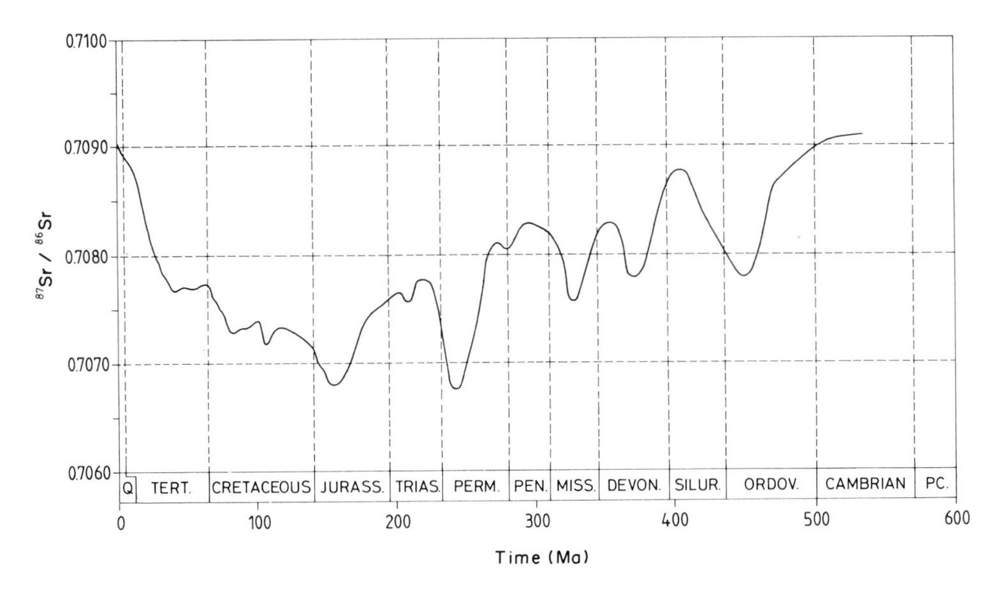

Fig. 4.4. Time-dependent variation of the $^{87}Sr/^{86}Sr$ ratio for ancient sea waters. (Burke et al. 1982)

1.2.1.2 Nd Isotopic Variations

The trend of secular variation of Nd isotopes in sea water is at present poorly known for several reasons. Isotopic homogenization has not been attained between the water masses of different oceans. Therefore, only samples representative of limited oceanic regions might provide information about the nature of the secular variation within a limited time scale for the particular oceanic region (Stille et al. 1992). Metalliferous sediments, Fe-Mn ores, carbonate fossils, marine phosphates, glauconites, and other authigenic clay minerals have been analyzed for information about the Nd isotopic variation in ancient ocean waters; but the interpretation of the data obtained from the authigenic minerals is not free of controversy, because some minerals might have acquired a strong signature from detrital precursors. Furthermore, sea waters across the depth can be potentially varied in their Nd isotopic composition, causing the Nd isotopic signatures of fossils of deep sea origin to be suspect.

The Nd secular isotopic variation in Precambrian sea water is constrained by only a very small data base which is supported mainly by information from banded iron formations (BIF) (Miller and O'Nions 1984; Jacobsen and Pimentel-Klose 1988a, b). These sediments presumably inherited the Nd isotopic composition of sea water at the time of their formation. The range of variation in Nd isotopic values for an individual BIF deposit was as much as 6 ε_{Nd}-units which is about the range of variation in present-day sea water in a single oceanic basin. A trend of increasing ε_{Nd} values with age indicates that the early Precambrian sea water was dominated by a much larger mantle component due to hydrothermal activity than the late Precambrian or Phanerozoic sea water. This conclusion is similar to the one based on the Sr isotope data (Veizer et al. 1983).

The Paleozoic sea water was varied in Nd isotopic compositions with ε_{Nd} values ranging from −3 to −20 (Fig. 4.5). Based on Nd isotopic analyses on brachiopods and conodonts, Keto and Jacobsen (1987) inferred that two separate ocean basins existed between North America and Baltica in the time span between 550 and 400 Ma and that they were separated by an island and/or a shoal circulation barrier. The Iapetus Ocean, which was the smaller of the two, had ε_{Nd} values ranging between −5 and −9 and the coeval Pacific-Panthalassa Ocean had ε_{Nd} values between −10 to −20. In the Late Ordovician to Silurian time (400–440 Ma), the Iapetus ocean was closed by the Taconic orogeny in North America. At the end of the Paleozoic, 300 to 250 Ma ago, the Nd isotopic composition of sea water sharply increased toward more radiogenic values (ε_{Nd} from −3 to −5), pointing to increased influence of volcanic activities and new generations of subduction zones around the Pacific/Panthalassa Ocean.

The Mesozoic and Cenozoic time span comprised different evolutionary trends of Nd isotopic compositions in the Pacific Ocean, the Indian-Tethys-

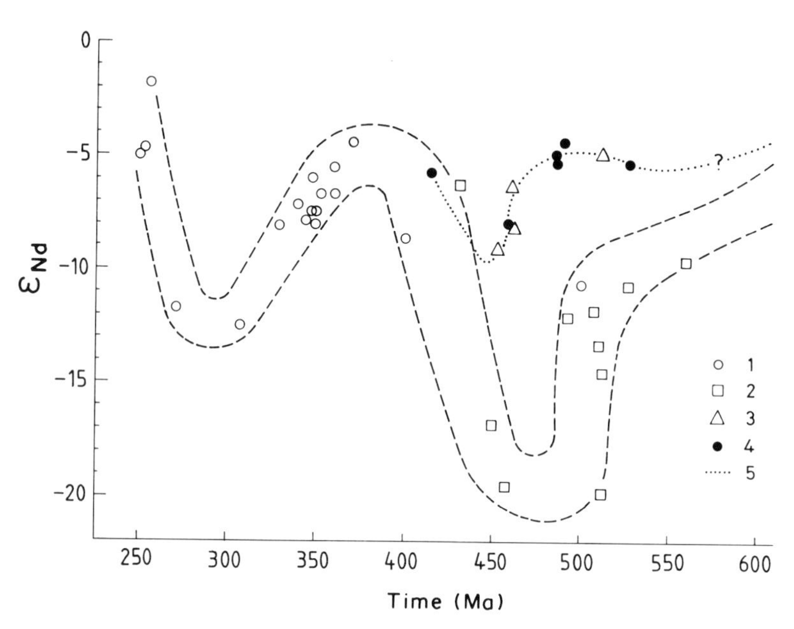

Fig. 4.5. Secular variation of the $^{143}Nd/^{144}Nd$ ratio of Paleozoic sea water. *1* and *2* North American samples; *3* samples from the southeastern United States; *4* European samples. The *dashed line* represents the trend in the Nd evolution of the Pacific-Panthalassa Ocean and the *fine dotted line* that of the Iapetus Ocean. (Stille et al. 1992)

Mediterranean Sea and in the Atlantic Ocean (Fig. 4.6). The ε_{Nd} values for the Pacific Ocean increased from −10 to −7 some 220 Ma ago and to −1 and −4 at the present day. The early Tethys sea water had Nd isotopic signatures identical to that of the Pacific Ocean, 200 to 180 Ma ago, but the influence of the continental runoff became more important, as indicated by a decrease in the ε_{Nd} values. No data are available for the early Atlantic Ocean, but about 100 Ma ago, a broad seaway existed between the Tethys and the north Atlantic Ocean, suggesting that the sea waters of both oceans had the same Nd isotopic composition. In the time span between 70 Ma and the present, the Tethys and the Atlantic sea waters evolved isotopically in different ways, probably caused by the slowly closing of the Tethys seaway.

1.2.1.3 Oxygen Isotopic Variations

The records for the secular variation of the oxygen isotope composition of sea water come from the analyses of carbonate rocks, cherts, and phosphorites of marine origin. These varied data all indicate that the $\delta^{18}O$ values decrease, although somewhat irregularly, as the ages of the analyzed minerals increase (Weber 1965; Perry and Tan 1972; Knauth and Lowe 1978; Shemesh et al. 1983; Veizer et al. 1986). The $\delta^{18}O$ value of sea water appears to have decreased at a rate of about 5 per mill per billion years, and three different explanations were offered for this trend in sedimentary rocks. One view holds that the decrease in the isotope composition with increase in the age of the rock is due to continuous post-depositional exchange with isotopically light meteoric water and hence the records in the rocks do not reflect the isotopic composition of the sea (Degens and Epstein 1962; Keith and Weber 1964). Two other views maintain that the isotope variations are reflections of the conditions of sea waters at the time of deposition of the different marine sediments, because these sediments have very different degrees of susceptibility to alteration by meteoric or groundwater and yet have the same general rate of decrease of approximately 5 per mill per billion years. These two views nonetheless are different, because one considers that the ocean water had a nearly constant $\delta^{18}O$ value and the trend represents precipitation of the marine sediments from sea waters whose temperatures, beginning as high as about 60 °C during the early Precambrian time, gradually decreased as the time decreased (Knauth and Epstein 1976; Knauth and Lowe 1978), and the other supports that the temperature remained nearly constant and the trend of increase in the isotopic value with decrease in the age is due to the continuous cycling of water through the mantle (Perry and Tan 1972). Focusing on the oxygen and hydrogen isotope compositions of granites of different ages, Taylor (1977) independently concluded that the ocean water isotope composition remained .generally uniform at least since 2.5 Ga ago. Muehlenbachs and Clayton (1976), Gregory and Taylor (1981) and Smith et al. (1984) advocated for an invariant

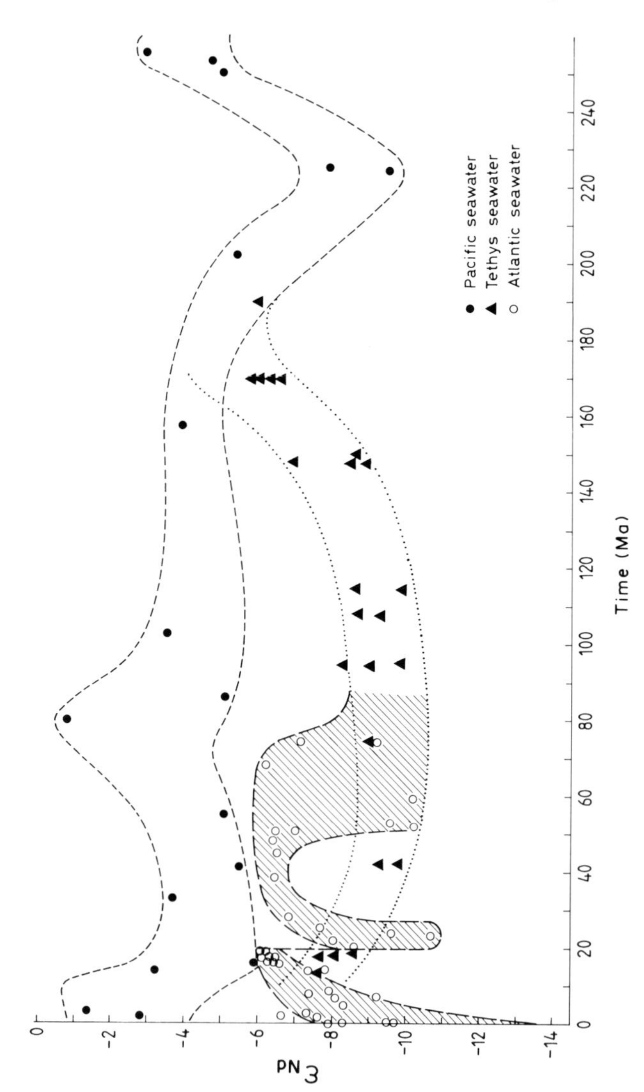

Fig. 4.6. Secular variation of the $^{143}Nd/^{144}Nd$ ratio of Mesozoic and Cenozoic sea water of different oceans. (Stille et al. 1992)

oxygen isotope composition for sea water through geologic time, in maintaining that the buffering of the sea water oxygen isotope composition is achieved by the two competing processes between ^{18}O depletion resulting from low-temperature submarine weathering and ^{18}O enrichment resulting from high-temperature hydrothermal activities. For additional discussions on this subject the readers are referred to the works by Muehlenbachs (1986), Veizer et al. (1986), Wadleigh and Veizer (1992), and Knauth (1992).

1.2.2 Gauconites and the Sedimentation Time

Isotopic investigations of glauconite minerals began with the K-Ar isotopic work by Wasserburg et al. (1956) and the Rb-Sr isotopic study by Cormier (1956), both reporting that isotopic dates of glauconites were lower than the time of sedimentation. Subsequent works during the late 1950s and early 1960s, which also found that many isotopic dates for glauconites are low as compared to stratigraphic ages of the corresponding sediments, explored the causes for these low ages (Amirkhanov et al. 1958; Herzog et al. 1958; Lipson 1958; Goldich et al. 1959; Hurley et al. 1960; Evernden et al. 1961; Polevaya et al. 1961). More recent works not only tried to determine the reasons why many of these minerals were prone to yield anomalous ages, but also attempted to establish mineralogical and chemical criteria which could serve as guides for discriminating between dates that are suspect and those that are reliable concerning the time of sedimentation (e.g., Odin 1975; Odin and Matter 1981; Harris and Fullagar 1989; Clauer et al. 1992a).

Although some examples exist showing agreement between isotopic dates of glauconites and stratigraphic ages of associated sediments, the K-Ar and the Rb-Sr isotopic dates of glauconites in many instances are 10 to 20% lower than the depositional age of the sediments containing these minerals (Keppens and Pasteels 1982). The low isotopic dates have been variously explained, attributing them to weathering (Keppens and Pasteel 1982), deep-burial diagenetic and low-grade metamorphic recrystallization (Bonhomme et al. 1969; Keppens and Pasteel 1982), emergence above sea level (Morton and Long 1980), and partial loss of radiogenic isotopes from "open" K sites in the expandable interlayer segment (Thompson and Hower 1973). In a few instances, the isotopic dates of glauconites were higher than the depositional ages, and these older dates have been commonly ascribed to the inheritance of isotopic memories of precursors by the glauconites, having found no indication that the anomalies in the dates could arise from reworking of older glauconitic minerals. The "anomalous" glauconites generally contain relatively low amounts of K_2O, commonly less than 6%, and could be described as nascent or less evolved glauconites, as Odin and Matter (1981) suggested. According to these authors, those glauconites with more than 6% K_2O, which indicates that they are evolved enough to have the isotopic memories of precursors obliterated, are likely to yield depositional ages of the host sediments.

The effect of the precursor minerals and the significance of the K_2O content on the isotopic dates of glauconites relative to their stratigraphic ages have been recognized in studies by Ghosh (1972), Adams (1975), Clauer (1976) and Tisserant and Odin (1979). Analyzing late Miocene glauconites from Morocco and also a sample of late Miocene glauconite from Algeria, Tisserant and Odin found that the K-Ar isotopic dates ranged between 8 and 37 Ma and that the dates decreased as K contents increased (Fig. 4.7), the trend of which was very similar to that observed by Odin et al. (1979) for the Recent glauconies in the Gulf of Guinea. The results suggested decreased influence of radiogenic ^{40}Ar from the precursors as a function of the degree of glauconitization. The most probable age of deposition of the sediments was deduced by the authors from the asymptote of the curve indicating a relationship of decrease in the dates with increase in the K contents.

Ghosh (1972), studying glauconites of the Eocene Weeches Formation in Texas, also noted that K-Ar dates decreased as the K contents increased

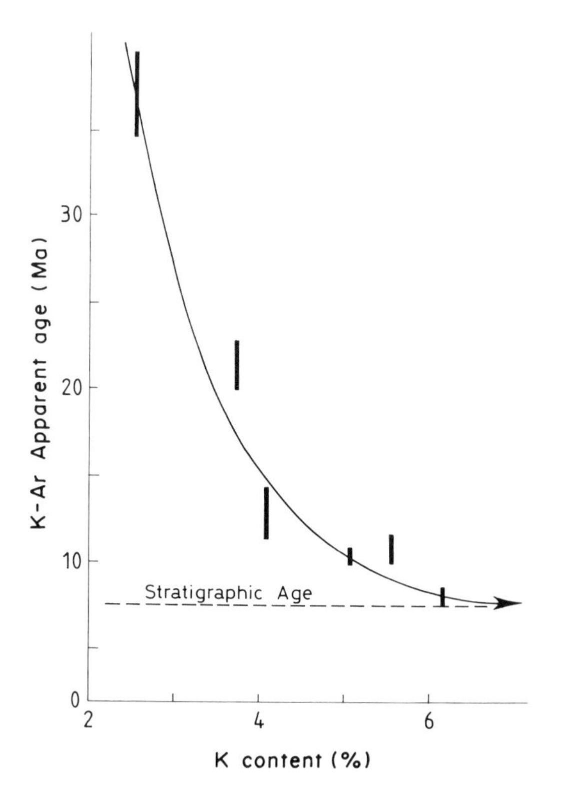

Fig. 4.7. Relationship between K-Ar dates and K_2O contents of Tertiary glauconites from Morocco and Algeria. (Tisserant and Odin 1979)

along a curvilinear trend. Adams (1975), analyzing some early Eocene glauconites, reported that glauconites with less than 2% K_2O yielded the highest K-Ar date of about 130 Ma, whereas the glauconite with more than 4% K_2O gave the lowest K-Ar date of about 30 Ma. The results of the study by Adams conveyed the same relationship between the K-Ar dates and the K contents of glauconites as that previously noted by Ghosh (1972) in his study of glauconites from Texas, and by Tisserant and Odin (1979) in their study of Moroccan and Algerian glauconites.

Analyses of the Rb-Sr system can also significantly increase understanding of both the timing and the process of glauconitization. Clauer (1976) analyzed some glauconites and associated clay minerals from Cretaceous and Tertiary rocks of the Paris and London basins. He found that the Rb-Sr dates of those glauconites with more than 4% K_2O and with initial Sr isotopic ratios identical to the Sr isotopic ratio of contemporaneous sea water were in accord with the stratigraphic ages. By contrast, the isotopic dates of associated clay minerals were significantly higher than those of the glauconites. He also observed that the isotopic dates of glauconies with less than 4% K_2O corresponded to the stratigraphic ages when the glauconies were considered to have had an initial Sr isotopic composition identical to that of the contemporaneous clay minerals dispersed in the same rocks. The initial Sr isotopic ratios of the glauconies had then to be significantly higher than the Sr isotopic ratio of contemporaneous sea water whose Sr isotopic signature was determined from analyses of associated carbonate minerals (Fig. 4.8). Based on this Rb-Sr evidence, Clauer argued for two distinct modes of glauconitization process, a "transformation mode" and a "neoformation mode". His suggestion was later reaffirmed by Bonnot-Courtois (1981), who studied the REE contents of the same aliquots of glauconites.

Odin and Hunziker (1982) presented very convincing evidence that isotopic analyses of very carefully selected glauconite minerals can provide reliable dates corresponding to the depositional age of the host sediments. The authors analyzed both the K-Ar and the Rb-Sr systems of high K-bearing glauconite fractions from Albian and Cenomanian units in Normandy, France. The K-Ar isotopic analyses of the Albian glauconites yielded a date of 98.6 ± 2.1 Ma and those of the Cenomanian glauconites gave a date of 93.0 ± 1.4 Ma. The Cenomanian glauconites also yielded a Rb-Sr date of 93.5 ± 1.6 Ma. Both the Rb-Sr and the K-Ar dates of the Cenomanian glauconites agreed well with the age of sedimentation.

Harris and Fullagar (1989) also demonstrated that carefully selected glauconite samples can yield reliable information about the time of deposition of sediments. Selecting only those glauconites that had K_2O contents of more than 5.6% and shallow burial history, Harris and Fullagar found agreement between the K-Ar and the Rb-Sr dates of glauconite samples from middle Eocene Castle Hayne Limestone in the southeastern Atlantic Coastal Plain of the USA. The Rb-Sr isochron dates were 45.3 ± 0.3 Ma

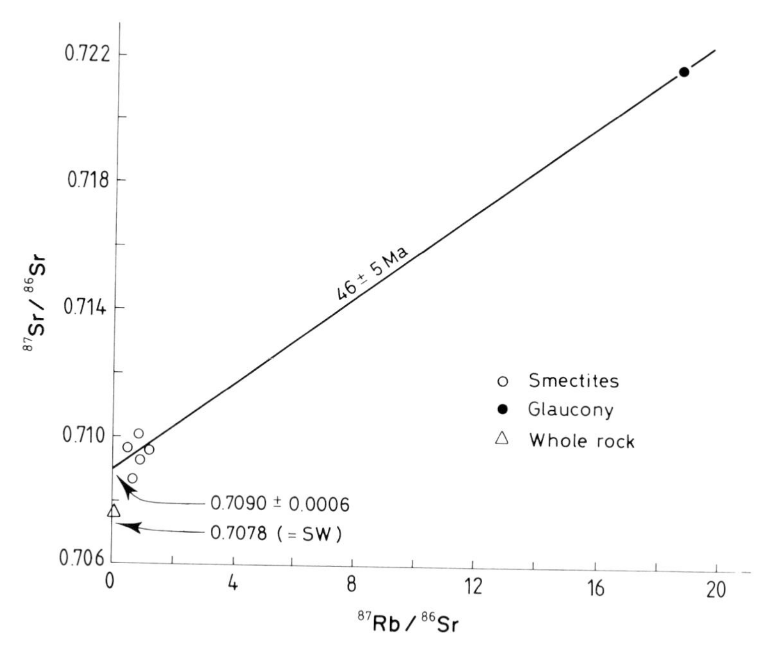

Fig. 4.8. Rb-Sr isochron data of Middle Eocene smectite and glaucony samples from Paris Basin. The *slope* yields a date identical to the stratigraphic age with an initial $^{87}Sr/^{86}Sr$ ratio different from that of the contemporaneous sea water given by the carbonate whole-rock sample. (Clauer 1976)

and 43.1 ± 1.2 Ma with initial Sr isotope values of 0.70764 ± 0.00015 and 0.70790 ± 0.00011, respectively (Fig. 4.9). The Rb-Sr and the K-Ar dates corresponded closely to the time of deposition of the sediments. The initial Sr isotopic compositions of the glauconite samples were very similar to the Sr isotopic compositions of contemporaneous sea waters.

The studies discussed above indicate that the depositional ages of sediments may be defined by isotopic dates obtained on glauconites in the sediments, provided these minerals are highly evolved in terms of their K contents and have had simple burial history. The retention of the isotopic memory of the precursors by glauconites is often a major cause of the failure of the minerals to indicate the time of deposition of sediments. Glauconies with less than 6.5% K_2O often yield K-Ar or Rb-Sr isotopic dates in excess of the stratigraphic ages, whereas those with more K_2O give dates that correspond very closely to the time of sedimentation, signifying that at the advanced evolutionary stages of glauconies, the residual memory of the precursor is obliterated within a very short time period, probably within several tens of thousand years after the deposition of the sediments. Clauer (1976) claimed bimodal evolutionary steps for many glauconites: a transformation growth for those glauconies with less than 4% K_2O and whose isotope compositions can still be traced to their precursors, and a neofor-

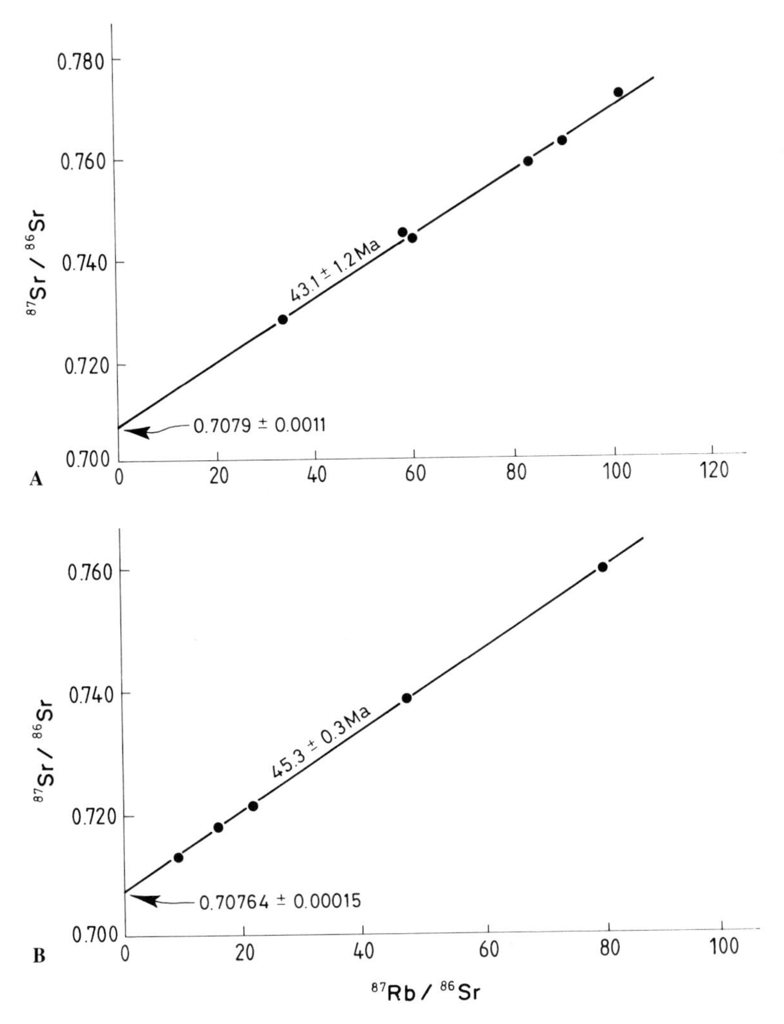

Fig. 4.9. Rb-Sr isotopic data of glauconites from Castle Hayne Limestone of Middle Eocene age. The samples of **A** were taken 30–35 cm below a bentonite unit, and those of **B** 25–30 cm above the same bentonite unit. (Harris and Fullagar 1989)

mation growth for those glauconites with more K_2O and whose isotopic compositions evolved to the point becoming untraceable to associated clays. The recent studies of Clauer et al. (1992a, b) and Stille and Clauer (1994) on Recent glauconies and glauconites have given further insight into the dynamics of the growth of these minerals and have emphasized that the two independent processes suggested by Clauer (1976) to explain glaucontization are in fact two sequential steps of the same process. These studies also presented a rational basis for Odin's (1975) conclusion that only those glauconites whose K contents are high are suitable for stratigraphic dating.

1.2.3 Clay Fractions in Shales and the Sedimentation Time

Clay fractions in shales are often enriched in smectites, illite/smectite mixed-layer minerals and, to a less extent, in illites. The common occurrence of these minerals made them an object of isotopic analyses for a number of studies that have been concerned with depositional and post-depositional histories of sedimentary rocks. Many inherent complexities in the history of these minerals prevent a simple straightforward interpretation of the isotopic data for clay fractions separated from a sample of sedimentary rock (Clauer 1979b).

1.2.3.1 K-Ar Data

Evernden et al. (1961), the first to report K-Ar dates of illite-rich clay fractions, noted that the isotopic dates decreased with decreases in size of the clay fractions and that the K-Ar dates of the fine fractions were lower than the stratigraphic ages of the sediments. To understand the origin of clay minerals in ancient sedimentary rocks, Bailey et al. (1962) determined the K-Ar ages of sedimentary illites in several cyclothemic Pennsylvanian shales and underclays from the midcontinent of the USA. Achieving the separation of coarse $2M_1$-rich fraction from fine $1M_d$-rich fraction, the authors found that the K-Ar dates of $2M_1$ illites were higher by as much as 75 million years than the depositional ages (280–310 Ma) of the sediments, but the K-Ar dates of $1M_d$-rich fine clay fractions were as much as 190 million years lower than the depositional ages. The authors attributed the low dates for the fine clay fractions to a number of causes such as preferential Ar loss because of the small sizes of the particles, reorganization and recrystallization with attendant K-fixation in the process of transformation of $1M_d$ to $2M_1$ illite minerals, and reconstitution of degraded micaceous minerals long after the deposition of the sediments during the Pennsylvanian time. Evidence gathered from subsequent studies does not fully corroborate these explanations, because 2M micaceous polytypes always originate at high temperature and are of detrital origin in most sedimentary environments, whereas the $1M_d$ illite polytypes are of authigenic origin and can form at various times, even long after the deposition of the sediments.

Like Evernden et al. (1961) and Bailey et al. (1962), Hower et al. (1963) also observed that K-Ar dates decreased as sizes of the clay particles decreased, and that the age of the coarsest fraction was greater, whereas that of the finest fraction was smaller than the stratigraphic age. The trend in decreases of K-Ar isotopic dates with decreases in the particle size was also observed by Hofmann et al. (1974) from analyses of samples of underclay from a Pennsylvanian cyclothemic sequence in Ohio (Fig. 4.10), but the authors found that even the very fine fraction ($<0.2\,\mu m$) gave K-Ar isotopic dates which were higher than the depositional age of the sediments.

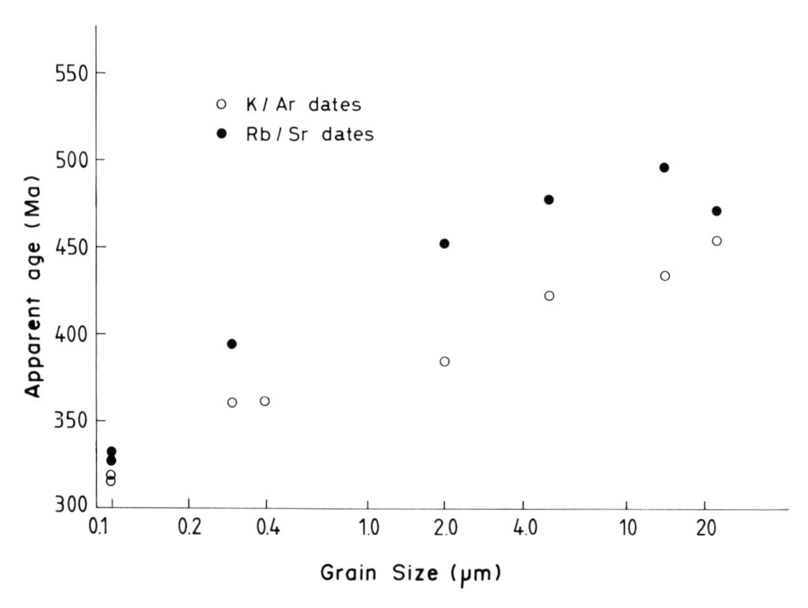

Fig. 4.10. Relationship between K-Ar dates and grain sizes of Pennsylvanian underclays. (Hofmann et al. 1974)

1.2.3.2 Rb-Sr Data

An assumption that clay minerals can undergo varied amounts of isotopic exchange with pore fluids soon after the deposition of sediments prompted many investigators to analyze the Rb-Sr isotopic compositions of clay fractions in sedimentary rocks for the purpose of determining ages of deposition. Bonhomme et al. (1966a) suggested that the authigenic clay minerals separated from shales may yield information about the time of sedimentation or diagenesis. Analyzing Rb-Sr isotopic compositions of clay fractions in some Cambrian shales, Bonhomme et al. (1966b) obtained a Rb-Sr isochron date of 373 ± 12 Ma which was clearly much lower than the sedimentation age. Bonhomme et al. (1968) also reported that clay fractions of shales in a Siluro-Ordovician sequence yielded a Rb-Sr isochron date of 308 ± 19 Ma which was much lower than the time of deposition. In contrast to the well-defined isochrons of clay fractions yielding ages which were lower than the age of deposition, the whole-rock isochrons had a high degree of scatter and yielded dates that were higher than the sedimentation time (Fig. 4.11).

Cordani et al. (1978) also reported that Rb-Sr isochron dates of clay fractions in several shales of different ages were lower than the times of sedimentation. Analyzing very fine clay fractions (<0.2 μm) separated from Upper Devonian shales at different localities in Texas, Morton (1985b) obtained a well-defined colinear trend among the data points in a Rb-Sr

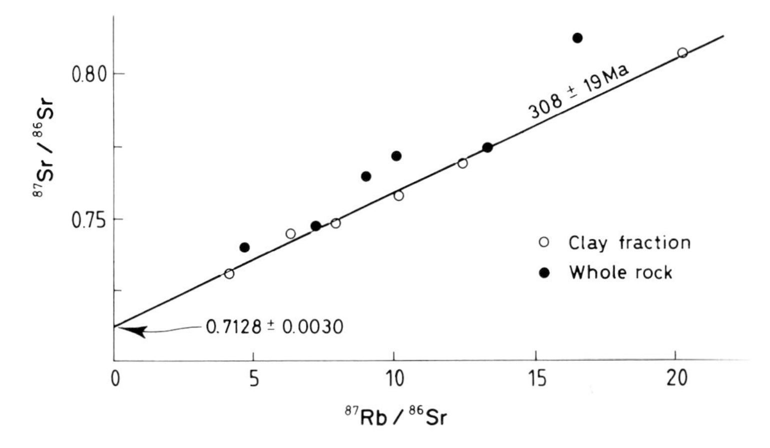

Fig. 4.11. Rb-Sr isochron data of whole-rock samples and <2 μm clay fractions of Siluro-Ordovician shales. (Bonhomme et al. 1968)

isochron diagram. This linear trend corresponded to a date of about 302 Ma, which was lower than the stratigraphic age. Clauer and Bonhomme (1970a) reported that data points for clay fractions of Ordovician shales and Precambrian slates and marls of the Vosges mountains in France were highly colinear in a Rb-Sr isochron diagram. The authors attributed the isotopic date of 358 ± 4 Ma, being lower than the time of sedimentation, to a recrystallization episode during the Hercynian tectonic activity.

Examples have also been found which illustrate that Rb-Sr isotopic dates of clay fractions are higher than the stratigraphic ages. Analyzing <2 μm clay fractions of a Permian shale unit at three widely separated localities in Kansas, Chaudhuri (1976) found three separate colinear trends in the Rb-Sr isochron diagram, yielding dates that ranged from 320 to 398 Ma. The dates were higher than the stratigraphic ages (Fig. 4.12). Morton (1985b) also found, based on the Rb-Sr isotopic analyses of 1–2-μm-sized clay fractions of some Upper Devonian black shales in Texas, that the isotopic date exceeded the stratigraphic age by as much as 45%.

Some studies on Rb-Sr isotopic analyses of clay fractions in sedimentary Paleozoic rocks have found that the isotopic dates corresponded to the time of sedimentation. Clauer and Bonhomme (1970b) obtained a Rb-Sr isochron age of about 425 Ma for the <2 μm clay fractions extracted from Ordovician shales of the Vosges mountains, France. Morton (1985a) reported a Rb-Sr isochron date of 23.6 ± 0.8 Ma for clay fractions <0.05 μm separated from shales of Miocene-Oligocene Frio Formation in the Texas Gulf Coast area (Fig. 4.13). Although this Rb-Sr isotopic date for clays in the Gulf Coast sediments was interpreted to be the time of a large-scale diagenesis, this event apparently occurred soon after the deposition of the sediments, which is estimated to be between 25 and 29 Ma.

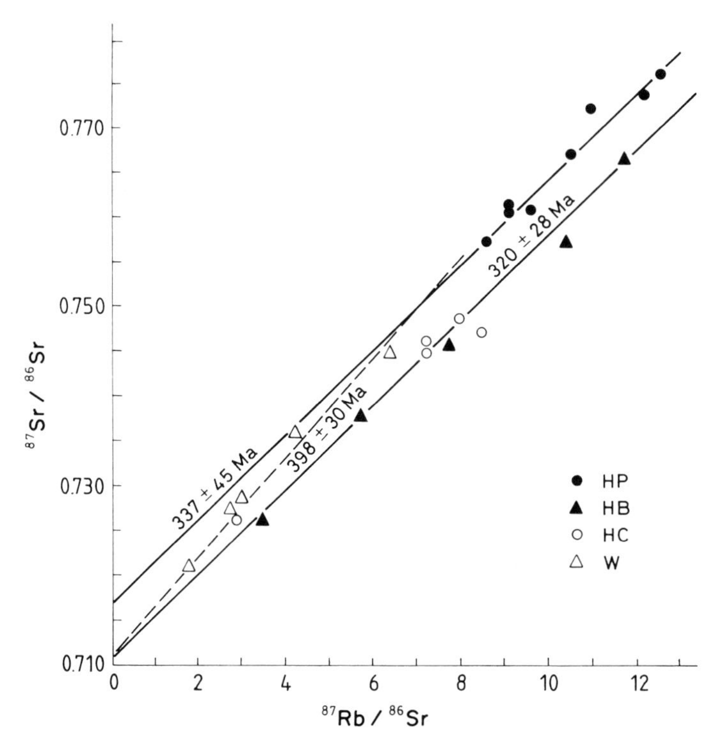

Fig. 4.12. Rb-Sr isochron dates of <2 μm clay fractions in a Lower Permian shale unit from Kansas and northern Oklahoma. The *letters* designate different sample locations. The *dates* exceed the stratigraphic age. (Chaudhuri 1976)

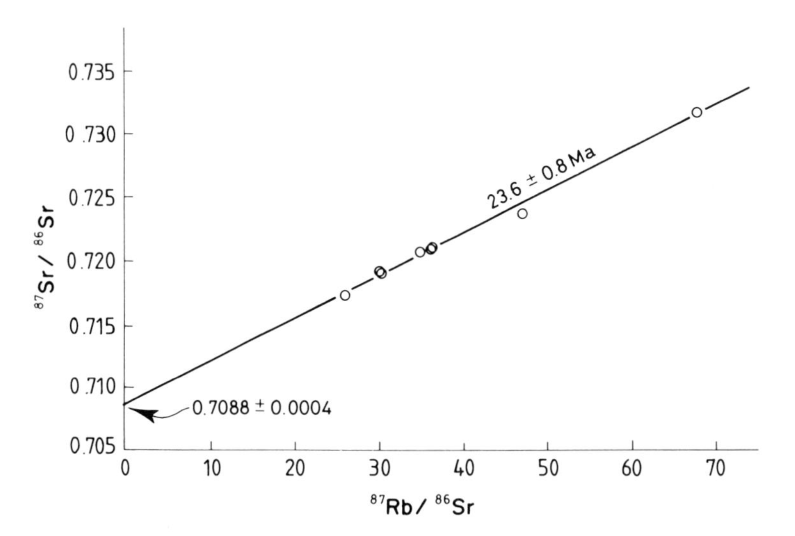

Fig. 4.13. Rb-Sr isochron data of <0.05 μm clay fractions of Miocene-Oligocene shales from the Texas Gulf Coast area. (Morton 1985a)

The complex evolutionary history of clay minerals in many sedimentary rocks poses a major obstacle in resolving the isotopic dates to reasonably good approximations of depositional ages. Simplified clay evolutionary histories seem to exist for clay sediments in carbonate rocks for apparent lack of long-term crystallization or multi-stage recrystallization. Hence the clay minerals in carbonate rocks may yield potentially valuable information concerning the time of deposition of sediments, but available studies are very few. The study of Kralik (1982) gave a Rb-Sr isotopic date of 1537 ± 52 Ma for clay fractions in some Proterozoic carbonate rocks in the McArthur basin in Australia. The author concluded, based much on speculation, that the isotopic date is a reasonably good approximation for the time of sedimentation of the sequence (Fig. 4.14).

The presence of different generations of diagenetic clay minerals in ancient sedimentary rocks may also stand as a major obstacle in establishing the time of sedimentation by isotopic analyses of clay separates. However, Gorokhov et al. (1994) have shown that by analyses of morphologically different clay fractions, the history of the clays in the sediments may be reasonably constructed to define with good approximation the time of deposition. Based on combined evidence of mineralogy, morphology, and chemical and isotopic compositions, Gorokhov et al. established three generations of illite clay minerals in some blue clays of Cambrian-Precambrian age of the East European platform. The earliest generation clay minerals were of 2M type and had a detrital origin with a Rb-Sr isotopic date of about 722 ± 13 Ma. The second generation of clay minerals were lath-shaped and well-

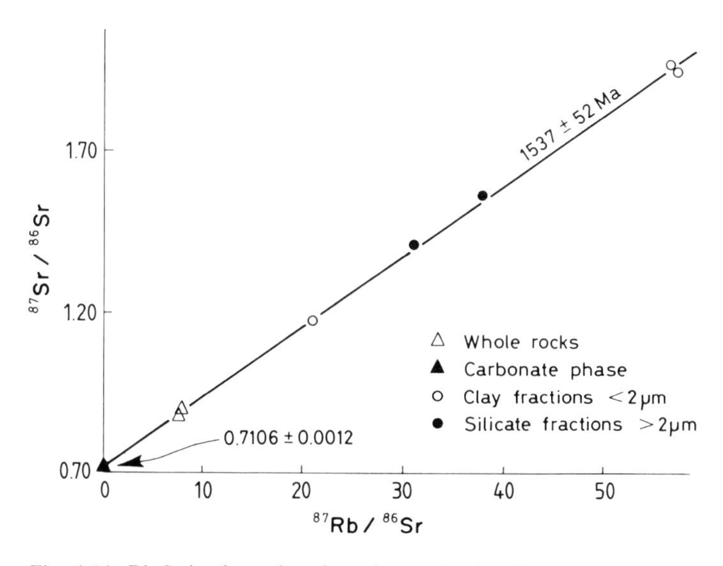

Fig. 4.14. Rb-Sr isochron data for <2 μm clay fractions of Proterozoic carbonate samples in McArthur Basin, Australia. (Kralik 1982)

crystallized, yielding a Rb-Sr date of 533 ± 8 Ma, which is regarded by the authors as an approximate age of the time of deposition of the sediments. A later generation of lath-shaped illite minerals yielded a date between 430 and 480 Ma. An acid leaching of these different generations of clays removed exchangeable Sr with $^{87}Sr/^{86}Sr$ ratios between 0.7120 and 0.7199, suggesting that environmental fluids during these evolutionary episodes were isotopically different.

In a study on Late Precambrian sediments of the Hammamat Series from the northeastern desert of Egypt, Willis et al. (1988) showed that these sediments recorded the uncovering of the evolving crust of this region. Petrographic, geochemical, and isotopic data showed that the lower part of the stratigraphic section consists of coarse detritus shed from nearby ensimatic terranes, while the upper part derived from rapid reworking of bimodal igneous rocks similar to those of volcanics and granites preserved within the region. The Rb-Sr whole-rock analyses of the Hammamat sediments gave an isochron date of 585 ± 15 Ma with an initial $^{87}Sr/^{86}Sr$ ratio of 0.70323 ± 0.00013, which the authors interpreted as being approximately close to the time of sedimentation. The Rb-Sr date of about 585 Ma was also in agreement with the K-Ar data range of 588–567 Ma obtained for the coarse (0.5–2-μm) clay fractions. The authors also leached the 0.2–0.5-μm clay fractions with dilute HCl and observed desorption of varied amounts of Sr with an $^{87}Sr/^{86}Sr$ ratio of 0.7124. The clay residues gave a Rb-Sr isochron age of 524 ± 17 Ma, which is in agreement with K-Ar dates ranging from 542 to 532 Ma. This Rb-Sr date was considered to be the time of a thermal episode. By acid leaching, the authors showed that the environmental Sr of the clay particles was unrelated to the genetic history of the clays and that its adsorption resulted in a mixing line in the isochron diagram (Fig. 4.15). Willis et al. interpreted this adsorption of Sr arising from interaction between the clay particles and genetically unrelated Sr-rich pore fluids, as occurring during or after the 525 Ma thermal episode.

Many Rb-Sr isochrons for clay fractions of Precambrian sedimentary rocks are found in the literature. Because of large uncertainties about the precise stratigraphic ages of many of these rocks, the merits of isotopic dates concerning predictability about the times of deposition remain in doubts. Clauer (1976) and Clauer and Deynoux (1987) investigated the Rb-Sr isotopic compositions of clays in several separate stratigraphic units of the upper Proterozoic sequence in Atar, Mauritania, and found that the order of the isotopic dates, which ranged from 964 ± 32 Ma to 575 ± 43 Ma, followed the relative stratigraphic order of the units (Fig. 4.16). The crystallinity indices and polytypes of the illites suggested that the sediments were not buried beyond the domain of diagenesis. Also, the initial $^{87}Sr/^{86}Sr$ ratios given by the intercepts of the isochrons were close to the isotopic composition of Sr of the depositional environments, as determined from the Sr isotopic composition of the associated carbonates. The authors, therefore, concluded

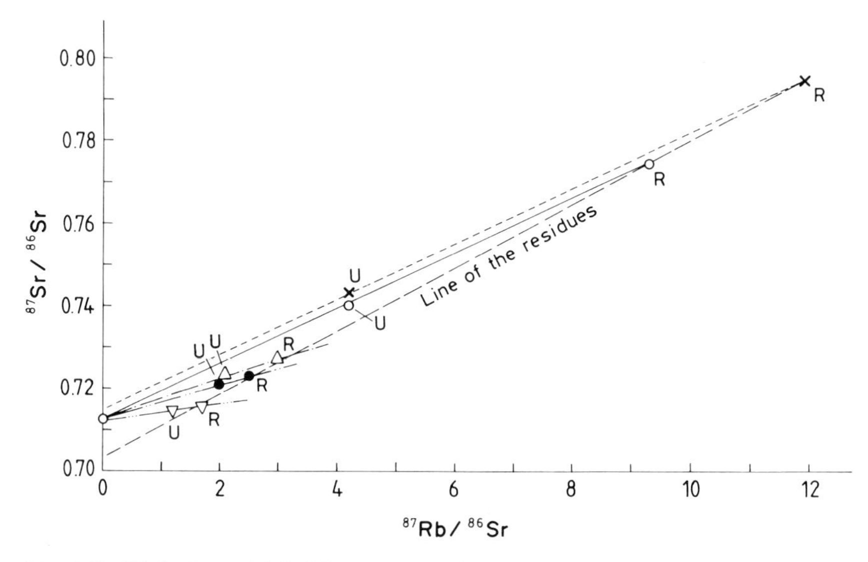

Fig. 4.15. Rb-Sr data of 0.2–0.5-μm untreated and acid-leached clay fractions of sediments from the Proterozoic Hammamat series in Egypt. The data points of *residues* plot on an isochron, whereas those of *untreated fractions* show trends toward an HCl-leachate value suggesting mixing with the soluble Sr. (Willis et al. 1988)

that the Rb-Sr isotopic dates of the clays may be considered as reasonable approximations of the times of sedimentation.

1.2.3.3 Sm-Nd Data

Because of the small decay constant of ^{147}Sm and the low Sm/Nd ratios for sedimentary rocks, the rate of growth of ^{143}Nd is small. Hence Sm-Nd isotopic dates can be measured accurately for those clays which are of Precambrian age. Little is known till now about the Sm/Nd isotopic date of clays and its relation to stratigraphic age of the sediment containing the clays. Stille and Clauer (1986) analyzed the Sm-Nd isotopic composition of a clay fraction of an argillite from Proterozoic Gunflint Formation in Canada. The isotopic data of this clay, in conjunction with the Sm-Nd isotopic data of some whole-rock argillite samples, yielded a date of 2.08 ± 0.56 Ga, which was interpreted by the authors to be approximately the time of sedimentation (Fig. 4.2). The Sm-Nd isotope date contrasted with the K-Ar and the Rb-Sr isotopic dates of about 1.4–1.5 Ga, suggesting that diagenesis or low-grade metamorphic conditions at a later period may have reset the Rb-Sr and the K-Ar systems but not the Sm-Nd one. Additional Sm-Nd isotopic studies of clays from the Gunflint Formation, as well as from other sedimentary rocks, are needed to evaluate the true potential of the Sm-Nd isotopic method in defining stratigraphic age of sedimentary rocks.

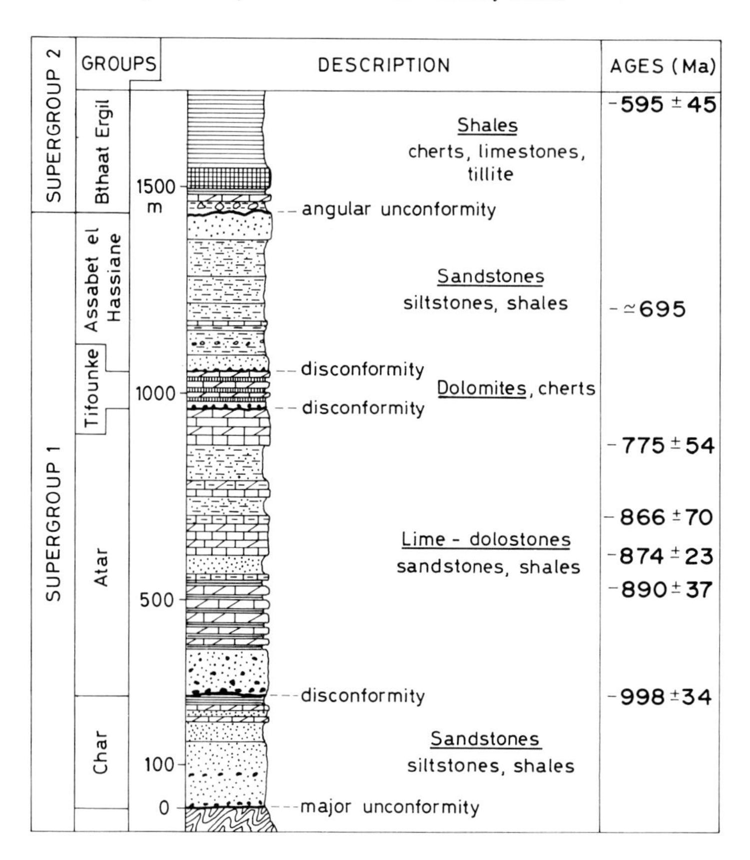

Fig. 4.16. Simplified stratigraphic succession of the sedimentary sequence of Mauritania with the Rb-Sr dates of the <2 μm clay fractions of seven units of the sequence. (Clauer et al. 1982c)

1.2.4 Syndepositional Isotope Evolution of Clay Minerals

Clauer's (1976) studies of two separate lacustrine deposits shed some light on the isotopic evolutions of syndepositional nonglauconitic clay minerals. One of these clay deposits belongs to an Oligocene lacustrine claystone sequence intercalated with carbonate beds occurring in the Hercynian Central Massif of France and the other belongs to a Tertiary-late Cretaceous lacustrine sequence of the Mormoiron basin in southeastern France.

The Oligocene lacustrine claystone sequence is about 250 m thick and is overlain by a volcanic ash bed. Illite and kaolinite are dominant clay minerals in the lower part of the claystone sequence. This illite-kaolinite clay mineral assemblage is succeeded upward in the sequence by an assemblage of illite, illite/smectite mixed-layer, and kaolinite minerals, and further upward by an assemblage of smectite and illite minerals, and then by an assemblage of essentially palygorskite minerals with occasional occurrences of smectite minerals. The entire clay series from illite-kaolinite association at the lower part of the sequence to palygorskite-smectite association at the upper part of the sequence is interpreted to be a prograde evolutive sequence, the deposition of which is followed by the deposition of a volcanic layer at the top of the claystone sequence. The illite in the lower part of the sequence is of 2M type, signifying a detrital origin, but the palygorskite in the upper part of the sequence was considered as being a product of recrystallization of smectite as it was in other lascustrian sediments (Trauth 1977) and in oceanic environments (Tazaki et al. 1986, 1987). Clauer noted that the Rb-Sr isotopic compositions of clays from different parts of the claystone sequence follow a curvilinear trend in a Rb-Sr isochron diagram (Fig. 4.17). This curve may be resolved into two linear segments, one with a positive slope defined by the illite-rich clay fractions, and the other with nearly a zero slope defined by the smectite-rich and the palygorskite-rich clay fractions. The steep part of the curve gives a date of about 102 Ma, which is far above sedimentation time, and an initial $^{87}Sr/^{86}Sr$ ratio of 0.7195 ± 0.0009, which is nearly the same as the isotopic composition ratio of $0.7182 + 0.0010$ for associated carbonate deposits (Clauer and Tardy 1971). The model age of the palygorskite-rich samples is 42 Ma, which is about the same as the age of the immediately

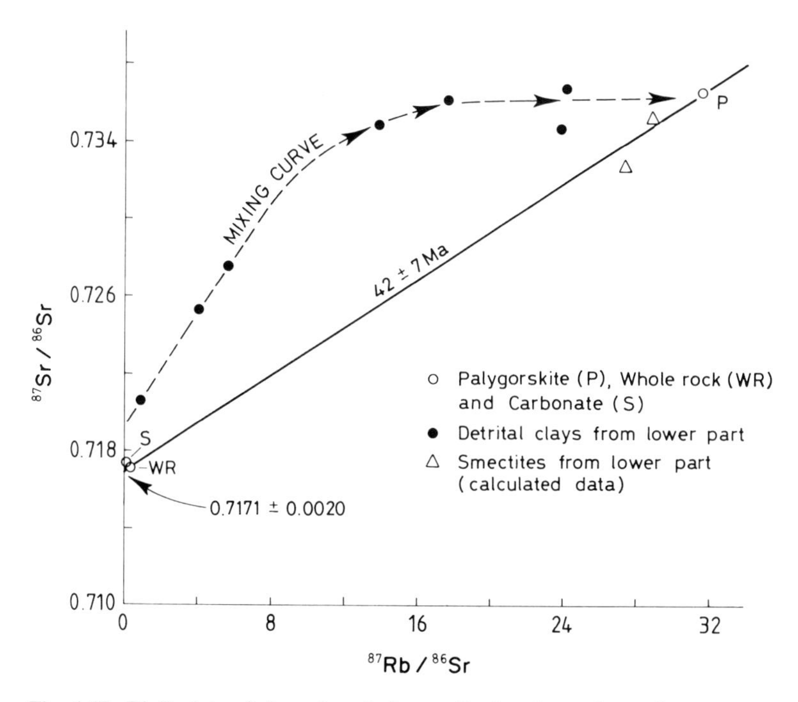

Fig. 4.17. Rb-Sr data of clay minerals from a Tertiary lacustrine sedimentary sequence in the French Massif Central. The data points of the pure smectites (*open triangles*) were calculated from illite-smectite mixtures which data points plot on the mixing curve. (Clauer 1976)

overlying volcanic beds, attesting to a near syndepositional origin for the palygorskite minerals.

Clauer (1976) also observed records of apparent syndepositional clay genesis in the lacustrine sediment of late Paleocene-Eocene age in the Mormoiron basin. The clay mineral compositions changed from an assemblage of kaolinite, illite, and smectite in the lower part of the sequence through an assemblage of smectite and palygorskite in the middle part of the sequence, to pure palygorskite in the upper part of the sequence. The clay minerals in the lower part have been thought to be of detrital origin, and those in the middle and the upper parts to be of authigenic origin. Analyzing clay samples from different segments across the sequence, most of which belonging to the upper and the middle parts of the sequence, Clauer obtained a Rb-Sr isotopic date of 59.1 ± 1.4 Ma with an initial $^{87}Sr/^{86}Sr$ of 0.70887 ± 0.00036 (Fig. 4.18). The initial Sr isotopic composition of the clay fractions, as defined by the isochron, was higher than the $^{87}Sr/^{86}Sr$ ratio of 0.70765 ± 0.00017 for a carbonate phase in an associated marl bed. Trauth (1977) convincingly presented a case for the palygorskite clays in the sequence, showing fibers of palygorskite that have grown around the smectite particles, as being products of recrystallization of detrital smectite in the lacustrine environment. Hence, the isotopic date of about 59 Ma may be considered as the time of transformation-crystallization of smectite to palygorskite. This date closely approaches the Paleocene-Eocene depositional age for the sediments. As the initial Sr isotopic composition of the clays was significantly different from the Sr isotopic composition of the depositional environment, as given by the Sr isotopic composition of the

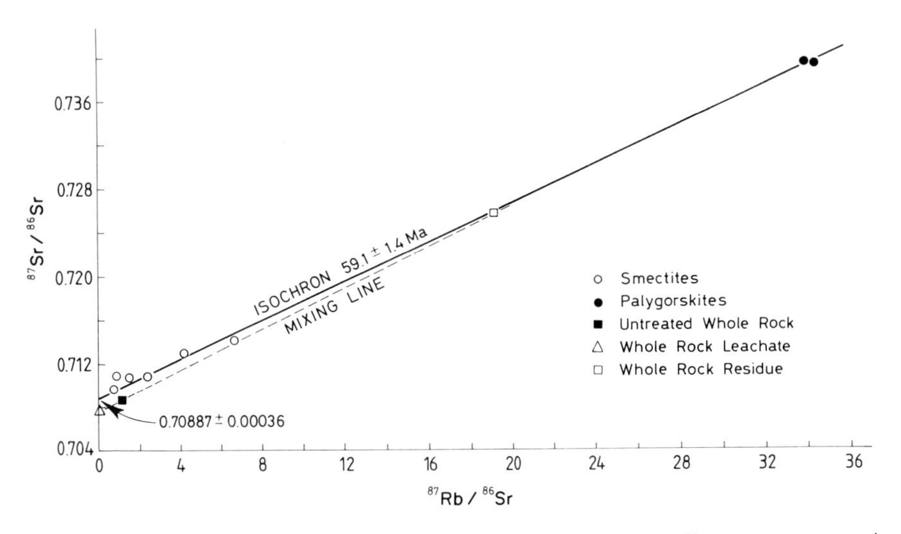

Fig. 4.18. Rb-Sr isochron data of clays from a Tertiary lacustrine sedimentary sequence in southeastern France. The initial value of the isochron is different from that of the whole-rock leachate, representing the environmental Sr. (Clauer 1976)

carbonate phase in the nearby marl bed, the recrystallization mechanism involving smectite to attapulgite conversion must have occurred in a pore-fluid environment without open communication with the depositional environment.

1.2.5 Duration of the Syndepositional Evolution of Clay Minerals

No direct evidence presently exists to suggest how long the evolution continues until the process of transformation-crystallization of nonglauconitic clays is completed. Studies on Recent glauconies suggest that it may be of short duration amounting to generally no more than several tens of thousand years (Odin and Dodson 1982). This time period is too short to be detected by any of the commonly used isotopic methods of dating geologically old materials.

Experimental studies by Eberl and Hower (1976) suggest that the time necessary to transform smectite to an illite/smectite mixed-layer mineral with low expandability may vary from about 100 million years at 20 °C, to 1 million years at 60 °C, and to 100 000 years at 80 °C. These estimates were based on extrapolation of experimental data obtained at 300 °C and 1.2 kbar pressure. Experiments by Decarreau (1983) demonstrated that dioctahedral smectite (nontronite) can form from gels in periods ranging from several weeks at 100–150 °C to about 100 years at 5 °C. Synthesis of Mg-saponite (stevensite) takes about half the time of the formation of nontronite, as the kinetics of crystallization of Al-bearing phases is extremely slow (Decarreau 1983). Harder (1974, 1976) succeeded in synthesizing some nontronites but also some illites with mixed octahedral population under surface temperature condition. Mattigod and Kittrick (1979) and Lin and Clemency (1981) were successful in crystallizing illites from aqueous gels. These experimental results suggest that clay minerals, including illite, are able to form within a few tens of thousand years under appropriate conditions.

2 Diagenetic Evolution of Clay Minerals in Deeply Buried Shales and Sandstones

Many different independent lines of mineralogical and chemical evidence have shown that illitization of smectite is a dominant process in the history of diagenesis of many argillaceous sedimentary rocks (Burst 1959; Powers 1959, 1967; Weaver 1960, 1961; Dunoyer de Segonzac 1969; Perry and Hower 1970; Hower et al. 1976; Boles and Franks 1979; and many others). Consensus has not yet been reached as to the details of the mechanism of illitization, although many recent studies have used Hower et al.'s (1976) model of conversion of smectite to illite. Perry and Hower (1970) reported from a study of Tertiary argillaceous sediments in the Gulf Coast area that, while the illitization increased with increase in depth, the K content of the whole-rock samples remained largely invariant (Fig. 4.19). The conservation of K meant that it was merely redistributed within the unit volume of the rock. Weaver and Beck (1971) suggested a solid-state reaction for the conversion of smectite to illite with some interlayered chlorite:

$$\text{Smectite} + \text{K-Spar} \rightarrow \text{illite/smectite/chlorite} + \text{Si} + H_2O.$$

Hower et al. (1976), presenting mineralogical and chemical data of bulk samples and different size fractions, gave a modified version of the above reaction by excluding the chloritic phase in the illite/smectite mixed-layer minerals. Considering that the formation of illite from smectite would require additions of K and Al and subtractions of Mg, Fe, and Si, they suggested that the source of K and Al are from dissolution of K-feldspar, and possibly also some from detrital mica, within the same unit volume of rock where illitization occurred. Hower et al. expressed the following reaction scheme:

$$\text{Smectite} + K^+ + Al^{3+} \rightarrow \text{Illite} + Si^{4+} + H_2O.$$

Boles and Franks (1979) proposed a reaction in which Al is conserved:

$$\text{Smectite} + K^+ \rightarrow \text{Illite} + \text{Chlorite} + H^+ + Si^{4+}.$$

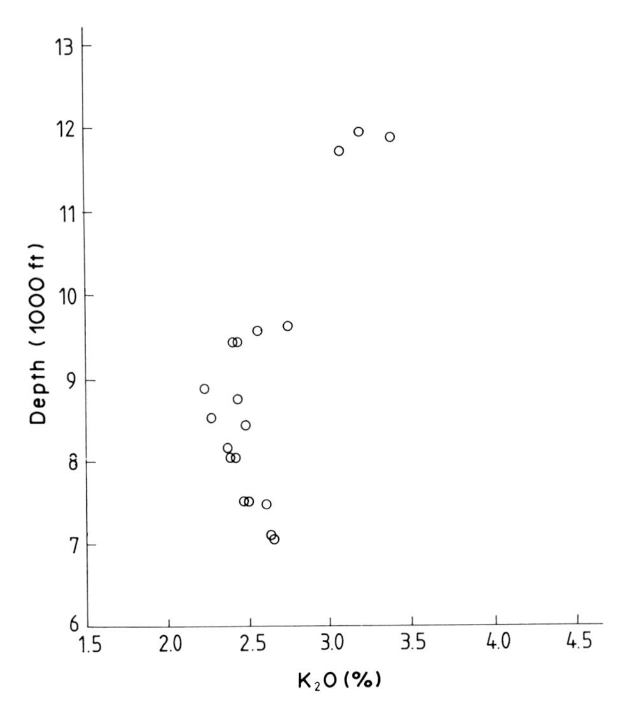

Fig. 4.19. Relationship between the K contents of shale whole rocks and their depth in the Gulf Coast sedimentary sequence. (Perry and Hower 1970)

Ahn and Peacor (1986) suggested an intermediate process between dissolution-precipitation reaction and solid-state transformation. More recently, Furlan et al. (1992) claimed that the K needed for the illitization of smectite in the Mahakam delta sediments had to be transported from a place outside the zone of illitization, thereby making the K transportation scheme unlike that suggested by Hower et al. (1976).

2.1 Diagenetic Clay Fractions in Buried Shales

Opinions about the mechanism of illitization of smectite are divided between neoformation and transformation processes. Little is known about the importance of different controlling factors which influence the course and the rate of the reaction. The factors which appear to have a major influence on the course of illitization of smectite are burial temperature (Perry and Hower 1970), time of exposure of the clay minerals to elevated temperature (McCubbin and Patton 1981), presence or absence of inhibitor ions in the pore fluids (Blatter 1974; Eberl et al. 1978), supply of K and Al (Hower et al. 1976), and the composition of the smectite (Eberl et al. 1978; Foster and Custard 1982).

2.1.1 K-Ar Isotopic Studies

The K-Ar isotopic data of both the bulk sediments and their clay fractions have proved extremely useful in defining diagenetic reactions involving illitization of smectite. Perry (1974) analyzed the K-Ar isotopic compositions of different clay fractions of some Miocene shale samples buried to depths between 1.5 and 5.5 km in the Louisiana Gulf Coast region. The temperature in the deepest part of this stratigraphic interval was estimated to be 165 °C. The $<0.5\,\mu m$ clay fraction consisted predominantly of the illite/smectite mixed-layer mineral and smaller amounts of kaolinite, detrital mica, and chlorite. In this clay fraction, the amount of illite in the mixed-layer phase increased from 20% at about 1.5 km depth to 80% at 5.5 km depth, whereas the K-Ar dates of the $<0.5\,\mu m$ fraction decreased from 164 Ma at 1.5 km to 100 Ma at about 5.5 km. All these isotopic dates were considerably higher than the corresponding stratigraphic age. The relatively coarse fraction ($0.5-2\,\mu m$) contained less illite/smectite mixed-layer mineral, but more detrital illite, feldspar, and quartz minerals, than the fine fraction. The K-Ar dates of this coarse fraction ranged between 312 Ma at about 1.7 km depth and 129 Ma at about 5.9 km depth. The differences in the isotopic dates between the two size fractions decreased as depths increased, but the dates of the two size fractions never became identical at any point within the 5.9-km stratigraphic interval examined (Fig. 4.20).

The trend of decreases in K-Ar date with increases in depth for the clay fractions in the Gulf Coast sediments was also observed by both Weaver and

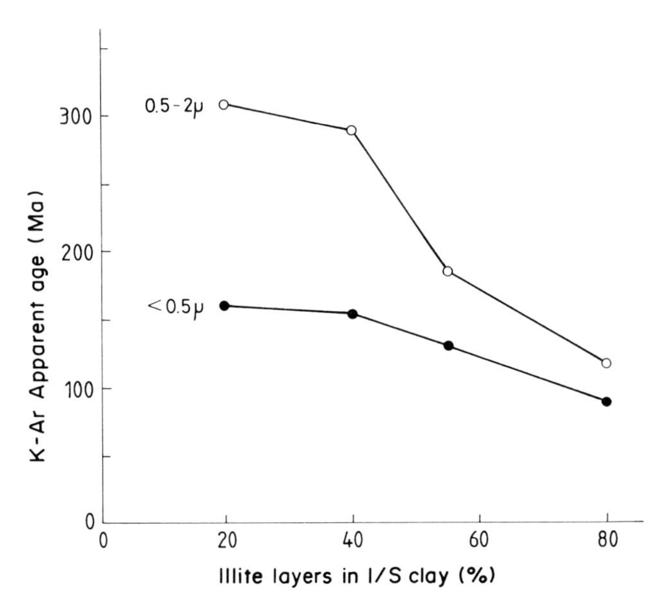

Fig. 4.20. Changes in the K-Ar dates of two clay fractions of shales with changes in illite contents in the illite/smectite mixed-layer minerals from the Gulf Coast. (Perry 1974)

Wampler (1970) and Aronson and Hower (1976). In their study, Weaver and Wampler (1970), examining the <0.2 μm clay fraction, noted that the K-Ar dates decreased from 165 ± 60 Ma at about 1.3 km depth to 125 ± 20 Ma at 5.1 km depth. Their analyses required large corrections for atmospheric Ar, introducing a high degree of analytical uncertainty that caused an apparent overlap in the dates. The data of Aronson and Hower clearly indicated that the K-Ar dates decreased as depths increased. Examining <0.1 μm clay fractions of Oligocene sediments in the Texas Gulf Coast, the authors reported that the illite contents increased and the K-Ar dates decreased, as the depths increased. The rate of decrease in the K-Ar value with depth was not uniform. Although the dates for the <0.1 μm clay fractions remained nearly constant at about 60 Ma across the top 2.47 km interval, the isotopic dates decreased rapidly from 60 Ma at 2.47 km to about 35 Ma at 3.7 km depth, following which any increase in depth produced no change in the date (Fig. 4.21). The K-Ar date-depth trend agrees well with the general trend in the evolution of illite layers in the illite/smectite mixed-layer minerals, yet the youngest K-Ar date was in excess of the stratigraphic age of the sediments. Aronson and Hower (1976) also noted that while the dates for the bulk samples remained nearly constant at about 150 Ma across the top 1.8-km stratigraphic interval, the dates decreased from 150 Ma to about 75–85 Ma as the depths increased from 1.8 to 4 km, beyond which the dates remained invariant. The decrease in the K-Ar date with increase in depths across the 1.8–4 km interval parallels to some degree the decrease in radiogenic ^{40}Ar content of the bulk shales, in contrast to the

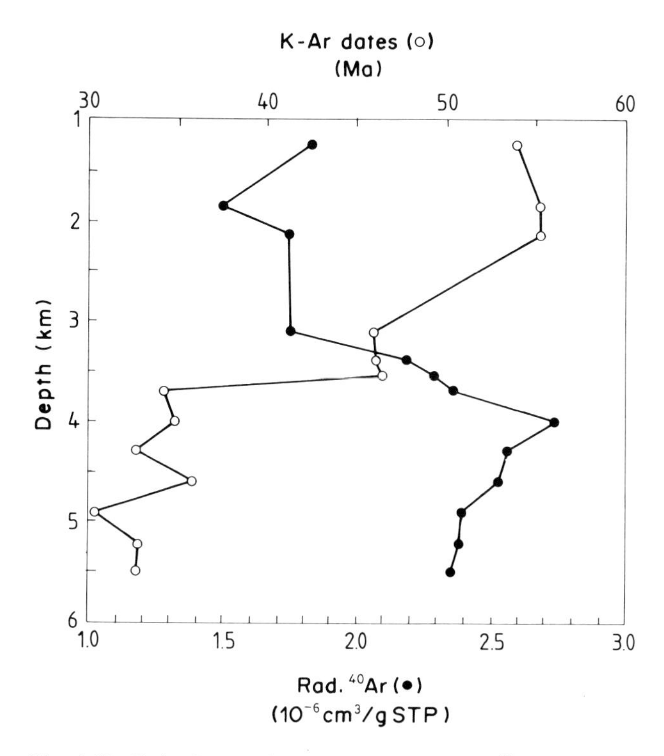

Fig. 4.21. K-Ar dates and amounts of radiogenic ^{40}Ar in the illite/smectite mixed-layer minerals in relation to depth for sediments of the Gulf Coast. (Aronson and Hower 1976)

gain in radiogenic ^{40}Ar by the fine clay fractions. Aronson and Hower explained that the decomposition of detrital components, such as K-feldspar and mica, would be a very logical cause for the radiogenic ^{40}Ar loss by the bulk samples.

Aronson and Hower (1976) considered that the variations in the radiogenic ^{40}Ar content of the fine <0.1 µm clay fractions in the zone of illitization were due to small variations in the proportion of discrete illite. Based on this assumption, the authors determined an average time of diagenetic formation of illite in the illite/smectite mixed-layer minerals in the Gulf Coast sediments. They noted that the amounts of radiogenic ^{40}Ar in the clay fractions remained nearly constant both across the top 3 km interval at about 1.7×10^{-10} mol/g, and below 4.5 km at about 2.5×10^{-10} mol/g (Fig. 4.22). Since illitization of the fine clay fraction at the very deep levels is by addition of K in the smectitic layer, the difference between the two amounts of Ar is then considered to be the amount of radiogenic growth by the decay of the amount of K which was added at the time of diagenesis. The amount of K added was computed to be about 3%, as determined from the difference in the K content between the shallow and the deep clay fractions (Fig. 4.22). Based on estimates of the amounts of radiogenic ^{40}Ar produced

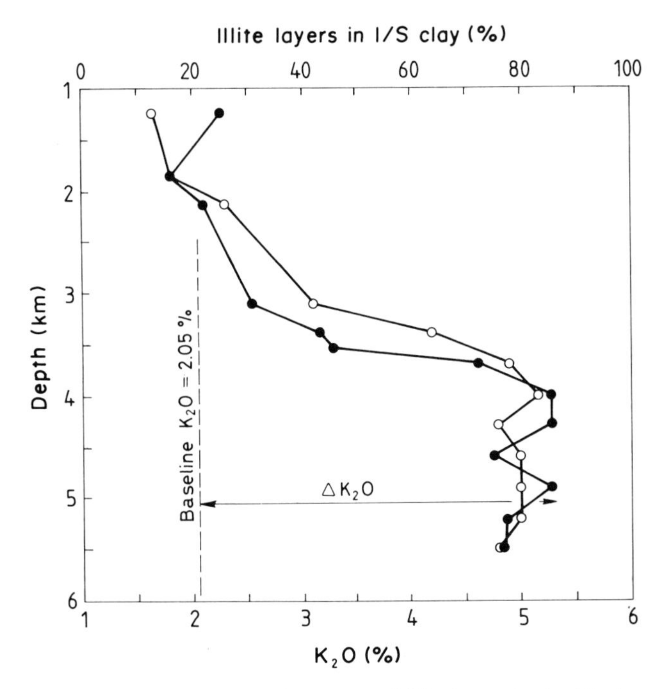

Fig. 4.22. K_2O contents and amounts of illite layers in the illite/smectite mixed-layer minerals in relation to depth for sediments of the Gulf Coast. The ΔK_2O value is considered to be the amount of K_2O incorporated by the illite-type clay minerals during diagenesis. (Aronson and Hower 1976)

from the decay of the amounts of ^{40}K added as a result of diagenetic illitization, Aronson and Hower (1976) calculated a K-Ar time of about 18 Ma. They claimed this time of 18 Ma to be a mean time of Ar addition due to the clay diagenesis, believing that illitization most likely occurred over an extended time interval. The calculated 18 Ma mean age appears reasonable, because it clearly post-dates the time of 25 Ma for the deposition of the sediments. Aronson and Hower's estimate of the time of diagenesis for the Gulf Coast sediments should be considered with some caution, because the discrete illite clay fractions between the shallow and the deep samples have been considered to be of the same age. Different studies on the Gulf Coast sediments have suggested that the K-Ar dates are widely varied, probably as a result of the isotopic variation of the detrital components.

Burley and Flisch (1989) presented an account of K-Ar dates obtained from analyses of <0.5 μm clay minerals from Upper Jurassic rock samples of the Piper and the Tartan fields of the Outer Moray Firth in the North Sea. The mudrocks are present in the interval between 2.75 km and 4.60 km depths. In these rocks, the illite/smectite mixed-layers of the clay fractions in mudrocks changed to increasingly illitic phase by increase in depth. The maximum change in the illitization of the illite/smectite mixed-layers occurred at the interval between 3 and 3.3 km depths. The variations of K contents, illite interlayer amounts, radiogenic ^{40}Ar contents, and K-Ar dates with depths are shown in Fig. 4.23. The model K-Ar dates for the clay separates from mudrocks buried to less than 4 km deep ranged from 153 to 125 Ma, whereas those from mudrocks buried to more than 4 km deep varied between 141 and 68 Ma. The age of sedimentation of the rocks in the Piper Formation is thought to be between 140 and 150 Ma. The authors believed that the high dates of about 150 Ma for the shallow samples reflect formation of illite/smectite clays in Jurassic weathering profiles which served as the sources of the deposited clay particles, and that the relatively low dates suggest presence of some amounts of newly formed illite-rich component. They arrived at this conclusion because the data points defined a linear trend in a K-Ar isochron diagram with a negative intercept on the radiogenic ^{40}Ar-ordinate (Fig. 4.24). Burley and Flisch held that the negative intercept in such a diagram means that either the smectite-rich sample lost ^{40}Ar during early burial or the illite-rich components inherited ^{40}Ar from their parent minerals. When the age pattern indicates a mixed age, as is the case here, the negative intercept can generate from a number of many different causes, and their discussions do not bring any dividend to the main objective of K-Ar dating of clays in shales. Nonetheless, the dates of the clays in the deep buried shales were significantly lower than those in shallow buried shales (Fig. 4.25). Burley and Flisch (1989) concluded that the date of 68 Ma may be designated as a maximum age for the illitization in the mudrocks.

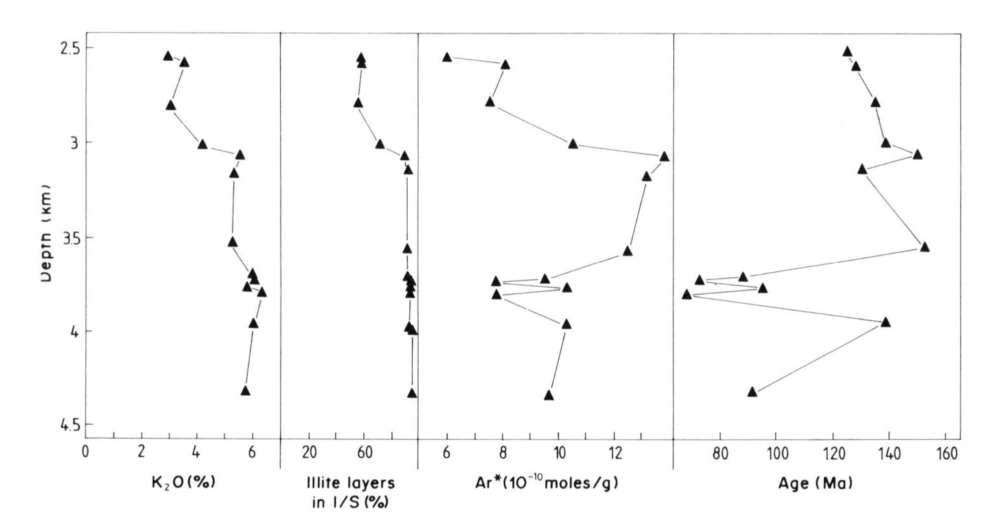

Fig. 4.23. Trends in K_2O contents, amounts of illite layers in the illite/smectite mixed-layer minerals, concentrations in radiogenic ^{40}Ar and K-Ar dates relative to depth for <0.5-μm clay fractions from mudrocks of the Piper and Tartan fields. (Burley and Flisch 1989)

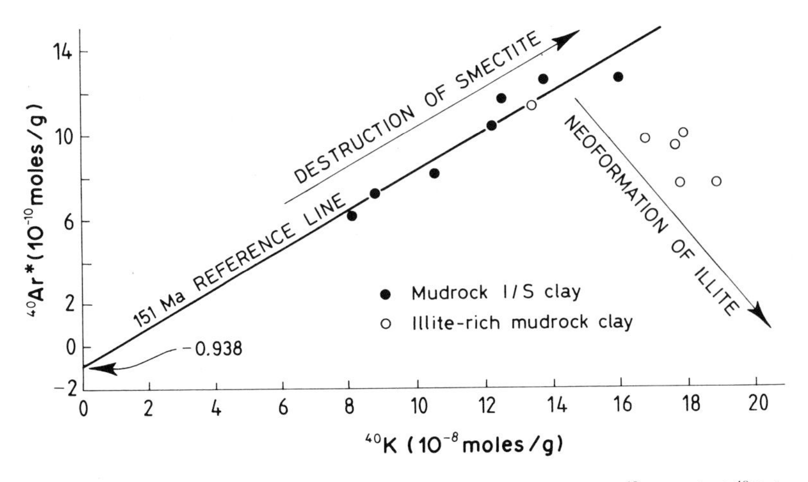

Fig. 4.24. Relationship between the concentrations of radiogenic ^{40}Ar and of ^{40}K for the <0.5-µm clay fractions from mudrocks of the Piper and Tartan fields. (Burley and Flisch 1989)

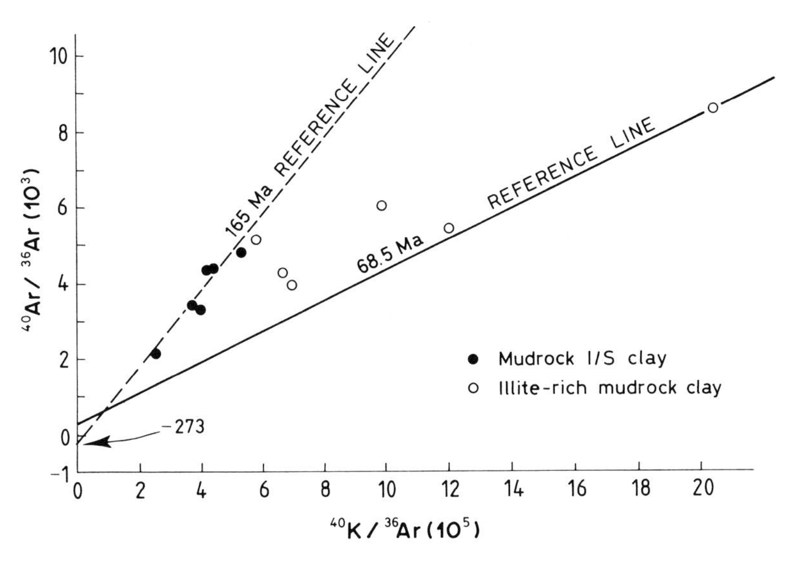

Fig. 4.25. K-Ar isochron data of the <0.5 µm clay fractions from Piper Formation mudrock samples. The *165-Ma reference line* is a best-fit regression through the I/S data points, whereas the *68.5-Ma reference line* represents the youngest age of illitized mudrock clay separates. (Burley and Flisch 1989)

The study by Burley and Flisch (1989) is a good illustration that even after making much effort in characterizing the analyzed clay materials, K-Ar isotope data of clay minerals in mudstones may not be successfully used in constraining illite clay diagenesis. The dates can potentially represent mixed ages because the clay fractions may consist not only of varied mixtures of authigenic minerals and detrital minerals of different ages, but also of authigenic minerals of different generations. The use of clay samples of <0.5 μm size, as has been done by Burley and Flisch (1989), offers little hope of defining the age of illitization of smectite during burial of mudstones, as it is an almost impossible task to physically separate coeval clay minerals in mudstones.

Since the presence of detrital components in clay fractions of shales distorts the measure of the time of diagenetic illitization, knowledge of change in the date due to fractional change in the amount of detrital components may be used in some situations to determine by extrapolation the period of illite diagenesis at zero amount of detrital minerals. This approach, which was successfully applied by Liewig et al. (1987a) to mixtures of illites and varied amounts of detrital feldspar in sandstones, was used by Pevear (1992) in K-Ar dating of diagenetic illites in deeply buried Albian-Aptian and Turonian (Cretaceous) shales interbedded with bentonites in Arkansas. He assumed that illite in the shales consisted of a two-component mixture of smectite-to-illite mixed layers and discrete detrital illite. The author quantified the fraction of detrital illite in the clay fraction using a modeling technique close to that developed by Reynolds (1980, 1985) and applied to the X-ray diffraction patterns. In a diagram in which the K-Ar dates were plotted against the amount of detrital illite present, the different size fractions of the two samples of the Aptian-Albian shale fitted along a line which yielded a K-Ar date of 30 Ma for the diagenetic end-member consisting in pure authigenic illite and a date of 355 Ma for the pure detrital end-member. The different size fractions of two samples of the Turonian shales, on the other hand, fitted along a second line, the extrapolation of which yielded a K-Ar date of about 125 Ma for the detrital end-member, that of the authigenic one being the same as for the previous sample (Fig. 4.26). The K-Ar isotopic analyses of the clay fraction separated from the bentonite bed intercalated between the shale beds yielded a K-Ar date of 27 Ma. The age of diagenesis of illites in the shales, as determined by extrapolation of the trends in dates of mixed components, agreed well with the age of the clay fraction of the bentonite, which is known not to contain detrital components. The approach taken by Pevear (1992) holds promises for estimating the isotopic age of clay diagenesis in those shales which have had simple burial history of detrital clay components.

2.1.2 Rb-Sr Isotopic Studies

Many studies on Rb-Sr isotopic age determinations of apparently unmetamorphosed whole-rock samples of shales and their separated clay fractions were carried out during the 1960s and early 1970s. As the Rb-Sr isochron dates from several of these studies appeared to have post-dated the time of

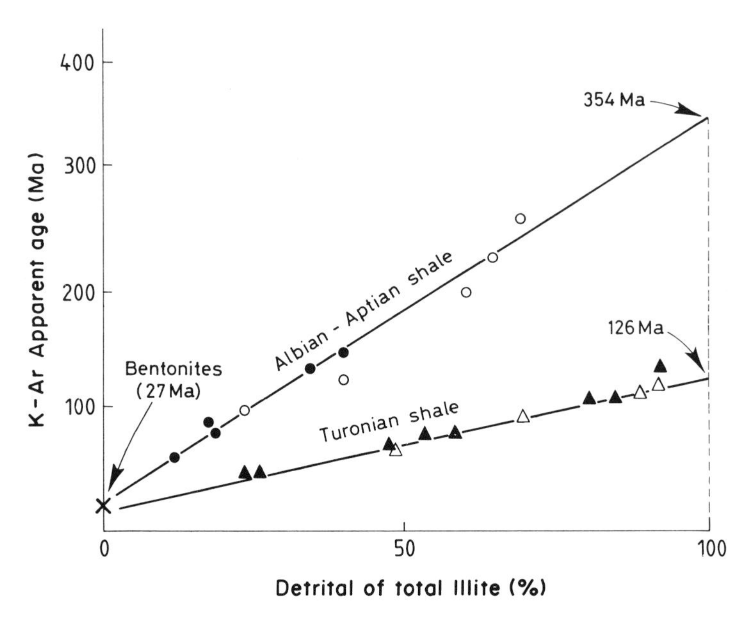

Fig. 4.26. Relationship between K-Ar dates and amounts of detrital illite for different size fractions of Turonian shale samples and Albian-Aptian shale samples from a well in Arkansas. The *filled* and *open symbols* represent the fractions of the two samples in each case. (Pevear 1992)

deposition, suggestions have been made that delayed diagenesis produced apparent Sr isotopic homogeneities. However, such an inference about the Sr isotopic homogeneity was soon realized to be highly questionable based on observations made by Perry and Turekian (1974), who analyzed the $^{87}Sr/^{86}Sr$ ratios of different clay size fractions of deeply buried Miocene shales in the Louisiana Gulf Coast. They provided indisputable evidence that, although a trend may be suggested toward Sr isotopic homogeneity among different size fractions of even the most deeply buried (5.5 km) shale sample, the data still remained so widely scattered that no meaningful isotopic date could be estimated by an Rb-Sr isochron diagram (Fig. 4.27).

As different independent lines of evidence demonstrated that diagenetic illitic clays in shales are most evident in the very fine fractions (generally <0.1 μm), several investigators focused on analyzing isotopic compositions of these very fine clay fractions separated from shales to determine the time of diagenesis (Clauer 1979a). One of such studies which has lately received much attention is that of Morton (1985a), who analyzed the Rb-Sr isotopic compositions of fine clays in deeply buried Oligocene shales in the Gulf Coast of Texas. Based on the Rb-Sr isochron analyses of <0.05 μm clay fractions of shales across a 2-km stratigraphic interval between about 3.5 and 5.5 km depths, Morton hypothesized that the illitization of smectite in the illite/smectite mixed-layer minerals was episodic, attendant with homo-

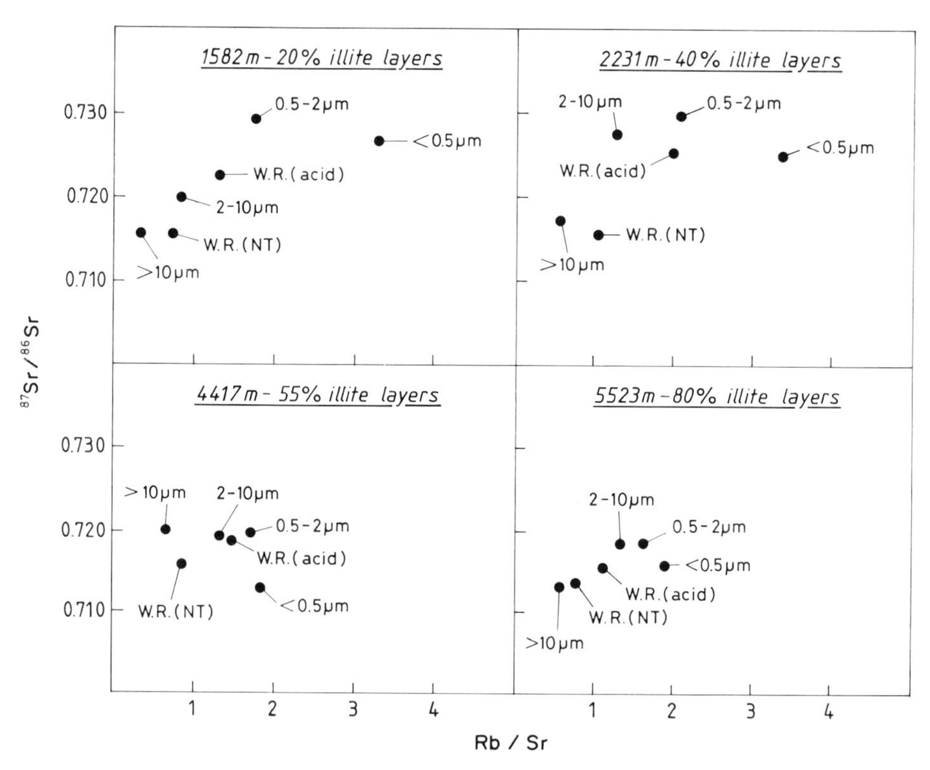

Fig. 4.27. Relationship between the $^{87}Sr/^{86}Sr$ and the Rb/Sr ratios of different size fractions in sediments from different depths in the Gulf Coast area. (Perry and Turekian 1974)

geneization of Sr isotopes among these fine fractions at about 23.6 ± 0.8 Ma. The depositional age of the sediments was estimated to be between 25 and 29 Ma. Morton, therefore, postulated that a sudden change in pore-water chemistry occurred 23.6 Ma ago, causing dissolution of K-bearing phases, detrital feldspar and mica, with attendant 80% illitization of smectite in the illite/smectite mixed-layer minerals mostly in the present deeply buried sediments.

Diagenetic illitization of the Texas Gulf Coast sediments was also investigated by Ohr et al. (1991). They analyzed samples of shales collected from a well which penetrated a sequence of shales and sandstones to a depth of about 3.65 km, covering a stratigraphic interval from lower Miocene to Paleocene. The well studied by Ohr et al. is located about 80 km to the northwest of the well studied by Morton (1985a), but the samples studied by Ohr et al. came from a lower stratigraphic interval than the ones studied by Morton. The analyses of illite/smectite mixed-layer clays of the shales studied by Ohr et al. indicated that illitization of the mixed-layer clays was more than 70% complete at depths greater than 2.4 km. The Rb-Sr isotopic data of <2 μm clay fractions of shales from the zone of intense illitization were highly scattered, but that of <0.1 μm clay fractions from this zone of illitization were linear in a Rb-Sr isochron diagram, giving a date of about 34 ± 3 Ma with an initial Sr isotopic ratio of 0.7147 ± 0.0003 for the fine clay fractions. The depositional age of the sediments being about 50 Ma, Ohr et al. concluded that the 34 Ma of the <0.1 μm clay date was the time of a single episode of diagenetic illitization covering a stratigraphic interval of more than 1.2 km. Some cautionary judgment should be made about this estimate, because the fine clay fraction of <0.1 μm size, which the authors analyzed, consisted of particles ranging from about 0.4 to less than 0.05 μm in diameter, and the influence of the coarse particles on the isotopic date has not been critically analyzed. The details have yet to be worked out about the spread in the Rb/Sr ratios attendant with a Sr isotopic homogeneity across a stratigraphic interval of more than one km.

2.1.3 Sm-Nd Isotopic Studies

Clay fractions of a sample of sediment can be measurably varied in both the Sm/Nd ratio and the Nd isotopic compositions. Sediments transported by modern rivers clearly reflect that such variations exist within the clay fraction. Stordal and Wasserburg (1986) found from analyses of both the <0.4 and the <0.1 μm clay fractions in suspended sediments of the Mississippi and the Amazon Rivers that the two clay fractions in the same suspended sediment are different in their Sm/Nd ratios and Nd isotopic compositions. The success of dating clay diagenesis by the Sm-Nd method hinges on sufficient spread in the Sm/Nd ratios of clays, while the Nd isotopic differences are erased by the process of diagenesis.

Chaudhuri and Cullers (1979), analyzing clays of both <2 and <0.5 μm sizes in Gulf Coast sediments, claimed that the diagenesis hardly has any effect on the fractionation of the REE, but this view is not shared by all. Mack and Awwiller (1990) and Awwiller and Mack (1991) compared the Sm/Nd ratios of <0.5 μm clay fractions to that of the bulk rock and of different diagenetic carbonate cements and concluded that mineralogical fractionation of Sm from Nd occurs during diagenetic illitization of smectite. The fractionation between the two elements attendant with illitization of smectite has been also claimed by Ohr et al. (1991). They reported that acid residues of very fine diagenetic clays of <0.1 μm size from the zone of illitization in sediments of the Gulf Coast area differ from corresponding leachates in their Sm/Nd ratios, while the individual residue and the corresponding leachates for four out of five samples from different depths within the zone of illitization have nearly identical Nd isotopic composition (Fig. 4.28). What should be stressed here is that the Nd isotopic homogeneity in diagenetic sediments may be seen only at a very small scale of a few centimeters at the most. Ohr et al. suggested, considering that the leachate involving very fine clay fractions (<0.1 μm size) came from an acid-soluble

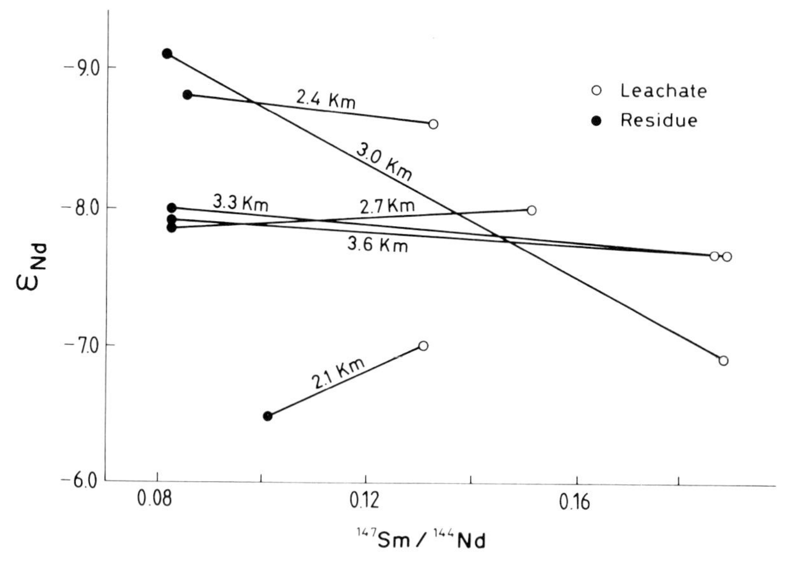

Fig. 4.28. Relationship between the Nd isotopic composition and the Sm/Nd ratio for both leachate and the corresponding residue of very fine clay fractions of shale samples at different depths in the Tertiary sediments of the Gulf Coast. (Ohr et al. 1991)

phase, that precipitation of phosphate minerals attendant with illitization of smectite may account for the spread in the Sm/Nd ratios between clay residues and corresponding leachates.

We have used the data of Ohr et al. (1991) to determine the potential K-bearing mineral which may have served as a source of K for the illitization of smectite in the Gulf Coast sediments. We have, therefore, recast in a Sm-Nd diagram the data given by Ohr et al. for three bulk samples and their size fractions from the zone of illitization (Fig. 4.29). As evident from data presented in the figure, the finest clay fraction ($<0.1\,\mu m$) has the lowest Sm/Nd ratio and the highest $^{143}Nd/^{144}Nd$ ratio. The trend in the Sm/Nd and the $^{143}Nd/^{144}Nd$ ratios for each sample and their clay fractions describes a two-component mixing line. By extrapolation, the trends for the three samples intersect at a Sm/Nd ratio of about 0.13, corresponding to a common size of slightly less than $0.1\,\mu m$. As the evidence of diagenesis is in the very small clay particles, the Sm/Nd ratio of 0.13 than belongs to the diagenetic illite phase. This extrapolation value of 0.13 is smaller than the average Sm/Nd ratio of 0.19 for many argillaceous sedimentary rocks or many granitoid rocks, but it approximates the values of K-feldspars in granitoid rocks (Faure 1986). The similarity between the Sm/Nd ratio of the diagenetic illite in the Gulf Coast sediment and that of average K-feldspar in granitoid rocks may be used to reinforce the suggestion that illitization of smectite in the illite/smectite mixed-layer minerals was accompanied by alteration of K-feldspar.

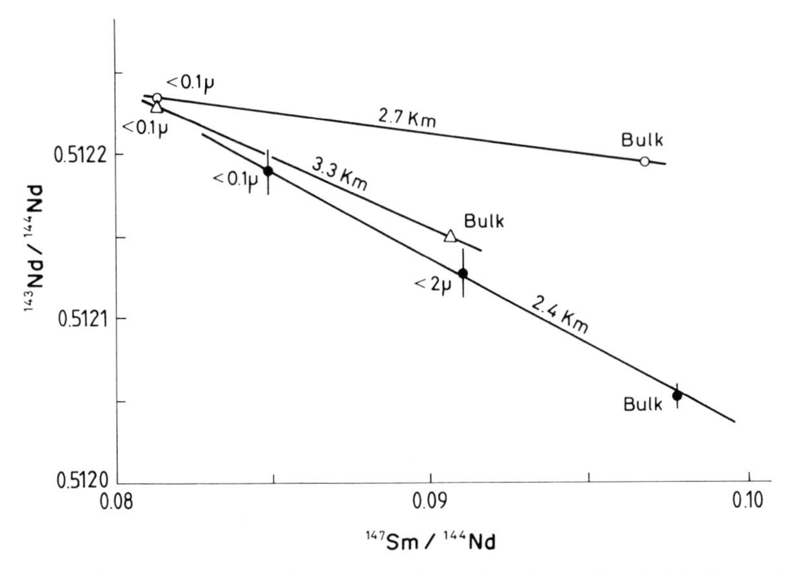

Fig. 4.29. Sm-Nd isochron diagram with data points for acid-leached bulk samples and corresponding clay size fractions of sediments at three different depths in the sequence of Tertiary sediments of the Gulf Coast area. (After Ohr et al. 1991)

The age of diagenesis of young sediments, such as those of the Gulf Coast, cannot be determined by the Sm-Nd method, but that of old sediments may be defined by it. The Sm-Nd study of Bros et al. (1992) on two nearby samples of black-shale sediments in the Lower Proterozoic sequence of the Francevillian basin in southeastern Gabon has shown that the Sm-Nd isotopic data of different size fractions and HCl acid leachates of the two samples each independently defined Sm-Nd isochron dates of 2099 ± 115 Ma and 2036 ± 79 Ma (Fig. 4.30). The authors interpreted the dates to be syndepositional or early diagenetic ages of the sediments. By comparison, the Rb-Sr and K-Ar dates of the same clay fractions were about 1.8 Ma, suggesting a later diagenetic event which obviously did not affect the Sm-Nd isotope records. What may be of much interest in the results of the study by Bros et al. is that the Sm-Nd data point of the shale which contained the most organic matter (about 13% kerogen) plotted between two isochrons defined by the fine clay fractions of the two samples with relatively lower organic matter contents. The agreement in the Sm-Nd isotopic evolution between the organic-rich sample and the clay fractions of the two relatively organic-poor samples suggests a Nd isotopic equilibrium between the clay minerals and diagenetic fluids at the time formation of the clay minerals. A fracture-filling bituminous sample had the highest $^{147}Sm/^{144}Nd$ ratio (0.15) and the highest Nd isotopic ratio. The Sm-Nd isotopic data of this bituminous material conformed to the isochron trends defined by the clay fractions of the two shales. The results suggest that Sm/Nd isotope studies may be used

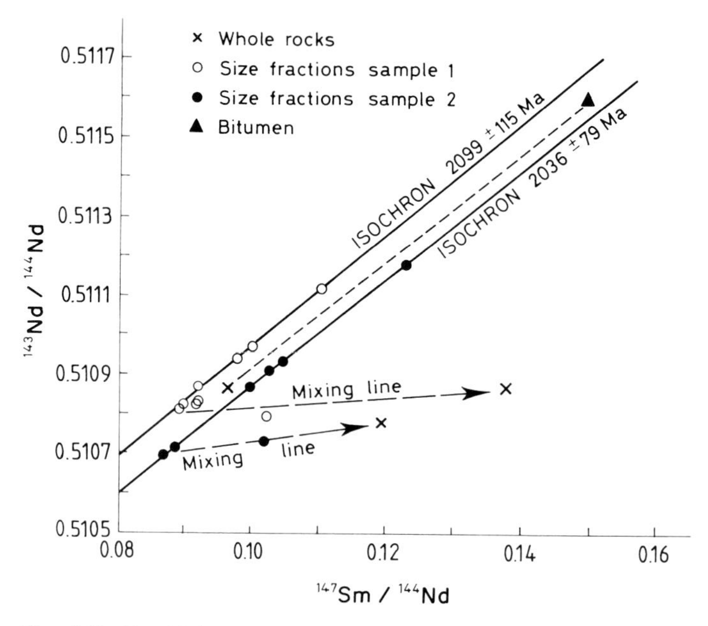

Fig. 4.30. Sm-Nd isochron data for different size fractions and the bulk samples of two Proterozoic black shales from Francevillian in Gabon. Also plotted are the data points for one bulk shale sample enriched in organic matter and one bitumen. (Bros et al. 1992)

to investigate clay-organic associations in the sediments, as Manning et al. (1991) also noted Nd isotopic links between kerogen and source rocks.

2.1.4 Oxygen and Hydrogen Isotopic Studies

The isotopic compositions of structural hydrogen and oxygen of a clay mineral at the time of its formation in a diagenetic environment are dependent on the isotopic compositions of water with which the mineral was in equilibrium at the time of the precipitation and the temperature influencing the isotope fractionation between the mineral phase and the water. Temperature, grain size or surface area, and crystal chemistry may variously affect these primary isotopic compositions by isotopic exchanges between the mineral and any associated water (James and Baker 1976; O'Neil and Kharaka 1976). A large number of studies have shown that analyses of oxygen and hydrogen isotopic compositions of diagenetically formed clay minerals may yield very useful information concerning both the processes and the physical and chemical characteristics of the diagenetic environment involved in the formation of the minerals (Yeh and Savin 1977; Yeh 1980; Sucheki and Land 1983; Longstaffe 1986; Longstaffe and Ayalon 1987, 1990; Ayalon and Longstaffe 1988; Tilley and Longstaffe 1989; Fallick et al. 1993).

Analyzing clay fractions of different sizes, ranging from 2 to <0.1 μm, in shales collected from three different Tertiary stratigraphic sections in the Gulf Coast area, Yeh and Savin (1977) noted that the $\delta^{18}O$ values of the finest (<0.1 μm) fraction clearly decreased with depth and that the ranges in the $\delta^{18}O$ values among the different clay fractions also decreased with depth (Fig. 4.31). At very deep levels with distinct evidence of illitization of smectite, the fine clay fractions consisted of essentially smectite/illite mixed-layer minerals and were presumably in isotopic equilibrium with pore waters at the time of the clay diagenesis, whereas the coarser fractions consisted of varied amounts of detrital illite, kaolinite, and chlorite and had experienced only partial amounts of oxygen isotope exchange with coexisting pore waters. The decreases in the $\delta^{18}O$ values with increases in depths for the very fine fractions followed a trend which can be explained in terms of isotopic exchange with increase in temperature accompanying the burial. Much of the diagenetic illitic clays in Tertiary shales of the Texas Gulf Coast were found to have $\delta^{18}O$ values between about +17 and +22 per mill. Figure 4.32

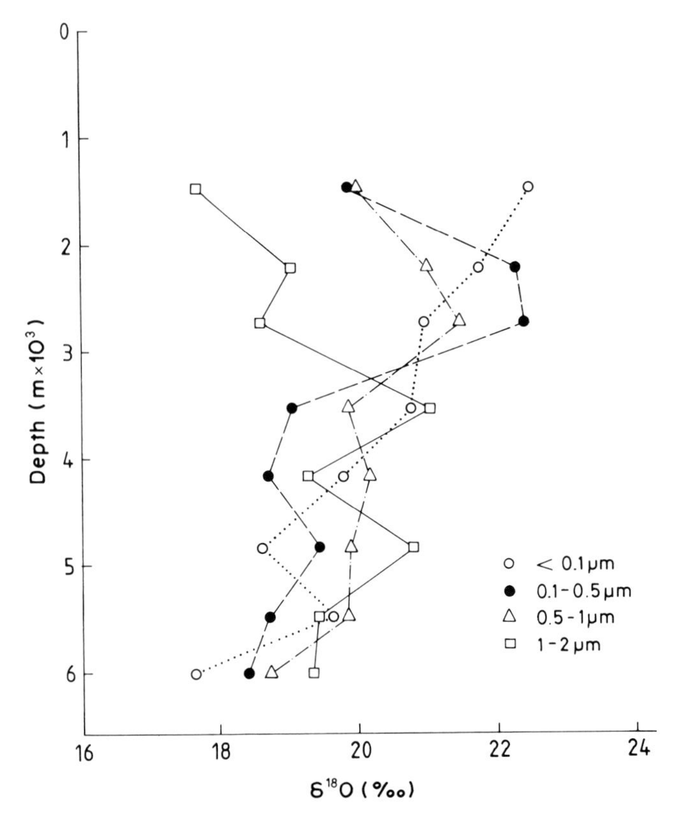

Fig. 4.31. Change in the $\delta^{18}O$ values as a function of depth for <0.1-μm clay fractions of Tertiary sediments from the Texas Gulf Coast area. (Yeh and Savin 1977)

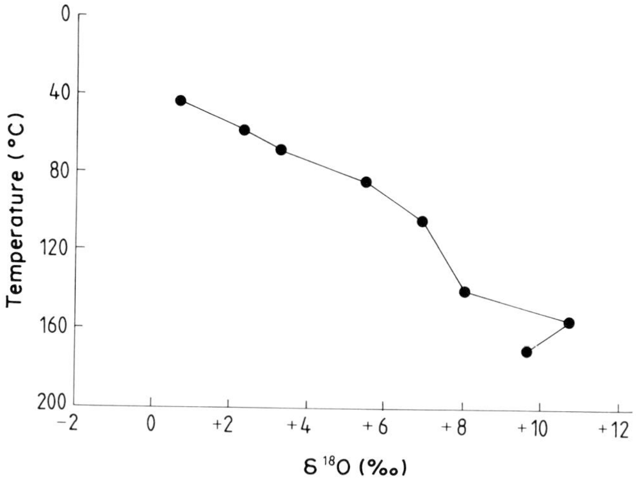

Fig. 4.32. Calculated O-isotope compositions of pore waters in equilibrium with the <0.1-μm clay fractions from CWRU Gulf Coast well 6, as a function of measured well temperatures. (Yeh and Savin 1977)

illustrates the changes in depth for the calculated $\delta^{18}O$ values of the pore waters which were in equilibrium with the finest <0.1 μm clay fractions. At the illitization zone with temperatures exceeding 80 °C, the pore waters which were in isotopic equilibrium with the diagenetic illites apparently had $\delta^{18}O$ values between about +5 and +11 per mill. These calculated $\delta^{18}O$ values for the pore waters in shales were later found to be broadly similar to the values of formation waters in associated sandstones as reported by Kharaka and Carothers (1986). The significance of the apparent variations in the calculated $\delta^{18}O$ values of pore waters in shales at depths, as noted by Yeh and Savin (1977), is that pore waters at different depths in a single well were not in communication with each other, suggesting strongly that the diagenetic reaction involving illitization in the Gulf Coast sediments occurred under approximately closed-system conditions. Morton (1985a) also analyzed the oxygen isotope compositions of clays from a well which was located a few kilometers southwest of the well that provided sediments for the study by Yeh and Savin. Like Yeh and Savin (1977), Morton observed decreases in $\delta^{18}O$ values of the very fine clay fractions (<0.05 μm) with increases in depths. He agreed with the general interpretation of Yeh and Savin that the decreases in the oxygen isotope values with increases in depths reflect increased illitization of the illite/smectite mixed-layer clays at progressively higher temperature accompanying burial.

Yeh (1980) reported hydrogen isotopic compositions of different size fractions of clays in the Texas Gulf Coast sediments from the same well which provided sediments for the study by Yeh and Savin (1977). Like the trends in the oxygen isotope variations with depths, the hydrogen isotope compositions of very fine <0.1 μm clays varied with depths and the isotope exchange among detrital and diagenetic clays appeared more pronounced at greater depths than at shallow depths (Fig. 4.33). Although the δD values of the very fine clays (<0.1 μm) from at least some wells generally increased with increases in depth, Yeh observed a marked shift toward higher δD values by an amount of about 20 per mill for the fine diagenetic clays in the zone of illitization. The hydrogen isotopic variations with depths suggest that the diagenetic illitization occurred in nearly closed-system conditions. Yeh concluded from the data that significant D/H fractionation occurred between residual and expelled pore water in the shales and that the conversion of smectite to illite in the illite/smectite mixed-layer minerals is an important mechanism of late-stage dehydration of shales.

Suchecki and Land (1983) analyzed oxygen and hydrogen isotopic compositions of diagenetic illitic clays in mudstones within volcanogenic sedimentary beds of the Upper Jurassic-Lower Cretaceous Great Valley

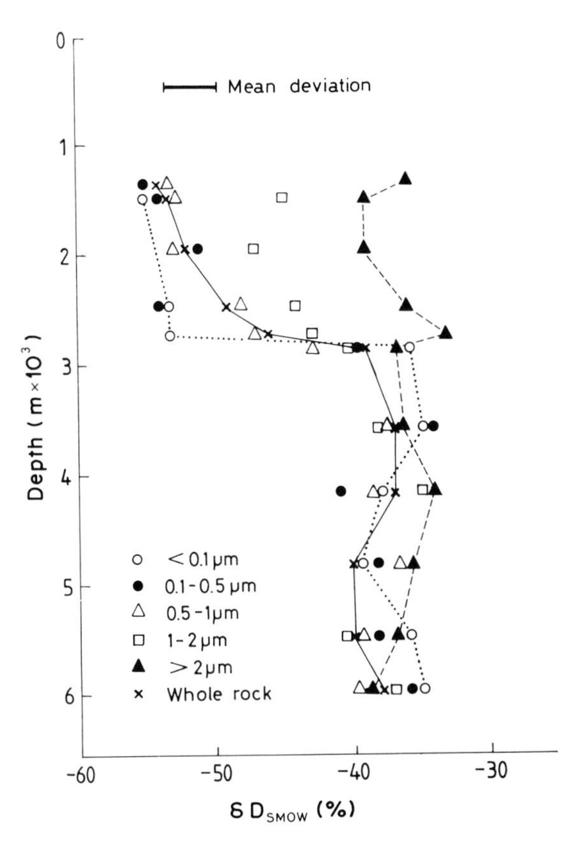

Fig. 4.33. Hydrogen isotope variations with depth of whole-rock samples and various size fractions in sediments from the Gulf Coast. (Yeh 1980)

sequence in northern California. The diagenetic illite content of the mudstones increased with increase in depth. The $\delta^{18}O$ values of the <1.0 μm diagenetic clays decreased with increase in depth, with a change from about +22 to +15 per mill within the analyzed stratigraphic interval. Suchecki and Land also noted, as did Yeh (1980) in his study of the Gulf Coast sediments, that the changes in the δD values, which ranged from about −69 per mill at the shallow part of the sequence to about −44 per mill at the deep part of the sequence, shifted by more than +20 per mill within a stratigraphic interval of few hundred meters between 4 and 5 km depths. The D-enrichment shift for the clay fraction approximately paralleled the increase in the relative proportion of illite in the illite/smectite mixed-layer phase. This isotopic change was explained by the authors as due to the effect of late-stage dehydration of shales, an explanation which was previously given by Yeh (1980) in his study of the Gulf Coast sediments. Using oxygen and hydrogen isotopic evidence on the northern Californian mudstones, Suchecki and Land suggested that the smectite underwent late-stage dehydration and that illite diagenesis in the mudstones occurred in nearly closed systems. Assuming that smectite to illite conversion occurred in rock-dominated systems, they calculated that model $\delta^{18}O$ and δD values of the formation fluids, whose initial $\delta^{18}O$ and δD values were equal to that of sea water before the reaction of illitization, were about +8 and −25 per mill, respectively.

2.2 Diagenetic Clay Fractions in Buried Sandstones

As the permeability of sandstones may be severely influenced by the formation of diagenetic clay minerals, their origin must be clearly known to define the development of reservoir properties of many hydrocarbon-bearing sandstones. Diagenetic clay minerals in sandstones occur as filling pores, replacement of feldspars and other detrital particles, and overgrowths on earlier diagenetic minerals or detrital particles. Kaolinite, discrete illite, illite/smectite mixed-layer, smectite and chlorite are the common diagenetic clay minerals in sandstones. Illite/smectite mixed-layer minerals in sandstones could contain more expandable clays than those in shale, but, like in shale, the smectite content of illite/smectite mixed-layer in sandstone commonly decreases with increase in depth or temperature (Boles and Franks 1979). The source of the chemical components of the diagenetic clays and their formation in a closed versus an open system have been much debated (Loucks et al. 1977; Hurst and Irwin 1982; Curtis 1983; Bjorkuum and Gjelsvik 1986).

In many sandstones, diagenetic clays are often more abundant than the detrital ones. This makes separation of the diagenetic phases in sandstones somewhat less difficult than that in shales. This potential advantage of getting a clean separation of the diagenetic phases from detrital components has generated much interest in the isotopic analyses of clays in sandstones focusing on understanding of clay diagenesis in sandstones.

2.2.1 K-Ar Isotopic Studies

2.2.1.1 The Jurassic Formations of the North Sea

Hamilton et al. (1987) analyzed K-Ar isotopic compositions of diagenetic illite in sandstones of the Jurassic Brent Formation in the North Sea basin, and the dates scattered between 59 and 46 Ma. The samples were supposedly prepared following a method of crushing. The mineralogic purity of each separate was checked by XRD, a method that is usually unable to detect the presence of any discrete contaminants occurring in minute quantities, which may be sufficient to influence the isotopic dates of the analyzed samples. Hogg et al. (1987) reported K-Ar dates of illites in sandstones of the Jurassic Brent Formation from the southern Alwyn field. They recognized three kinds of morphology among the illites. The K-Ar dates of the illites ranged between 45.5 and 60.3 Ma. Hogg et al. noted, as did Hamilton et al., that the samples from shallow depth had the relatively young dates. All these authors suggested that illitization began and was completed at shallow depths as soon as the sediments reached specific temperature through burial. Jourdan et al. (1987) claimed that the diagenetic history of the Brent Formation in the Alwyn area is long and complex. Based on their own K-Ar isotopic data of clay fractions and on that of other studies, especially that of Liewig et al. (1987a), the authors identified four separate episodes of diagenesis within the time span ranging from the end of the Lower Cretaceous (100 Ma) to the Eocene (50–40 Ma). Liewig et al. (1987a), as the discussion below emphasizes, showed that at least part of the scatter in the K-Ar dates can be ascribed to the presence of varied amounts of detrital K-feldspar contaminant in the clay fractions. The different ages for illite authigenesis claimed by Jourdan et al., although possible, are questionable simply for lack of convincing evidence on the degree of purity of the analyzed fractions.

Illite fractions from Brent sandstones of the Alwyn area were also dated by Liewig et al. (1987a), who followed a rigorous procedure for separation of the clay fractions from sandstones. Liewig et al. carefully monitored the mineralogy and the morphology of the separated phases by combined X-ray diffraction and electron microscopic observations. They determined the K-Ar and Rb-Sr isotopic dates of $<0.6\,\mu m$ clay size fractions which were separated by adopting both a classical crushing technique and a repetitive freezing-thawing technique for disaggregations of pieces of sandstone samples. The clay fractions separated, by using the freezing-thawing technique, from sandstone samples belonging to the same core consisted of 100% illite for one sample and about 90% illite and 10% dickite for the two others. The XRD analysis and electron microscope observation revealed that these clay fractions were totally free of any detrital component. The K-Ar dates, ranging narrowly between 35 and 44 Ma, averaged 41 Ma. On the contrary, the dates for the clay fractions separated from several sandstone samples following a conventional crushing technique varied widely between 50 and 129 Ma. The relatively higher dates and the wide spread in the results can be explained by the presence of varied amounts of detrital feldspar in the clay fractions (Table 4.1). This becomes evident by comparing the K-Ar dates with the illite-to-feldspar ratios determined by XRD semi-quantitative measurements using an internal standard (Fig. 4.34). An extrapolation from the trend to pure illite gave an isotopic date of about 40 Ma, which is identical to the average age obtained directly on the pure illitic fractions separated by the repetitive freezing-thawing technique. Cocker et al. (1988) followed the same gentle freezing-thawing disaggregation technique to separate illite fractions of Brent sandstones in Hutton field of the United Kingdom region and obtained a limited scatter in the K-Ar dates of 43 Ma for <0.4 μm clay size fractions from the watersaturated zone in the sandstones. These K-Ar values were found to be about 2 million years lower than that of the illite fractions from the oil-bearing zone, suggesting that the oil entrapment occurred shortly before the diagenetic evolution of illite in the underlying water-saturated zone.

Liewig (1993) also analyzed K-Ar dates of different size fractions of Brent sandstone samples from depths between 3.3 and 3.4 km in Heimdal field of the Norwegian area. These size fractions, which were extracted following a thawing-freezing disaggregation technique, consisted in a mixture

Table 4.1. K-Ar and mineralogic data of clay fractions from Alwyn sandstones, North Sea. (Liewig et al. 1987a)

Samples (μm)	Depth (m)	Feldspar/illite ratio	K_2O (%)	K-Ar dates (Ma ± 2σ)
3-1 <1	3261.2	34.4	4.00	100 ± 3
3-2 <1	3273.6	22.8	3.83	129 ± 6
4-1 <1	3419.4	13.0	6.07	72 ± 3
4-2 <1	3424.0	9.9	7.15	52 ± 2
5-1 <1	3549.5	9.2	5.12	54 ± 2
5-2 <1	3565.6	7.9	3.59	50 ± 3
2-5 0.4-1	3325.5	13.2	7.81	77 ± 2
1–2		20.1	7.56	89 ± 2
2-6 1–2	3327.1	24.9	6.37	79 ± 2
2-7 0.4–1	3331.7	10.3	5.86	58 ± 2
1–2		11.4	5.52	70 ± 2
6-1 <0.6	3238.5	0	6.68	44 ± 2
6-2 <0.6	3239.7	0	5.53	44 ± 2
6-3 <0.6	3242.0	0	7.45	35 ± 1

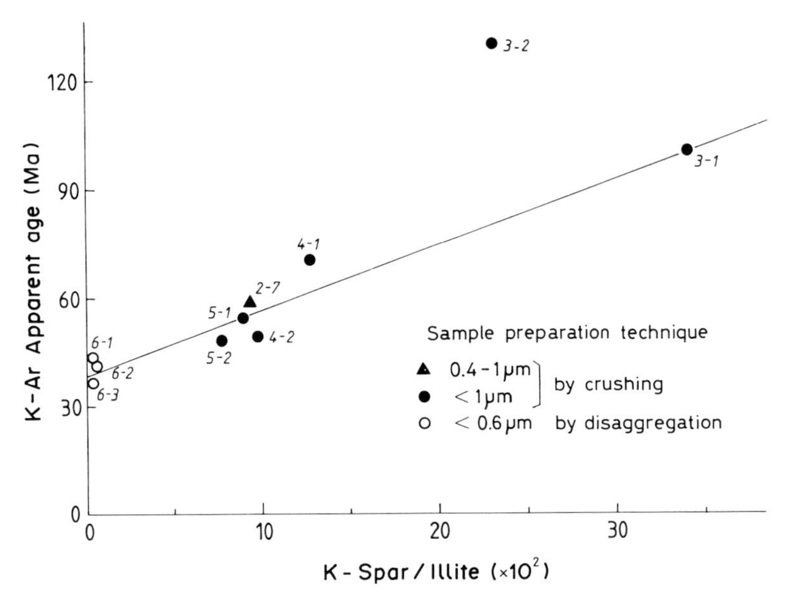

Fig. 4.34. K-Ar data of clay fractions relative to the K feldspar/illite ratio as indicated by the use of varied techniques for sample preparation. (Liewig et al. 1987a)

of illite and kaolinite. Illite was mainly concentrated in the small <0.4 µm size fraction, whereas kaolinite was in the coarser 0.4–1 and 1–2-µm size fractions. The K-Ar dates of the small size fractions were scattered between 46 and 54 Ma, whereas those of the coarser size fractions ranged from 72 to 162 Ma. The younger dates agree well with the results from similar materials by several other investigators, whereas the high dates certainly reflect the occurrence of detrital components in the coarse clay fractions.

Burley and Flisch (1989) analyzed the K-Ar dates of illite-rich clay separates from Upper Jurassic sandstones occurring within a depth interval between 2.8 km and 5 km in the Piper and Tartan fields in the United Kingdom region. Illitic minerals occurred in a variety of forms. Detrital clay was present sporadically in the matrix and also found as small mudflakes and as replacements of rock fragments, feldspars, and micas. The origin of these illites is in doubt, either they refer to inheritance from source areas or to diagenetic crystallization in the sandstones; but illites occurring as replacements of highly altered K-feldspar and mica grains were certainly of diagenetic origin. The authors noted that the distribution of clay mineral species in sandstones was varied between the Piper and the Tartan fields and also between the oil-saturated and the water-saturated zone in each field. Several species of illite/smectite mixed-layer clays were found in both fields. In the Piper field, discrete authigenic illite was limited to less than 1%, but in the Tartan field, the discrete authigenic filling was often more than 5%. In the Piper field, the ordered illite/smectite mixed-layer clays were abundant throughout the entire unit and were often with 50% expandable phase. In the shallow part of the unit in the Tartan field, the clay assemblages contained an ordered illite/smectite clay with about 40% expandable phase and in the deeper part of the unit, the clay assemblages increased in illite content and contained less than 5% of expandable phase. The clay minerals were more illitic in the water zone than in the oil zone in both the Piper and the Tartan fields. Burley and Flisch (1989) observed that the K-Ar dates of the <0.5 μm clay separates from Jurassic sandstones fall into two groups, differentiating the samples of the oil-saturated zones from that of the water-saturated zones. In the Piper field, the K-Ar dates for the <0.5 μm authigenic-rich clay separates from oil-saturated zone ranged from 131 to 144 Ma, whereas that from water-saturated zone varied between 103 and 129 Ma. In the Tartan field, the dates for the clay separates from the oil-saturated zone were between 65 and 93 Ma, but that from the water-saturated zone were between 29 and 65 Ma (Fig. 4.35). The relatively low K-Ar dates for clay separates from the water-saturated zones imply that authigenic illitization continued in the water-saturated zone even after the oil accumulation in the zone above. However, these dates cannot be used to constrain either the time of accumulation of hydrocarbon or the time of diagenesis of illite, because the clay separates contained both authigenic phases and inestimable but influential amounts of detrital minerals. The problems of contamination with fine-grained detrital micas and feldspars and of mixing of authigenic illite phases of different generations require extreme care to obtain pure separates of individual generations of authigenic components. These aspects were recently discussed in detail by Clauer et al. (1992d), because failure to

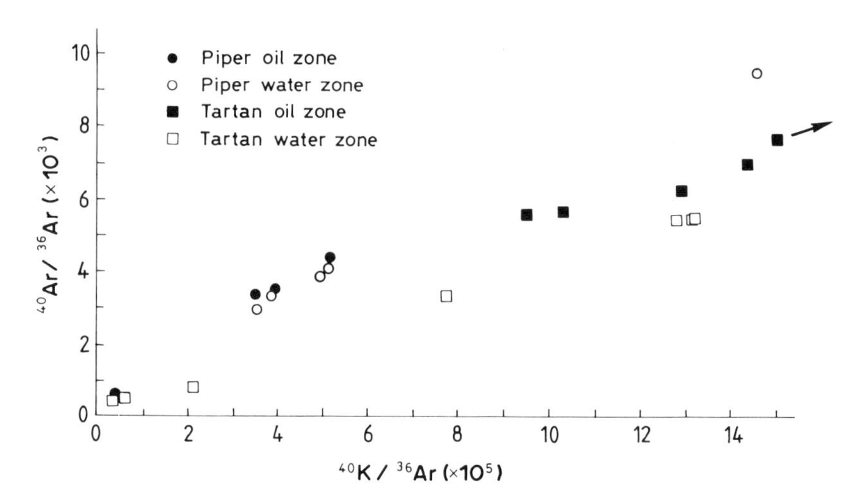

Fig. 4.35. K-Ar isochron data of <0.5-μm clay fractions from sandstones of the Piper Formation collected from different wells in the Piper and Tartan fields. The data are uncorrected for the K feldspar content. (Burley and Flisch 1989)

obtain a pure phase often leads to meaningless, erroneous interpretations of the data. The problem of multiphase diagenetic illites may be avoided by very careful mineralogical characterization of the clay separates involved prior to the isotopic analyses.

Ehrenberg and Nadeau (1989) investigated the timing of formation of diagenetic illite in subarkosic sandstones of the Middle Jurassic Garn Formation in the Haltenbanken area located on the mid-Norwegian continental shelf. The Garn Formation, a principal hydrocarbon reservoir unit of the Haltenbanken province, occurs at a depth of about 3.7 km beneath the sea floor. The temperature at this burial depth has been estimated at about 140 °C. The authors noted that the abundance of illite increases with increase in depth and also that the crystallinity index increases with increase in the abundance of illite. The shallow cores of the sandstones contained a small degree of illitization amounting to between about 10 and 35%, whereas the deep cores had extensive illitization amounting to between 80 and 95%. The nonillite clay fraction consisted of kaolinite with subordinate amounts of chlorite. The K-feldspar constituted the principal nonclay mineral. In general, extensive illitization was accompanied by decreases in both K-feldspar and kaolinite. The K-Ar isotopic dates of the $<0.2\,\mu m$ clay fraction ranged between 31 and 55 Ma, no discernible relation existed between the dates and the depths of the samples. Electron microscopic observations indicated that the illite was a mixture of two end-member morphological types. XRD analyses showed no detectable amount of feldspar in the illite-rich clay separates that were analyzed for the dating. The K-Ar dates did not correlate with the indices of thermal maturity (i.e., vitrinite reflectance), although the degrees of illitization correlated well with the indices of thermal maturity. The authors, therefore, concluded that the dates reflected an unspecified contamination problem.

Glasmann et al. (1989b) studied the diagenetic evolution of hydrocarbon-bearing sandstones in a deltaic sequence of the Middle Jurassic Brent Formation in the Heather field of the United Kingdom region. The diagenetic history of the sequence in this field is similar to that of many other North Sea fields, although they vary in the timing and the intensity of the diagenetic events. The major diagenetic processes which affected the reservoir properties are, from the earliest to the latest, calcite cementation, K-feldspar dissolution, kaolinite precipitation, quartz precipitation, and illite precipitation. Illite is present both as illite/smectite mixed-layer and discrete illite minerals. The K-Ar dates of the $<0.1\,\mu m$ clay fractions, ranging from 27 to 57 Ma, generally decreased as the depths increased (Fig. 4.36). Glasmann et al. concluded that the decrease in age with increase in depth can be related to progressive filling of the reservoir with hydrocarbons, beginning in middle Eocene and continuing into late Oligocene. Comparing the K-Ar date of the $<0.1\,\mu m$ clays to that of the $0.1-0.2\,\mu m$ particles at the same depth in the Heather field, Glasmann et al. noted that the two dates differ by about 5–10 Ma, the coarser having the higher values. This pattern of age difference

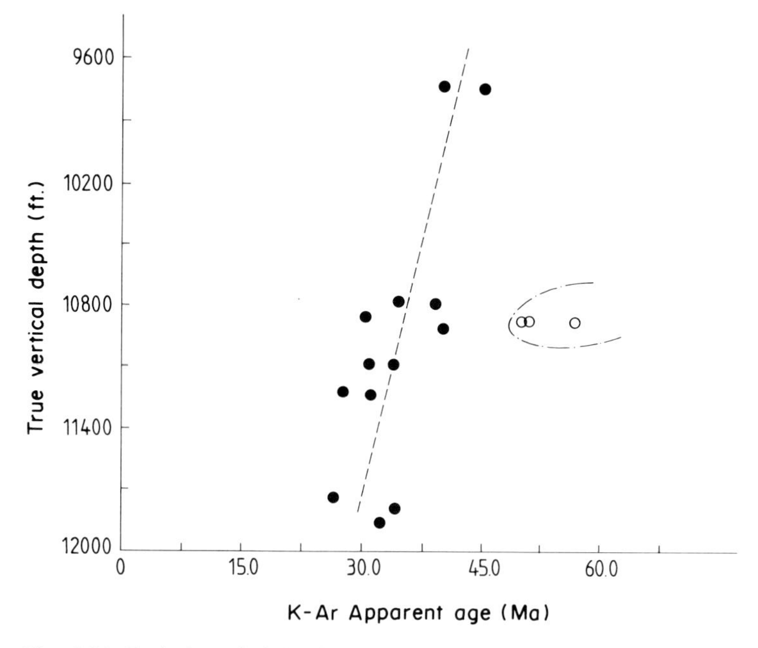

Fig. 4.36. Evolution of K-Ar dates of <0.1-μm illite fractions of sandstone samples from Jurassic Brent Formation relative to depth. The *filled symbols* belong to samples from Main Heather field, the *open symbols* to those from 2/5–9 well. (Glasmann et al. 1989b)

is commonly explained by the presence of relatively higher amounts of detrital minerals in the coarser materials, but the authors suggested that the size-date pattern, as given by these North Sea samples, reflect a record of early growth history in the coarse crystals of illite.

2.2.1.2 The Rotliegendes Formation of the Southern North Sea and Western Europe

The Permian Rotliegendes sandstones in the southern part of the North Sea basin have been the subject of some recent geochemical studies focusing on the time of hydrocarbon emplacement in the reservoirs. Lee et al. (1985, 1989) analyzed K-Ar dates of different clay fractions (0.1 to 0.5 μm) which ranged in composition from pure illite to illitic-rich illite/smectite mixed-layers of sandstones of the Permian Rotliegende Formation from several different subbasins in the southern North Sea and in northeastern Netherlands. The authors observed that newly formed illites enveloped older illite in intergranular pore spaces of the sandstones and that the amounts of diagenetic illite increased with increase in depths. The correlation may reflect increase in the rate of crystal growth due to increase in temperature or increase in the amounts of constituents necessary for illite growth from sources outside the Rotliegendes. The formation temperatures for illite were estimated to be 95 and 135 °C, whereas that of the mixed-layer illite types with expandable component was somewhat lower. The K-Ar dates, which mostly ranged between 100 and 180 Ma, were thought to indicate that the time of diagenesis varied from one basin to another. For example, the K-Ar dates of the illitic clays from central Netherland basin were mostly in the range between 103 and 115 Ma, those from Broad Fourtees basin ranged mostly from 125 to 165 Ma, those from much of northeastern Netherlands basin ranged widely between 106 and 175 Ma, and those from Indefatigable gas field ranged between 130 and 165 Ma. Several K-Ar dates from northeastern Netherlands basin were between 39 and 42 Ma, but the illites with such low dates came from a fault zone, suggesting a localized effect due to movements of fluids along the fault. Lee at al. argued that the formation of diagenetic illite was closely related to the Kimmerian orogeny and also the Jurassic and Late Cretaceous inversion movements. The authors held that the duration of illitization could have been either as short as about 8 Ma, as illustrated by much of the data from the Laman gas field in the southern North Sea, or as long as about 70 Ma, as suggested by much of the data from the northeastern Netherlands.

Robinson et al. (1993) published K-Ar analyses of illites in the lower Lenan Sandstone Formation of the Lower Permian Rotliegendes Group in the Village field of southern North Sea. These rocks are principal hydrocarbon reservoir units and contain illite as a common diagenetic cement post-dating chlorite, some kaolinite, and carbonate minerals. The K-Ar dates of the illites belonging to samples from burial depths between 2.8 and 3.4 km below the sea floor, ranged between 138 and 177 Ma, the fine fractions generally yielding 10% lower dates than the coarse clay fractions (0.2–0.5 μm) (Fig. 4.37). The authors maintained that the clay separates involved were essentially free of any detrital K-bearing component, arguing that the non-radiogenic Ar incorporated into the clays had the isotopic composition of the present atmosphere. The authors also held that, as the correlation is extremely poor between the K-Ar dates and the maximum burial depth estimated from burial history modelling, any suggestion of diffusive Ar loss from most of the fine clays could not be substantiated. Hence, Robinson et al. (1993) concluded that the evidence of the finest clay fractions having slightly lower K-Ar dates than the coarser fractions could be a reflection of the duration of the illite growth wherein the finest clays were the last to grow. According to the authors, the most reasonable interpretation of the K-Ar dates is that they represent the Middle-Late Jurassic time interval of about 35 Ma for the growth of illite across a large region (1450 × 100 km) which was subjected to uplift due to thermal doming and rifting. Compared to the data published by Lee et al. (1989), the K-Ar data of Robinson et al. (1993) have a much wider dispersion, suggesting that they could have been influenced to some degree by the presence of detrital K-bearing minerals.

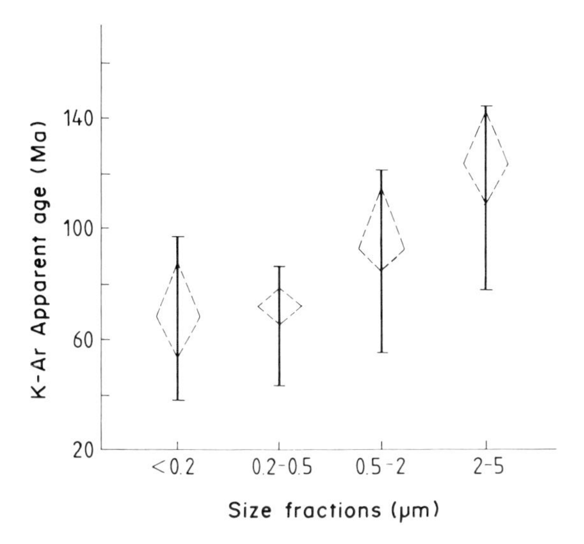

Fig. 4.37. K-Ar dates for different size fractions of sandstone samples from Jurassic Brent Formation in the North Sea. (Robinson et al. 1993)

Liewig (1993) studied different size fractions of gas-bearing Rotliegendes sandstones from several drill holes next to a salt dome in northwestern Germany. The measured drill-hole temperature at 5 km depth was about 150 °C, which corresponds to a thermal gradient of about 30 °C/km. The observations by optical and electronic microscopies suggested two distinct populations of illite particles: tangential ones developed in compacted lamellar pore spaces and post-compactional ones newly formed in large intergranular pore spaces. The TEM observations revealed that the illite particles had very constant morphologic characteristics: well-crystallized, lath-shaped, sharp edges with pseudo-hexagonal endings. In the coarser fractions (>0.4 μm), illite also occurred as moire flakes with irregular edges corresponding to a detrital micaceous component. The 2–6-μm fractions of three out of the five disaggregated samples and one intermediate fraction (1–2 μm) yielded K-Ar dates between 217 and 275 Ma. The fine clay fractions (<0.4 μm) yielded K-Ar dates with narrow scatter between 175 and 203 Ma. The author found no discernible relationship between radiogenic ^{40}Ar loss and decrease in size of the separated fractions. On the contrary, for one sample, the K-Ar dates increased from 175 to 190 Ma, as the size of the particles decreased from 1–2 to <0.4 μm. This is, to the best of our knowledge, the only time that such an age-size trend may give support to the idea of recrystallization by the process of Ostwald ripening (Eberl et al. 1990). As the present-day temperature is still about 150 °C at a depth of 5 km and no loss of radiogenic ^{40}Ar has been recorded by the author, one may suggest that the illite minerals formed at a significantly higher temperature 200 to

180 Ma ago. The author suggested that the high temperature event may correspond to the time of the breakdown of the western European craton at the beginning of the formation of the northern Atlantic Ocean. Liewig (1993) also noted that the illites which appeared more or less compacted and recrystallized yielded 180 Ma, whereas the noncompacted illites which grew in the secondary porosity gave dates of about 200 Ma. To explain this relationship, the author suggested two explanations. One, neoformation of illite and dissolution-crystallization of illite/smectite mixed layers occurred during the same event, which could have spanned over about 20 Ma. The age difference could have been induced by either a kinetic factor for the two processes or introduction of gas at about 200 Ma in the traps which stopped authigenesis of illite in the macropores, whereas brines remaining in the micropores favored continuation of the dissolution-crystallization of the illite/smectite mixed-layer minerals for another 20 million years. The other, illitic cementation in the pores and the dissolution-crystallization of pre-existing illite/smectite mixed-layer minerals occurred during two different events. As the recrystallized illites were found between compacted quartz grains and along pressured contact grain surfaces, this process could have happened during a specific event 180 Ma ago, after the neoformation process was blocked 20 million years earlier by gas entrapment. In this case, the event at 180 Ma could have been related to either the emplacement of the Bramsche plutonic massive during early Cretaceous, the effect of which on the maturation of the organic matter has been studied by Teichmüller et al. (1979) and Altebäumer et al. (1981), or late tectonic events and plumes in this region which have been suggested by Kettel (1983). This scenario would also explain the younger K-Ar ages for the illites of the Rotliegendes sandstones in the Groningen area of the Netherlands (Lee et al. 1989).

2.2.2 Rb-Sr Isotopic Studies

Literature is abundant on K-Ar isotopic dates on diagenetic illitic clay fractions in sandstones. In contrast, information about Rb-Sr results is scanty. A critical factor in the dating of any clay fraction by the Rb-Sr isotopic methods is knowledge of the initial $^{87}Sr/^{86}Sr$ ratio during the crystallization, the initial Sr isotopic ratio of the clay fraction being the same as that of the fluids from which the mineral accumulated the initial Sr. Several cogenetic samples which might have formed from fluids with nearly the same Sr isotopic composition and had a reasonable spread in their Rb/Sr ratio, can define a Rb-Sr isochron from which the age and the initial isotopic composition of the Sr integrated by the minerals during formation can be obtained. A Rb-Sr model age can also be calculated for each individual clay sample, but the ages thus calculated can be far from the true diagenetic age if the initial isotopic composition of Sr is wrongly assumed, or the clay materials analyzed are not free of any detrital or secondary impurity. All these reasons certainly explain why the Rb-Sr method is far less used for dating of diagenetic clay materials of sandstones.

The Rb-Sr data published by Liewig et al. (1987a) on illite of the Middle Jurassic Brent sandstones from the Alwyn area give a perspective on the complexity in the interpretation of such isotopic dates. The results available from this study were determined on clay fractions separated following the classical method of crushing for disaggregation of the rocks. The $^{87}Sr/^{86}Sr$ ratios of the size fractions ranged from 0.71425 to 0.72621, except for an ankerite-enriched fraction whose $^{87}Sr/^{86}Sr$ ratio was 0.70913. As the Sr isotopic compositions and the Rb/Sr ratios were found to be poorly correlated, no meaningful isochron age could be obtained. Therefore, the Rb-Sr model ages for the individual clay fractions ranged widely between 220 and 350 Ma. These model ages were found to be considerably higher than the range of 50 to 130 Ma for the K-Ar dates obtained on the same aliquots. Like the K-Ar dates, the Rb-Sr dates for the coarser fractions (1–2 μm) were somewhat higher than for the finer ones (<0.4 μm) of the same sample, reflecting increased influence of detrital components in the coarser fractions. As preferential diffusional loss of radiogenic ^{40}Ar out of K-bearing clay particles is no longer a convincing argument for the low K-Ar dates relative to the Rb-Sr values, a reason for the difference between the two ages could be that the Rb-Sr model ages are calculated based on the assumption of a too low initial $^{87}Sr/^{86}Sr$ ratio for the fluid from which the clay material formed.

Although the Rb-Sr isotopic dates may not be reliable for the calculation of the age of a diagenetic illite, reliable ages obtained by other independent means, such as by the K-Ar method, may be used to investigate the Sr isotopic evolution of clay minerals at the time of their formation. To obtain such information for the illites in the Alwyn reservoirs, initial $^{87}Sr/^{86}Sr$ ratios were calculated for the different size fractions by taking the 40 Ma K-Ar date as being the formation age of the diagenetic illite in the Alwyn reservoirs. The ratios ranged from about 0.7136 to 0.7239, with values within narrower limits for samples from the same drill hole. For example, the initial $^{87}Sr/^{86}Sr$ values for ten fractions, taken within a 30-m interval from a core at a depth of 3.31–3.34 km, ranged between 0.7136 and 0.7169. Part of the variations in these Sr isotopic compositions can certainly be attributed to the presence of varied amounts of Rb-bearing detritals carrying radiogenic ^{87}Sr. The overall variations of the initial $^{87}Sr/^{86}Sr$ ratio of the measured size fractions may also arise from isotopic variations of fluids within some limited scales. The difference in the initial $^{87}Sr/^{86}Sr$ ratio of authigenic illite may be suggested by the two trends in the relationship between the Sr isotopic compositions of illitic clays and the detrital feldspar to authigenic illite ratios of the clay assemblages from the same aliquots (Fig. 4.38). By extrapolating the trends, the $^{87}Sr/^{86}Sr$ ratios of the end-member component pure in illite phase may be defined as 0.712 and 0.709. The initial value of 0.709 for some illites, as determined by the extrapolation, corresponded to the Sr isotopic

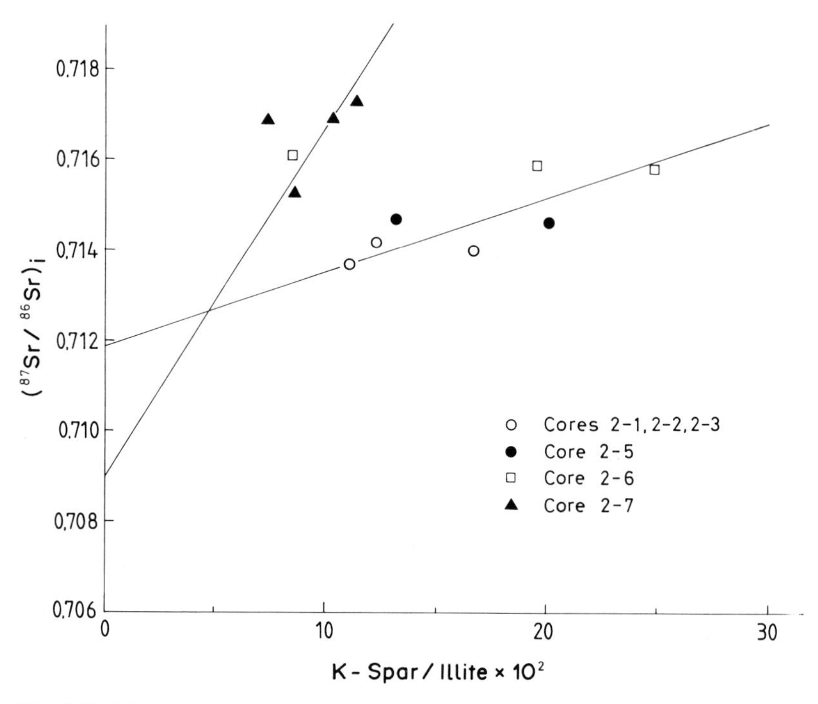

Fig. 4.38. Rb-Sr isochron data of clay samples from several cores in Alwin field. (Liewig et al. 1987a)

ratio of associated ankerite cements. The data may be interpreted to suggest that two generations of brines or two types of nearly contemporaneous brines were involved in the crystallization of the illites in the Brent reservoirs of Alwyn field.

2.2.3 Oxygen and Hydrogen Isotopic Studies

Diagenetic clay minerals in sandstones may be separated individually without significant contamination by detrital and nonclay diagenetic clay minerals. Exchange of oxygen isotopes between water and framework of clay minerals, especially illite, is considered to be extremely slow at temperatures of less than about 100 °C (James and Baker 1976; O'Neil and Kharaka 1976). By contrast, the rate of hydrogen isotope exchange can be rapid at most surficial temperatures (Lawrence and Taylor 1972; Bird and Chivas 1988). Hence, numerous efforts have been made to analyze the oxygen isotope compositions of diagenetic clay minerals in sandstones to reconstruct the history of pore fluids and uplift histories of strata in sedimentary basins (Land and Dutton 1978; Milliken et al. 1981; Longstaffe 1983, 1986; Sucheki and Land

1983; Dutton and Land 1985; Longstaffe and Ayalon 1987, 1990; Tilley and Longstaffe 1989; Longstaffe et al. 1992).

Land and Dutton (1978) observed that diagenetic mineralization in a Pennsylvanian deltaic sandstone sequence in north-central Texas consisted of early chlorite growth which was followed in order by syntaxial quartz overgrowths, cementation by calcite, dissolutions of calcite cement and feldspar, and finally cementation by ankerite and kaolinite. They considered that the Fe-rich chlorite formed as a result of reaction between amorphous aluminosilicates and Fe-oxy-hydroxides during shallow burial soon after deposition. The $\delta^{18}O$ value of the chlorite was about +14.3 per mill. Assuming a reasonable temperature under the burial condition, the authors calculated that the $\delta^{18}O$ value of the pore water was between +2 and +4 per mill. Quartz overgrowths which followed chlorite mineralization had $\delta^{18}O$ values of approximately +24 per mill. The quartz overgrowths formed presumably from silica derived from adjacent compacted shale beds undergoing conversion of smectite layers into illite layers during the maximum burial of the sequence of sediments. The temperature of burial corresponding to the quartz overgrowth was estimated and the $\delta^{18}O$ value of the fluid in the sandstones at this time was computed to be between +3 and +5 per mill. The latest major diagenetic event was likely to have happened when ankerite formation and kaolinitization of feldspar occurred more or less simultaneously as a result of late maturation of hydrocarbons. The $\delta^{18}O$ value of the kaolinite was about +18.8 per mill. The temperature was estimated at about 50 °C, and this suggested that the fluid $\delta^{18}O$ value was about −2.5 per mill. Thus the oxygen-isotope composition of pore waters in sandstones changed little through the periods covering the chlorite formation, the smectite to illite conversion and the quartz overgrowth in the sandstones, as the isotopic values of the fluids varied between +2 and +5 per mill. The path of the pore water oxygen-isotope composition shifted to significantly low ^{18}O values only during the very late stage of clay mineralization.

Milliken et al. (1981) reported that progressive burial of sandstones of the Oligocene Frio Formation in Texas resulted in intense diagenesis below 2.6 km depth. The diagenesis produced extensive reaction between the pore fluids and the detrital clays and feldspars in volcanic rock fragments, and also resulted in precipitations of Ca-carbonate with kaolinite at the upper zone and albitization at the lower zone. Abundant quartz overgrowth was found only below 3.6 km. This active zone of diagenesis, which is characterized by kaolinite formation, corresponded to the early smectite/illite transition at a temperature of about 100 °C, and the zone with albitization corresponded to a relatively more advanced illite/smectite transition at temperatures between about 120 and 150 °C. The $\delta^{18}O$ value of kaolinite averaged at about +20 per mill, making the corresponding $\delta^{18}O$ of water about +5 per mill. The average $\delta^{18}O$ of albite was about +17 per mill, which amounted to pore water with a $\delta^{18}O$ value between about +1 and +4 per mill. Lack of quartz overgowth in the zone of albitization or kaolinite formation has been explained as due to utilization of the silica in pore fluid by the formation of albite. The calculated pore fluid $\delta^{18}O$ values between +1 and +5 per mill, corresponding to the formations of albite and kaolinite in the sandstones, were found to be within the range of $\delta^{18}O$ values reported by Kharaka and Berry (1980) for the present-day formation waters in these sandstones.

Ayalon and Longstaffe (1988) examined the oxygen isotope compositions of diagenetic minerals in the Upper Cretaceous basal Belly River sandstone in southwestern Alberta and reconstructed changes in the pore-water oxygen isotope compositions during the diagenesis. The basal Belly River sandstone unit is a complex sequence of deltaic and nearshore fine to medium-grained sandstones interbedded with mudstone, shale, and siltstone. A general paragenesis of diagenetic minerals in the basal Belly River sandstone includes growth of chlorite in the early diagenesis and that of kaolinite, smectite, illite, and illite/smectite, preceding feldspar dissolution, in the late diagenesis. Calcite growths were dominant during both early chlorite formation and later kaolinite-illite-illite/smectite formations. Quartz overgrowths followed chlorite formation, but preceded the kaolinite-illite-illite/smectite formations. The $\delta^{18}O$ values of chlorite ranged from +6.1 to +8.3 per mill and that of quartz overgrowths ranged from +13.1 to +18.0 per mill. The $\delta^{18}O$ values of kaolinites ranged from +7.8 to +11.9 per mill, whereas that of the illite/smectite minerals ranged from +10.9 to +13.1 per mill and that of smectite from +12.6 to +13.7 per mill. Limiting the temperatures of burial to the range between 90 and 120 °C, based on the reconstruction of the burial history of the sediments, Ayalon and Longstaffe estimated that the pore water oxygen isotopic compositions ranged between 0 to −10 per mill. The oxygen isotope value of the pore water increased due to increased water-rock interaction attendant with burial. Quartz overgrowths began crystallizing at or near maximum burial. Burial diagenesis terminated by Laramide orogeny in the early Eocene. This was followed by recharge of the basin by low ^{18}O-bearing meteoric water and continued crystallization of quartz, kaolinite, illite/smectite, and smectite at lower temperatures during the post-Eocene cooling and erosion. Hence, low ^{18}O-bearing meteoric water played an important role basinwide during deposition and early diagenesis of these Upper Cretaceous sandstones and also following uplift of the basin in early Eocene time. Figure 4.39 is a summary of the changes in pore-water oxygen isotope compositions that accompanied diagenesis of the Upper Cretaceous basal Belly River sandstone.

Longstaffe and Ayalon (1987) reported the $\delta^{18}O$ values of diagenetic clay minerals in sedimentary rocks from Lower Cretaceous Viking Formation in the western Canadian basin. The Viking Formation consists of interbedded sandstones, conglomerates, and shales of a regressive marine sequence. Early, shallow diagenesis resulted in the formation of kaolinite, calcite, and chlorite. The early diagenetic kaolinite formed probably at

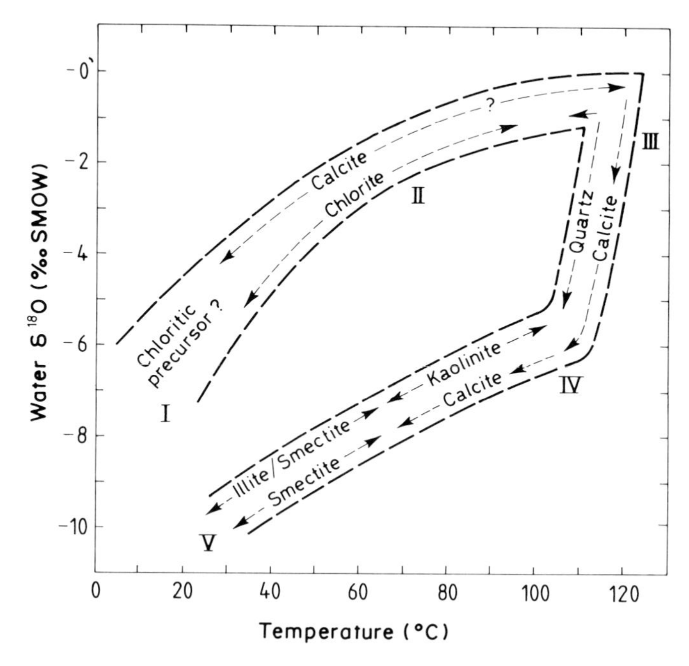

Fig. 4.39. Model-based evolution of the oxygen-isotope composition of pore fluids and diagenetic minerals in the basal Belly River sandstone, Alberta. (Ayalon and Longstaffe 1988).

temperatures of 10–30 °C. As burial diagenesis proceeded, diagenetic chlorite and quartz overgrowths formed. The burial diagenesis ended with uplift of the basin and influx of meteoric water which caused dissolution of feldspar deep inside the basin and formation of additional kaolinite at pore-water temperatures of about 120 °C. Pore waters declined in temperature and the final major stage of diagenetic clay mineralization involved formation of illitic clays at temperatures of about 45–65 °C. The early diagenetic kaolinites were found to have an average $\delta^{18}O$ value of $+26 \pm 2$ per mill. The calculated $\delta^{18}O$ value of the pore waters was about -1 per mill. Later kaolinites were found to have an average $\delta^{18}O$ value of $+14 \pm 3$ per mill. The illitic clay minerals had an average $\delta^{18}O$ value of $+14 \pm 2$ per mill. The calculated values of the pore waters at the time of formation of the illites ranged between -3 and -7 per mill. Thus, the pore waters seem to have evolved into ^{18}O depletions from early diagenetic kaolinite to late diagenetic illite formation. The authors found that the diagenetic kaolinite minerals in the Viking Formation, as in other Cretaceous sandstones in the basin, are not in oxygen equilibrium with each other and also not in oxygen isotopic equilibrium with present-day formation waters in the subsurface. Lack of oxygen isotope equilibrium also exists between the illitic clay mineral and the present-day formation water. This contrasts with the oxygen isotope relation between the illitic clay minerals and the present-day formation waters in the Upper Cretaceous Belly River sandstone where some of the diagenetic illitic clay minerals appear to have formed in isotopic equilibrium with the present-day formation waters (Longstaffe 1986; Ayalon and Longstaffe 1988). The oxygen isotope equilibrium between the illitic clay and the formation water in the Belly River sandstone may suggest neoformation of illite.

More recently, Longstaffe and Ayalon (1990) reported hydrogen isotope data of diagenetic clay minerals in the Viking Formation and the Belly River Formation. The δD values for the diagenetic kaolinites and illitic minerals in each stratigraphic unit were limited within a narrow range. For example, late kaolinites in the Belly River sandstone and the Viking Formation had δD values between -128 and -137 per mill and between -112 and -131 per mill, respectively. Similarly, the δD values for illitic clays ranged from -129 to -138 per mill in the Belly River sandstone and from -110 to -132 per mill in the Viking Formation. The limited range in the δD values contrasts with the considerable range in the $\delta^{18}O$ values. Unlike the lack of oxygen-isotope equilibrium, hydrogen-isotope equilibrium exists between diagenetic clay minerals and present-day formation waters, suggesting that the hydrogen isotope re-equilibration had occurred since the precipitation of the clay minerals. The authors maintained that hydrogen-isotope re-equilibration between diagenetic clays and formation waters in sandstones can occur independently of oxygen-isotope exchange at temperatures as low as 40 °C.

Although many have found poor correlations between oxygen and hydrogen isotopic values for diagenetic clay minerals in sedimentary basins, Fallick et al. (1993) reported significant positive correlations between $\delta^{18}O$ and δD values for both kaolinite and illite in the Kimmeridge Magnus sandstone in the northern North Sea province. The results of the study by Fallick et al. are in sharp contrast with that of others who have found no significant correlation between the two isotopic values for diagenetic minerals from other northern North Sea oilfields. As post-formation hydrogen isotope exchange is the main cause for poor correlation of δD and $\delta^{18}O$ values of clays in many fields, Fallick et al. suggested that the positive correlations for the data, suggesting the isotopic compositions to be inherited at the time of formation of the minerals, is best explained by a model of precipitation at a more or less constant temperature from pore fluids whose isotopic compositions varied accros the studied oilfield.

The sandstones of the Middle Jurassic Brent Formation constitute a major reservoir unit in the North Sea. Glasmann et al. (1989b) reconstructed the paleo-fluid dynamics in the Heather field from analyses of diagenetic minerals in Middle Jurassic Brent Group sandstones. Calcite cementation, K-feldspar dissolution, and kaolinite, quartz, and illite precipitations constitute the major diagenetic records in the paragenetic sequence. Illites, which post-date kaolinite, had K-Ar dates within the range of 55 to 27 Ma, indicating that precipitation of illite occurred through much of the

Paleogene. The average oxygen-isotope composition of kaolinite in the Heather field was +13.8 per mill, and the average hydrogen-isotope composition was −53.2 per mill. The kaolinite isotope composition was uniform across the field, in contrast to the varied isotopic compositions for the early formed calcite cements. Considering geologically reasonable temperature, the authors concluded that the oxygen isotope composition of pore fluids from which the kaolinite minerals precipitated ranged between −6 and −8 per mill. The field-wide uniformity in the oxygen isotope compositions of kaolinite minerals has been interpreted as extensive flushing of the reservoir by meteoric water at temperatures between 45 and 60 °C. The important period of kaolinitization preceded late Cretaceous deep burial (pre-90 Ma). Fluid inclusions in quartz suggest that entrapment temperatures ranged between 75 and 120 °C. The $\delta^{18}O$ values of illite were varied between +13.2 and +17.3 per mill, suggesting that the fluids from which the minerals formed at this temperature had an oxygen isotope composition between −3 and +5 per mill. According to Glasmann et al., such fluids probably evolved through time from partial mixing of trapped meteoric water in the Brent sandstones with saline compaction water from adjacent sub-basins.

Permian Rotliegendes sandstones, known as important hydrocarbon bearing reservoir units in southern North Sea, have been recently investigated for paleofluid environments during diagenetic formations of illite minerals. Lee et al. (1989) analyzed oxygen isotopic compositions of illites in the Permian sandstones of the northeastern Netherlands. Depending on the time of formation of the illitic clays, they divided the clay separates into two broad groups: the early diagenetic clays with K-Ar dates between 150 and 170 Ma, and the later clays with dates between 107 and 120 Ma. The $\delta^{18}O$ values of the older diagenetic illite clays were at about +20.7 per mill, whereas those of later illitic clays were between +17.6 and +19.9 per mill. The authors estimated that the $\delta^{18}O$ values of formation waters involved in the growth of illites were mostly between 0 and +5 per mill, although some waters could have values as low as −5 per mill, and thus that the diagenetic fluids involved in the growth of most of the illitic clays were slightly enriched in $\delta^{18}O$ values relative to sea water. The authors observed an apparent trend in the increase of the $\delta^{18}O$ values of the formation waters with increase in depths, suggesting increased mineral-water interaction with depth or less influence of meteoric water component.

Robinson et al. (1993) analyzed oxygen isotope compositions of illite-rich clay fractions, ranging in size from about <0.2 to 5 µm, separated from Permian Rotliegendes sandstones in Village Fields area in southern North Sea basin. The K-Ar dates of the illite samples averaged 158 Ma and suggested that illitization occurred across a large region during a relatively short time interval at Middle-Late Jurassic time. The period of illite growth appeared to be synchronous with rifting and subsidence of the basin. The $\delta^{18}O$ values of the clay separates ranged from +14.1 to +17.8 per mill (Fig. 4.40). The values of the finest fractions (<0.2 µm) were significantly lower

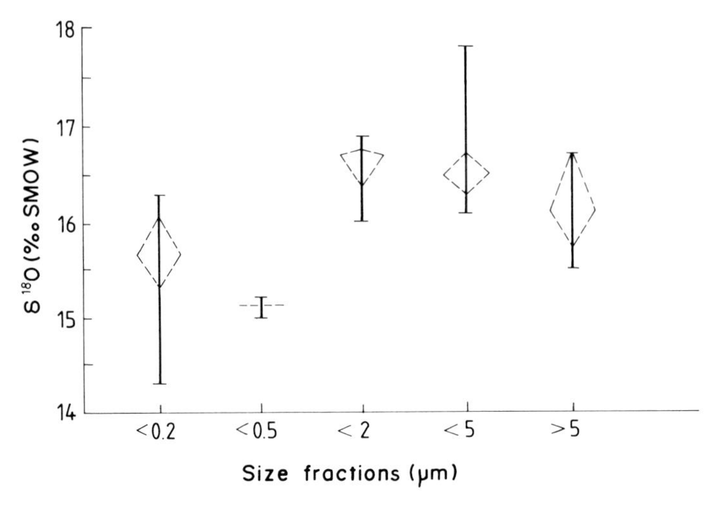

Fig. 4.40. Oxygen-isotope composition of different size fractions of sandstone samples from the Jurassic Brent Formation in the North Sea. (Robinson et al. 1993)

than that of the coarser fractions (0.5–2 μm). Correlation between the $\delta^{18}O$ values and current depths of the samples is extremely poor, but a trend seems to exist between the $\delta^{18}O$ values and Middle Jurassic burial depths (Fig. 4.41). This trend may be explained in terms of the precipitations of illite at different temperatures from fluids with a nearly uniform isotopic composition. This temperature-dependent isotopic variation of the clays suggests that the isotopic composition of the fluids had to be about +4 to +5 per mill.

2.3 Comparative Evolution of Diagenetic Clay Minerals of Buried Shale-Sandstone Associations

Isotopic investigations of clays in both shales and sandstones in association with each other are few, although potentially much can be learnt from the comparison of their isotopic signatures about the sources and movements of solutes involved in the clay diagenesis, the compaction history of the sediment, and the thermal and tectonic history of an entire sedimentary basin. The paucity of investigations seems to stem from the common difficulty in separating diagenetic clay minerals from the detrital ones, particularly in shales, producing a condition for a very low degree of success in relating the isotope transfers between shale and sandstone units. However, the works of Rinckenbach (1988), Glasmann et al. (1989a), Clauer et al. (in progress) and Furlan et al. (in progress), which are discussed below, illustrate that carefully controlled analytical data may be integrated to derive

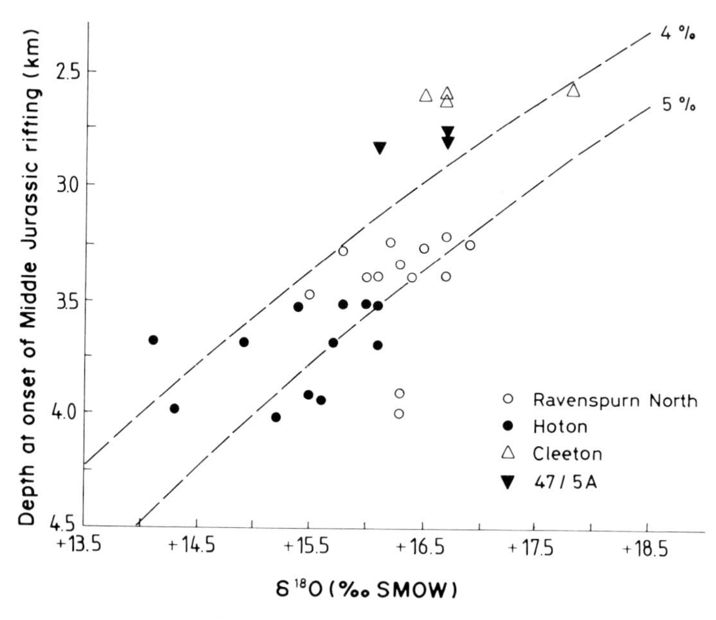

Fig. 4.41. The $\delta^{18}O$ values as a function of depth for clay separates of sandstone samples from the Jurassic Brent Formation in the North Sea. (Robinson et al. 1993)

important information about the differential clay diagenesis in shale and sandstone units and the paleofluid movements in a sedimentary basin.

2.3.1 The Mahakam Delta Deposits

Rinckenbach (1988) and Clauer et al. (in progress) examined the <0.4 μm clay fractions which were extracted, following a gentle thawing-freezing technique for disaggregation of the pieces of rock, from shales and associated sandstones of a 4-km-thick Miocene to Pliocene sedimentary succession in the bassin of the Mahakam Delta in eastern Kalimantan (Borneo). The analyses by XRD indicated a classical pattern of evolution of illite/smectite mixed layers with depth, changing from a randomly oriented mixed-layer phase across the top 1-km interval, through a short distance ordered mixed-layer clays at intermediate depths, to a long distance ordered phase at depths below 3 km. The percentage of illite layers in the mixed-layer phase relative to depth was described by an S-shaped curve like the ones found for similar clays in many other buried sedimentary sequences (Fig. 4.42). The XRD data also indicated that the compositions of the mixed-layer clay materials were identical in both the shales and the adjacent sandstones across the entire sequence in the Mahakam Delta basin.

Microthermometric analyses of the fluid inclusions contained in the quartz overgrowths gave two sets of values: one with a temperature range of

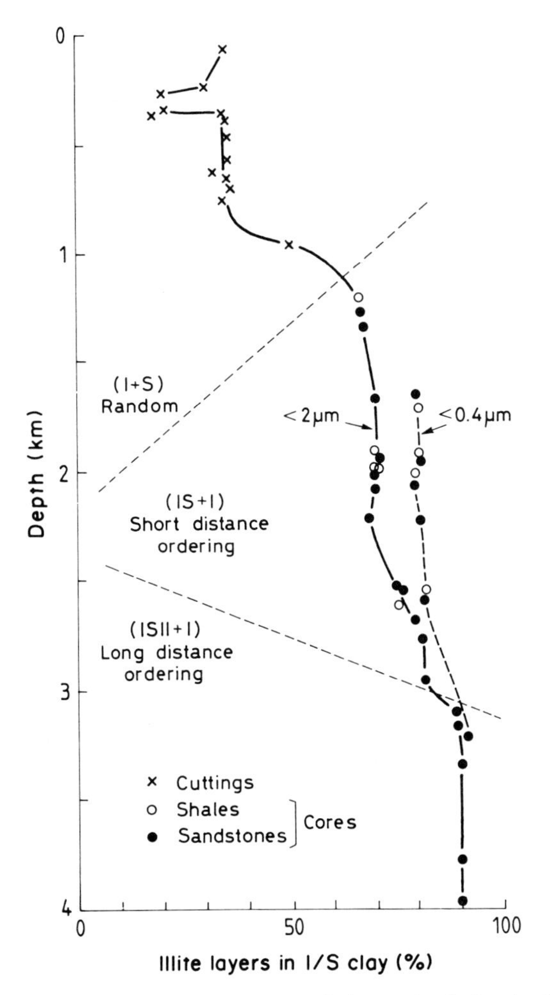

Fig. 4.42. The illite contents relative to depth for the illite/smectite mixed-layer minerals of shale and sandstone samples from the buried sedimentary sequence in the Mahakam Delta basin. (Rinckenbach 1988)

140–170°C and 3–6.5 wt% eq. NaCl and the second set with a temperature of about 130°C and 1.2 wt% eq. NaCl which corresponds to the values of the present-day brines (Chaudhuri et al. 1992c). The K-Ar dates were reasonably similar between the clay fractions of the shales and those of the sandstones in the very upper part of the sequence, but differed considerably in more deeply buried sediments. The isotopic dates of the <0.4-µm clay fractions of shales decreased from 80 Ma near the surface to 52 Ma at 3.16 km, whereas those from associated sandstones decreased from 80 Ma near the surface to 25 Ma at 2.6 km (Fig. 4.43). Observations by TEM revealed that the illite-type particles of the deeply buried sandstones were

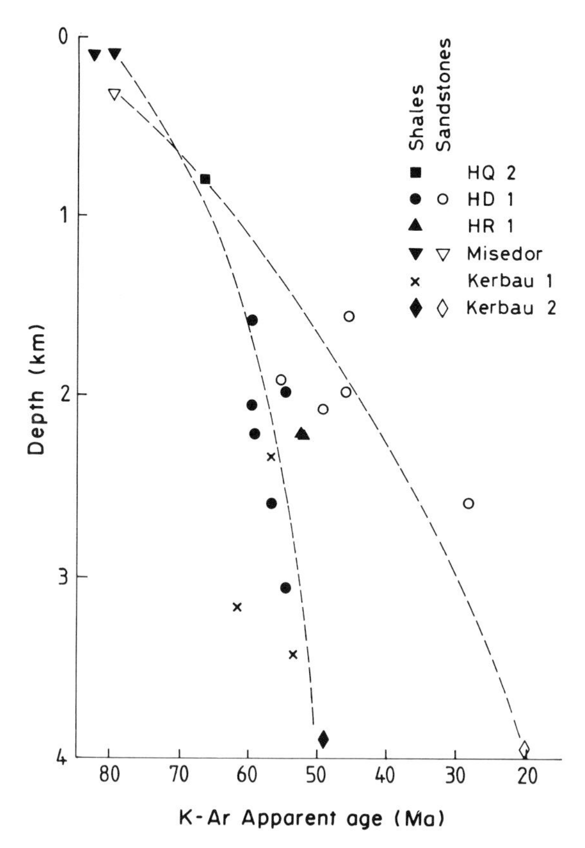

Fig. 4.43. Relationship between the K-Ar dates of the <0.4-μm clay fraction of shales and sandstones and depth in the buried sedimentary sequence of the Mahakam Delta basin. (Rinckenbach 1988)

euhedral lath-shaped, whereas those of the associated shales were very similar to that of the nonburied sediments: rounded particles with ill-defined edges. The $\delta^{18}O$ values of the <0.4-μm clay fractions from shales decreased from +14.7 to +12.6 per mill at a depth of about 3 km, whereas those from the associated sandstones decreased from +15.6 to +10.5 per mill at the same depth (Fig. 4.44). In assuming a temperature of about 170 °C for illitization at a depth of 4 km, which is the maximum temperature found in the fluid inclusions of the quartz overgrowth, the calculated $\delta^{18}O$ of the brines would be about +4.5 per mill in the shales and about +2.5 per mill in the sandstones. Clauer et al. postulated that the isotopic and the morphological differences between the clays in the two lithologically distinct rock types required a distinctly separate mechanism for the clay formation during burial. Based on the morphologic and isotopic evidence, they proposed that the mixed-layer clays in the shales evolved in a process of transformation of pre-existing clay structures in a relatively closed chemical system, whereas

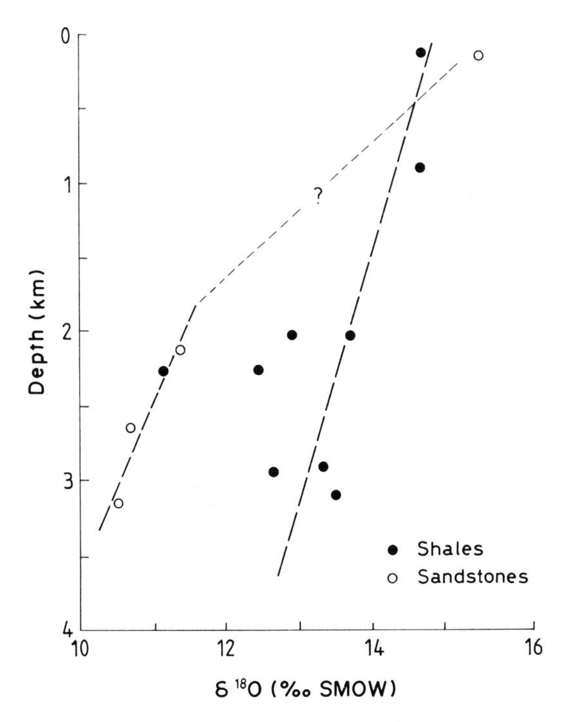

Fig. 4.44. Relationship between the $\delta^{18}O$ of the <0.4-μm clay fractions of shales and sandstones relative to depth in the buried sedimentary sequence of the Mahakam Delta basin. (Rinckenbach 1988)

similar illite/smectite mixed-layer clays in the sandstones grew by a process of dissolution and reprecipitation of pre-existing clay particles in a relatively open chemical system.

The illitization process during burial of sedimentary sequences is definitely dependent upon K availability. Hower et al. (1976) suggested that for shales, which seem to act as closed systems, the K for the illitization mechanism seems to have come from the dissolved K-feldspars in the same volume of rock, suggesting a direct internal transfer of K. Based on petrographic estimates on sandstone rocks progressively buried in the Mahakam delta, Furlan et al. (submitted) found that 1% of K-feldspar was present in the sandstones of the upper zone of illitization between 0.35 and 2 km, and that less of 5% of the feldspar volume was altered. In the same zone, the authors found that 5% of illite/smectite mixed-layer minerals were present and that the amount of change in the expandability of the mixed-layer phase was as much as 35%. The calculated amount of K available from alteration of the K-feldspar was found to be 20–25 times less than the amount of K needed for the illitization process. Furlan et al., therefore, suggested that additional supply of K had to come from a place outside the illitization zone, and they claimed that this supplementary K came by upward migration from a zone as deep as 4.2 km where dickite and silica precipitation were visible.

2.3.2 The North Sea Deposits

In the Norwegian sector of the northern North Sea, Glasmann et al. (1989a) analyzed the fine clay fractions (<0.1 μm) of both the Tertiary and Mesozoic shales and the underlying Jurassic sandstones of the Brent Group from three different fields (Huldra, Veslefrikk and Oseberg). The clay minerals were separated in different size fractions after disaggregation of the samples by gentle crushing and mild sonic treatment. In the shales, the mineralogic evolution of the illite/smectite mixed-layer minerals relative to depth gave an S-shaped trend from randomly ordered smectite-rich fractions above 2.2-km depth to regularly ordered illite-rich fractions below 2.4-km depth (Fig. 4.45). Electron microscopic observations revealed that the illite-type minerals consisted of predominantly euhedral lath-shaped crystals with small amounts of platy crystals and also that some very minor amounts of detrital mica were visible in the analyzed samples. The relationship between the lath and the platy illites could not be clearly explained. Inoue et al. (1987) suggested that lath-shaped illites transform through platy illites to $2M_1$ mica with increase in the temperature of alteration. The K-Ar dates on the

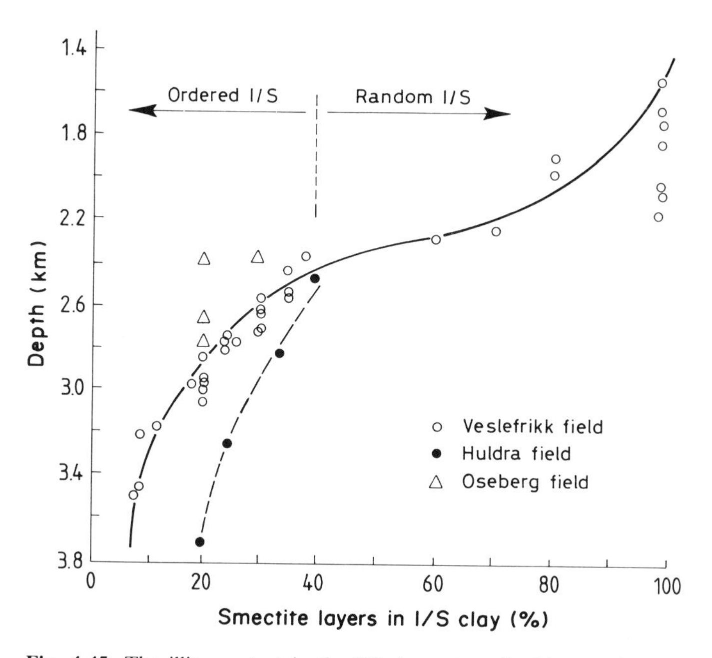

Fig. 4.45. The illite content in the illite/smectite mixed-layer minerals in shales of the Brent Formation in three fields of the Bergen High area. (Glasmann et al. 1989a)

<0.1 μm clay fractions of Veslefrikk field increased from about 64 Ma at 2.55 km to about 87 Ma at 3.5 km (Fig. 4.46). In Huldra field, the K-Ar dates of the fine clay minerals remained constant at about 77 Ma across the 2.47–3.9 km interval, with a slight increase to 82 Ma at 4.13 km. In Oseberg field, the K-Ar dates decreased from 125 Ma at 2.73 km to about 100 Ma at 2.77 km. The measured isotopic dates of the fine clay fractions of the shales were lower than the corresponding stratigraphic ages, suggesting that the illitic mixed-layer minerals were primarily diagenetic in origin. However, the scatter in the dates suggested the presence of small amounts of detrital K-bearing components, and the measured dates were corrected by using a

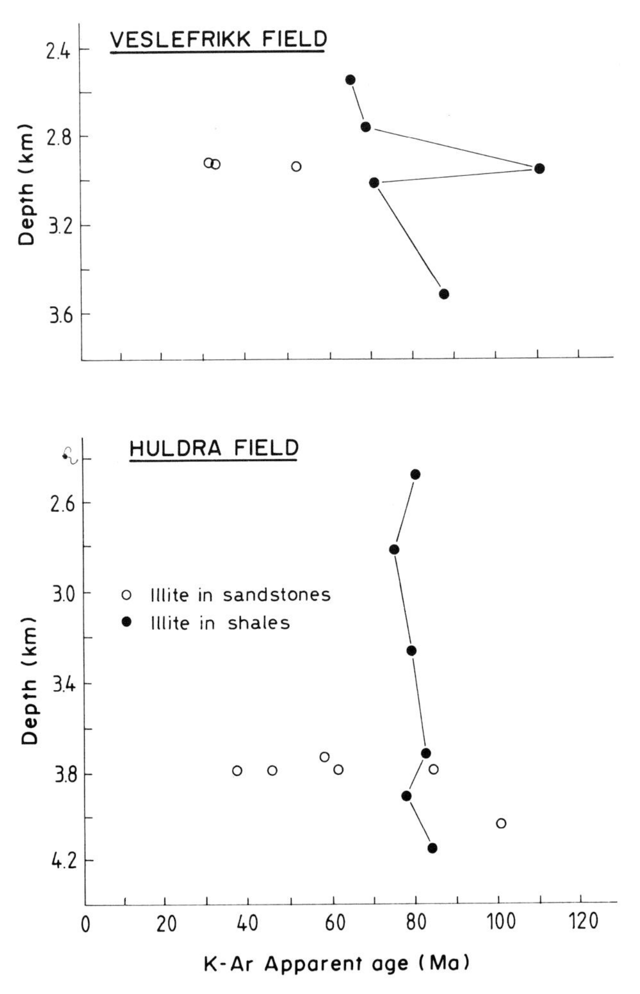

Fig. 4.46. Relationship between K-Ar dates and depth for the <0.1-μm clay fraction in Brent shales and sandstones of two fields of the Bergen High area. (Glasmann et al. 1989a)

mean age of the detrital component and an estimate of its abundance. The corrected date appeared to be 65 Ma.

Data of fluid-inclusion microthermometry obtained by Glasmann et al. on quartz overgrowths of the Jurassic Brent sandstones which underlie the Tertiary-Mesozoic shales yielded a maximum temperature of 180 °C at about 4-km depth in Huldra field. The fluid-inclusion temperatures were systematically higher than the measured temperatures at the same depths, and hence the authors concluded that either the thermal gradients were different during the formation of the clay particles, or "thermal pulses" might have occurred during the burial history of the sediments in the North Sea. The illite particles of the Brent sandstones very commonly were euhedral lath-shaped, suggesting that their growths were unconstrained by the size of the pores. The K-Ar dates of these illitic fine fractions ranged between 38 and 98 Ma. The authors acknowledged that the high dates were also due to increasing amounts of detrital materials, but the dates were significantly lower than the time of deposition. A lath-shaped <0.1 µm illite from hydrocarbon saturated zone gave a K-Ar age of about 58 Ma, and similarly shaped illite between water-saturated and hydrocarbon-saturated zones yielded a significantly lower age of 38 Ma. The authors suggested that the difference between the two values might be due to illitization terminating sooner in the hydrocarbon level, immediately after accumulation of the hydrocarbons, whereas diagenesis continued in the water zone. The lowest age of 38 Ma reported by Glasmann et al. (1989a) agrees well with the average of 41 Ma obtained by Liewig et al. (1987a) for the clay fractions from the Alwyn field and by Cocker et al. (1988) for those from the Hutton Field. But blocking of the diagenesis at 58 Ma in the zone where hydrocarbons accumulated remains questionable when compared to the 43 Ma obtained by Cocker et al. (1988).

The comparison in the K-Ar dates of diagenetic illite-type minerals between the sandstones and the shales in the Bergen High area led Glasmann et al. (1989a) to suggest that the illitization in the Brent sandstones continued over a long period, beginning about 90 Ma ago in the Late Cretaceous time and continuing until 38 Ma in the Neogene time. Accordingly, migration of hydrocarbon fluids from the deep Viking graben into the Brent sandstones may have began as early as Late Cretaceous. The K-rich warm fluids which were accumulated in the Brent sandstones leaked continuously through Late Tertiary time to the overlying shales, until deep burial caused considerable decrease in the porosity of these fine-grained sediments. The leakage of the warm K-fluids from sandstones into the overlying shales presumably altered the chemical and thermal environments of pore waters in the poorly compacted shales, thereby promoting dissolution of smectites and precipitation of illites during the late Cretaceous to early Tertiary time. The porosity change in the shales due to deep burial during the Late Tertiary brought a halt to the precipitation of illite in the shales much earlier (at 65 Ma) than in the underlying sandstones (at 38 Ma).

2.4 Duration of Diagenetic Illite Formation

The formation of authigenic clay minerals in diagenetic or low-temperature metamorphic environments is thermodynamically controlled in response to the fluid chemistry, temperature, and pressure. The growth of a particular mineral is kinetically controlled by a number of factors including transport of ions to the reaction surface or site, process of surface reactions manifested as ion exchange, dehydration, etc., removal of the reaction products, and temperature (Fisher and Lasaga 1981). In addition to these factors, several others that appear important for the drive of the smectite to illite reaction are pore fluid chemistry, presence of detrital components providing K and Al to drive the reaction, initial composition of the smectite, and the length of time that clay sediments have been subjected to the critical burial temperature (Perry and Hower 1970; Hower et al. 1976; Foster and Custard 1982; Anjos 1986). Opinions may differ about relative importance of these different factors for the change from smectite to illite, probably the most important diagenetic clay reaction in progressively buried sediments. Most would consider that temperature, coupled with the length of time the clay materials have been subjected to the critical temperature, is probably the dominant factor in this mineralogical change in deeply buried sediments (Eberl and Hower 1976; Hoffman and Hower 1979; Roberson and Lahann 1981; Inoue 1983; Howard and Roy 1983; Jennings and Thompson 1986; Ramseyer and Boles 1986; Whitney and Northrop 1988; Pollastro 1989).

Hoffman and Hower (1979) suggested that the change from random to ordered illite/smectite mixed-layer minerals occurs at some critical temperatures over several million years by residence in long-term burial settings with average geothermal gradient. The critical temperatures in these settings are about 100–110 °C for the change from random ($R = 0$) to short-ordered ($R = 1$) and about 170–180 °C for the change from short-ordered ($R = 1$) to long-range-ordered ($R = 3$) illite/smectite mixed-layer minerals. On the other hand, studies by Jennings and Thompson (1986) and Ramseyer and Boles (1986) have indicated that the change from random to ordered illite/smectite mixed-layer clays occurs also at some critical temperatures over a short period of less than 2–3 million years in short-lived geothermal settings with elevated geothermal gradient. The critical temperatures for the change in these geothermal systems are about 130–140 °C for the transformation from random ($R = 0$) to short-ordered ($R = 1$) and about 170–180 °C for the conversion from short-ordered ($R = 1$) to long-range-ordered ($R = 3$) illite/smectite mixed-layer minerals. Eberl and Hower (1976) have suggested from experimental data that the time involved in the change from smectite to illite/smectite mixed-layer minerals with low expandability may vary from about 1 million years at 60 °C to about 100 000 years at 80 °C. The study of Cocker et al. (1988) for K-Ar dates of illites in oil reservoirs of the Brent Formation has suggested that illite precipitation occurred in a short time of less than 1 million years at temperatures of about 80 °C. Small (1993) predicted from experimental results that illitization at threshold temperature of about 60 °C would correspond to a precipitation time of about 2 million years, and that at a temperature range of 80–90 °C it could correspond to less than 100 000 years.

The work of Hower et al. (1976) provided a very convincing argument to develop a common notion that the illitization of smectite is highly dependent on the rate of dissolution of K-feldspar that is providing the K necessary for the illitization process. Altaner (1986) compared potential rates of the illitization of smectite with that of the dissolution of K-feldspar based on some estimates for activation energy and frequency factor for different orders of reaction at different temperatures. He determined that between 50 and 220 °C the rates of K uptake by the smectite in the process of illitization were much slower than the rates of dissolution of K-feldspar, suggesting that the K-feldspar dissolution rate does not influence the rate of illitization of smectite in deeply buried sediments.

2.5 Reconstruction of Thermal Histories of Sedimentary Basins

2.5.1 The Paris Basin, France

Part of the thermal history of the Paris Basin has been unraveled through isotopic analyses of authigenic clay minerals in upper Triassic (Rhaetian) sandstones in north-central France (Mossmann et al. 1992). The sandstones consist of sandy tidal flat-type sediments alternating with pelitic and silty layers in the eastern part of the basin, but change progressively to open marine silty sandstones toward the central part of the basin. The Paris basin has had an apparently simple tectonic history with the pre-Triassic basement having a major fault zone in the northern part of the basin and a series of roughly E-W-trending echelon faults and being covered by a saucer-shaped accumulation of post-Triassic sediments. The major tectonic features of the basin were established in Hercynian times, but some later Mesozoic and Cenozoic activities affected both the basement and its sedimentary cover (Pomerol 1974; Curnelle and Dubois 1981).

Mossmann et al. (1992) analyzed K-Ar dates of clay fractions in Rhaetian sandstone from samples taken along a SW-NE traverse across the basin. Along this traverse, the sandstone unit occurs at depths ranging from outcrop to 2.7 km below the surface. Quartz, feldspar, and sericitized to kaolinitized lithic fragments were the major detrital grains in the sandstone samples. Feldspar grains occurred mostly as highly corroded K-feldspar at the shallower part of the basin. Siliceous, argillaceous, calcareous, and sulfatic cements were commonly observed in the sandstone samples. Kaolinite, chlorite, illite/smectite mixed-layers, and smectite in various amounts constituted the essential clay mineral compositions of the rocks. Kaolinite, common in the shallow part of the basin, made up a major fraction of the coarse clay, whereas illite/smectite mixed-layer clay was most abundant in the fine fractions. Chlorite occurred in the deep part of the basin. Only in the outcrop samples was discrete smectite abundant. Based on diffraction patterns of <0.2 µm clay fractions of sandstone samples, Mossmann et al. (1992) recognized three distinctly different types of illite/smectite mixed-layer minerals: illite/smectite with less than 10% smectite which is present in dominant amounts in all samples from shallow (1.2 to 2.7 km) to deep levels, illite/smectite with about 20% smectite and illite/smectite with about 35%, the latter two being relatively minor. Analyzing the K-Ar dates of <0.2 µm clay fractions of samples from different depths and deconstructing the composite array of the dates, the authors established three different growth periods for the three different types of illite/smectite minerals. The mixed-layer minerals with less than 10% expandable layer yielded K-Ar dates between 189 and 192 Ma. These ages were assumed to be close to the age of crystallization of this illite-rich mineral. The mixed-layer minerals containing 20% expandable layer gave a mean calculated K-Ar date of about 150 Ma, and that of the mixed-layer with about 35% expandable layers yielded about 80 Ma (Fig. 4.47).

Several of the size fractions that were dated by the K-Ar isotopic method were also analyzed for their oxygen isotope compositions (Clauer et al. 1994). The 190-Ma-old illite/smectite mixed-layer phase, that is the one with less than 10% expandable layers, were distinguishable into two groups, based on the oxygen isotope, although both formed at the same time and at nearly the same depth of about 1 km (Fig. 4.48). Of these two groups, one

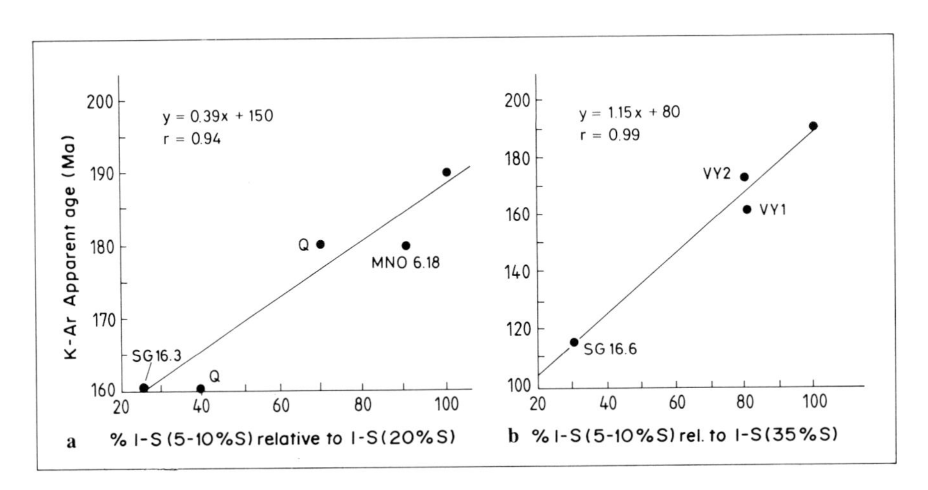

Fig. 4.47. K-Ar dates of different generations of illite/smectite mixed-layer minerals in Triassic sandstones from the Paris Basin. **a** Relationship between phases 1 and 2; **b** between phases 1 and 3. (Mossmann et al. 1992)

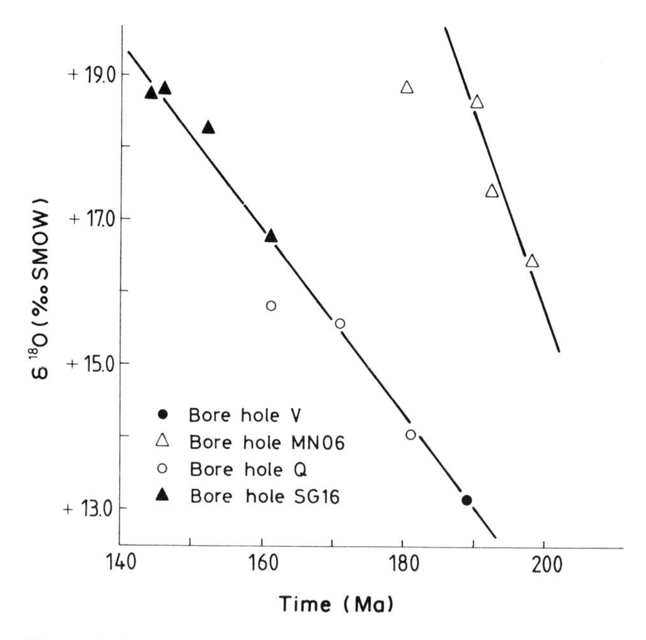

Fig. 4.48. Relationship between K-Ar dates and $\delta^{18}O$ values of <0.4 μm illite/smectite mixed-layer minerals in Triassic sandstones from the Paris Basin. (Clauer et al. 1994)

consisted in $\delta^{18}O$ values of about +17 per mill, belonging to those illite-rich minerals occurring in rocks near or within a major fault system in the basin. The other consisted in $\delta^{18}O$ values of +13 per mill, belonging to those illite-rich minerals occurring in rocks in the deeper parts of the basin. The difference in the oxygen isotope compositions between the two groups of illite of the same age can be related to temperature, composition of the reactant fluids, water-to-rock ratio during mineralization, or to all of these. The temperature-dependent difference would suggest that the illite-rich minerals in close proximity to the fault system formed at a lower temperature than that in the presently deep part of the basin. The difference in temperature between the two areas within the basin was probably not sufficient to account for the difference in the $\delta^{18}O$ values of the two groups of illite-rich fractions. On the other hand, the influence of the water-to-rock ratio cannot be ignored in view of the fact that the isotopic difference is attributable whether the minerals formed near or away from the fault.

As the 150- and the 80-Ma-old illite/smectite phases occur in association with the 190-Ma-old phase, and not as a single phase in any of the clay separates, the $\delta^{18}O$ values of the two later phases had to be extrapolated from the trends in changes in the $\delta^{18}O$ values with changes in the amounts of each phase. The trend is well constrained for the $\delta^{18}O$ values described by varied mixings between the 190-Ma-old phase and the 150-Ma-old phase.

The extrapolated $\delta^{18}O$ value of the 150-Ma phase, as determined from trend is approximately +17.6 per mill. By comparison the trend in $\delta^{18}O$ values by varied mixings between the 190-Ma-old phase and the 80-Ma-old phase is poorly constrained, giving a roughly approximate $\delta^{18}O$ value of about +23 per mill for the 80-Ma-old phase. The extrapolated $\delta^{18}O$ values of the two later illite/smectite phases suggest that they probably formed at a lower temperature than the earliest one.

Clauer et al. (1994) also found about +17.5 per mill for the $\delta^{18}O$ values of quartz overgrowths in association with the formation of the 190-Ma-old mixed-layer phase. Based on calculated oxygen-isotope fractionation coefficients of coprecipitating quartz and illite, the authors obtained a crystallization temperature of 230 ± 20 °C for the illite-rich mixed-layer phase occurring presently in the deep part of the basin. This temperature is significantly higher than the present temperatures in the wells. Assuming a mean thermal gradient of 35 °C/km, comparable to the thermal gradient today, a burial of about 6 km would be needed to explain the paleotemperature of about 230 °C. This is not in agreement with the tectonic history of the basin, and thus the illite-rich clay fractions must have originated from a hydrothermal fluid.

Based on Huon et al.'s (1993) program for Ar diffusion related to time, temperature, and size of the particles, Clauer et al. (1994) estimated how much time would be needed for the complete erasing of the inherited Ar memory in the illitic clay minerals in Paris Basin if the clays were assumed to have derived from a source of Hercynian age and were exposed to the various estimated temperatures that existed in the basin at different times. They calculated that a detrital <0.4 μm illitic clay fraction with a Hercynian age of 360 Ma would have lost all its Ar in less than 1 million years at a temperature condition of about 230 °C, which presumably existed in the basin 190 Ma ago. Further calculation shows that the <0.4 μm illite/smectite fractions, formed at 190 Ma and exposed at an estimated temperature of 180 °C during the thermal event at 150 Ma, would have lost their Ar in less than 1 million years. Additional calculation shows that should the 150-Ma-old illite/smectite mixed-layer phases become exposed to a temperature of 150 °C about 80 Ma ago, these clays would loose all their Ar in a time span of about 37 million years. According to Mossmann et al., the two former thermal events in the Paris Basin are of hydrothermal origin, implying high temperatures and short durations. Clauer et al.'s calculations show that Mossmann et al.'s inferences are plausible. The calculations suggest that the latest event could have been a long duration diagenetic process, which is also suggested by evidence from apatite fission track dates ranging from about 90 to 50 Ma.

In summary, the paleo-temperature dates for the two former events give very high thermal gradients of 60 to 105 °C/km, and therefore indirectly support the evidence of the occurrence of hydrothermal activities possibly related to basement tectonic activities or readjustments. Mossmann et al.

(1992) and Clauer et al. (1994) suggested that the abnormally high formation temperatures of the two early authigenic clay mineral generations, at least of the Triassic units, were associated with migrations of hydrothermal fluids along fault systems acting as pathways. Convective heat transfer at this time appears to be reasonable, considering that contemporaneous tectono-hydrothermal activity was recorded at several other places of Western Europe related to the breakdown of the west-European platform which generated the northern Atlantic Ocean (Bonhomme et al. 1983; Bonhomme et Millot 1987; Liewig et al. 1987b; Liewig 1993). R. Worden (pers. comm.) reconstructed patterns of sedimentation rates through time for the Paris Basin (Fig. 4.49). These patterns show that the sedimentation rates were greatly accelerated during the 200–180 Ma and the 150 Ma periods and also, to a lesser degree, during the 100–80 Ma period. The paleothermal history of the Paris Basin reconstructed from the age-temperature relationship for the clay minerals underscored three major thermal events, at periods of around 190, 150, and 80 Ma. Hence, the thermal events closely correspond in time to the three accelerated sedimentation rates. We believe that this relationship reflects basement readjustments which could have induced high heat flow and accelerated subsidence; but the true significance of such a correspondence has yet to be worked out and requires similar studies in other basins.

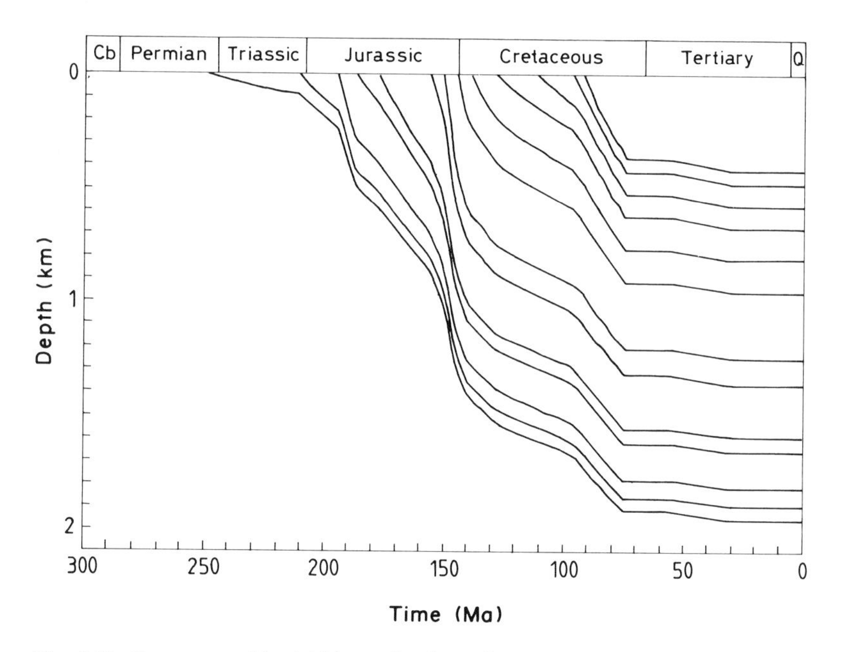

Fig. 4.49. Decompacted burial history for the sedimentary sequence in the Paris Basin. (After R. Worden, pers. comm.)

2.5.2 The Cuanza Basin, Angola

The Lower Cretaceous arkose sandstones in the Cuanza Basin of the west equatorial African margin were well studied during the offshore petroleum exploration. Girard et al. (1989) recently completed a study of mineralogic, isotopic, and fluid inclusion analyses of these rocks to determine the timing and formation temperatures of varied diagenetic cements which included clay minerals. The arkosic sediments of the basin were typically of a rift margin transgressive-upward sequence. Feldspars such as orthoclase, microcline and plagioclase, and biotite made up the major framework minerals. Illite/smectite mixed-layer phases and illite together with small amounts of chlorite constituted the bulk of the diagenetic clay fractions. Other diagenetic minerals included K-feldspar, quartz, calcite, dolomite and a few others. Clay cements appeared late in the diagenetic evolutionary history, and their K-Ar dates typically decreased with decreases in size. The K-Ar dates suggested that illite/smectite mixed-layer minerals formed at about 50 Ma at depths between 1.0 and 1.5 km during the Eocene, and a pure illite at depths greater than 2 km about 16 Ma ago during the Miocene. The temperature of formation for the illite/smectite phase was estimated to be in the range of 65–125 °C from a fluid with $\delta^{18}O$ values between +2 and +9.5 per mill. The estimated temperatures of formation for the pure illite was significantly higher than that of the mixed-layer phase, at a range between 140 and 195 °C, from a fluid with a $\delta^{18}O$ between +5.6 and +10.5 per mill.

We find the temperature and the age relationship among the clay minerals reported by the authors to be troublesome. It is difficult to explain, to our best knowledge, how the K-Ar system of the illite/smectite mixed-layer phase formed 50 Ma ago in a temperature range of 66–125 °C remaining unaffected by the subsequent formation of the illite that occurred 34 Ma later in a temperature range of 140–195 °C, knowing that this illite is stratigraphically located less than one hundred meters above the illite/smectite mixed-layer phase in the same well.

3 Isotopic Records of Clay Diagenesis in Formation Waters

Isotopic and chemical signatures of present-day deep subsurface waters in sedimentary basins may carry records of clay diagenesis in connection with the deep burial of the sediments. These deep subsurface waters have been variously termed as brines, formation waters, oil-field waters, or connate waters. Formation waters in sedimentary basins are commonly of Na-Cl to Na-Ca-Cl type, with salinities ranging from a value near the salinity of average sea water to as much as ten times that of average sea water (White 1965; Collins 1975). The various processes by which the solutes in these waters could have originated have been summarized in several recent articles (Hanor 1987; Chaudhuri and Clauer 1992b; Kharaka and Thordsen 1992). Deep subsurface waters are believed to be most often mixtures of waters of different origins, although several examples of predominantly local meteoric origin have been recognized (Clayton et al. 1966; Kharaka and Carothers 1986; Knauth and Beeunas 1986; Sheppard 1986).

A potentially major impact of clay fractions on the composition of the formation waters is the fractionation of the oxygen and hydrogen isotopes of water moving updip through clay-rich sediments behaving like a membrane in a deeply subsiding basin. Laboratory evidence and theoretical considerations suggest that the membrane filtration effect, while causing decrease in the concentration of solutes in effluent water, lowers the δD and the $\delta^{18}O$ values of the effluent water relative to that of the influent water (Graf et al. 1965; Coplen and Hanshaw 1973; Phillips and Bentley 1987; Demir 1988). Results of laboratory experiment forcing water through compacted montmorillonite by Coplen and Hanshaw have shown that the effluent waters are depleted in δD by 2.5 per mill and in $\delta^{18}O$ by 0.8 per mill relative to the residual or influent waters.

Stable isotope compositions of pore waters in sediments may be modified by reaction with clay minerals at basinal temperatures (Savin 1980; Milliken et al. 1981; Kharaka and Carothers 1986; Longstaffe 1989). The extent of modification is dependent upon the compositions of the two reacting species and the temperature of the reaction. The shift in the oxygen isotopic composition of the water due to reaction with clay minerals is difficult to differentiate from that due to reaction with other common minerals such as carbonate or non-clay silicate minerals undergoing dissolution. However, since the clay minerals are important reservoirs of hydrogen in sedimentary rocks, reaction between clay minerals and waters may be potentially seen through the δD of the formation waters (Kharaka and Carothers 1986). Relation to changes in the hydrogen isotopic compositions of formation waters due to clay reactions should be evaluated with some caution. Hydrogen isotope compositions of water may be shifted due to large isotope-exchange fractionation between water and hydrogen sulfide gas. The isotope exchange between water and hydrogen sulfide can be locally important (Hitchon and Friedman 1969). Also, hydrogen gas may be generated in significant amounts in sedimentary basins containing a large amount of organic deposits. The exchange reaction between hydrogen gas and water is rapid and the fractionation factor is also very large, potentially causing a large shift in the hydrogen isotopic composition of the water.

Studies of isotopic compositions of Sr in formation waters have proved useful in understanding the chemical evolutions of these waters (Starinsky et al. 1983; Stueber et al. 1984, 1987; Chaudhuri et al. 1987, 1992b; McNutt et al. 1987). The $^{87}Sr/^{86}Sr$ ratios of formation waters may be influenced by reactions with clay minerals, especially illitization of smectite minerals, but other common mineral reactions such as dissolution-replacement of alkali feldspar minerals and dissolution-recrystallization of marine carbonate and sulfate minerals are also important (Chaudhuri and Clauer 1993). Illitization of smectite minerals will not greatly increase the Sr isotopic compositions of pore waters, much of the elevation being most likely due to alteration of K-feldspar and mica minerals which provide the K necessary for the clay transformation or its precipitation. Many formation waters are enriched in Rb but depleted in K relative to evaporated sea water. Chaudhuri and Clauer explained that the K/Rb ratios of the waters generally relate to the conversion of smectite to illite under conditions of deep burial diagenesis.

Ion-exchange interactions with clay minerals can potentially influence the Sr isotopic compositions of formation waters. Experiments on leaching of clay minerals with dilute HCl acids give a perspective of the limits of the isotopic compositions of Sr that can be derived from different clays. In the case of illite and illite/smectite mixed-layer minerals, the $^{87}Sr/^{86}Sr$ ratios of the exchangeable Sr, which includes mainly the surface-adsorbed and some interlayered Sr, are seldom higher than 0.714 and often less than 0.712, and for smectite even less than 0.710 (Clauer 1976; Chaudhuri and Brookins 1979; Clauer et al. 1984, 1990). These studies also indicate that the amount of exchangeable Sr in illite-rich clay minerals is small, between a tenth and a few micrograms per gram of the minerals, but in smectite it is about five to ten times that in illite. These results clearly show that formation waters with Sr isotopic values much above 0.715 derived Sr from reaction with at least detrital K-feldspars or micas.

4 Summary

Clay minerals in sedimentary rocks bear important records for many events in a long and complex history of these rocks since the time of their deposition. Different kinds of isotopic analyses of these minerals have proved extremely useful in finding information, not only about the periods of these different events, but also about chemical dynamics in fluid-rock interactions in a sedimentary basin at these various times.

Clay aggregates in an ancient sedimentary rock are products of components having diverse origins of detrital, syndepositional, and diagenetic types. The determination of the time of deposition of a sedimentary rock by isotopic analyses of its clay minerals requires extremely careful separation of the syndepositional clays from the rest and critical information about any potential post-depositional isotopic modification of the clay minerals, the failure of which can result in an erroneous interpretation of the isotopic data. Early K-Ar and Rb-Sr isotopic studies for the purpose of determining the time of deposition have produced mixed results, yielding isotopic dates that ranged widely from excess to deficient relative to the stratigraphic ages.

Many early Rb-Sr isotopic studies were carried out on whole-rock samples. The whole-rock data for stratigraphic purpose is often meaningless, unless the geologic evidence strongly suggests that the sediments derived from very young contemporaneous crystalline source rocks and that the diagenetic modification of the constituents of the rocks occurred essentially very shortly after the deposition. Some early efforts focused on analyses of only clay fractions of the sediments, but little or no care was taken in many of these studies to determine whether or not the analyzed samples composed of clay minerals differed widely in their history of formation. Where stratigraphic ages have been poorly constrained, such as in many Precambrian sedimentary rocks, because of lack of well identifiable fossil records, isotopic dates produced in these studies were uncritically assumed to correspond to a good approximation of the time of deposition. The claim of success in defining the time of deposition for ancient sedimentary rocks may be justified enough for those studies that have carefully integrated varied isotopic data with detailed mineralogic and chemical data. Few recent studies have emphasized a large potential of the Sm-Nd isotopic method, as compared to that of the Rb-Sr and the K-Ar isotopic methods, for dating Archean and Proterozoic rocks, because the REE seem to be much less sensitive to post-depositional changes than the alkali and alkali-earth elements.

A highly successful line of investigation for stratigraphic dating is the study of glauconites whose evolutions may be completed in environments that are in close proximity to sediment-water interface. Those studies which have endeavored to analyze only very carefully selected samples of glauconites have demonstrated that isotopic dates from glauconite can be interpreted as the dates of the sediments containing these minerals. Highly evolved glauconites with K_2O content more than 4.5% appear to yield K-Ar and Rb-Sr isotopic dates that are not only in agreement with each other, but also in close agreement with the time of sedimentation determined by independent means. The low K_2O-bearing glauconitic materials often reflect varied degrees of residual isotopic memories of replaced precursors from which glauconites evolved. Both Rb-Sr and Sm-Nd isotopic data have revealed substantially the course in the chemical evolution of glauconites from replacement of smectitic precursors, beginning with a process confined to a system that allowed limited chemical and isotopic exchanges with the ambient sea water, and later proceeding with a process that brought a highly evolved state for the glauconites under a condition that allowed more or less unrestricted chemical and isotopic exchanges with sea water. The span of time involved in the entire replacement phenomenon, culminating in highly evolved glauconitic minerals, may cover a period of only a few tens of thousand years. Besides glauconite, some clay minerals could also be found to have formed from replacement of precursors within a period shortly after deposition. The analyses of such minerals can yield dates that can be reasonably interpreted as the time of sedimentation. Palygorskite minerals replacing smectitic minerals in some lacustrine deposits are known to yield

Rb-Sr isotopic dates that seem in very good agreement with the stratigraphic age of the deposits.

Illitization of smectite is a major process in the burial diagenetic history of many sedimentary rocks. In recent years, isotopic tools have been used extensively to understand both the timing and the mode of formation of these diagenetic illites. Significant amounts of both stable isotope, particularly oxygen isotope, and radiogenic isotope, most commonly K-Ar isotope, data of diagenetic illite minerals now exist in the literature. The record of illitization in shales is most often evident in the $<0.1\,\mu m$ clay fractions. As very careful studies on shales by many have shown, even in this very fine size fraction, minor to trace amounts of detrital minerals can be found, the amounts of which may escape detection by X-ray diffraction analyses. The influence of the detrital phase on the isotopic dates of the clays by either the K-Ar or the Rb-Sr method can be significant, especially when the detrital phase is geologically very old relative to the diagenetic phase. The degree of influence on the isotopic dates for fine clay fractions due to the presence of varied amounts of detrital components having varied ages has been quantified by Hamilton et al. (1989). The data for the diagenetic minerals may be extracted from mixed dates by extrapolation, following knowledge of the variations in dates relative to the amounts of the diagenetic phases. Such extrapolations can yield meaningful diagenetic dates for bimodal clay assemblages. The timing of illite diagenesis can be accurately defined only when the illitization occurred in a very short interval of time or, as some would call it, in a "punctuated" one. If the duration is a prolonged one, the calculated dates can only yield, as some suggested, an "average" age for illitization.

Many inherent complications often hinder the separation of specific diagenetic illitic clay minerals from others in shales. These problems have contributed to varied isotopic dates for diagenetic clay minerals in the same or equivalent rocks analyzed by different investigators, generating disagreements about precise definition of the time of any single episode of illite diagenesis. The discord in the opinions about the time of illitization is well illustrated in studies of illitization in the Jurassic sediments of the North Sea and in Tertiary sediments of the Texas Gulf Coast. For instance, a K-Ar isotopic study of illitic clays from the Gulf Coast sediments claimed an "average" age of 18 Ma for the diagenetic illitization of smectite, whereas another study analyzing the Rb-Sr isotopic systematics of equivalent illitic clays, concluded that the same process occurred in a very short time about 23 Ma ago. The apparent disagreement in the dates emphasizes the necessity of complete documentation of the analyzed clay samples so that any bias in the dates due to potential contamination by any unwanted mineral can be fully avoided.

Recently, some attention was given to Sm-Nd isotopic analyses of diagenetic illites in shales. Studies on such minerals from the Gulf Coast Tertiary sediments have shown that by leaching a young diagenetic illite with dilute acid, the leachate had a higher Sm/Nd ratio than the acid-leached residue, while both had the same Nd isotopic composition. This result, suggesting that Sm-Nd isotopic dating may hold great promises for determining the time of illitic diagenesis in old sedimentary rocks by analyzing the Nd isotopic compositions of both acid-leachates and acid leached residues, was confirmed by analyses of illite-rich clays from Archean to Proterozoic sediments, which yield critical information about the time of diagenetic illitization.

From knowledge of a reasonable range of temperature during the formation of diagenetic illite in a rock, oxygen isotope data of the pore fluid in equilibrium with this illite can be estimated. At present, some data are available in the literature for pore fluids in equilibrium with diagenetic illitic minerals in shales. Results from studies of diagenetic illitic clays in Tertiary shales from the Texas Gulf Coast area and Cretaceous mudstones from northern California area have indicated that the $\delta^{18}O$ values of the pore fluids ranged between +3 and +11 per mill, suggesting that the pore waters at the time of illitization were considerably chemically evolved, if they were either meteoric waters or sea water. As the estimated $\delta^{18}O$ values for pore waters related to illitization in shales appear to be varied across the stratigraphic column, diagenetic fluids at different depths were evidently not in communication with each other.

Although in many situations separation of diagenetic illite or illite/smectite mixed-layer minerals free of other minerals from sandstones is less difficult than that from shales, yet K-Ar isotopic dates reported by different investigators were found to be considerably varied for illites and illite/smectite mixed-layer minerals from stratigraphically equivalent sandstone beds in the same sedimentary basin. The difference in the reported ages is well illustrated by studies made on illitic minerals from the Jurassic sandstones in the Alwyn field, or elsewhere, in the North Sea basin. The opinions have been divided because results of individual studies were variously biased by the presence of different amounts of detrital or diagenetic minerals of another generation in the analyzed clay samples. Some recent work, finding that crushing and grinding processes for disaggregation of sandstones contribute to the artificial creation of clay-sized detrital particles that mix intimately with the diagenetic clays, have recommended the adoption of a method of freezing and thawing for the disaggregation allowing the separation of the intended clays with no or little detrital components of negligible consequence.

The K-Ar isotopic dates of diagenetic illites and illite/smectite mixed-layer minerals in many sandstone sequences have been found to decrease not only with increase in depth, but also with decrease in particle size. The relationship between the K-Ar dates and the particle size may easily be explained in terms of the increased influence of detrital components in the coarser clay fractions. The trend in age-depth relationship has been explained by some as an effect of progressive downward shift of the interface between hydrocarbon-saturated zone and water-saturated zone by gradual filling of the reservoir with hydrocarbon-rich fluids. As the K-Ar dates depend on the relative behavior of the K and the radiogenic ^{40}Ar during the illitization process, the evolution of the age-depth trend depends mainly on the availability of K to enter the mineral structures and the possibility of the radiogenic ^{40}Ar to escape from mineral structures during the process at any place in the sedimentary sequence. A possible reason for the trend could also be that the reaction kinetics for the formation of illites are varied with a slow crystallization process in small pores and a relatively rapid one in large pores.

The calculated $\delta^{18}O$ values of pore waters in equilibrium with illite and illite/smectite minerals in sandstones from different sedimentary basins range between −7 and +5 per mill. This range of values appears to be somewhat different from the values between +3 and +11 per mill for pore waters in equilibrium with diagenetic illites in shales. This general oxygen isotopic difference suggests that chemical dynamics in the evolution of illite are different in these two contrasting lithologic environments. Several studies on diagenetic illites from associated shale-sandstone sequences have noted isotopic differences between the two sets of illites. Although mineralogic compositions of illite/smectite mixed-layer minerals are nearly similar for shale and sandstone in close proximity to each other in a sequence alternating between the two, the K-Ar dates for illites in shales are substantially higher than those in sandstones at comparable depths. The two sets of illitic minerals also differ in their oxygen isotope compositions. To explain the difference in the K-Ar dates, some have suggested that increased availability of K in relatively more porous sandstones permits a prolonged period of illitization, in contrast to early cessation of illitization in shales with rapidly decreasing porosity (Glasmann et al. 1989a). This explanation does not provide any clue for the cause for the difference in the oxygen isotope compositions between the two sets of clays, unless one admits that it could be induced by a different water-to-rock ratio. Rinckenbach (1988) and Clauer et al. (in progress) offer an alternative interpretation in claiming that differences in the K-Ar dates and the oxygen isotopic compositions relate to differences in the process of illitization, the clays in shales having formed by a transformation-type process in somewhat closed-system conditions and that in sandstones having formed by a dissolution-crystallization process in relatively open-system conditions. This interpretation explains the differences in K-Ar dates, taking into account the specific behavior of the K and the radiogenic ^{40}Ar of the illitic mineral phases in the two types of rocks.

Chapter 5

Isotope Geochemistry of Mica-Type Minerals from Low-Temperature Metamorphic Rocks

The systematics of common isotopes in sedimentary masses are often difficult to interpret because of the complex history of evolution of the rocks which very often consist in mixed suites of detrital and authigenic minerals. The chaotic isotopic relationship that is very common among different components in sedimentary rocks begins to give way to a relatively more coherent relationship among evolved components with the onset of metamorphism. Many metasedimentary rocks afford excellent opportunities to study the characters of dynamism of different isotopes in clays subjected to a wide variety of low-temperature metamorphic environments.

Even a low degree of metamorphism of sedimentary rocks can have pronounced effects on the Rb-Sr isotope systematics of the sedimentary mineral components; an effect among many could be the Sr reequilibrations of minerals and rocks at various scales (Peterman 1966; Bofinger et al. 1968, 1970; Bonhomme and Prévôt 1968; Bonhomme et al. 1968; Montigny and Faure 1969; Clauer and Bonhomme 1970a, b; O'Nions et al. 1973; Bath 1974a, b; Clauer 1974, 1976; Gebauer and Grünenfelder 1974; review in Clauer 1979a). Many early Rb-Sr isotopic studies made on whole-rock samples indicated that the isotopic data points in several occasions defined linear trends in isochron diagrams. Such common occurrences of linear trends among the Rb-Sr data suggest that metamorphism can induce Sr homogenization among minerals within at least a small rock volume. Clauer (1974) and Clauer et al. (1980) observed in studies of metasediments from Morocco that whole rocks consisting of quartz, plagioclases, and clay minerals but no K-feldspars, gave isochrons whose slopes yielded dates that have been believed to indicate the age of the metamorphism. How the homogenization occurs among minerals in a complex assemblage as a result of metamorphism still remains virtually unknown.

Isotopic dates alone may not be sufficient in understanding metamorphism-related processes of reorganization of the chemical constituents resulting in the formation of equilibrium mineral assemblages. The significance of isotopic dates remains in doubt unless mineralogy and morphology of the analyzed materials are thoroughly investigated. This is best illustrated by the results of geochronologic investigations on apparently unmetamorphosed shales in the Precambrian Belt Supergroup in Montana. Based on the Rb-Sr and K-Ar isotopic data of whole-rock samples, clays, and glauconites in the Precambrian Belt Supergroup in Montana, Obradovich and Peterman (1968) obtained three sets of ages of about 900, 1100, and 1300 Ma. The authors concluded that these dates represented three periods of sedimentation separated by two hiatuses of 200 Ma or more. Eslinger and Savin (1973) calculated from $\delta^{18}O$ values of quartz and illite in a series of outcrop samples of the Belt Supergroup in Montana that the temperatures were in the range between 225 and 310 °C. These temperatures are consistent with an estimated burial of 5.5 to 6 km based on stratigraphic and structural considerations. Moreover, a paleomagnetic study by Elston and Bressler (1980) suggested that the entire sequence was deposited during middle Proterozoic.

1 Isotope Geochemistry of Mica-Type Minerals in Different Metamorphic Environments

Mica-like minerals are a common component in both sedimentary and low-grade metasedimentary rocks. A study of the transformation of sedimentary illite to a mica-like entity as a consequence of low-temperature metamorphism provides an insight into chemical and isotopic readjustments that must occur in a rock as a result of change at the onset of metamorphism. Hunziker et al. (1986) recognized the following mineralogical and chemical changes across the transition from diagenetic illites to metamorphic micas: (1) platy minerals with irregular or ill-defined borders modified to euhedral-shaped particles, (2) interstratified illite/smectite changed into well-crystallized illite, (3) the crystallinity index of illite decreased from greater than 7.5 in the diagenetic zone to less than 4.0 in the epimetamorphic zone, (4) illite minerals transformed from $1M_d$ forms in the diagenetic environment to $2M_1$ forms in the anchizone and epizone environments, (5) K_2O contents increased from 6–8% in the diagenetic zone, through 8.5–10% in the anchimetamorphic zone, to 10–11.5% in the epimetamorphic zone, (6) the total layer charge increased from about 1.2 in the diagenetic zone to about 2.0 in the epimetamorphic zone, (7) the average value of b_o of the layering decreased from diagenetic to metamorphic zones, and (8) the temperature at which a prominent endothermic peak appears in the differential thermal curve progressively shifted from 510 °C for diagenetic illites to 750 °C for the epimetamorphic micas.

1.1 Regional Thermal Metamorphism

An influence of increased temperature related to orogenic activities on the isotopic systematics of sedimentary clay minerals is best illustrated by the results of two studies, one on the French Oxfordian "Terres Noires" and the other on the Triassic-Permian claystones of the Swiss Molasse. The rocks in these two areas have undergone various degrees of metamorphic evolution, and the isotopic results of mica-like minerals in these rocks shed much light about the physical and chemical conditions for the growth of micas in response to these varied metamorphic conditions.

1.1.1 A Geotraverse in the "Terres Noires" of the French Alps

The Jurassic "Terres Noires" consist in monotoneous, clay-rich, gray to black shales which crop out in the western French Alps. The sequence was deposited during the middle Jurassic and was presumably metamorphosed during the early Alpine orogeny. Dunoyer de Segonzac (1969) described the petrography, mineralogy, chemistry, and structure of these rocks. The mineralogic and structural evolutions of metasedimentary rocks in the "Terres Noires" sequence, having pronounced eastward schistosity, are illustrated by an east-west transect (Fig. 5.1). The eastward variations are marked by: (1) disappearances of kaolinite and irregular illite/smectite mixed layers, (2) replacement of the 1Md polymorphic illite by the 2M type, (3) decrease in the illite crystallinity index, (4) increase in the Al/Mg ratio of the octahedral composition of illite, (5) decrease in the cation exchange capacity, and (6) morphologic change from ill-formed illite particles to straight-edge mica particles. A microthermometric study of fluid inclusions in quartz overgrowths indicated that thermal intensity reached about 170 °C in the western part and about 250 °C in the eastern part (Barlier et al. 1974).

The <2-μm clay fractions of the samples from the nonmetamorphic western side consisted of a mixture of illite, chlorite, and illite/smectite mixed-layer minerals with occasionally small amounts of kaolinite (Clauer 1976). Illite crystallinity indices suggested that the evolution of the illites belonged to the diagenetic domain. About 45% of the illite were detrital 2M types. On the other hand, the samples from the eastern side contained a regular illite/smectite mixed-layer phase with discrete illite and chlorite. Illite crystallinity implied that the minerals formed in an upper anchizonal environment. As much as 50 to 75% illite were of 2M type.

The Rb-Sr whole-rock data plot along a line in an isochron diagram (Fig. 5.2). The slope of the line yielded a date of about 250 Ma which is significantly higher than the depositional time between 170 and 155 Ma. The initial $^{87}Sr/^{86}Sr$ ratio of 0.70691 ± 0.00024 is similar to the Oxfordian (Middle Jurassic) sea water $^{87}Sr/^{86}Sr$ ratio of 0.7067 ± 0.0002, suggesting that these rocks contained Sr of marine origin, probably in a carbonate phase. The Rb-Sr dates of two clay fractions from the nonmetamorphic western part were similar to the date obtained for the whole rocks. The

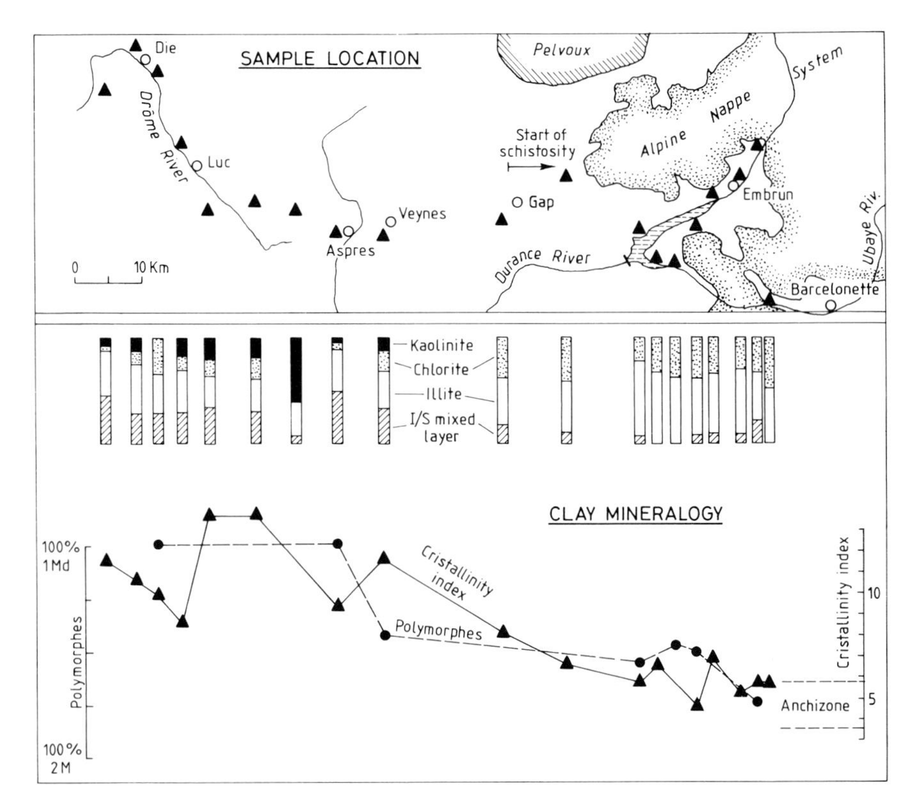

Fig. 5.1. Mineralogic composition and both crystallinity index and polymorphic type of the illitic component of the clay fraction for each different sample of "Terres Noires" shale at various location in the French Alps. (Clauer 1976)

dates of the clays, therefore, were far in excess of the sedimentation time. By contrast, the Rb-Sr dates of the clay fractions from the eastern part were found to be as low as about 60 Ma (Fig. 5.2). As the temperature of the metamorphism for these samples from the eastern part was about 250 °C, the Rb-Sr date of 60 Ma may be interpreted as the response to the period of westward thrusting during the Alpine episode.

The data show that the trend of the decrease in the Rb-Sr dates is largely due to change in the $^{87}Sr/^{86}Sr$ ratio and not to that in the Rb/Sr ratio. The decrease in the $^{87}Sr/^{86}Sr$ ratio with respect to recrystallization of illite during metamorphism is accompanied by an increase in Sr content, suggesting that Sr with a low $^{87}Sr/^{86}Sr$ ratio had to be either incorporated by the illite during the recrystallization or adsorbed on the particle surfaces because these minerals recrystallized in a Sr-enriched environment in the presence of carbonate phases. It may be suggested that the illite remained open to both the Rb and the Sr at the time of recrystallization.

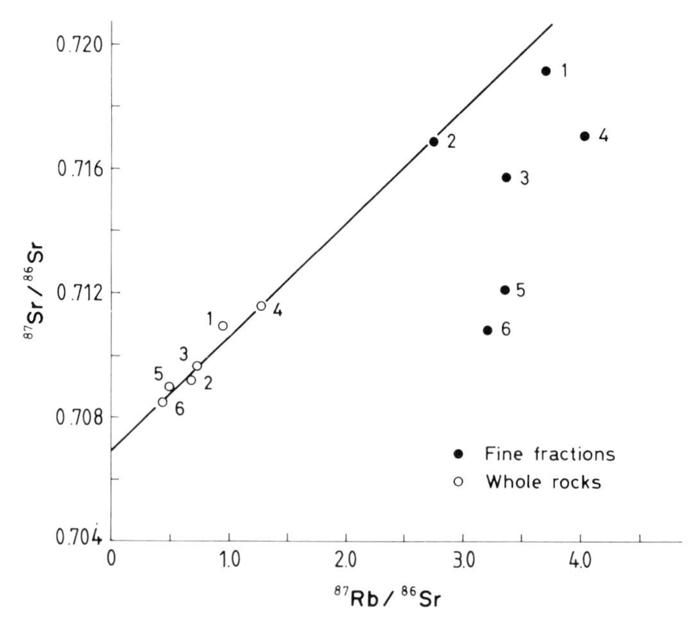

Fig. 5.2. Rb-Sr data for fine fractions and whole-rock samples of the "Terres Noires" shales. (Clauer 1976)

1.1.2 A Geotraverse in the Glarus Region of the Swiss Alps

Triassic and Permian lithologic associations of claystones, shales, slates and phyllites in the Helvetic region of the Central Alps reflect changes from diagenetic, through anchimetamorphic, to epimetamorphic recrystallizations. Hunziker et al. (1986, 1987) analyzed K-Ar, $^{40}Ar/^{39}Ar$, Rb-Sr, and stable isotope compositions of clay separates from these rocks and observed the following trends: (1) a 1Md illite in the diagenetic zone changed to a 2M polytype in the epimetamorphic zone, (2) K_2O increased from about 6–8% in the diagenetic zone to about 10–11.5% in the epizone, (3) an endothermic peak in differential thermal curves shifted from 500 °C for the diagenetic illite to 750 °C for the epimetamorphic mica, (4) the morphology of the particles changed from irregular flakes with ill-defined borders in the diagenetic zone to euhedral particles with sharp borders in the epimetamorphic zone, (5) decreases in both the K-Ar and the Rb-Sr dates from the diagenetic zone to the epimetamorphic zone, and (6) the K-Ar dates were generally lower than the Rb-Sr dates for the clays in the diagenetic zone, but the two sets of isotopic dates became nearly concordant in the epimetamorphic zone. The authors estimated that the critical temperature for total Ar resetting of detrital illite was about 260 ± 30 °C (Fig. 5.3).

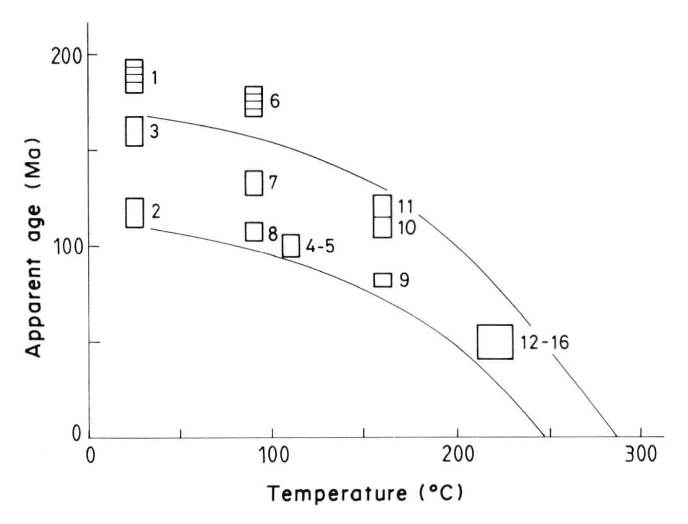

Fig. 5.3. K-Ar dates in relations to the present-day borehole temperatures for illite fractions in the Glarus region. (Hunziker et al. 1986)

We have re-examined the data of Hunziker et al. to analyze changes in the Rb, Sr, and K contents of the illites accompanying the change from diagenesis to epimetamorphism. Both the Rb and the K contents increased as the intensity of metamorphism increased. The average Rb content increased from 228 μg/g for diagenetic illite, through 367 μg/g for anchimetamorphic illite, to 424 μg/g for epimetamorphic illite. On the contrary, the average Sr content decreased from 86 μg/g for nonmetamorphic illite, through 76 μg/g for anchimetamorphic illite, to 46 μg/g for epimetamorphic illite. The average Rb/Sr ratio increased from 2.8 for diagenetic illite, through 8.3 for anchimetamorphic illite, to 18.4 for epimetamorphic illite. The average K/Rb ratio decreased from 226 for diagenetic illite, through 174 for anchimetamorphic illite, to 166 for epimetamorphic illite. The present-day average $^{87}Sr/^{86}Sr$ ratios ranged from 0.724 for diagenetic illite, through 0.726 for anchimetamorphic illite, to 0.732 for epimetamorphic illite. The average $^{87}Sr/^{86}Sr$ ratios of acid leachates from illites ranged from about 0.714 for epimetamorphic illites to about 0.710 for both the diagenetic and anchimetamorphic illites. The leachate with a relatively high $^{87}Sr/^{86}Sr$ ratio also had a relatively high Rb/Sr ratio, as the leachates from epimetamorphic illites had an average ratio of 0.21, whereas those from diagenetic or anchimetamorphic illites were nearly identical in their values between 0.01 and 0.03. These results suggest that the trend in the decrease in Rb-Sr isotopic date, as a function of increase in the intensity of recrystallization of illite, is related more to the increase in the Rb/Sr ratio than to the decrease of the $^{87}Sr/^{86}Sr$ ratio.

The $^{87}Sr/^{86}Sr$ ratios increase as a function of increase in the intensity of metamorphism for the illites from both the Glarus region and the "Terres

Noires" region, whereas the Sr contents of the minerals concomitantly decreased in the former and increased in the latter. The decrease in the Sr content with increase in the intensity of recrystallization in the Glarus region was apparently accompanied by removal of some amount of radiogenic ^{87}Sr by the effect of epimetamorphism. This is suggested by the higher average $^{87}Sr/^{86}Sr$ ratio of 0.714 for leachates from the epimetamorphic illites as compared to that of the 0.710 for the diagenetic and epimetamorphic illites. No explanation can be given at this time for the two opposite trends in the Sr contents found for illites between the two regions. Additional studies are evidently needed for a clear understanding of the Rb-Sr systematics of illites in their evolution from diagenesis to metamorphism.

The oxygen-isotope data were useful to distinguish the process of recrystallization during the evolutionary path of the illites from diagenesis to epimetamorphism in the Glarus region. Hunziker et al. (1986) reported that no discernible change in the oxygen-isotope composition by the change from diagenetic to metamorphic conditions. On the basis of these data, Hunziker et al. suggested that the evolution from 1Md illite to 2M muscovite involves a continuous lattice "restructuration" without the rupture of the tetrahedral and octahedral bonds and the change of the hydroxyl radicals. This no change in the oxygen isotope composition in the process of the transformation from diagenetic illites to metamorphic micas contrasts with changes in both the K-Ar and the Rb-Sr isotopic compositions and the morphology of the particles. The K-Ar and Rb-Sr isotopic data indicate that significant losses of radiogenic ^{40}Ar and ^{87}Sr occurred even in the absence of lattice reorganization (Fig. 5.4). The reorganization process apparently caused continued repartition of Ar and K in the mineral lattices, as suggested by the $^{40}Ar/^{39}Ar$ data which showed a progressive change from a "staircase" pattern typical for an inhomogeneous mineral fraction toward a "plateau" pattern typical for a homogeneous authigenic mineral fraction (Fig. 5.5). Thus, changes in the K-Ar and the Rb-Sr isotopic compositions without having any change in the oxygen isotopic composition suggests that a dissolution-crystallization of the illite particles could have occurred in a very low water-to-mineral ratio environment.

1.1.3 The Eocambrian Slates of the Mulden Group in Namibia

Clauer and Kröner (1979), studying pelitic sedimentary rocks of the Eocambrian Mulden Group of northern Namibia using petrographic, mineralogic, and isotopic means, differentiated two major mineral parageneses in the rocks: an illite-chlorite-quartz-albite paragenesis with occasional amounts of smectite of possibly sedimentary origin, and an illite-chlorite-quartz-albite-microcline-pyrite paragenesis with occasional amounts of stilpnomelane. These two parageneses were considered to belong to two separate regional metamorphic events of very low-grade intensity with

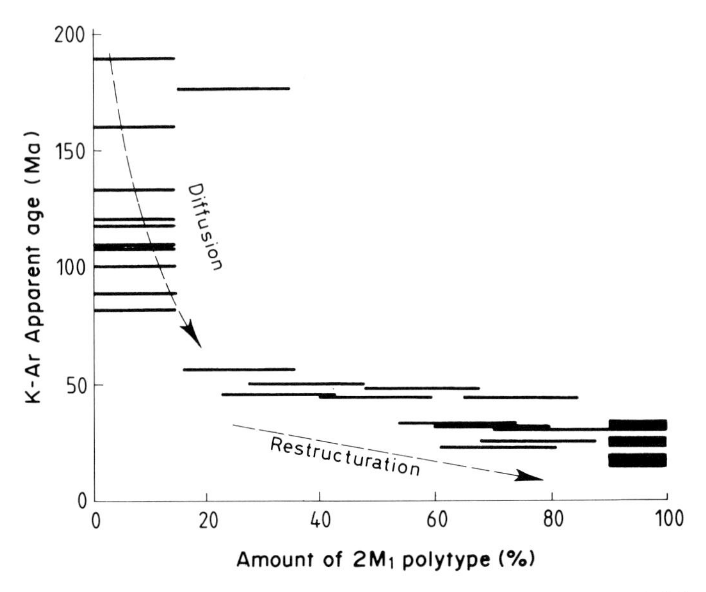

Fig. 5.4. Relationship between the K-Ar dates and the amounts of $2M_1$ polytype in the illite fractions of samples from the Glarus region. (Hunziker et al. 1986)

temperatures between 250 and 300 °C and pressures up to 2 kbar. The <2-µm clay fractions for most samples belonging to the first paragenesis consisted mainly of illite with subordinate chlorite and trace amounts of quartz and albite and the clay fractions of some samples contained 5–15% smectite. The <2-µm clay fractions belonging to the second paragenesis constituted of similar amounts of illite and chlorite with trace amounts of quartz and microcline. Some of these clay fractions also contained 5 to 10% stilpnomelane.

The Rb-Sr analyses of the clay fractions of the first paragenesis gave an isochron date of 537 ± 7 Ma with an initial $^{87}Sr/^{86}Sr$ ratio of about 0.7115. Five of the six clay fractions belonging to this first paragenesis without any smectite had a mean K-Ar date of 538 ± 12 Ma, but one fraction containing smectite gave a slightly lower K-Ar date of 501 ± 13 Ma. The Rb-Sr analyses of the clay fractions of the second paragenesis gave an isochron date of 457 ± 12 Ma with an initial $^{87}Sr/^{86}Sr$ ratio of about 0.7171. The samples belonging to this second paragenesis without any stilpnomelane gave K-Ar dates of 457 ± 12 Ma, while the three samples containing stilpnomelane gave higher K-Ar dates grouped at 493 ± 13 Ma. Evidently, presence of smectite in clay fractions of samples belonging to the first paragenesis and presence of stilpnomelane in clay fractions of samples belonging to the second paragenesis have had effects on the K-Ar dates, suggesting an unrelated post-metamorphic disturbance for each of the minerals.

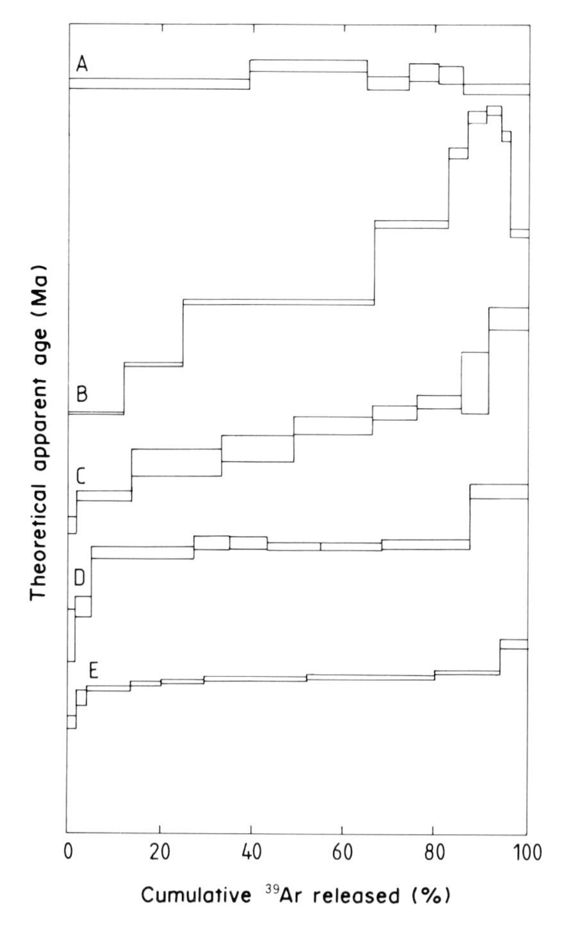

Fig. 5.5. $^{40}Ar/^{39}Ar$ incremental release ages for illite-rich clay fractions from the Glarus region. The evolution from *pattern A* to *pattern E* represents progressive change in the recrystallization from initial $1M_d$ to authigenic 2M type. (Hunziker et al. 1986)

1.1.4 The Archean Carbonates of the Barberton Greenstone Belt in South Africa

The Sm-Nd analyses of clay minerals from carbonates of the Barberton Greenstone Belt in South Africa by Toulkeridis et al. (1994) gave an isochron, defined by untreated fractions, acid leachates, and acid-treated residues, yielding a date of 3104 ± 63 Ma with an initial $^{143}Nd/^{144}Nd$ ratio of about 0.508557 (Fig. 5.6). The clay fractions consisted mainly in 2M illite with a crystallinity index characteristic of the greenschist facies. The morphology of the illite particles was typical of in situ crystallized forms. Based on U-Pb zircon ages of volcanoclastic rocks at the base and the top of the

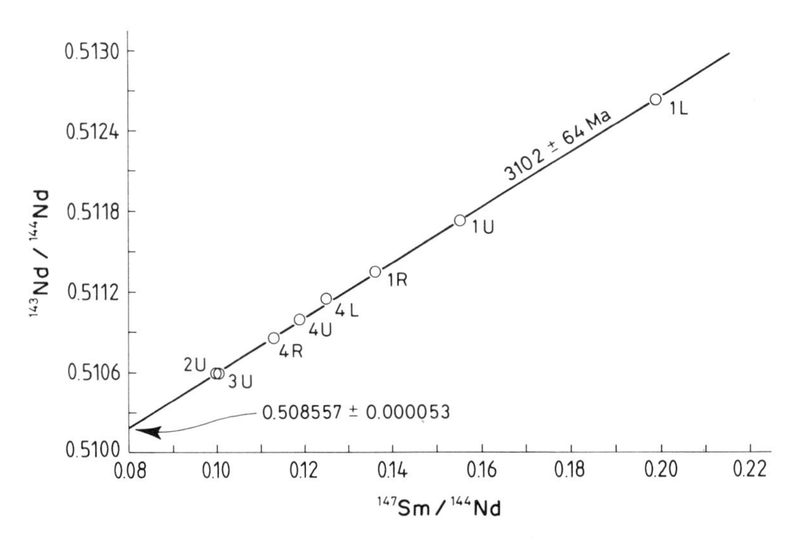

Fig. 5.6. Sm-Nd isochron of <2-μm mica-type fractions of Fig-Tree carbonate rocks in the Greenstone belt in South Africa. (Toulkeridis et al. 1994)

sequence, the sedimentation time of the Fig Tree carbonate sequence has been estimated to be between 3259 ± 4 and 3227 ± 4 Ma (Kröner et al. 1991). Toulkeridis et al., therefore, claimed that the Sm-Nd date of about 3.1 Ga for the sediment is a record of a thermal event, although no other independent evidence has yet been found in this part of the innermost Barberton Greenstone Belt.

1.2 Dynamothermal Metamorphism

Many different works on cleavage formation in metasedimentary rocks have shown that cleavage dissects the rocks and isolate microlithons of different sizes. In these microlithons are preserved sedimentary rock fabrics with detrital micas, whereas in the domains that are disturbed detrital micas are mechanically rotated onto the cleavage plane. During the cleavage development, the size of the microlithons progressively diminishes and the cleavage zones correlatively increase until being continuous (McPowell 1979). Cleavage morphologies, therefore, can be qualitative indicators of the amount of strain that accumulated in metasedimentary rocks. Cleavage development is often contemporaneous with thermal activity of a metamorphic event. Hence, investigations of the systematics of K-Ar isotopes in mica-type minerals can give an insight into chemical dynamics of recrystallization for the minerals subjected to combined effects of heat and deformation, in connection with cleavage development.

The eastern part of northern Morocco represents an internal domain of the Hercynian chain and is characterized by a low-grade metamorphism with contemporary development of flat-lying cleavage. The main deformation presumably occurred between the Middle Devonian and the Upper Visean (380–330 Ma). During this time, a low-grade metamorphism affected the Lower Carboniferous rocks in northern Morocco and developed a generally steeply-dipping axial-plane cleavage in the rocks (Piqué 1981). Huon et al. (1987) and Huon et al. (1993) made isotopic studies of mineralogically uniform Lower Carboniferous shale and slate samples containing clay minerals, micas, and minor amounts of quartz, feldspars, and hematite. The authors separated six fractions of different size from a sample of slate from the Oulmes area in the western part of the Moroccan Hercynian belt. The micas of this rock sample had a high degree of orientation which amounted to as much as 90%. The mica-like minerals, among the size fractions, consisted of 70 to 90% 2M polymorphic illite and 30 to 10% chlorite/vermiculite mixed layers. The illite-crystallinity indices suggested that the minerals belonged to anchi-to-epimetamorphic varieties. The K-Ar dates of the fractions ranged between 291 Ma for the $<1\,\mu m$ fraction to 332 Ma for the 6–10 μm fraction (Fig. 5.7). The three smallest fractions (<1, 1–2, and 2–6 μm) yielded nearly identical dates (290.9 ± 6.5, 294.7 ± 6.7, and 300.5 ± 6.5 Ma), amounting to an average value of 295 ± 5 Ma. The authors interpreted this date as being the time of a major recrystallization event for the micaceous materials, because the nearby 300-Ma-old Oulmes granite is considered to be synkinematic with the cleavage development (Piqué 1976). By contrast, the 6–10-μm coarse fraction yielded a K-Ar date of 332 Ma that was significantly higher than the K-Ar dates of 308 and 314 Ma for the 10–30 μm and 30–50 μm fractions. The authors attributed the relatively low ages for the two coarsest fractions to the loss of ^{40}Ar by mechanical bending and orientation of the larger particles within the cleavage planes. Electron microscopic observations indicated that the coarsest fractions were not aggregates of smaller particles, thereby ruling out the possibility of small particle size as a cause for the low ages.

The micaceous minerals of different size fractions of a diagenetic-to-anchimetamorphic slate sample from the Azrou region in the oriental part of the same Hercynian belt in Morocco, consisted of 60 to 80% 2M illite and 20 to 40% chlorite. The K-Ar date of 330 Ma for the <0.5-μm fraction was lower than that of 410 Ma for the 6–10 μm fraction and of 377 Ma for the 30–50-μm fraction. The relatively high K-Ar date of 330 Ma for the <0.5 μm fraction of the Azrou region, in contrast to the date of 295 Ma for the <0.5 μm fraction of the Oulmes region, was thought to reflect a detrital K-Ar isotopic memory which could not be totally erased by the weak diagenetic-to-metamorphic condition. The evidence for low intensity of metamorphism of the rocks in the Azrou region is visible in the low degree of orientations of the micaceous particles, amounting 10 to 70% (Fig. 5.8). The difference in the K-Ar dates between the <0.5-μm and the 6–10-μm

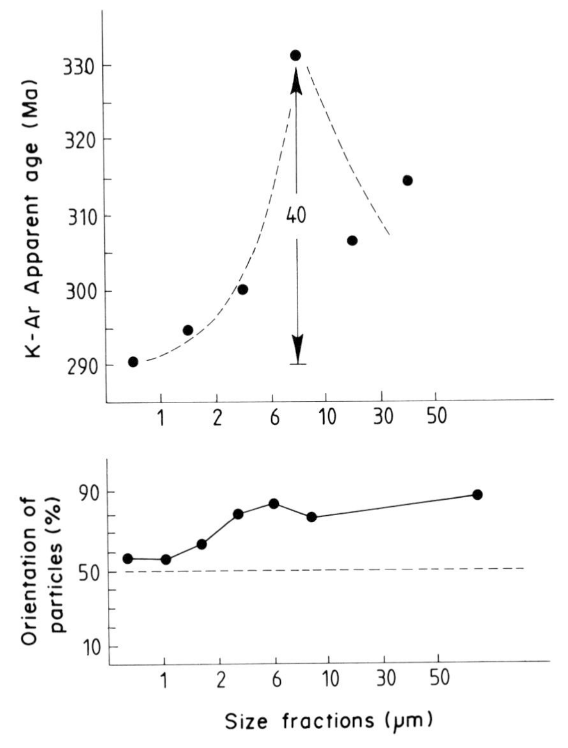

Fig. 5.7. K-Ar dates in relation to the orientation of particles for different size fractions of a sample of slate from the Oulmes region in Morocco. (Huon et al. 1987)

micaceous fractions from the Azrou region was about 80 Ma, whereas that from the Oulmes region, where the intensity of recrystallization was relatively higher, was about 40 Ma. The difference in these dates, coupled with a smaller degree of orientation of the particles in the schistosity of the Azrou sample, favors the hypothesis that the K-Ar system of the large detrital micaceous flakes is affected by mechanical granulation with attendant rotation into the cleavage planes. The data on mica-like minerals from the two regions make evident that fine fractions of shale samples can yield K-Ar dates that approximate the age of metamorphism when it reaches the epimetamorphic grade with a high degree of orientation of the particles. The observed decrease in the K-Ar dates of the large particles (>10 µm) presumably results, to some degree, from mechanical rotation of the particles onto the cleavage planes.

1.3 Cataclastic Metamorphism

The process of recrystallization can become significant in areas which experienced large-scale fault or overthrust movements. The isotopic records

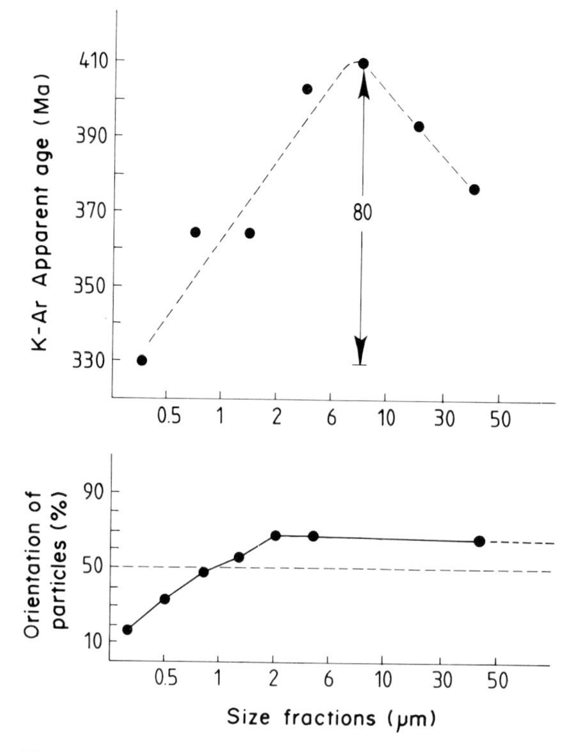

Fig. 5.8. K-Ar dates in relation to the orientation of particles for different size fractions of a sample of shale from the Azrou region in Morocco. (Huon et al. 1987)

of these recrystallized minerals may therefore be useful to describe the dynamics of tectonic processes. The temperature of an overridden section may be rapidly increasing in a very short period of time. The amount of increase in the temperature is a function of the thickness of the overriding plate and its basal temperature. Oxburgh and Turcotte (1974) calculated that underneath a 10- to 20-km-thick thrust sheet, the temperature of strata a few km thick underneath the thrust plane could attain values between 100 and 250 °C within a period of a million years. The effects of overthrust can be particularly evident in ash beds within a sedimentary sequence beneath the thrust plane, because of the high potential of some fraction of the constituents of the ash beds undergoing recrystallization or devitrification into the formation of clay minerals by the tectonic condition.

The Lewis overthrust belt in Montana with several Cretaceous bentonite beds in the underthrust plate has been examined by Hoffman et al. (1976) to determine the impact of overthrusting on the volcanogenic sediments. The authors noted, based on "fixed" K content of acid-insoluble residues of 0.2–0.5-μm clay fraction of each bentonite sample, that the K-Ar dates ranged from about 57 to 72 Ma. They also observed that the K-Ar dates of different size fractions within a sample are nearly identical, suggesting that particle size-dependent diffusional Ar loss is negligible. The authors further suggested that the K-Ar dates represented the time of conversion of potassic enrichment of bentonites in connection with the thrusting. The oldest K-Ar dates were found to be nearly in agreement with the time of thrusting, as determined by field and structural evidence. The duration of the thrusting could not be determined from the field evidence, but the youngest K-Ar dates may suggest that the recrystallization may have continued for about 15 Ma.

Clauer et al. (1991) investigated the time of thrusting of a late Proterozoic sedimentary succession on the early Paleozoic sedimentary cover of the north-central Mauritanide of Western Africa. The <2-μm clay fractions of the underthrust plate consisted mainly of illite with subordinate chlorite, but the samples from the thrust zone itself also contained some smectite. The crystallinity indices of the illites corresponded to values characteristic of anchi-to-epimetamorphic conditions. The K-Ar dates of the clays away from thrust zone (>5 m) were scattered between 472 and 337 Ma, whereas the clays from samples collected within 1 m below the thrust zone yielded K-Ar dates between 301 and 312 Ma. The authors suggested that the dates of the samples away from thrusting were considered to reflect the influence of detrital components. The dates of the samples within the thrust zone and very close to it most likely recorded the time of thrusting.

1.4 Contact Metamorphism

The effect of temperature on K-Ar and Rb-Sr isotopic compositions of clays in shales is probably best evident in the vicinity of igneous intrusions. To determine the course of clays in shales in such conditions, Aronson and Lee (1986) analyzed the K-Ar dates of clays in shales and associated bentonite beds in the Cretaceous sedimentary sequence intruded by a Tertiary stock near Cerrillos, New Mexico. Although some questions may exist about the author's interpretation of the dependence of the K-Ar dates on the size of the clay fractions, the results of their study indicate that the K-Ar dates of the clay fractions of bentonite beds are distinctly lower than that of the associated shale beds. For example, the K-Ar dates of clay fractions of different size in a sample of bentonite ranged between 28 and 30 Ma, which is very similar to the age of the intrusion, whereas the K-Ar dates of clay fractions of different size in a sample of shale ranged between 91 and 119 Ma. An important aspect of the results is that the very high K-Ar dates of the clay fractions in the shale are clear evidence that illitic minerals in a shale can retain their detrital isotopic memories by various degrees, even when the minerals were strongly heated in the proximity of an intrusive igneous body.

The Ordovician Nara sequence in Western Africa was cut by numerous dolerite dikes whose K-Ar dates ranged between 175 and 207 Ma (N. Clauer, unpubl. data). Clauer (1976) studied the clay fractions of shales both close to and away from contacts with the dikes. The <2-μm clay fractions in the shales away from the contacts with the igneous intrusions consisted of illite, smectite, and chlorite. The crystallinity indices of illite suggested that the shales away from the intrusive rocks were not metamorphosed. The Rb-Sr isotopic data points of the clay separates defined a line in an isochron diagram, yielding a date of 423 ± 19 Ma with an initial $^{87}Sr/^{86}Sr$ ratio of about 0.7093 (Fig. 5.9). The initial Sr isotopic value given by the isochron being the same as that of the contemporaneous sea water, Clauer interpreted the Rb-Sr date to approximate the depositional age of the sedimentary sequence. By contrast, the <2-μm clay fractions of shales taken near the intrusive contacts consisted mainly of illite and less chlorite, crystallinity indices of illite having values that correspond to anchizonal and epizonal metamorphic conditions. The Rb-Sr data points of these clay fractions were scattered variously below the previous isochron obtained with the clay fractions from unmetamorphosed shales. Four out of six samples that were analyzed define a poorly linear trend with a date of 267 ± 52 Ma and an intercept of about 0.7112. The intrusion apparently lowered the Rb-Sr isotope date, but the previous isotopic memory was retained in sufficient amounts by the clays to cause the date to be in excess of the time of intrusion.

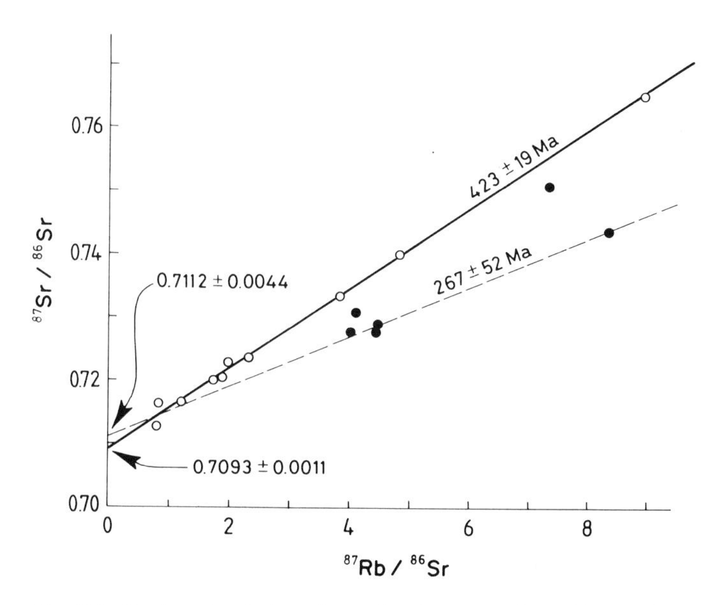

Fig. 5.9. Rb-Sr data of the <2 μm clay fractions of the Nara sedimentary sequence. *Filled circles* are for dolerite-intruded contact metamorphic samples, whereas *open circles* are for samples from outside the contact metamorphic zone. (Clauer 1976)

1.5 Retrograde Metamorphism

Some evidence has been produced suggesting that the isotopic records of micaceous minerals should be carefully evaluated, because the clay minerals may experience changes in the original isotopic records due to the effects of later retrogressive metamorphism. Any change in the original metamorphic record depends not only on temperature, time, and mineral characteristics such as crystallinity and surface area, but also on chemical and isotopic compositions of the fluids, as well as on the ratio of the mineral to the fluid involved.

Wilson et al. (1987) investigated the influence of retrograde metamorphism on isotopic compositions of clay minerals by analyzing some associated with uranium deposits in sandstones of the Proterozoic Athabasca Group in Saskatchewan, Canada. The authors observed that illites from alteration halos surrounding the uranium deposits in the sandstones had δD values of about −70 per mill and $\delta^{18}O$ values of about +12 per mill which corresponded to a formation temperature of about 200 °C, based on the assumption that the water in equilibrium with the clays had the same isotope composition as many present-day mid-latitude basinal brines. The K-Ar dates of these illites from alteration halos surrounding the uranium deposits were found to be about 1200 Ma which is close to the U-Pb ages of the uranium ores. By contrast, many illites within the uranium ore deposits had δD values as low as −169 per mill and $\delta^{18}O$ values as high as +15 per mill. The XRD data did not indicate the presence of any expandable lattice in the illites occurring within the uranium deposits, yet these illites contained as much as 15.4 wt. % water, exceeding the theoretical amount of 9 wt. % water in a unit cell of pure illite mineral. Wilson et al. observed that the K-Ar dates decreased, and that the water contents increased, as the δD values decreased (Fig. 5.10). The authors also observed that, after degassing under vacuum at 110 °C, the illites released waters at two distinct temperature ranges, one between 350 and 400 °C accompanied by 20 to 50% of the total water release, and the other between 600 and 1300 °C yielding the remaining water which derived from the hydroxyl sites. The δD values of the waters released at the low temperature range were lower than that of the waters released at the high temperature range. Wilson et al. considered the water released at the temperature range betweeen 350 and 400 °C as being "interlamellar" water molecules either trapped in nonexpandable layers as lenses of water or occupying vacant potassium sites. The authors also found that the "interlamellar" water once extracted at 400 °C could not be replaced again by the immersion of the clays in water, suggesting that this water is not very labile. A comparison of the isotopic compositions of the clay particles with that of the local meteoric waters showed that the "interlamellar" waters had a somewhat lower range of δD values, but a higher range of $\delta^{18}O$ values than the local meteoric water. The enrichment in ^{18}O was explained as being the result of isotopic exchange between oxygen of the "inter-

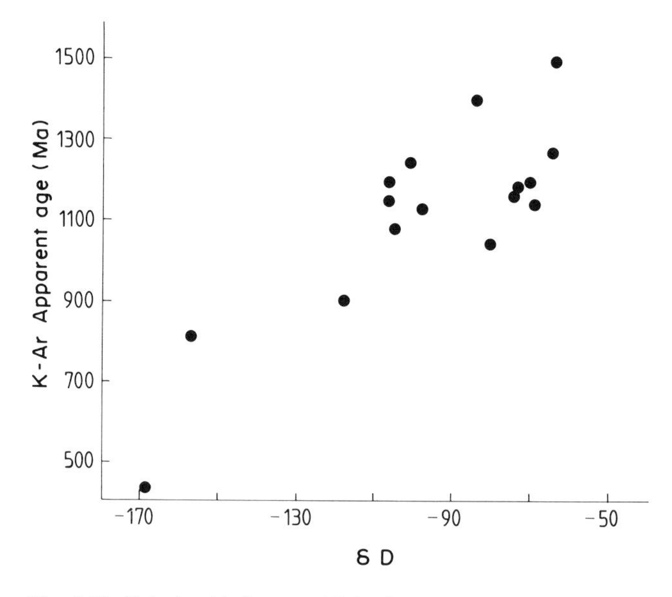

Fig. 5.10. Relationship between K-Ar dates and δD values of illite samples from unconformity-type uranium deposits. (Wilson et al. 1987)

lamellar" water and that of both the hydroxyl ion and the silicate framework. Wilson et al. concluded that continued isotope exchange interaction resulted in concomitant loss of Ar causing the lowering of the K-Ar dates, and maintained that the low δD and high $\delta^{18}O$ values of the altered illites represented the onset of the formation of an illite/smectite mixed-layer from pure illite by retrograde metamorphism.

The views of Wilson et al. (1987) are not supported by Halter et al. (1987), who claimed that the stable isotopic changes, and the attendant decreases in the K-Ar dates for clay minerals associated with uranium mineralization in the Athabasca deposits of Saskatchewan, are related to specific radiation-catalyzed retrograde metamorphism as a result of structural alteration produced by natural irradiation from decay of uranium isotopes, and not to a general case of retrograde metamorphism. Halter et al. observed that the illite and chlorite minerals that were in intimate association with the uranium mineralizations were strongly depleted in δD, with values between −90 and −170 per mill as compared to clay minerals extracted from rocks containing low amounts of uranium and which had δD values between −48 and −62 per mill. The K-Ar dates of about 750 Ma for the clay minerals associated with the uranium mineralizations were found to be low, as compared to the K-Ar dates of 1260 Ma for the unaltered clay fractions. Both the δD values and the K-Ar dates decreased as the uranium content in the whole rocks increased (Fig. 5.11). Halter et al. suggested that the low δD values are compatible with radiation catalyzed retrograde

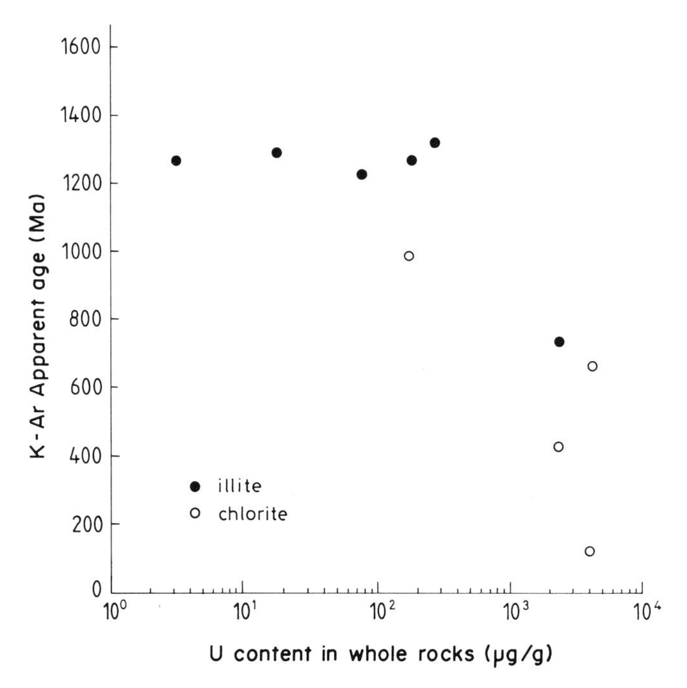

Fig. 5.11. Relationship between K-Ar dates of the <2-µm clay fractions and uranium contents of the Athabasca sandstone rocks in the Cluff area. (Halter et al. 1987)

exchange of the clay minerals with post-Cretaceous meteoric waters. Additional supporting evidence to this interpretation was provided by Philippe et al. (1993), who obtained K-Ar dates as low as 38 ± 10 Ma for clay fractions in sandstones with uranium mineralization in the Cigar Lake area in eastern Saskatchewan.

Kotzer and Kyser (1991) also investigated the stable isotopic compositions and K-Ar dates of illitic clay minerals associated with uranium mineralization in the Athabasca basin of Saskatchewan. They noted that the illites had a narrow range of $\delta^{18}O$ values but a very wide range of δD values. They also found, as did Wilson et al. (1987), that the hydrogen isotope compositions of molecular water differed from those of hydroxyl groups in illites. Although Kotzer and Kyser observed that the K-Ar dates decreased and the weight percentage of H_2O increased, as the δD values decreased. However, unlike Halter et al. (1987), they found no correlation between the uranium contents and the δD values of the illites (Fig. 5.12). Halter et al.'s radiation catalyzed model would produce changes in both the oxygen and the hydrogen isotope compositions of the clay minerals, but that is not what Kotzer and Kyser observed, as they found a wide range in the δD values and a narrow range in the $\delta^{18}O$ values. Kotzer and Kyser claimed that the process of low-temperature retrograde alteration in the isotopic composi-

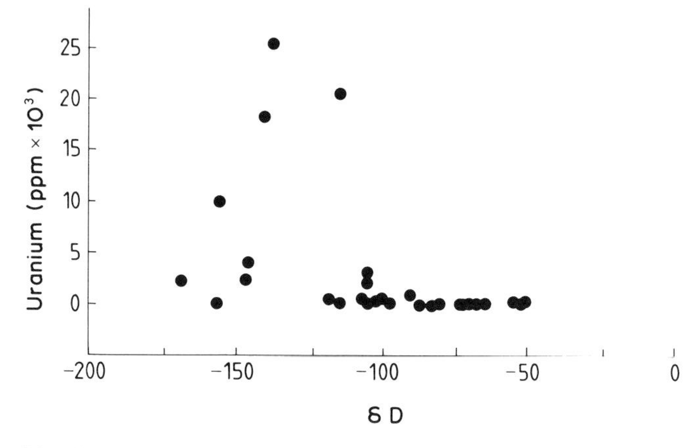

Fig. 5.12. Relationship between uranium contents in rocks and δD values of their illite fractions for different samples of uranium-bearing sedimentary rocks in Canada. (Kotzer and Kyser 1991)

tions of illite should be common under conditions of high water-to-mineral ratios.

1.6 Multiple Metamorphism

Metamorphic overprinting is a common phenomenon in the history of many sedimentary rocks. The records of overprinting may become obvious from evidence on mineralogical associations, but the times of different metamorphic events may not be easily determined by a single mode of isotopic approach, because a specific type of isotopic memory may be variously affected depending upon the temperature, the mineralogy, and the degree to which the system was open during the recrystallization processes. The use of multiple isotopic tools can be fruitful in defining various episodes of metamorphism in the history of a metasedimentary rock.

In the southwest of Casablanca in Morocco, Middle Cambrian metagraywackes were subjected to a progressive Hercynian metamorphic overprint ranging from nonmetamorphic, through anchizonal, to epizonal conditions (Piqué 1981). Different fractions ranging in size from 10–63 to <0.2 μm belonging to samples of different grades of metamorphic recrystallization were investigated for their mineralogy and their Rb-Sr, Sm-Nd and K-Ar isotopic systems by Schaltegger et al. (1994) and Clauer et al. (accepted). The clay fractions of the nonmetamorphic and the anchimetamorphic samples consisted mainly in illite with subordinate amounts of chlorite and that of the epimetamorphic samples contained in addition smectite the amount of which at times was as much as 75%. The smectite originated from a hydrothermal activity and was mainly concentrated in the smallest fractions of the epimetamorphic samples, suggesting that the thermal event responsible for its crystallization was less intense than that responsible for the formation of illite and chlorite (Rais 1992).

As Schaltegger et al. (1994) noted that the Rb-Sr systems of the clay fraction were dominated by detrital and authigenic illite and chlorite minerals, as well as carbonate phases. The Rb-Sr leachate-residue regressions of the two smallest fractions of a nonmetamorphic sample gave internal dates of about 343 and 352, whereas that of the coarser fractions gave higher internal dates up to 437 Ma. The leachate-residue regressions of the five size fractions ranging from <0.2 to 2–6 μm of an anchimetamorphic sample gave dates between 298 and 346 Ma, while the leachate-residue regressions of the 0.4–1- and 1–2-μm size fraction for each of the two epimetamorphic samples gave 304 and 331 Ma. The authors, therefore, assumed that isotopic homogenization of Sr occurred during Hercynian metamorphism with Rb-Sr leachate-residue ages between 298 and 349 Ma, the untreated <0.2 μm clay fractions, residues, and leachates giving an isochron at 342 ± 21 Ma with an initial $^{87}Sr/^{86}Sr$ ratio of about 0.715 (Fig. 5.13).

The Sm-Nd isotopic system was dominated by cogenetic apatite and Fe-oxi-hydroxides, both of which are known for their high contents of REE. The REE in the cements were leached with acids. The data points of the different leachates fitted a Sm-Nd line in an isochron diagram which yielded a date of 523 ± 72 Ma. Schaltegger et al. claimed that the isochron evolved

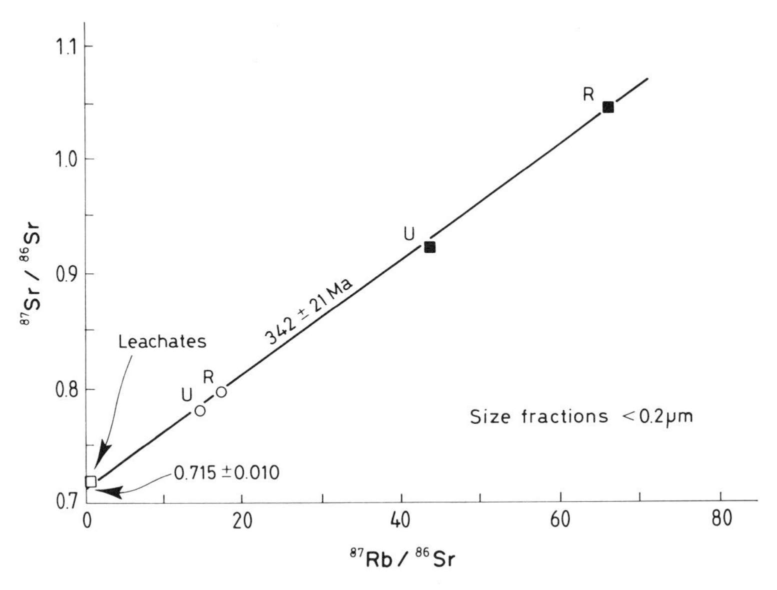

Fig. 5.13. Rb-Sr data of leachates and residues of <0.2-μm size fractions of non-metamorphic graywackes from a transect in the southwestern Hercynian region of Casablanca in Morocco. (Schaltegger et al. 1994)

as the result of a post-depositional, probably shortly after the deposition, diagenetic Nd isotopic equilibrium between apatite and oxi-hydroxides (Fig. 5.14). The Sm-Nd isotopic data points of <0.2-μm size fractions of the two nonmetamorphic samples plotted on the leachate isochron, suggesting that Nd of the clay minerals was in isotopic equilibrium with Nd of the diagenetic apatite and oxi-hydroxide mineral phases (Fig. 5.15). The Sm-Nd systems of all different analyzed size fractions of the high-grade metamorphosed samples were variously disturbed by, at least, the Hercynian metamorphic episode.

The K-Ar dates of the different fractions of the nonmetamorphic samples indicated that the dates decreased from values of about 430 Ma to values of 362 and 387 Ma, as the size decreased (Clauer et al., accepted; Fig. 5.16). The K-Ar dates of most of the size fractions of an anchimetamorphic sample scattered within 300 ± 10 Ma, as did those of the coarsest fractions of the three epimetamorphic samples. In these three samples, a relationship exists between the amount of smectite in the small size fractions and the K-Ar dates (Fig. 5.17). Clauer et al. suggested that the occurrence of smectite was the signal of a mineral paragenesis related to a late hydrothermal activity about 210 Ma ago.

In summary, combined evidence of the Sm-Nd, Rb-Sr and K-Ar isotopic results of different size fractions from samples of the same facies and stratigraphic age helped to reconstruct the nonmetamorphic and metamor-

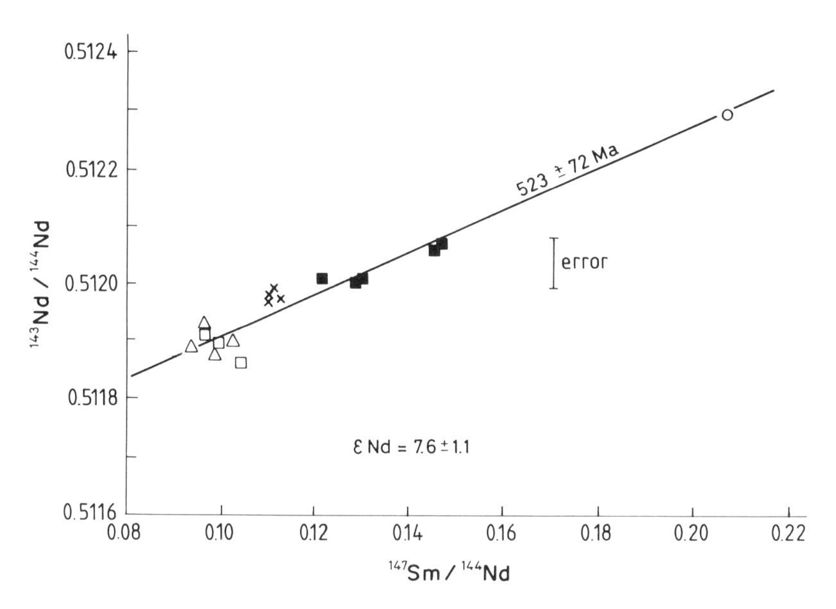

Fig. 5.14. Sm-Nd data of HCl-treated leachates for different size fractions of samples from a transect in southwest Morocco. Leaches of different fractions from the same sample are indicated by the *same symbol*. (Schaltegger et al. 1994)

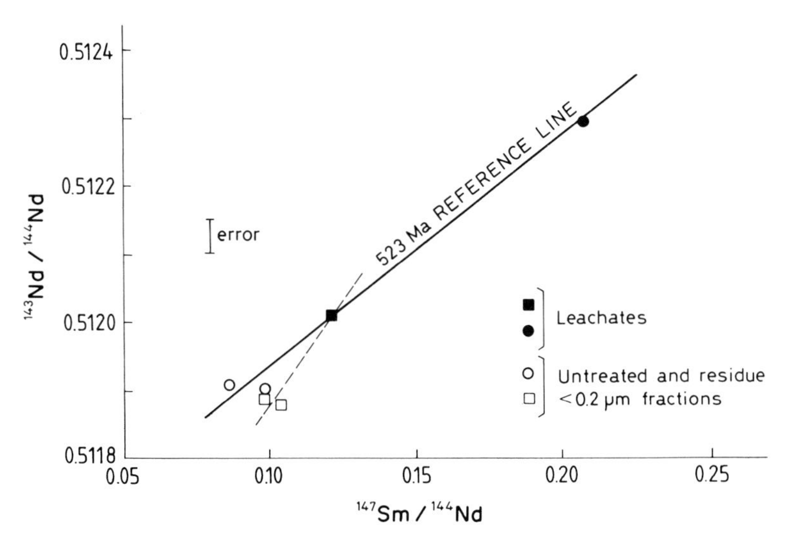

Fig. 5.15. Sm-Nd data of untreated <0.2-µm clay fractions, their HCl-treated leachates, and acid-leached residues for the two least recrystallized samples from the transect in southwest Morocco. (Schaltegger et al. 1994)

phic events recorded in a complex mineral assemblage. The Sm-Nd analyses of the leachates from clay minerals of the nonmetamorphic samples allowed identification of very early diagenetic interactions, the records of which seem to have been kept preserved despite later intense complex events. The Rb-Sr isotopic data of the clay minerals of rocks from the anchi- to epimetamorphic domains revealed the occurrence of a Hercynian event. The K-Ar dates suggested a late hydrothermal activity which was probably related to the appearance of smectite. The relationship between K-Ar dates and amounts of smectite content was found to be either linear or assymptotic. Clauer et al. suggested that the hydrothermal event at 210 Ma, which may have caused the formation of a smectite-rich mineral paragenesis, either was responsible for a diffusive loss of radiogenic ^{40}Ar out of the Hercynian illite particles and thereby resulted in an asymptotic relationship between the K-Ar dates and the smectite contents, or had no effect on the Ar budget of the previously formed illitic material and thereby resulted in a linear relationship between isotopic dates and amounts of smectite in the clay fractions.

2 Isotope Geochemistry of Mica-Type Minerals Related to Metamorphic Rock Lithology

Considerable lithologic changes can occur quite frequently across stratigraphic sequences. The different rock units in the same sequence can be

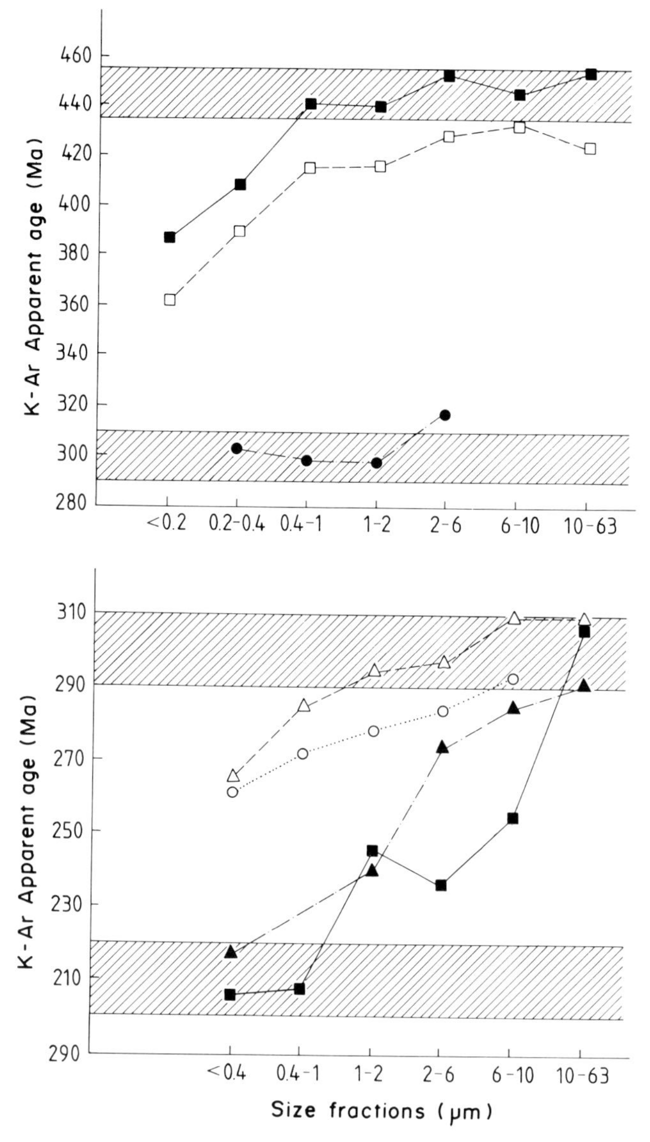

Fig. 5.16. K-Ar dates of the different size fractions for seven metagraywacke samples from a transect in southwest Morocco. (Clauer et al., accepted)

expected to respond differently if they were subjected to various metamorphic constraints. The conditions of varied juxtaposed lithologic units produce a favorable situation for chemical transfers across the different units. Examinations of the isotopic systems of mica-like minerals in different rock associations subjected to the same forces of metamorphic temperature and pressure can help in gaining a broad perspective about the relative influences of neighboring and host rocks on the isotopic compositions of the

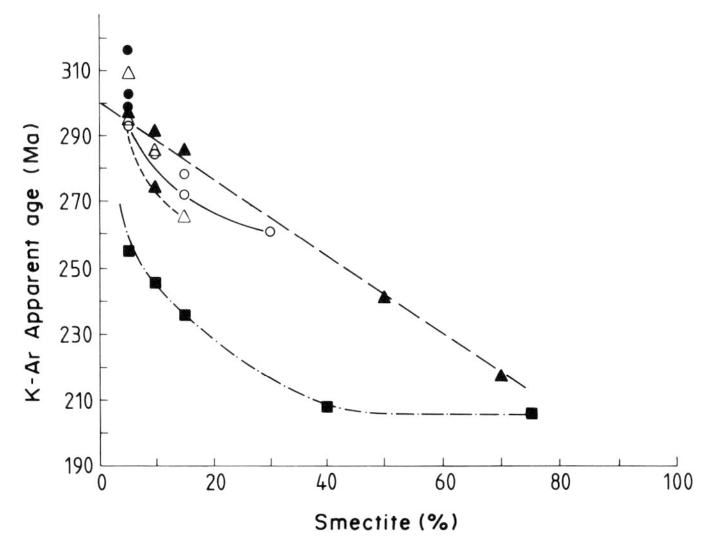

Fig. 5.17. Relationship between K-Ar dates and smectite contents of different size fractions of hydrothermally altered metagraywacke samples from a transect in southwest Morocco. (Clauer et al., accepted)

clay minerals in a given rock of a sequence with rapidly changing lithologies. Such studies can reveal a great deal about the nature of chemical dynamics during the metamorphism.

2.1 Metapelite-Metavolcanic Association

Clauer et al. (1985a) made an isotopic study of clay minerals from an 8-m-thick sedimentary unit intercalated between spilitic pillow lavas in the upper part of the Cambrian Spilitic Group of Erquy in Brittany (France). Chemical data of the clay fractions and the whole rocks of the sedimentary unit indicated a strong volcanoclastic input at the bottom of the sequence that was superceded by progressively increased supply of terrigenous components until the deposition of the next pillow lavas. The <2-μm clay fraction of the sedimentary unit consisted mainly of illite and chlorite with subordinate but varied amounts of smectite or illite/smectite and chlorite/smectite mixed-layer minerals. The illite crystallinity indices suggested that these clays crystallized in an epimetamorphic environment. The Rb-Sr isochrons of the <2-μm clay fractions and also the whole rocks of the sediments gave a date of 494 ± 11 Ma with an initial $^{87}Sr/^{86}Sr$ ratio of 0.7052 ± 0.0005 (Fig. 5.18). This date for the clay fractions is identical to the Rb-Sr isochron date of 482 ± 10 Ma with an initial $^{87}Sr/^{86}Sr$ ratio of 0.7055 ± 0.0002 for the overlying and underlying spilitic rocks, as reported by Vidal (1980). The K-Ar isotopic

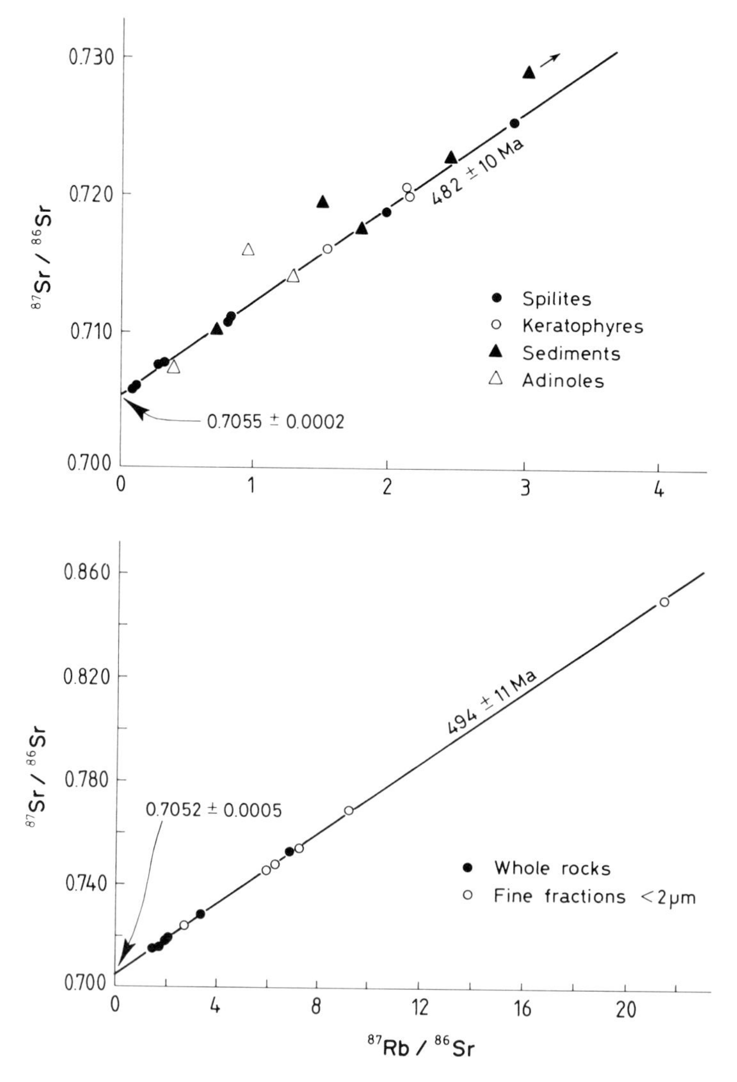

Fig. 5.18. Rb-Sr dates of metavolcanic and metasedimentary whole-rock samples and <2-μm fine fractions of the Erquy sequence in Brittany (France). The *upper isochron diagram* is from Vidal (1980), the *lower* from Clauer et al. (1985a)

analyses of the <2-μm fractions, which consisted of 100% clay minerals, gave dates of about 485 Ma. Some clay fractions that were also analyzed for their K-Ar dates, contained varied amounts of nonclay, glassy materials. These fractions with the glassy materials yielded K-Ar dates as low as 400 Ma. The above Rb-Sr and K-Ar dates on the clay minerals in the sedimentary unit and the Rb-Sr data on the enclosing spilitic rocks, suggest that the formation of illitic minerals was highly chemical dependent on the spilitization process and that both occurred the same time about 490 ± 10 Ma ago. Structural studies have demonstrated that the Erquy volcanosedimentary sequence was tilted by a tectonic event long after spilitization and associated clay mineral crystallization. Although it might have caused some scatter in the K-Ar dates of the glass-bearing clay fractions, this tectonic event which probably occurred about 285 ± 16 Ma ago (Martineau 1976), apparently did not affect the Rb-Sr and the K-Ar dates of the clay materials.

A sequence of alternate layers of metapelite and metatuff, each layer about a few meters thick, occurs in the eastern Rheinisches Schiefergebirge in Germany. The sequence was deposited during the Devonian, but the two rock types were variously metamorphosed between anchimetamorphic and epimetamorphic grades. The rocks differ in the intensity of the slaty cleavage: it is discontinuous in the metapelites, but is continuous in the metatuffs. The <2-μm size fractions of both rock types consisted almost exclusively of illite with small amounts of chlorite and quartz (Reuter 1985, 1987). The coarse fractions, between 2 and 20 μm in size, contained large amounts of quartz, amounting to as much as 40%, with subordinate amounts of albite in both rock types, but metapelites contained in addition some calcite. The illite-type minerals in the metapelites were essentially of 2M polytype, whereas that in the metatuffs were of both 2M and 1Md polytypes.

The K-Ar dates of different size fractions of metapelites from both the anchimetamorphic and the epimetamorphic zones distinctly made evident that the dates decreased as a function of the decrease in size. The decrease in the dates for the entire size range amounted to changes from about 400 to 345 Ma for the anchimetamorphic samples and from about 390 to 314 Ma for the epimetamorphic ones. Also, for the same size fraction, especially for the fractions in the clay size, the K-Ar dates of the anchimetamorphic samples were higher than that of the epimetamorphic ones (Fig. 5.19). By contrast, the K-Ar dates for the fractions ranging between 20 and <0.4 μm of the metatuffs in the anchimetamorphic zone remained almost invariant between 336 and 319 Ma (Fig. 5.20). Also, considerable overlap existed in the dates of different fractions between the anchimetamorphic and epimetamorphic metatuffs, although slightly lower dates by about 10–15 Ma, may be traced for the epimetamorphic fractions of <1 μm size. The apparent invariant nature in the K-Ar dates of the metatuffs, with respect to changes in the particle sizes and also similarity in the dates among the fractions >1 μm, signify that a distinct recrystallization of the metatuffs occurred during the period 325 ± 5 Ma, which has been related to a metamorphic event by Reuter (1985, 1987).

The internal difference in the dates between the metapelite and the associated metatuff is a reflection of the difference in the nature of metamorphic recrystallization within these rocks of different lithologies. Based on the K-Ar data of the metatuff, what becomes clear is that only the very fine (<0.6 μm) fraction of metapelite in the epimetamorphic zone has the same record of the disturbance as the one held by the clay minerals in the

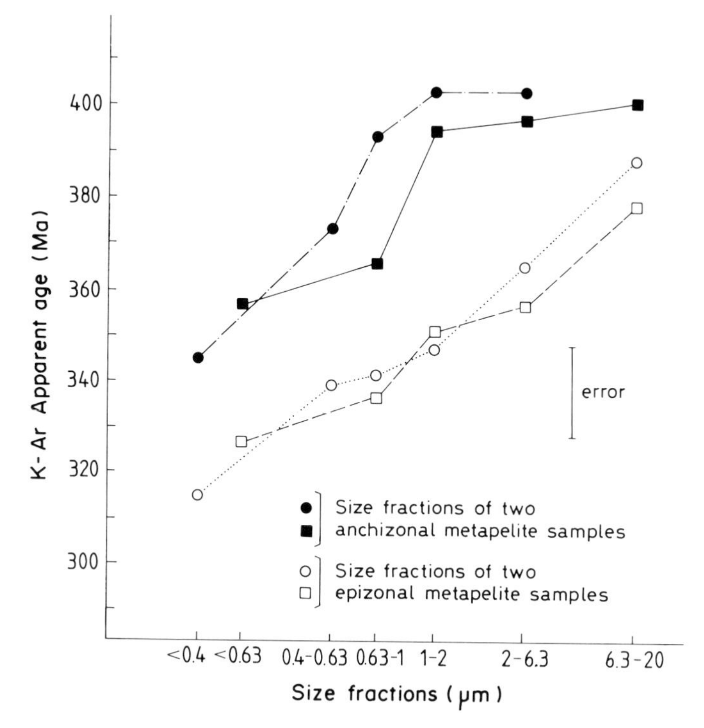

Fig. 5.19. Relationship between K-Ar dates and the size of different fractions for two anchizonal and two epizonal metapelite samples from Rheinische Schiefergebirge in Germany. (Reuter 1985, 1987)

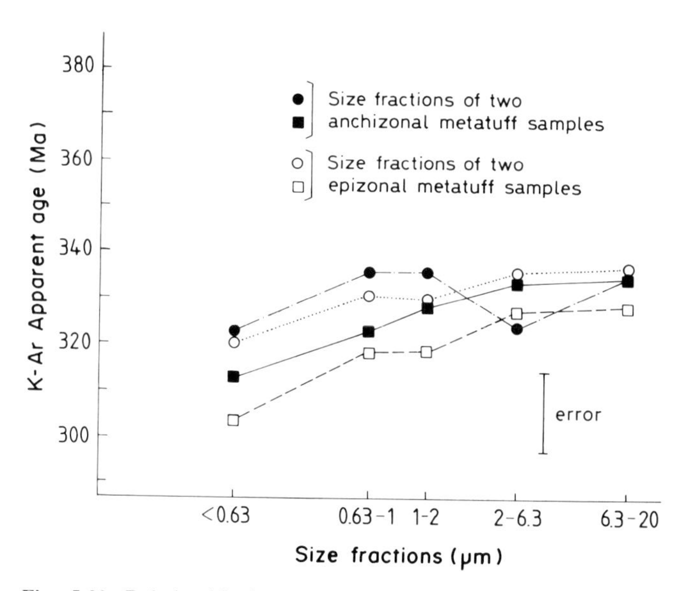

Fig. 5.20. Relationship between the K-Ar dates and sizes of different fractions for two anchizonal and two epizonal metatuff samples from Rheinische Schiefergebirge in Germany. (Reuter 1985, 1987)

metatuff. The anchimetamorphic condition was insufficient to obliterate completely the previous isotopic signatures in the metapelite. This is not to say that no new illite-type or mica-like minerals formed in the metapelite under both anchi- and epimetamorphic conditions, because the dates decreased as a function of decrease in the size of the fraction, signifying that very fine fractions smaller than the ones analyzed potentially can have the unambigous isotopic record of the metamorphism.

2.2 Metaarkose-Slate Association

Isotopic records of mica-like minerals in arkoses that are associated with shales or siltstones in a sequence can provide some indications of a past hydrothermal event, where the evidence for such an activity is not at all apparent in the rocks. The study by Brockamp et al. (1987) on the Upper Carboniferous arkose-shale host rocks of uranium deposits in the Baden-Baden area (Germany), gives valuable information about the pattern of changes in the isotopic signatures of mica-like minerals in both arkose sandstones and associated shales affected by the mineralization. The host rocks of the uranium deposit consist of irregular alterations of shales and arkoses of Upper Carboniferous age. The uranium deposit is considered to be epigenetic-hydrothermal in origin. The temperature of mineralization was estimated to be between 240 and 290 °C. The age of mineralization, based on a U-chemical study, is about 150 Ma during the Jurassic. The uranium deposit is surrounded by zones of decreasing hydrothermal alteration. Sericitization occurred in the most altered zone about 2 km wide which is surrounded by a less intense alteration zone about 2 km wide marked by albitization. The zone of albitization is itself surrounded by a zone of weak alteration. The sericite zone contains micas of phengite composition.

The <2-μm size micas in the arkosic samples of the three zones from ore-associated zone to the albitized zone yielded dates of about 155 Ma, whereas that in an arkosic sample from the least altered zone yielded a date of about 175 Ma. By contrast, the <2-μm micas in shale samples yielded progressively increased dates from about 170 Ma for the clay minerals in the ore-associated zone, through about 180 Ma for that in the sericitized zone, and about 185 Ma for that in the albitized zone, to about 190 Ma for that in the least altered zone. Thus, the data show that records of even coarse-sized micas in arkoses at a distance of several km away from the ore can signify the time of a hydrothermal event. By contrast, the dates for the micas in the shales are systematically higher than those for the same in the arkoses within the same alteration zone. The lack of record of hydrothermal event in <2-μm micas in shales does not indicate that no mineralization occurred in these rocks, because analyses of <0.63-μm clays from the sericite ore-zone and the albite zone yielded dates of 151 ± 3, 162 ± 4, and 172 ± 4 Ma, respectively. The date from the altered ore zone only is very close to the time of mineralization. What one can conclude from these age patterns in shales and associated arkoses hydrothermally altered, is that chemical potential and fluid dynamics are the most important factors in the recrystallization of micas. Also the K-Ar isotope dates of clays in shales are useful in identifying hydrothermal events only when the studies are made on the very fine-grained particles.

2.3 Coal Bed-Slate Association

The impact of organic-rich beds on mineralization in associated sediments has been the subject of an investigation by Daniels et al. (1994), who examined the mica-type minerals in an anthracite bearing sequence in eastern Pennsylvania. Fractures in these anthracite coals often carry illites with small amounts of kaolinite and chlorite. Occurrences of NH_4-illite along with K-illite in fractures in the anthracite are common. The illite in the host shales is very often a K-bearing illite, although some NH_4-illite was found in the organic-rich shale. Daniels and Altaner (1990) concluded that NH_4-bearing illite in anthracite beds in eastern Pennsylvania formed exclusively during late-stage coalification at temperatures between 200 and 275 °C from organically derived N and a kaolinite precursor:

$$3\text{Kaolinite} + 2NH_4 \rightarrow 2NH_4\text{-illite} + 2H^+ + 3H_2O.$$

The K-Ar and the Rb-Sr isotopic data were collected by Daniels et al. (1994) from different types of illite in both the coal beds and the associated shale beds. The K-Ar isotopic analyses of both the NH_4-bearing and the K-bearing illites in the coals gave an isochron date of 254 ± 6 Ma. The same K-Ar date is given by the NH_4-bearing illite in organic-rich shale. By contrast, the K-bearing illite in the shale yielded an age of 281 ± 6 Ma.

The Rb-Sr isotopic analyses of both the NH_4-bearing and the K-bearing illites in the coal units yielded an isochron with an age of 260 ± 44 Ma with an initial $^{87}Sr/^{86}Sr$ ratio of about 0.7097. The isotopic data of the NH_4-bearing illite from the organic-rich shale bed conformed to the isochron defined by the illites from the coal bed, suggesting similar coeval history. By contrast, the Rb-Sr isotopic data of the K-bearing illite from the shale deviated significantly from the isochron trend (Fig. 5.21). Thus, the Rb-Sr data of illites in both the coal beds and the organic-rich shales provided a similar chronologic information as the K-Ar dates on the same illites.

The authors interpreted that the isotopic dates at about 260 Ma represented a thermal and deformational pulse of the Alleghanian orogenic

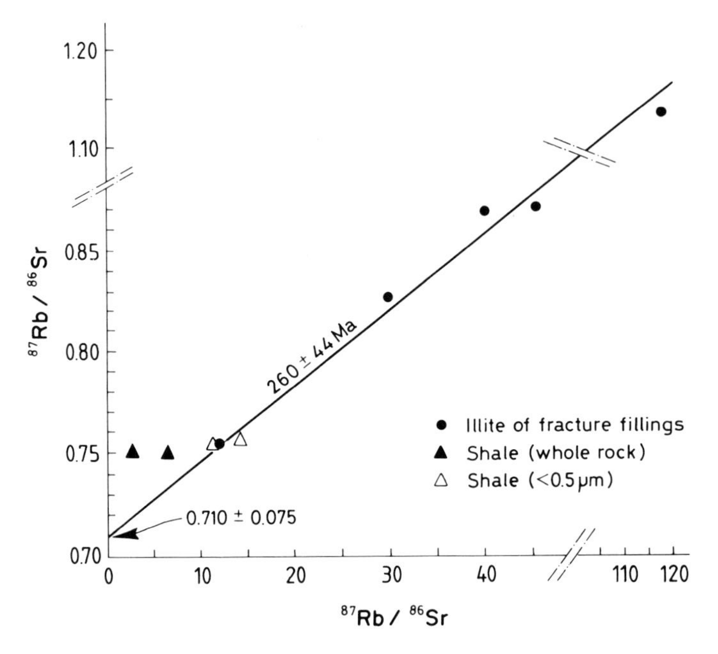

Fig. 5.21. Rb-Sr dates of fracture-filling NH_4-bearing illites in anthracite coals, as compared to that of whole-rock and clay-fraction samples from associated shales in Pennsylvania. (Daniels et al. 1994)

episode. The data clearly suggest that new illites were created in an organic-rich environment, and that the organic-rich environment under the anthracite metamorphic condition seems to have not been intense enough to obliterate completely the previous isotopic signatures of all the clays fractions in the barren pelitic host rocks.

3 Isotope Geochemistry of Mica-Type Minerals Under Hydrothermal Conditions

Hot saline fluids brought from outside and flowing through faults, joints, and pores in rocks, and also along unconformities and bedding planes, may promote varied degrees of recrystallization and neoformation of minerals in the rocks. Phyllosilicate minerals are often common products of these processes. Chemical and isotopic analyses of phyllosilicate minerals can be potentially useful means for delineating low-temperature hydrothermal related mineralizing events in a sedimentary basin.

3.1 Isotopic Dating of Hydrothermal Activities

The geothermal system in the Rhine graben in northeastern France is presently active and affords an excellent opportunity to examine the hydrothermally related isotopic records imprinted in the clay minerals of the Triassic Buntsandstein sandstones. The Rhine graben in northeastern France is about 300 km long and 35 km wide. The relatively undisturbed Mesozoic sedimentary sequence with the Triassic Buntsandstein at the bottom was deposited over the deformed Permian strata. In the period between late Cretaceous and Paleogene, about 135 to 46 Ma ago, the entire Mesozoic and the underlying Paleozoic sequences were affected at various times by pre-rift volcanic activities. The rifting with some associated volcanic activities began in the Eocene time about 37–46 Ma ago. The period of Oligocene, about 34–23 Ma ago, saw a considerable amount of deposition of sediments in the basin as a result of continued subsidence and associated igneous activities. The distension and fracturation activities at the rift during the Miocene-Pliocene time, about 23 to 2 Ma ago, brought renewed continental sedimentation and associated volcanic activities. In the period since the Pliocene, the graben has been tectonically active at various times and the site for the deposition of alluvial sediments. As a result of the tectonic effect in connection with the formation of the graben, the Triassic Buntsandstein sandstones, which deposited in fluvial to deltaic environments in northeastern France were variously displaced. At the present time, the sandstones crop out at the surface and are also found at depths to about 3 km below the surface in the central part of the graben. The present temperature at 3.2 km depth, as measured in a drill hole, is at least 140–160 °C. Jeannette et al. (1988) reported on depth-related porosity developments by analyzing two cores of Buntsandstein sandstones recovered from a depth of about 3 km, with respect to the equivalent surface samples. Fritz et al. (1983) analyzed the chemical and isotopic compositions of the present-day hydrothermal waters recovered from Buntsandstein sandstones deeply buried in the graben.

Liewig (1993) analyzed the mineralogy and K-Ar dates of various size fractions separated from these rocks. The <2-μm fractions consisted essentially of illite, whereas the fractions >2 μm also contained trace amounts of quartz and feldspar. The very fine fractions were devoid of any feldspar. The illite crystallinity indices corresponded to anchizonal characteristics. Observations by TEM revealed three types of illites: lath-shaped particles up to 10 μm long, euhedral hexagonal particles, and irregularly shaped particles. The first two were considered to be authigenic in origin and the third detrital. The K-Ar dates of the bulk <2-μm fraction clays of several samples ranged between 84 and 188 Ma. Five different size fractions, ranging from <0.4 to 10 μm in size, of three other samples yielded a wide range of K-Ar dates, the dates decreasing as the particle size decreased and ranging from about 115 Ma for the coarsest fraction (6–10 μm), through about 101, 95 and 78 Ma for the intermediate size fractions, to about 54 Ma for the finest fraction (<0.4 μm). Quite evidently, the dates are mixed records of multiple events. However, by plotting all the K-Ar dates against the corresponding K_2O contents, Liewig was able to define the youngest illitization period which appeared to be 40 Ma based on a 10% K_2O content for the youngest illite-type minerals (Fig. 5.22). The 40 Ma date is geologically reasonable because about this time the rifting already began with considerable thermal activities. The highest K-Ar date of 188 Ma observed by Liewig for illite in the sandstones certainly includes records of pre-rifting events. The potential of a pre-rifting metamorphic event about 190 Ma old has been previously been described by Bonhomme et al. (1980, 1983) and Liewig et al. (1987b) in other west European regions.

The K-Ar dates of the illites in the Buntsandstein sandstones from the Rhine graben are important in several respects. They clearly indicate that at least the records of the hydrothermal activities some 40 Ma ago in the illitic clays are still very much preserved in the midst of the continued activities, suggesting that for the last 40 Ma, temperature of the hydrothermal system never exceeded the temperature of formation of the illite. Furthermore, although the calculation by Fritz et al. (1983) shows that the present composition of the hydrothermal fluid in the Rhine graben is in thermodynamic

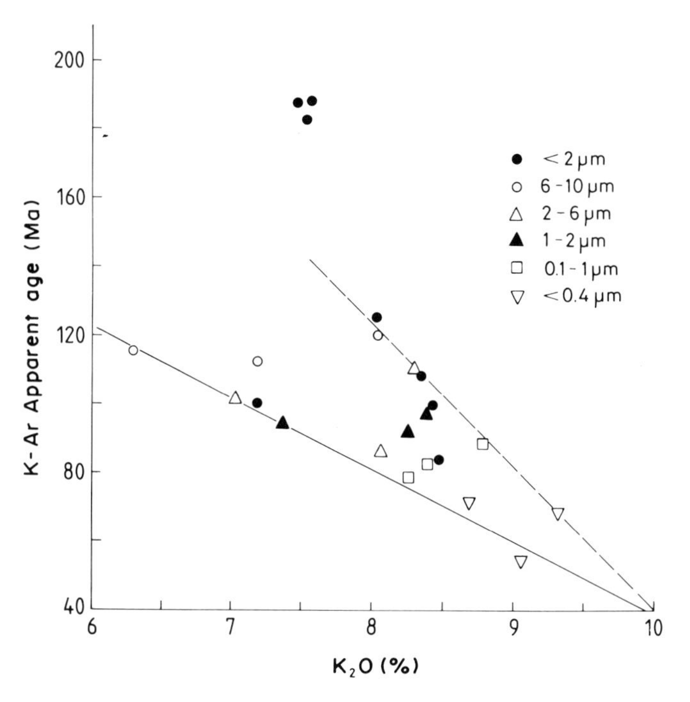

Fig. 5.22. Relationship between K-Ar dates and K_2O contents of different size fractions of Buntsandstein sandstones in the Rhine graben. (After Liewig 1993)

equilibrium with respect to an illite, the K-Ar isotopic dates do not suggest that illite is presently forming in the sandstone. This could mean either that no illite is presently precipitated or dissolved, or that illite neogenesis would be found only in size fractions much smaller than those studied by Liewig (1993).

The Middle Ordovician bentonitic beds in the eastern United States illustrate an effect of hydrothermal activity on metastable volcaniclastic sediments. These bentonitic beds are generally a few centimeters to a few meters thick, and are enclosed in limestone beds. Trace element analyses suggest that the precursors of these bentonites were mostly of trachyandesite composition with minor amounts of dacite and rhyolite (Huff and Turkmenoglu 1981). The K-Ar isotope data of clay minerals in these K-bentonites are of much interest because the clays are composed of highly K-rich illitic minerals and had a shallow burial history. The temperatures of the hydrothermal alteration were in the range between 125 and 60 °C (Elliott and Aronson 1987). The shallow burial and the widespread occurrence of these beds with K-rich illitic minerals suggest involvement of tectonically driven hydrothermal fluids. Elliott and Aronson analyzed K-Ar dates of clay minerals from some bentonite beds at several different outcrops in the Appalachian basin of Alabama, Tennessee, Kentucky, and Virginia. The separated clay fractions of 1–2, 0.2–0.5, and <0.125 μm sizes typically consisted of overwhelming amounts of illite/smectite mixed-layer minerals with very minor amounts of chlorite and kaolinite, and occasionally trace amounts of K-feldspar in the coarse fractions. The fraction of the expandable component in the mixed-layer minerals ranged beween 25 and 35%. The K-Ar dates dispersed narrowly between 272 and 303 Ma, indicating that the illitization occurred in a short time interval during the period between the late Carbonifeous and the early Permian, which was apparently coincident with the Alleghanian orogeny. Such a widespread illitization is most likely owed to westward subsurface expulsion of hot saline fluids from the deeply buried parts to the edges of the foreland basin during the orogeny. The nonclay K-bearing minerals, such as biotites and K-feldspars, are abundant in the K-bentonites. As these minerals are present in the bentonites as well-preserved crystals, they could not have served as the source of K for the formation of illitic minerals in these rocks, and hence K had to come from outside the bed. An outside source for the K is further suggested by the Al-normalized comparison in the chemical compositions between an average K-bentonite and its presumed average trachyandesitic precursor (Huff and Turkmenoglu 1981). The evidence is strong that the illitization of smectite in the K-bentonite beds occurred in a very open-system condition at temperature no higher than 100 °C that is commonly attendant with the illitization in deep burial sediments.

Deutrich et al. (1993) and Zwingmann et al. (1993) made petrographic, geochemical and K-Ar isotopic investigations on illite-rich clay fractions of Rotliegendes (Late Permian) and Late Carboniferous sandstones in northwestern Germany. The clay fractions in the samples located in the vicinity of faults yielded REE patterns that were enriched in LREE with pronounced positive Eu-anomalies, suggesting intense leaching of feldspar precursors by the mineralizing fluids. By contrast, the clay fractions in the samples that were located further away from faults were depleted in LREE and showed lower Eu anomalies. The Li and Cs contents and the Si/Al ratios of the illite minerals were also distance-dependent to the faults. The K-Ar analyses of all the illites of different sizes gave a range of dates between 220 and 160 Ma with two mean maxima in the frequency distribution pattern, one at 210 for the coarser fractions (2–20 μm), and the other at 180 Ma for the finer fractions (<2 μm). These K-Ar dates overlap those obtained by Liewig (1993) on clay fractions in similar rocks from the same region. It is not yet clear if the two maxima in the dates correspond to one continuous episode of 30 Ma duration, or to two separate pulses of illitization process. The K-Ar dates are believed to reflect an early-to-mid Kimmerian extensional tectonic activity with mobilization of hydrothermal fluids along some reactivated Late Paleozoic faults.

3.2 Indirect Isotopic Dating of Hydrothermal Ore Deposits

The origin of many hydrothermal ore deposits hosted by sedimentary rocks remains highly controversial, largely because direct dating of ore minerals is often difficult, if not impossible, by common isotopic methods for lack of suitable materials or methods. In recent years, the limited available means for direct dating of ores have prompted studies of minerals from barren zones close to ore concentrations for defining the period of mineralization (reviews by Clauer and Chaudhuri 1991, 1992). Among the minerals examined for such isotopic dating purposes are the clay minerals which are potentially capable of recording tectono-thermal events in connection with the ore depositions. The isotopic systems of the clay minerals are useful in documenting both the time and the duration of hydrothermal processes.

3.2.1 Clay-Dating of Uranium Deposits

Lee and Brookins (1978) and Brookins (1980) reported a Rb-Sr isotopic date of 139 ± 10 Ma for <2-μm clay separates from uranium mineralized sandstones of the Upper Jurassic Morrison Formation of New Mexico. The date was based on an isochron defined by smectite-rich clay fractions from barren rocks and chlorite-rich clay fractions from ore zones. Lee and Brookins also obtained low Rb-Sr isochron dates from 110 to 115 Ma for clay minerals, which they interpreted as subsequent periods of mineralization. The U-Pb dates on the same ore deposit yielded lower discordant values from 87 to 113 Ma. In contrast to the Rb-Sr dates, the K-Ar values of the clay minerals were found to be widely scattered between 138 and 49 Ma, which suggest some additional hydrothermal events in the area.

Clauer et al. (1985b) reported K-Ar results of <2-μm clay fractions associated with uranium-ore deposits from Cluff Lake in Saskatchewan, Canada. Clay fractions selected from mylonitized parts of the sedimentary sequence which was affected by the same tectonothermal event as the basement and consisted largely of illite with smaller amounts of chlorite, provided K-Ar dates averaging 1293 ± 36 Ma. The analyzed clay samples were located no farther than 40 m from the massive uranium ore deposits and were considered to have experienced the tectonothermal event responsible for the major uranium concentration in the area. Philippe (1988) obtained a U-Pb date of 1278 ± 31 Ma for truly unaltered pitchblende samples from the same area. To elucidate the extent of the thermal event in this region, Clauer et al. (1985b) also analyzed several <2-μm clay fractions from both tectonically undisturbed barren sandstones close to the ore pit and sandstones outside the Carswell structure, away from any uranium show. This second set of clay fractions gave a K-Ar isochron date of 1222 ± 52 Ma which is similar to the age of the clay fractions associated with the ore-bearing zone. Based on the similarity in the isotopic dates, the authors suggested that the hydrothermal event near the tectonically disturbed units was part of a tectonothermal event widespread across the region at this time, with uranium concentrations only at the contact between basement and sedimentary cover. The temperature of this event was estimated at 200 ± 50 °C, based on microthermometric determinations in the fluid inclusions of the quartz overgrowths (Pagel et al. 1980). Similar dates were also obtained on clay fractions and uranium minerals in sandstones from the uranium deposit of Cigar Lake in eastern Saskatchewan (Philippe et al. 1993).

Lower Carboniferous sandstones in the Akouta region in central Niger are host rocks for a major uranium ore deposit. The temperature of the fluids from which the uranium precipitated could have been as high as 160 °C (Forbes et al. 1987). Turpin et al. (1991) contended that the uranium deposition occurred between 260 and 150 Ma, based on U-Pb determinations on well-characterized pitchblende samples. Illite-corrensite clay assemblages in proximity to the ore deposit gave K-Ar dates of 148 ± 6 Ma, whereas the illite-chlorite assemblages yielded a K-Ar date of 199 ± 8 Ma, suggesting the existence of an earlier hydrothermal event (Fig. 5.23). By contrast, clay fractions from the ore body itself yielded low K-Ar dates of 134 ± 4 Ma and 117 ± 6 Ma. Additional age determinations are certainly needed to constrain better the U-Pb age and to determine the relationship between the hydrothermal activity and the uranium deposition.

The uranium ore deposit of Lodève in southeastern France is believed to be either Upper Permian or Lower Permian in age. Lancelot and Vella (1989) dated the massive pitchblende concentrations of the basin at 183 ± 4 Ma by U-Pb and Pb-Pb determinations. Mendez Santizo (1990) found,

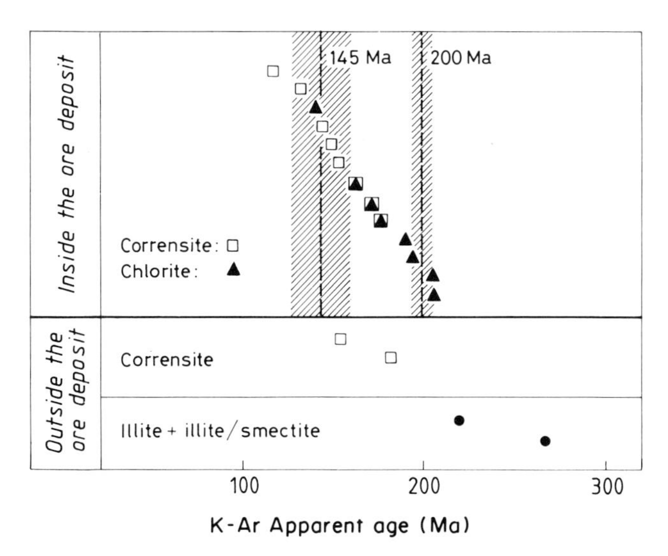

Fig. 5.23. K-Ar dates of both corrensite and chlorite assemblages for samples within the uranium ore deposit and corrensite-illite assemblages for samples outside the ore deposit of Akouta in Niger. (Turpin et al. 1991)

based on analyses of inclusions in quartz within and around the fault system of the basin, that temperatures for the mineralization ranged from 150 to 300 °C, with a maximum frequency at 170–180 °C. Mendez Santizo et al. (1991) made K-Ar determinations on different sizes of clay fractions from samples collected in a mine galery on a 20-m-long transect across a major fault. The clay fractions consisted mainly of illite and chlorite away from the fault, whereas illite was replaced by an illite/smectite mixed-layer mineral within the fault where chlorite was crystallized the best. Different size fractions of seven samples yielded K-Ar dates ranging from 254 to 103 Ma, depending on their location relative to the fault (Fig. 5.24). The lowest date of 103 Ma was found in the finest fraction (<0.2 μm) of the sample within the fault, whereas the coarser fractions (<0.4, 0.4–1, and 1–2 μm) of the same sample yielded dates between 110 and 165 Ma. The highest date of 254 Ma was found in a coarse fraction (1–2 μm) of a sample located about 10 m away from the fault. This K-Ar dates obtained here suggest a high sensitivity of the K-Ar isotopic systems in very fine clay particles to very discrete and late hydrothermal activities.

3.2.2 Clay-Dating of Mississippi Valley Type Pb-Zn Deposits

The Pb-Zn sulfide deposits in the midcontinent of the United States are probably the most extensively studied Mississippi Valley-type (MVT) ore deposits. These deposits are now accepted to have formed from hot basinal

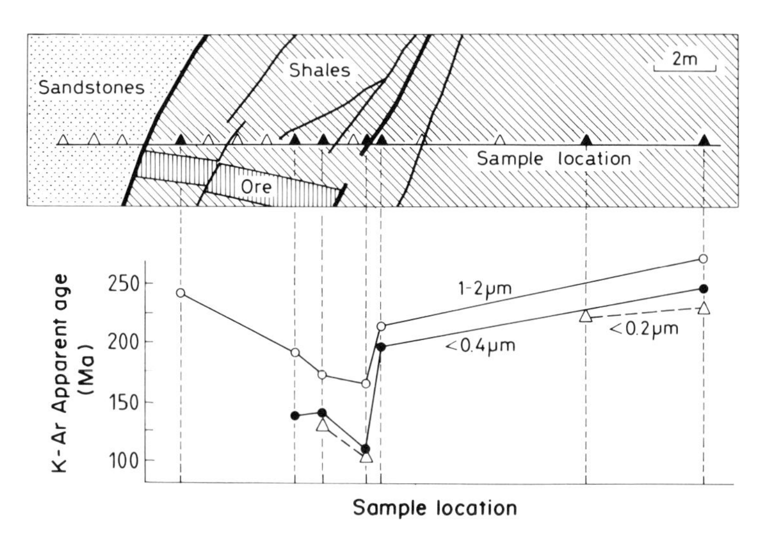

Fig. 5.24. Location of hydrothermally altered samples from the uranium ore deposit of Lodève in southern France. Samples were taken from a 20-m transect across a major fault. (Mendez Santizo et al. 1991)

brines (Hall and Friedman 1963; Viets and Leach 1990). In the midcontinent of the United States, three major MVT ore districts are central Missouri, southeast Missouri, and Tri-State-northern Arkansas, covering a region of about 24 000 km^2 . The Pb-Zn deposits in southern Missouri area are mainly in upper Cambrian rocks, whereas those in the Tri-State-Northern Arkansas area occur in Ordovician and Mississippian rocks and those in central Missouri in Ordovician rocks. The widespread occurrences of MVT deposits in Paleozoic rocks throughout the North American midcontinent suggest a common origin for these deposits. Leach and Rowan (1986) and several others suggested that the ore deposits in the region were probably linked to the Ouachita tectonism in Late Pennsylvanian-Permian time and that the temperatures of the ore fluids ranged between 80 and 170 °C, causing potential widespread heating of the Paleozoic rocks to about this temperature range. Despite many different geochemical, structural, and stratigraphic studies on the MVT deposits, many fundamental questions remain unresolved. The source of the fluid or the fluids, the time or the times of mineralization, and also the flow paths for the fluids, remain virtually unknown. Because several stratigraphic units within the ore bearing sequences contain glauconite minerals, some isotopic studies on these glauconites have been made to address the time of ore mineralization.

Kish and Stein (1979) and Posey et al. (1983) obtained Rb-Sr isochron dates of about 358 Ma for a suite of acid-leached glauconites separated from ore zones in southeast Missouri. Grant et al. (1984) determined both Rb-Sr isochron dates and K-Ar conventional dates from a suite of glauconites both away from ore zones and near ore zones in Upper Cambrian Bonneterre and Davis Formations in southeast Missouri. They found that the suite of glauconites from the Bonneterre Formation had a Rb-Sr date of about 423 Ma with an initial $^{87}Sr/^{86}Sr$ ratio of about 0.7095 and an average K-Ar date of about 423 Ma, and that the suite of glauconites from Davis Formation had a Rb-Sr isochron date of 387 Ma with an initial $^{87}Sr/^{86}Sr$ ratio of about 0.7136 and an average K-Ar date of about 368 Ma. Grant et al. explained that the difference of about 20–30 Ma in the results between their study and that of Kish and Stein (1979) and Posey et al. (1983) was due to an acid leaching of the minerals in the other studies. Conventional K-Ar analyses by N. Clauer and S. Chaudhuri (unpubl. data) on several glauconite separates from ore zones in the Ozark region also gave two sets of dates, one at 385–390 Ma and the other at less than 355 Ma. Many of these isotopic dates on glauconites are somewhat in agreement with the Rb-Sr isochron date obtained by Lange et al. (1983) from analyses of galena samples from Viburnum Trend in Missouri. If these ages can be accepted as a period of ore mineralization, then the isotope resetting of the glauconites about 360 to 390 Ma ago might be related to ore mineralization from hydrothermal fluid related to middle Ordovician uplift in the Ozark region, or to a thermal event contemporaneous with the emplacement of the Avon diatremes in the region about 380 Ma ago (Zartman et al. 1966).

The relationship of the isotopic dates of glauconites to mineralization is in dispute with conclusions from structural and stratigraphic considerations and also paleomagnetic and isotopic studies. Structural and stratigraphic considerations by some support the idea that the widespread ore mineralization in the region occurred following the late Pennsylvanian and early Ouachita tectonic uplift. A paleomagnetic study by Wu and Beales (1981) concluded that the ore mineralization occurred during Carboniferous-Permian time. A recent paleomagnetic study of Symons and Sangster (1991) on barite deposits which are thought to be genetically related to Pb deposits in central Missouri district, also concluded that the Pb-Zn mineralization in the region occurred during late Pennsylvanian to early Permian time, supporting the Ouachita uplift model of Leach and Rowan (1986) for the origin of the MVT ore deposits in the central North American continent. These paleomagnetic dates are duplicated by Rb-Sr results of sphalerite samples from the West Hayden ore body which yield two independently indistinguishable isochrons of Permian age at about 270 Ma (Brannon et al. 1992) and by widespread authigenic feldspar growths of about 280–315 Ma in several Paleozoic sandstones in Pennsylvania, Maryland, Virginia, and Tennessee (Hearn and Sutter 1985). These dates also agree with the K-Ar dates of illitization of Ordovician bentonites in the same area as reported by Elliott and Aronson (1987) and with some of the K-Ar dates of 265 Ma obtained by Hay et al. (1988) on illite/smectite mixed layers of Ordovician tuffs. Hearn et al. (1987) suggested that the authigenic feldspar and the Pb-Zn deposits were probably formed by the same brines in late Paleozoic time.

The migration of brines was probably a regional phenomenon associated with the tectonic activity that accompanied the Alleghanian orogeny. The debate about the dates continues as Nakai et al. (1990) obtained a Rb-Sr date of about 380 Ma for sphalerites from the east Tennessee MVT district, which they linked to a middle Paleozoic Acadian orogeny, and Hay et al. (1988) obtained Rb-Sr dates of about 400 Ma for authigenic feldspars and K-Ar dates of about 360 Ma for illite/smectite mixed-layer minerals from tuffs of the Upper Mississippi Valley. Possibility still exists that the MVT deposits may have formed at several different times. Should the mineralization happen to have occurred from a single tectonically related hydrothermal episode during late Pennsylvanian to early Permian time, as suggested by some investigators, then the failure of the glauconites to record this event needs further investigation.

3.2.3 Clay-Dating of Other Pb-Zn Deposits

Many of the Pb-Zn ore deposits of the Cévennes deposits in the southeastern French Massif Central are located within sedimentary formations (Michaud 1980). Several genetic models have been provided to explain these ore deposits, invoking syngenetic concentrations, diagenetic migrations, karstic processes, or hydrothermal circulations (see Le Guen et al. 1991 for review). Le Guen et al. (1991) determined Pb isotope compositions of Pb-Zn minerals from both ore and barren zones, and concluded that no evidence existed for inherited Pb from host Cambrian dolomite. They assumed that most of the metal stock generated during Triassic time and remained homogeneous despite successive phases of mobilization and concentration. A preliminary study of the mineralogy and K-Ar system of the <2-μm fractions and subsize fractions of nine shaly samples collected in the Malines mining district showed that these clay fractions were illite rich with minor amounts of kaolinite or chlorite (Toulkeridis et al. 1993). Illite crystallinity indices were characteristic of upper greenschist facies conditions. The K-Ar analyses of these clay fractions yielded dates that ranged from about 298 to about 177 Ma, having six of these values between 298 and 279 Ma and six others between 203 and 177 Ma, and values of six more scattered between these end-member values. The high dates at about 280 Ma correspond to a metamorphic overprint induced by the emplacement of nearby Hercynian granites which were dated at 280–290 Ma (see Le Guen et al. 1991). The low dates of about 190 Ma are believed to be records of the hydrothermal mineralizing event that was responsible for the deposition of most of the metal stock.

To determine the potential of dating ore deposits by the Pb isotope analyses of clay materials in ore-hosted rocks, Toulkeridis et al. (1993) compared the Pb isotopic compositions of clay fractions to that of the ore minerals in the Malines district in the Cévennes region. The clay fractions

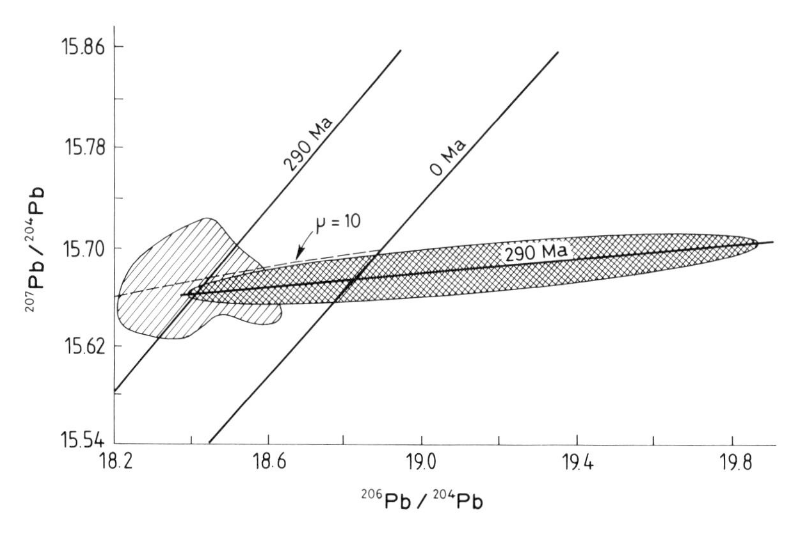

Fig. 5.25. Pb-Pb isotope data of clay minerals associated with the Pb-Zn ore deposit in the Malines district in southern France. The *dashed area* is for Pb isotopic dates for various galena samples measured by other authors, whereas the *hatched area* is for the Pb isotope dates for the clay fractions. (Toulkeridis et al. 1993)

were examined by SEM to ensure that no galena grains were associated with the clay particles in the samples selected for the analyses. The Pb in the clay materials had $^{206}Pb/^{204}Pb$ ratios that ranged between 18.48 and 18.60, $^{207}Pb/^{206}Pb$ ratios that ranged between 15.66 and 15.68 and $^{208}Pb/^{204}Pb$ ratios that ranged between 38.36 and 38.45. These values plot on a μ evolution curve of 10 and are ankored well within the Pb isotope ratios of the nearby ore minerals (Fig. 5.25). These preliminary results confirm the existence of a genetic link between the ore and the clay minerals and also support the existence of a mineralizing event during the Liassic time in the southeastern Massif Central. More similar studies are needed to determine the full potential of Pb isotope analyses of clay minerals.

4 Summary

The K-Ar isotopic records of clay-sized mica-type minerals in low-temperature anchimetamorphic to epimetamorphic metasedimentary rocks clearly demonstrate that recrystallization of these minerals attendant with the metamorphic processes is regulated by the permeability of the rocks involved. For clay minerals of nearly identical size, those in low permeability rocks such as shales have higher premetamorphic memory than those in relatively high permeability rocks such as sandstones. Furthermore, the effect of recrystallization on the premetamorphic isotopic memory is dependent on the size of the clay particles involved. The temporal history of a metamorphic event may be identified in several different size fractions within the clay fraction, but if the event has to be identified in shales, the selected clay fraction must be of very fine size. The characteristics of the isotopic records during the recrystallization of a rock producing clay minerals depend on the original composition of the materials. Clay minerals in metavolcanics, because of the high amounts of unstable components in the parent rock, provide more reliable indication of the time of recrystallization than those of associated metasediments. This effect is most evident in metabentonites. Clay minerals of these rocks from a wide variety of tectonothermal conditions have provided isotopic dates that are concordant among several different size fractions in a single sample, as well as among different samples. The isotopic dates of such minerals are also geologically compatible with a clearly discernible tectonic element.

The Rb-Sr isotopic records of clay minerals may also serve as a guide in identifying metamorphic recrystallization events, whereas more investigations with the Sm-Nd method are needed before reliable routine applications can be expected. In shales, metamorphic recrystallization causes some preferential expulsion of radiogenic ^{87}Sr along with expulsion of some amounts of Sr from clay minerals, resulting in decrease of their $^{87}Sr/^{86}Sr$ ratios, while producing enrichment in Rb and a decrease in the K/Rb ratio. The recrystallization-induced decrease in the $^{87}Sr/^{86}Sr$ ratio is more apparent than the increase in the Rb/Sr ratio.

The process of recrystallization of clay minerals into mica-like minerals in shales or other low permeability rock due to a metamorphic event may be decribed, as some called, as a "restructuration" process in a limited fluid-to-mineral ratio environment. The $\delta^{18}O$ values for the minerals change very little from diagenetic environment to the epimetamorphic environment. Examples are known illustrating recrystallizations in environments, closed to some degree, by the evidence that oxygen isotope records remained nearly unchanged, yet the K-Ar and the Rb-Sr isotope records were greatly modified.

The isotopic dating of clay or mica-type minerals can help resolve age problems of a wide variety of tectonic and metamorphic events such as regional, cataclastic, and hydrothermal metamorphisms, but the success would again require very careful selections of samples possibly analyzed along several different lines of isotopic determinations. A potentially very useful area of investigation is the analyses of clay minerals from rocks which host ore deposits from unidentified hydrothermal sources. Isotopic dating of clay minerals from metasedimentary rocks hosting uranium minerals can also help to constrain better the U-Pb dating of these ore minerals, as they are very sensitive to any post-formational alteration.

Chapter 6

The Frontiers of Clay Isotope Geochemistry

Most will agree that clay minerals are one of the major components of the materials in the shallow crustal segment of the Earth. For us, clay minerals also represent the most fascinating component of this complex "chemical world", because they carry records of many discrete evolutionary events in the constantly changing history of the upper crust. Isotopic data gathered from analyses of clay minerals can, therefore, serve as very useful tracers for understanding many aspects of the Earth's history. However, the potential is high for clay minerals and clay-like minerals to be of multiple origin in sedimentary and metasedimentary rocks. Furthermore, these minerals can experience pervasive changes in their chemical compositions, especially in the interlayer compositions, due to interactions with ambient fluids. Consequently, isotopic records of clay minerals must be carefully analyzed and not be taken as straightforward information about the history of their formation. Because the growth of clay minerals often results from replacement of silicate precursors, the reaction rate is an important control in determining the isotopic signatures of the newly formed clay phases.

Protocols for appropriate separation and detailed identification of clay minerals from rock samples have to be carefully considered prior to the isotopic analyses. The failure to meet these requirements clearly contributes to high degrees of uncertainty about the results and even to misleading conclusions. Many studies in the past have taken straightforward approaches in isotopic analyses for clay minerals only to find that the results could not be used to achieve the main objectives. Isotope systems of clay minerals in rocks of shallow crustal environments were often found to have been variously disturbed. Recognition of these disturbances can be important because such knowledge can provide powerful clues in discovering and defining important geological events in the history of the upper crustal rocks. Very detailed knowledge of the analyzed materials is also necessary to ensure that any trend in the isotopic data is due to natural and not artificial causes. Where care has been taken in gathering independently as much information as possible, isotopic results have proved extremely useful in gaining valuable insights into the time, process, and environment of formation of the clays in question. Although the early isotopic studies might have collected data under less prudent considerations, they have been extremely useful in providing the foundation for later development of a comprehensive framework relating isotopic parameters to various physical and chemical conditions during the growth of clay minerals.

Differentiation and characterization of clay minerals in question are essential steps that must be taken to ensure that the isotope data provide well-constrained and enlightened information about the history of the minerals. The practice of classical crushing and grinding techniques for disaggregation of the rocks should be discarded, because such a procedure leads to artificial creation of clay-sized particles from the mechanical breakdown of unwanted minerals such as K-feldspars, micas, and quartz that may be present in the rocks. Unrestrained disaggregation of rocks is a sure path to achieve equivocal or enigmatic isotopic results. Instead, the use of a repetitive freezing and thawing technique, or any other gentle procedure that produces only insignificant amounts of artificial clay-sized particles, is recommended. The separated clay fractions must always be mineralogically defined by X-ray diffraction analyses, but also be morphologically described by electron microscopic (both SEM and TEM) observations, and chemically characterized by major chemical and necessary trace element compositions including the REE distributions. Leaching of clay minerals by appropriate reagents and chemical and isotopic analyses of both leachates and acid-leached clays should be followed because such information may provide valuable insights into the environment in which the clay particles formed. The acid leaching should be carefully controlled to ensure total recovery of the acid-leached clay phase during elutriation of the samples, as the failure to do so can result in loss of potentially valuable information or even in an erroneous interpretation of the data.

1 Isotope Geochemistry as a Dating Tool for Clay Minerals

The objectives of isotopic dating of clay minerals can be varied, ranging from determining the provenance history of sediments to resolving the time of sedimentation or defining the time and duration of post-depositional events. The K-Ar method and, less commonly, the Rb-Sr method are the two principal lines of isotopic analyses that have been hitherto sought for these purposes. The Sm-Nd method of dating clay minerals has just begun to come into use, indicating a high potential for this method as a geochronologic probe in studies of ancient sediments. The $^{40}Ar/^{39}Ar$ and Pb-Pb methods have yet to be fully tested for their usefulness.

In many sedimentary systems, the growth of a clay mineral or any other secondary silicate mineral may be found to be related to the replacement of a precursor silicate mineral. However, chemical equilibrium between bulk fluids and secondary mineral products may not always be attained in the sedimentary systems. Suggestion has been made that secondary products which are formed from alteration of silicate precursors, especially volcanic glasses, may evolve through a series of rate-dependent, interface-controlled irreversible reactions leading to the formation of more thermodynamically stable phases (Dibble and Tiller 1981). As discussed in the previous chapters, an important insight may be gained from the isotopic data of clay minerals about the control a bulk solution may have on the formation of a secondary mineral in a given system. In soil profiles or sedimentary environments, for instance, the isotopic signatures of secondary clay minerals are clearly dependent on the size of the formational environment.

1.1 Depositional Time Indicator

Isotopic tools that have been used for a variety of sedimentologic purposes, have had to date only limited success in determining the time of deposition of sedimentary sequences because the results appear equivocal and even unrealistic in some instances. Inability to decipher the isotopic signatures of the clays at the time of their deposition is part of the cause for this limited success.

The Rb-Sr and K-Ar isotopic "model" ages for clays in young ocean sediments are commonly far in excess of the "zero" sedimentation age. The high dates for the clays are due to the retention of isotopic memory from a previous phase in their evolutionary history. Vast amounts of clays in the ocean sediments are detrital in origin, and laboratory studies of the reactions of well-crystallized illitic clay minerals in sea waters at low temperatures have shown that these minerals remain essentially nonreactive to sea water. "Degraded" illites and other clay minerals with expandable layers are known to absorb K and other dissolved ions from sea water. What is not clearly known is the degree by which isotopic compositions of terrigenous clays are modified by K fixation or ion exchange as a result of their transport to ocean basins.

Studies of clays from various DSDP Sites in different ocean basins have shown that smectites and, to a smaller degree, celadonites are very common as autochtonous clays in young ocean basins. Hydrothermal smectites and those clay minerals formed at sites away from hydothermal activities are very commonly of saponite-nontronite type, whereas smectitic clays of terrigenous origin are commonly of montmorillonite-beidellite type. As hydrothermal smectites and other autochtonous clay minerals are easily reworked in ocean basins and mixed in various amounts with terrigenous smectitic and illitic clay minerals, much caution is needed in the interpretation of the isotopic data of a clay mineral assemblage in any ocean sediment. Independent geochemical evidence must be integrated with the isotopic evidence to determine the fraction of authigenic clays in the ocean sediments. Isotopic investigations of clay minerals for the purpose of determining their sedimentation time require intensive scrutiny of testing the relative importance of factors that may have influenced the isotopic composition of the clay minerals in question.

K-rich illitic clays are ideally suitable candidates for age determinations. But evidence is common that these minerals do not form at least in sufficient amounts until long time after the deposition of the sediments. In rare instances, such as in Precambrian sediments of Western Africa and Northern America, K-bearing clays appear to have formed shortly after the deposition of the sediments. If such claims are valid, additional studies are needed to determine the conditions under which they could have formed.

Theoretically, glauconites are ideal minerals for isotopic determination of the time of sedimentation because they are known to be about synsedimentary in origin and they can be easily separated from other minerals in a rock. Different studies on mineralogic and chemical compositions of recent glauconites have shown that these minerals could evolve from replacement of smectitic clay precursors in some few tens of thousand years after the deposition of the sediments in a marine environment. Several studies have concluded that the isotopic records of only those glauconites that have K_2O contents of at least 6.5%, have the potential to yield information about the time of deposition, suggesting that such "mature" glauconitic minerals originating shortly after the deposition of the sediment probably experience very little isotopic disturbance under most normal burial conditions since the time of deposition. Nevertheless, possibilities exist that glauconites may also attain the "high" K values long after the time of deposition by some processes under deep burial conditions, which may explain why many glauconites have been found to have low isotopic dates. Analyses of carefully selected pure glauconite samples have shown that low dates for any pure glauconite samples are due not as much to preferential loss of radiogenic isotopes (Ar and Sr) as to gain of alkali elements (K and Rb). Additional studies are needed to establish independently the geochemical or isotopic criteria that can be used to differentiate between early evolved glauconites and late evolved or recrystallized ones, as the information is of critical importance to the interpretation of isotopic data of glauconites concerning the time of their formation.

1.2 Diagenetic Time Marker

Clay diagenesis is commonly considered to commence with the time the sediment becomes detached from the water-sediment interface subsequent to the deposition. The boundary between syngenesis and diagenesis is difficult to draw because sediments may be overturned many times across the sediment-water interface by physical and biological forces at the site of deposition. Diagenetic evolution of clays may be classified arbitrarily into early and late types. Early diagenesis is less influenced by temperature and pressure than late diagenesis, which is commonly attendant with significant amount of burial of the sediments. No specific value for the magnitude of burial has been agreed upon to distinguish an early diagenetic phase from a late diagenetic phase, although many consider the division between the two at a burial of about 2 km.

1.2.1 Early Diagenesis

Many have considered that terrigenous clay mineral assemblages in marine sediments are generally stable under shallow burial conditions. Studies of pore waters in shallow buried sediments in the floors of modern oceans have shown the existence of concentration-depth gradients for different elements. The trends have been attributed primarily to the mass transfer between underlying altered basalt and overlying sea water, while the intervening sediment column in most places remains more or less passive, especially when the sediment consists of illite, chlorite, kaolinite, and various mixed-layer minerals. Many have recognized the potential of formation of early diagenetic smectites, and to a lesser degree palygorskites, from alteration of volcanogenic detritus in shallow buried sediments. Some have even entertained the idea of fixation of K by degraded illitic minerals during the early burial stage in the history of sediments. Evidence consisting of identification, crystallinity variation, and abundance of minerals alone cannot be used to draw any specific conclusion about the genetic history of the clay minerals in marine sediments because the same mineralogic parameters can be equally influenced by paleotectonic, paleoclimatic, and diagenetic factors.

Isotopic evidence has proved useful to detect the early diagenetic formation of clay minerals in some young modern ocean sediments. Analyses of oxygen, hydrogen, and Sr isotopic compositions of very fine ($<0.1\,\mu m$) clay fractions, primarily of smectitic composition, in sediments on the floors of modern oceans have indicated records of authigenic clay formation. These fine clays appear to constitute no more than 5% of the bulk clay material. As interlayer cation exchange with ambient waters is an important mode of clay mineral modification, many traditional mineralogic analyses and microscopic observations are not sensitive indicators for the recognition of whether or not clay mineral assemblages have experienced ion exchange related modifications as a result of shallow burial diagenesis. Various isotopic data for clay minerals in shallow buried ocean sediments from many different locations are needed to obtain more insights into the process and extent of clay mineral modification as a result of early diagenesis.

1.2.2 Late Diagenesis

A major trend in the change in the clay mineral composition of sediments, attendant with an increase in burial depth, is the increase in the illite content in the illite/smectite mixed-layer phases. Numerous studies have addressed various factors that can be responsible for the decrease in expandability of illite/smectite mixed-layer minerals with increase in depth. Besides many disagreements about physical and chemical factors in the process of illitization, major controversies exist as to the time and duration of this illitization related to burial in the sedimentary basins. Results from studies of illitization in Tertiary shales of the Texas Gulf Coast area ended up in sharply divided opinions between a prolonged continuous process and a discrete or "punctuated" event about the burial-induced illitization of illite/smectite mixed-layer clays.

As already discussed, the K-Ar data of very fine <0.1-μm illite/smectite clay minerals in Oligocene-Miocene shales in the Gulf Coast region have been interpreted to represent the records of a long continued process with a suggested average duration of about 18 Ma (Aronson and Hower 1976), whereas the Rb-Sr data for nearly identical minerals from a 1.6 km stratigraphic interval have been used to claim that the illitization occurred in a short period at about 23–24 Ma ago (Morton 1985a). Following Morton's claim, the illitization occurred when the sediments were buried to a depth of about 2.5 km, but with no more illitization through an additional 3 km burial for the sediments. How the Sr isotopic homogeneity occurred for clays within a 1.6-km-thick interval remains obscure, especially in view of the Gulf Coast shales being known for their detrital components varied in origin. A convincing explanation for Sr isotopic homogeneity is even more needed in light of the fact that both oxygen and hydrogen isotopic data for the same clays signify lack of hydraulic communication within the zone of illitization.

Interpretations of isotopic dates for clay minerals separated from shales are difficult. Both separation of diagenetic minerals from abundantly present detrital components and high potential for inheritance of radiogenic isotopic memories from potential precursors by the diagenetic clays can have significant influence on the isotopic dates for the clays in question. The meaning of a Rb-Sr isochron date can become more clear by analyzing "internal" isochrons for several clay samples across a zone of illitization of smectite. An "internal" isochron for each clay sample is based on data from acid leachates and both acid-leached and untreated materials. The dates from these "internal" isochrons should be congruent, even though the initial Sr isotopic ratios defined by the isochrons may be found to be varied. Isotopic dates for clay minerals in shales can also be influenced by minute amounts of very old detrital components, which may escape detection by X-ray analysis or even electron microscopic observation. Clays in shales are known to yield high K-Ar dates relative to the dates of sedimentation established by some other independent means. Incorporation of radiogenic isotopes by clay minerals from their immediate precursors can be a potential cause for the anomalously high K-Ar dates for clays in shales.

The duration of illitization in deeply buried sediments may be defined straightforwardly by analyzing clays in sandstones from sedimentary sequences having records of illitization of smectite. Clay separates from sandstones generally contain much smaller amounts of detrital components than those from shales, because of the relative ease with which authigenic clay minerals can be separated from sandstones. Also, the potential is low for incorporation of radiogenic ^{40}Ar by the diagenetic clay minerals during their growth in sandstone. Studies of illitization of smectite in illite/smectite mixed-layer clays in Jurassic sandstone-shale sequences in the North Sea basin have well documented the fact that temporal records for illitization in sandstones are geologically reasonable compared to those in shales. The K-Ar dates for diagenetic illites in these sandstones indicate that widespread illitization occurred about 40–45 Ma ago. By contrast, the K-Ar dates for clays in shales in the North Sea basin have been found to be consistently high. These apparently high dates for illite-type clay minerals in shales relative to those for the same minerals in sandstones have also been observed in the Miocene sequence in the Mahakam Delta basin, where the K-Ar dates for clay minerals in a sandstone sample buried to about 4000 m were found to be about 4 Ma, which corresponds to the date of a structuration event in the area (Furlan 1994). Hence, the evidence from the data for clays in sandstones from the two sedimentary basins suggests that burial-related illitization of smectite in illite/smectite mixed-layer clays can be episodic in nature. Many studies are needed to determine whether the high dates and often wide variations in the isotopic dates for clays in shales are due to a problem associated with the separation of clays or to the evolution of the minerals from a transformation process that occurred in conditions of a closed chemical system.

The formation of clay minerals in many bentonitic beds also stands as a major unresolved subject. Bentonitic beds are known to be useful stratigraphic marker, and geologic records have shown that these clay-rich beds are often derived from alteration of volcanic ash debris that deposited in both marine and continental settings (Fisher and Schmincke 1984). Smectite and illite/smectite mixed-layer minerals are the dominant clay minerals in many bentonitic beds, but kaolinite and illite, as well as zeolites, can also be found in considerable amounts especially in the continental deposits (Gundogdu et al. 1989). Studies of altered volcanic ash in ocean-floor sediments that are as old as Miocene have shown that the clay minerals are essentially smectitic in composition. The high abundance of smectite minerals produced from alteration of volcanic ash in the floors of modern oceans contrasts with the high abundance of illite and illite-enriched illite/smectite mixed-layer clays in many ancient K-bentonite beds. Thus, the formation of many ancient K-bentonite beds appears to be a two-step process having an episode of smectite formation from alteration of volcanic ash in an early diagenetic period and a subsequent episode of illitization of smectitic clays in a late diagenetic period. Several important questions surround the phenomenon of formation of ancient bentonite beds. Why do unaltered volcanic materials occur alongside or in close proximity to altered volcanic materials that resulted in the formation of clay minerals? In this presumed two-step evolution for many ancient bentonite beds, when was the early diagenetic alteration? What influence did adjacent sedimentary beds have on the rate of alteration of the volcanic ash materials or on the change from smectite to illite/smectite mixed-layer minerals? What is the source of K for illitization of the smectite in many bentonite beds? Important clues for answers to these questions are hidden in the isotopic records of clay minerals in bentonitic beds. Investigations focusing on various isotopic studies, combined with detailed studies on mineralogical, morphological, and both major and trace chemical characteristics for clay minerals in bentonite deposits, are needed to determine many complex steps in the origin of ancient bentonite beds.

1.3 Provenance Tracer

A large body of literature has recognized many difficulties that stand in the way of reconstructing such major aspects in the history of many argillaceous sedimentary rocks as the source of clays, the paleocirculation pattern of the sediments and the tectonic factors at the time of deposition of the clays. Mineralogical changes in the composition of clays can be accurately determined using standard analytical procedures. Although such information is vital, it alone is not sufficient to define fully the causes for any spatial and temporal variations in the clay mineralogy, because a given magnitude of the variation can result equally from changes either in climate, provenance, or paleocirculation.

A few studies on oxygen, Rb-Sr, and Sm-Nd isotopic compositions of very young ocean sediments have demonstrated that isotopic evidence can be used to obtain information about the source of the sediments and their dispersal pattern in parts of the modern ocean basins. Much progress needs to be made in isotopic studies of modern sediments in oceanic and continental environments to fully explore how spatial changes in isotopic signatures relate to those in the mineralogy of clay sediments. This knowledge could be useful in applying isotopic variations in clay minerals in order to understanding the influence of various sedimentologic factors such as provenance, climate, tectonics, and paleocurrents in the history of deposition of ancient sediments.

1.4 Isotopic Dates and Particle Size

Many studies on K-Ar and Rb-Sr isotopic dates for clay minerals in sedimentary rocks have recognized positive covariations between isotopic dates and particle size of the analyzed clay materials. This common trend has also been seen for clay minerals or clay-like minerals in metasedimentary rocks. Increase in diffusive loss of the radiogenic isotopes from the fine clay minerals due to increased surface area is not a strong argument for their lowered ages relative to coarser particles. K-Ar data for mica-type minerals in the Buntsandstein sandstones buried to a depth of 3 km in the Rhine graben have shown that continuous loss of radiogenic ^{40}Ar has yet to occur, as the 40-Ma-old isotopic memory of these minerals is still preserved today under current hydrothermal conditions with a temperature of about 160 °C. The most likely cause for the size-date trend for clay minerals in sedimentary rocks is the relatively decreased amount of detrital components in the diagenetic finer clay fractions. The trend of decrease in isotope dates with decrease in size may also be influenced by increased recrystallization of fine clays due to an increase in temperature attendant with increased burial. Another possible cause for this size-date trend could be that the process of nucleation and crystallization of illite in very small pores was much slower than that in relatively large pores.

Some evidence has been found that suggests an inverse relationship between size and isotopic date for clays. For example, K-Ar isotopic dates for illitic clay minerals in a sandstone sample of the Rotliegendes Formation in Germany have been found to increase as a function of decrease in the particle sizes. This result may support the concept of Ostwald ripening, a process by which coarser particles grow from the reuse of solutes derived from dissolution of fine particles. The validity of this concept requires more studies with chemical and isotopic analyses of different clay fractions, especially of particles having sizes much smaller than those studied routinely, almost down to the "fundamental particle size".

2 Isotope Geochemistry as a Record of the Physical and Chemical Conditions of Clay Formation

The growth process of many common clay minerals other than those of illitic and glauconitic type is often little in dispute because the evidence in many cases is such that the newly formed minerals resulted from either replacement of a structurally unrelated precursor or precipitation from a solution. On the contrary, the processes of evolution of K-bearing clay minerals in sediment or sedimentary rocks is often unclear, because in many instances the minerals are intimately associated with replacement of smectitic minerals, following a reaction course which may be described as either a transformation reaction path or a dissolution-crystallization path. How can isotopic data be useful in deciphering the inside story of the change from smectite to either illite, or glauconite? Some insight into the process may be gained by information about factors related to the crystallization history of clay minerals such as sources of solutes, replacement-precipitation process related to the crystallization, origin of water, temperature, solid-to-water ratio, and degree of hydraulic communication both across a bed and among beds during the crystallization of clay minerals.

As post-crystallization processes can obliterate by various degrees the isotopic memories inherited by different clays about the time of their formation, detailed knowledge of such post-crystallization isotopic disturbance is useful for a reliable interpretation of the isotopic data regarding physical and chemical conditions of the systems in which the clays grew. The paths of K-Ar or Rb-Sr isotopes in clay assemblages may be variously influenced by disturbances related to diffusive loss of the radiogenic isotopes, restructuration of clays that followed a course of diffusive loss, and having more than one generation of clays. These potential isotopic paths for a natural clay assemblage may be depicted by Fig. 6.1, and are well illustrated by the course of illite-to-muscovite transition in the Glarus Alps. Diffusion dominated change will mainly induce a subvertical displacement of the dates for the first generation material (Fig. 6.1, arrow 1). The modification process of the first generation material being the response to a more pervasive change, the diffusion effect will be followed by an internal restructuration of the first generation material, which will thereby produce a negative curvilinear trend in the isotopic date-crystallinity relationship (Fig. 6.1, curve 2). On the other hand, the strict addition by precipitation of a second generation material to the preexisting first generation material will be seen

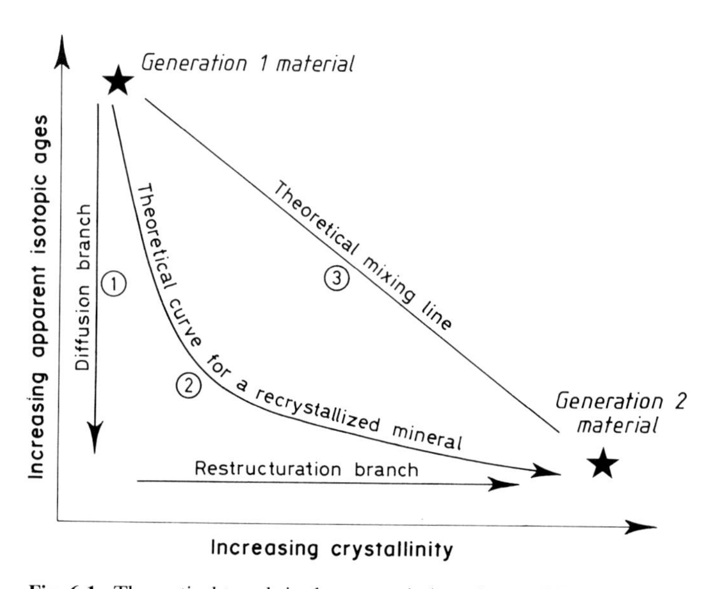

Fig. 6.1. Theoretical trends in the age variations due to different generations of clay minerals in an assemblage

by the arrangement of the data points along a mixing line joining the areas of the two populations (Fig. 6.1, line 3).

Clay minerals physically or chemically modified record a decrease in their isotopic dates along one of three primary courses in an isochron diagram. These courses with decreased isotopic dates result from either a decrease in the isotopic composition (Fig. 6.2, line 1), an increase in the elemental Rb/Sr or K/Ar ratio (Fig. 6.2, line 2), or a combination of both (Fig. 6.2, line 3). A diffusion process will be mainly characterized by the loss of the radiogenic isotopes, which will induce a major decrease in the isotopic composition and an accompanying increase in the elemental ratio. In the case of a late process which affects preexisting clay mineral phases, the data points of the resulting material will move towards the right of the diagram depending on the behavior of the isotopic system: towards the right in a closed system and towards the right but also downwards in an open system. In the case of an independent precipitation of a new generation of sheet silicates, the data points will plot on a horizontal line at t = 0, describing the initial ratio to be equal to the isotopic ratio of the aqueous environment. The scale of the displacement of the data points on this line depends on the

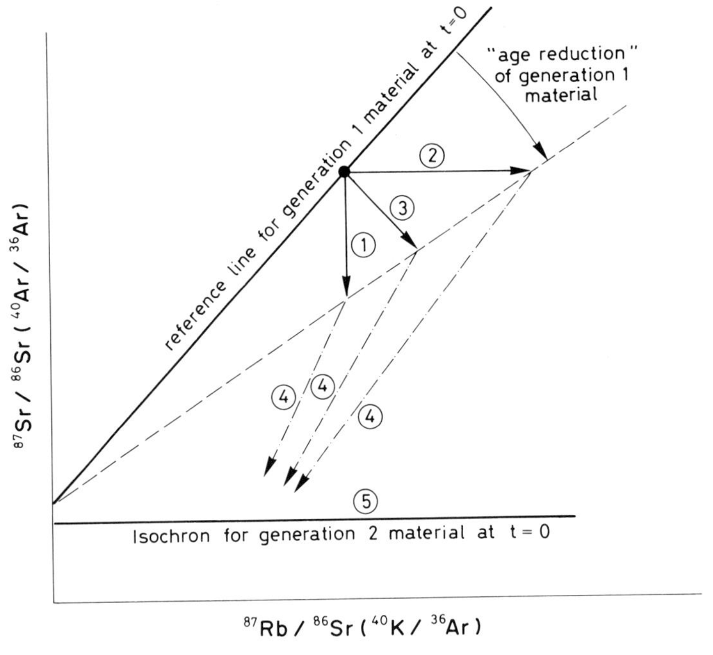

Fig. 6.2. Theoretical behavior of Rb/Sr and K/Ar isotopic systems of clay minerals induced by post-formational reorganizations. Line *1* refers mainly to a diffusion behavior; line *2* to a recrystallization behavior in a closed system; line *3* to a recrystallization behavior in an open system; line *4* to mixing lines consisting in mixtures of generation 1 and 2 materials; and *5* to an isochron of generation 2 material at t = 0

available elements in the mineralizing fluids (Fig. 6.2, line 5). If the mineral separation procedure has led to an assemblage of materials of two generations, the data points are likely to plot along one of several mixing lines with apparently realistic initial ratios (Fig. 6.2, line 4). The study on hydrothermally altered mica-type clays in Hercynian metasedimentary rocks of Morocco, presumably metamorphosed during the Triassic time (Clauer et al., accepted), provides a good case in which the clay and clay-type fractions plot either along line 1 or line 4 or line 5 depending on the samples considered.

2.1 Isotope Geochemistry and the Process of Clay Formation

Whitney and Northrop (1988) experimentally investigated the illitization process for K-saturated smectite phases and examined in detail the fractionation of the oxygen isotope composition between the fluids and the solid phases. The authors found that the $\delta^{18}O$ compositions of the fluids and of the solids paralleled the trend in the change of expandability of the clay minerals. Both the mineralogy and the changes in the oxygen isotope compositions as a function of the illitization process demonstrated that the reaction proceeded in three stages, including two different mechanisms. The first stage corresponded to a dominantly transformation reaction, leading to random illite/smectite mixed layers with approximately 65% of the oxygen in each layer having been reset when illitized. The second stage of the reaction corresponded to a transition stage during which random and ordered illite/smectite phases seem to have coexisted as separate phases. The third stage of the reaction produced ordered illite/smectite mixed-layer phases through a neoformation reaction mechanism during which the degree of isotopic resetting was directly proportional to the degree of illitization. These experimental results demonstrate that the illitization of smectite proceeds along a complex path involving initially a transformation reaction followed by a neoformation reaction.

Isotopic results obtained from a study of recent glauconite minerals have shown that the reaction involving glauconitization of smectitic minerals requires a two-step evolutionary process, starting with replacement by dissolution-precipitation for precursor minerals, occurring as pelletal aggregates (Clauer et al. 1992a, b). This first step of the process happens while the different pellets act as microenvironments that are more or less insolated from the condition of an open sea water environment. This initial alteration phase seems to terminate when the K_2O content reaches about 4.5% and is then followed by a process which is regulated by alkali enrichment from sea water, resulting in the growth of "glauconite" crystals. Part of this sequence in the two-step process has also been recognized in smectite authigenesis in some black shales. In this case, smectite mineralization resulted from dissolution of detrital clay precursors in a semi-closed system with some Fe that was supplied by associated metastable Fe-oxy-hydroxide phases. Thus, the initial dissolution-precipitation process for the formation of glauconite pellets and also authigenic smectites in black-shale rocks are reactions equivalent to that of the transformation type observed in the experimental study by Whitney and Northrop (1988). This reaction phase is dominated by a condition of small water-to-solid ratio with a limited amount of resetting of the oxygen isotopes for the replaced phase. By contrast, the second step in the glauconitization process, with a high water-to-solid ratio, is similar to the neoformation step in Whitney and Northrop's experiment.

The illitization process of smectite appears to be lithology-dependent. Illites in associated sandstone and shale rocks have been found to be varied in their morphology, K-Ar dates, and oxygen isotopic compositions. The results from various studies have suggested a restricted system for illitization of smectite in shales, as compared to a relatively open system for illitization of smectite in sandstones. Furlan et al. (accepted) have recently noted, in support of illitization of smectite in sandstones under open system conditions, that K necessary for that process in the Mahakam Delta basin in Indonesia had to come from a stratigraphic interval that was far below the zone of illitization.

The source of K for the illitization of smectite has been traced to dissolution of K-feldspar. Chaudhuri et al. (1992a) used Rb/K ratios of diagenetic fluids and Sm/Nd ratios of clay-sized sediments to suggest that major solutes necessary for the illitization process were related to dissolution of K-feldspars. Based on Sm-Nd data of clay minerals in Middle Precambrian shales, Stille et al. (1993) corroborated the relationship between illite crystallization and K-feldspar dissolution. Furlan (1994) found a similar genetic link between illitization and K-feldspar dissolution based on the REE patterns for clays and K-feldspars in the sandstones and shales of the Mahakam Delta basin.

2.2 Isotope Geochemistry as a Record of Closed Versus Open System Behavior

Knowledge of the source of solutes provides clues to the physical and chemical conditions of the mineral-fluid reactions involved in the formation of a given clay mineral assemblage in a rock. Information about whether the clay mineral reaction occurred in a closed or an open system with respect to certain components is critical to obtain information about any evaluation of paleohydrology, paleoclimate, or paleotemperature from data on clay minerals. Isotopic data can reveal the conditions that might have existed at the time of formation of a clay mineral assemblage.

At the heart of understanding of the mechanisms of clay formation at different environments in the shallow crustal region lies knowledge about the transport history of solutes for the formation of any particular clay or clay-like mineral. Chemical analyses of assemblages of clay minerals in paleosols have received some attention because such data can be potentially useful for information on past climates. Information about the openness of the chemical system in which a given clay mineral formed is very important in making any interpretation about climatic conditions based on chemical or isotopic evidence. Any degree of inheritance of the isotopic memory from precursor by the newly formed clay minerals is to a large degree dependent on the boundary condition for the reaction system. Many documents on isotopic compositions of clay minerals from different geologic settings suggest that a nearby closed system condition is common for much of the solutes that form clay minerals by replacement of precursors.

2.2.1 Soil Profiles

Isotopic studies of clay minerals in soil profiles are few. The kinetics of many chemical reactions in soil environments are dependent on the transfer of solutes from reaction sites at solution-solid interface. Nonequilibrium, heterogeneous reactions are expected to occur in soil profiles, so that clay minerals as products of such environments, can be expected to have considerable isotopic variations. The potential for diversity in the isotopic compositions of clay minerals formed in soil profiles has been documented by a Rb-Sr study on clay minerals across a soil profile that developed on a crystalline parent rock in the Congo Republic. The authigenic clay minerals were found to have Sr isotopic ratios between 0.7604 and 2.3130. Variations in the Sr isotopic ratios have also been found for clay minerals across several other soil profiles. Acid leachates from clay minerals of these different soil profiles were also varied in the Sr isotopic composition, suggesting that ambient fluids related to the clay authigenesis were extremely varied in their Sr isotopic compositions. Furthermore, combined study of untreated pure authigenic clay fractions, acid leachates, acid-leached solids seem to be potentially useful in dating the paleosols.

Several early isotopic studies of clay minerals in soils documented that oxygen and hydrogen isotope compositions of these minerals correlated well with those of local meteoric waters. The results of these studies have been taken to interpret that the growth of clay minerals in soils occurs in open systems with a high water-to-mineral ratio. However, such an observation about the stable isotope composition of soil clay minerals relating to that of local meteoric water has not always been corroborated. A few recent studies on clays in soils have shown that solutions in equilibrium with the clay minerals were depleted in both the ^{18}O and the D contents relative to the local meteoric waters, suggesting a strong influence in the isotopic compositions of the pore waters rather than of the bulk meteoric waters. In these cases, the initial growth of the soil clay minerals may have been kinetically controlled by the formation of metastable "protomineral" phases in microenvironments. Later, slow, continued infiltration of meteoric waters through the soil profile, causing some hydrogen and oxygen isotope exchanges with the clays but no significant purging of the radiogenic isotopes, may have stabilized the newly formed clay mineral phases. What is also clear from evidence gathered in a few other studies is that the clay minerals can inherit radiogenic isotopes at the time of crystallization in a soil environment. Anomalously high K-Ar isotopic dates were obtained for clay minerals across two independent soil profiles developed on volcanic rocks. The evidence for inheritance of a significant amount of radiogenic isotopes by clay minerals forming in soil environments requires that provenance studies based on K-Ar and Rb-Sr model ages for clays in young ocean sediments or old sedimentary rocks should be carefully considered. Additional studies are needed to gain insights into the relationship between the isotopic compositions of clay minerals and that of ambient waters in soil profiles to understand the growth mechanism of clay minerals with inheritance of apparently large amounts of radiogenic isotopes.

2.2.2 Depositional Environments

The early phase in the process of glauconitization of clay minerals in young ocean sediments appears to be a good illustration of the concept that the solutes can be derived locally for the formation of at least some clay minerals. As has been discussed earlier, collective K-Ar, Rb-Sr, and Sm-Nd isotopic data suggest that the formation of most evolved glauconitic minerals, that is those in excess of 4–5% K_2O contents, entails a two-step sequence. The initial step consists of formation of "immature" glauconitic minerals with limited K_2O content and involves a dissolution-crystallization process in an environment with limited accessibility to sea water for the solutes, and the final step consists of formation of "mature" or highly evolved glauconitic minerals with K_2O content exceeding 6% by having considerable exchanges with sea water for the solutes. In an isochron diagram, the data points move first almost horizontally to the right, starting from a line of detrital memory (Fig. 6.3A). The maturation step is characterized by a downwards trajectory of the data points until reaching the marine $^{87}Sr/^{86}Sr$ ratio (Fig. 6.3A).

The isotopic data on glauconitic minerals suggest that some clay minerals could also form by replacement of terrigenous clay minerals in shallow buried sediments. This appears to be illustrated by the Tertiary lacustrian shale samples of the French Central Massif. In these rocks, the data points of the clay fractions could only be traced along a right downward trajectory in an isochron diagram, while the mineralogy changed from detrital illite to authigenic palygorskite through an intermediate smectite stage resulting in

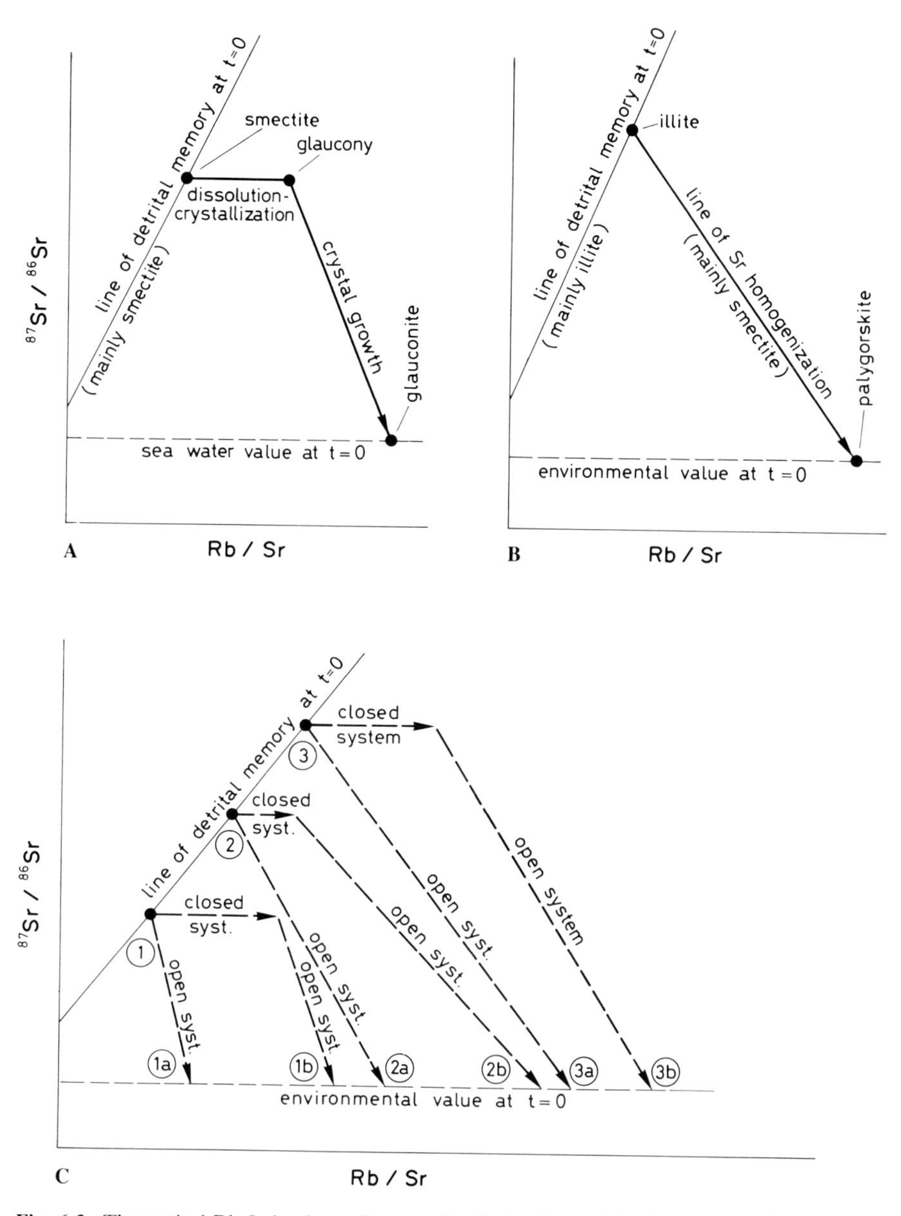

Fig. 6.3. Theoretical Rb-Sr isochron diagrams for **A** the glauconitization process off shore west Africa. **B** The illite-smectite-palygorskite transition in Oligocene lacustrine claystones from French Central Massif. **C** A combined effect of both

the Sr homogenization (Fig. 6.3B). Obviously, the process stopped when the $^{87}Sr/^{86}Sr$ ratio of the newly formed clays reached the value of the environmental Sr, but this progressive homogenization could terminate at any time when the conditions become unfavorable. An isochron-diagram presents a general framework combining the two trends of the homogenization (Fig. 6.3C) with following steps: (1) three studied samples deposited in a sedimentary environment have their data points plotting on or close to a line of detrital memory which characterizes the deposited detrital clay component of the sediments, (2) depending on the immediate environmental conditions, the samples are altered in either a restricted (closed) chemical system and the data points move horizontally, or a renewable (open) chemical system and the points move downwards to the right, (3) the data points will move as long as the process is going on, (4) when the conditions change (a closed one to an open one), the process either changes or stops, and (5) the ultimate stage is an isotopic equilibrium with the environmental Sr.

The consequence of having the formation of new minerals in a closed system is that the recognition of their formation may escape detection by routine chemical and isotopic analyses, because many chemical and isotopic characters of the newly formed minerals would most likely be similar to that of the precursors. In this case, isotopic dating of such clays is achievable only when the isotopic characteristics at the time of formation can be determined independently from those of the clay materials. The evidence for growth of clay minerals under closed system conditions requires many studies which integrate isotopic, chemical, and mineralogical data and do not depend on isotope evidence only, and which focus especially on determining the extent of growth of clay minerals by replacement of precursors in young ocean sediments.

2.2.3 Burial Diagenetic Environments

Increase in the illite content in illite/smectite mixed-layer clays with increase in depth is a well-known diagenetic phenomenon for many deeply buried sediments. Isotopic data can provide important clues as to the mechanism for change from smectite to illite in the mixed-layer clays. The depth-related decrease in the ^{18}O contents of clays in deeply buried shales has been attributed by some to increased influence of temperature. The data further suggest that hydraulic communications across the strata in the zones of illitization were often of limited extent. The restricted hydraulic communications across the sedimentary beds during the illitization of smectite in deeply buried sediments have also become apparent by the contrast between K-Ar isotope data of illite/smectite clays in sandstones and those in adjacent shales from the zone of illitization. The K-Ar isotope dates for the clays in shales have been found to be higher than those in adjacent sandstones. By analyzing clay minerals in deeply buried sediments in the Mahakam Delta basin, Furlan (1994) recently elaborated on this lithology dependent K-Ar isotopic trend. The $^{87}Sr/^{86}Sr$ ratios and the ^{18}O contents of illite/smectite clays in shales are also higher than those in adjacent sandstones. The collective isotopic data suggest that the illite/smectite evolution in shales occurred continuously in a restricted, or closed system, whereas that in sandstones occurred in a relatively open system with high water-to-mineral ratio.

Many combined mineralogical and isotopic studies have shown that isotopic homogenization is strongly dependent on the degree of recrystallization affecting the clay minerals. The isotopic behaviors of clay minerals in both shales and sandstones that have been affected similarly by thermal activities during burial vary; they are illustrated in Fig. 6.4, where the degree of isotopic homogenization is thought to equal the degree of recrystallization. At a given depth, the degree of recrystallization (or homogenization) of the illitic clay phase is more enhanced in the sandstone lithology (continuous

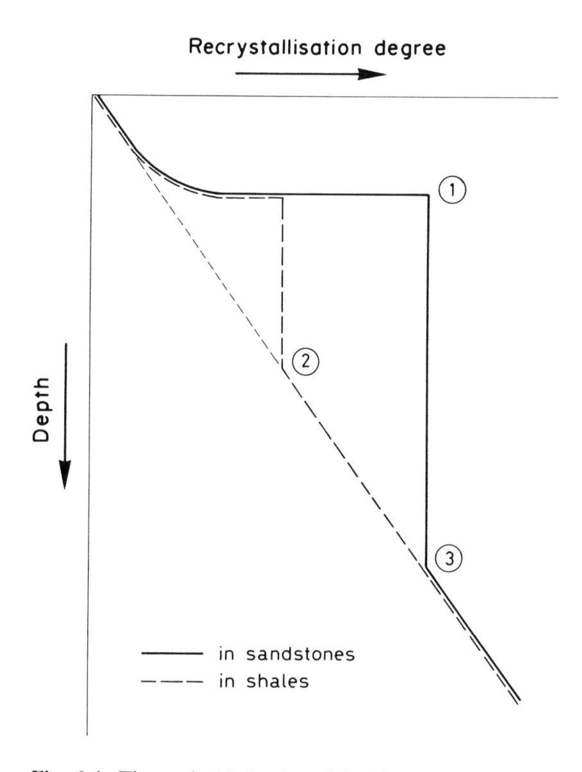

Fig. 6.4. Theoretical behavior of the K-Ar isotopic system of illitic clay fractions in a shale and a sandstone subjected to an intense recrystallization episode during the burial process

line) than in the shale lithology (dashed line), due to a thermally aided chemical effect by migrating fluids (Fig. 6.4, point 1). At greater depths (Fig. 6.4, point 2), the thermally induced recrystallization which occurred at depth 1 is still recorded in the sandstone lithology, but by this time it is erased in the shale lithology. The effect of the episodic influence of point 1 is erased in the sandstone lithology only at great depth (Fig. 6.4, point 3). As long as the clay fraction that is altered at the depth defined by point 1 is not buried to point 3, the time at which event 1 occurred can be determined by analyzing the clays of the sandstone rocks, while those of the corresponding shale rocks will give the "reference burial effect", as the record of event 1 has been erased rapidly by the burial effect. Additional isotopic studies of illitic clays in deeply buried sediments are needed to determine the relative importance of the various factors which may have some control on the isotopic configuration of diagenetic illites in deeply buried sediments.

2.2.4 Metamorphic and Hydrothermal Environments

Analyses of illites and micas from relatively unaltered and metasedimentary rocks of anchimetamorphic and epimetamorphic grades in the Alpine Glarus region have shown that change from illite to mica-like mineral attendant with increased thermal metamorphism produced no significant change in the oxygen isotope composition, but losses in both radiogenic Sr and Ar isotopes, similar to those described in Fig. 6.1. Anchimetamorphic and epimetamorphic recrystallizations of mica-like minerals in many metasedimentary rocks are believed to have occurred in low water-to-mineral ratio environments, approximating to closed-system conditions during the crystallization of the minerals.

Isotopic records of clay minerals are also useful markers of the process of clay mineral formation from hydrothermal reactions in a rock. While clay minerals in hydrothermally altered rocks of two lithologically different types can be nearly identical in mineralogical composition, these minerals can be very different in isotopic composition. This has been documented by a study on the isotopic compositions of illitic clay minerals in uranium-bearing hydrothermally altered shale and adjoining arkose beds near Baden-Baden in Germany. Illitic clay minerals in the arkose samples yield K-Ar dates that correspond to the time of uranium mineralization, but those in adjoining shale samples yield K-Ar dates that were found to be significantly higher than the time of mineralization. The relatively high dates for the clay minerals in shales can be attributed to the presence of various amounts of detrital components in this case, but the possibility also exists that high dates relate to inheritance of radiogenic isotopes by the newly formed clays evolving from replacement of precursors in an approximately closed system environment.

3 Can Isotope Geochemistry Elucidate the Concept of Clay Genesis?

As early as about 30 years ago, clay minerals in sediments and sedimentary rocks were genetically grouped into inherited, neoformed, and transformed types (Millot 1970). Theoretically, the distinction between neoformed and transformed clays is clear, as the latter are considered to have formed from a process of ionic replacement or substitution of an existing mineral, but in a practical sense the evidence for recognition of these two processes in a natural clay mineral assemblage is hard to gather. During the past three decades or more, vast improvements have been made in the analytical techniques for studying clay particles of extremely fine size, facilitating the understanding of mechanisms of formation of clay minerals, in a complex environment, such as that in burial diagenesis. Despite the use of many sophisticated analytical tools, we are still a long way from testing different hypotheses that have been suggested for the growth mechanism of diagenetic illitic clay minerals in sedimentary rocks.

Since the time of Millot (1970), the ideas about transformation and neoformation have been used in a variety of ways to explain the details concerning diagenetic illitization of smectite in illite/smectite mixed-layer minerals in deeply buried sediments. Mechanisms that have been suggested for the formation of diagenetic illite in the illite/smectite minerals include: (1) transformation of smectite into illite in the mixed-layer clays, essentially through ionic substitution and fixation through the replacement of ions from outside the immediate periphery of the crystallites that underwent illitization (Weaver and Beck 1971; Hower et al. 1976); (2) discontinuous, but progressive, transformation of smectite to illite for particles with varied thickness, structural dislocations facilitating diffusive transfers of ions, and compositions due to the detrital smectite particles being of different genetic sources (Ahn and Peacor 1986); (3) a process of dissolution-crystallization, where dissolution of small detrital smectite particles occurs and precipitation of illite follows, resulting in aggregates of a few nm-thick fundamental illite particles (McHardy et al. 1982; Nadeau et al. 1984, 1985); and (4) a process of dissolution-crystallization characterized by cannibalization of smectitic minerals where part of the illite/smectite clay minerals are dissolved and the elements released are then reused in part to form illite with an attendant increase in the illite content for the illite/smectite minerals (Pollastro 1985). Many of these suggested mechanisms for the growth of illitic minerals in deeply buried diagenetic environments may also be valid for the growth of clay minerals in other environments. The distinction between these different processes becomes diffusive if one considers varied influences of physical and chemical characteristics of the formational environments. Any success in distinguishing these processes would require that not only different sorts of isotopic evidence (Rb-Sr, K-Ar, Sm-Nd, oxygen) be collected for the same clay sample, but also that the isotopic evidence be integrated with that from detailed complementary chemical and isotopic analyses of the immediate environment. An area of research that promises to shed much light on the mechanism of the formation of clays is the integration of the isotopic and chemical data of leachates derived from clays, by following a course of analyses facilitating progressive removal of the adsorbed and interlayer cation populations, with the isotopic, chemical, and mineralogic data of the corresponding acid-leached clays.

Suggestions have been made that "interlamellar" and "interparticulate" domains with differing chemical compositions can exist in clay aggregates (Tessier 1984; Touret 1988). The chemical constituents in these "interlamellar" and "interparticulate" domains may belong to the environmental fluid related to the mineral crystallization, and to some post-formation migrating fluids, respectively. The potential of extracting the chemical history of the fluids from these domains exists by means of a selective dissolution that is followed by analyses of chemical and isotopic compositions of both the leachates and the leached residues. The studies on glauconite by Clauer and Stille (1993) and Stille and Clauer (1994) have demonstrated that considerable promise exists in describing the genetic history of clays by following the paths of changes in the isotopic compositions of clays and their leachates through a course of progressive sequential leachings of the clays by various reagents.

References

Aagaard P, Helgesson HC (1983) Activity/composition relations among silicates and aqueous solutions: II. Chemical and thermodynamic consequences of ideal mixing of atoms on homological sites in montmorillonites, illites, and mixed-layer clays. Clays Clay Min 31:207–217

Aberg F, Wickman FE (1987) Variations of $^{87}Sr/^{86}Sr$ in water from streams discharging into the Bothnian Bay, Baltic Sea. Nord Hydrol 18:33–42

Adams CJD (1975) New Zealand potassium-argon age list – 2. N Z J Geol Geophys 18:443–467

Ahn JH, Peacor DR (1986) Transmission and analytical electron microscopy of the smectite-to-illite transition. Clays Clay Min 34:165–179

Albarède F, Michard A (1987) Evidence for slowly changing $^{87}Sr/^{86}Sr$ in runoff from freshwater limestones of southern France. Chem Geol 64:55–65

Albarède F, Michard A, Minster JF, Michard G (1981) $^{87}Sr/^{86}Sr$ ratios in hydrothermal waters and deposits from the East Pacific Rise at 21 degrees N. Earth Planet Sci Lett 55:229–236

Allison GB, Hughes MW (1982) The use of natural tracers as indicators of soil-water movement in a temperate semi-arid region. J Hydrol 60:157–173

Altaner SP (1986) Comparison of rates of smectite illitization with rates of K-feldspar dissolution. Clays Clay Min 34:608–611

Altebäumer FJ, Leythaeuser D, Schaeffer RG (1981) Effect of geologic rapid heating on maturation and hydrocarbon generation in Lower Jurassic shales from NW Germany. Adv Org Geochem 80–86

Amirkhanov KI, Brandt SB, Bartnitskh EI, Gurvich VS, Gasanov SA (1958) On the question of preservation of radiogenic argon in glauconite. Dok Acad Sci SSSR 118:328–330

Amirkhanov KI, Brandt SB, Bartnitsky EI (1961) Radiogenic argon in minerals and its migration. In: Geochronology in rock systems. Ann N Y Acad Sci 91:235–275

Amouric M, Parron C (1985) Structure and growth mechanism of glauconite as seen by high resolution transmission electron microscopy. Clays Clay Min 33:473–482

Anand RR, Gilkes RJ, Armitage TM, Hillyek JW (1985) Feldspar weathering in lateritic saprolite. Clays Clay Min 33:31–43

Anati DA, Gat JR (1989) Restricted marine basins and marginal sea environments. In: Fritz P, Fontes JC (eds) Handbook of environmental isotope geochemistry. Elsevier, Amsterdam, pp 29–73

Anjos SMC (1986) Absence of clay diagenesis in Cretaceous-Tertiary marine shales, Campos basin, Brazil. Clays Clay Min 34:424–434

Aronson JL, Douthitt CB (1986) K/Ar systematics of an acid-treated illite/smectite: implications for evaluating age and crystal structure. Clays Clay Min 34:473–482

Aronson JL, Hower J (1976) Mechanism of burial metamorphism of argillaceous sediment. 2. Radiogenic argon evidence. Geol Soc Am Bull 87:738–743

Aronson JL, Lee MC (1986) K/Ar systematics of bentonite and shale in a contact metamorphic zone, Cerrillos, New Mexico. Clays Clay Min 34:483–487

Awwiller DN, Mack LE (1991) Diagenetic modification of Sm-Nd model ages in Tertiary sandstones and shales, Texas Gulf Coast. Geology 19:311–314

Ayalon A, Longstaffe FJ (1988) Oxygen-isotope studies of diagenesis and porewater evolution in the western Canada sedimentary basin: evidence from the Upper Cretaceous basal Belly River sandstone, Alberta. J Sediment Petrol 58:489–505

Bailey SW (1963) Polymorphism of the kaolin minerals. Am Min 48:1196–1209

Bailey SW (1980) Summary of recommendations of AIPEA Nomenclature Committee. Clay Min 15:85–93

Bailey SW (1988a) Polytypism of 1:1 layer silicates. In: Bailey SW (ed) Hydrous phyllosilicates (exclusive of micas). Rev Miner 19, Min Soc Am, Washington, pp 9–27

Bailey SW (1988b) X-ray diffraction identification of the polytypes of mica, serpentine, and chlorite. Clays Clay Min 36:193–213

Bailey SW, Brown BE (1962) Chlorite polytypism: regular and semi-random one-layer structures. Am Min 47:819–850

Bailey SW, Hurley PM, Fairbairn HW, Pinson WH (1962) K-Ar dating of sedimentary illite polytypes. Geol Soc Am Bull 73:1167–1170

Baker PA, Stout PM, Kastner M, Elderfield H (1991) Large-scale lateral advection of sea water through oceanic crust in the central equatorial Pacific. Earth Planet Sci Lett 105:522–533

Balashov YA, Ronov AB, Migdisov AA, Turanskaya NV (1964) The effect of climate and facies environment on the fractionation of rare earth during sedimentation. Geochem USSR 10:951–969

Banfield JF, Eggleton RA (1988) Transmission electron microscope study of biotite weathering. Clays Clay Min 36:47–60

Banks H (1972) Iron-rich saponite: additional data on samples dredged from the Mid-Atlantic Ridge 22 °N. Smithson Contrib Earth Sci 9:39–42

Barlier J, Ragot JP, Touray JC (1974) L'évolution des Terres Noires subalpines méridionales d'après l'analyse minéralogique des argiles et la réflectométrie des particules carbonatées. Bull Rech Géol Min II 6:533–548

Barrett TJ, Friedrichsen H (1982) Elemental and isotopic compositions of some metalliferous and pelagic sediments from the Galapagos mounds area, DSDP Leg 70. Chem Geol 36:275–298

Barshad I, Kishk FM (1970) Factors affecting potassium fixation and cation exchange of oil vermiculite clays. Clays Clay Min 18:127–137

Bashour I, Carlson RM (1984) Rubidium as a controlling factor in potassium release from micaceous minerals. Soil Sci Soc Am J 48:1010–1012

Bass NM, Moberly R, Rhodes JM, Shih CY, Church SE (1973) Volcanic rocks cored in the Central Pacific, Leg 17, Deep Sea Drilling Project. In: Winterer EL, Ewing JI et al. (eds) Init Rep of DSDP, vol 17. US Gov Printing Office, Washington, pp 429–503

Basset WA (1960) Role of hydroxyl orientation in mica alteration. Geol Soc Am Bull 71:449–456

Bath AH (1974a) Strontium isotope studies on sedimentary rocks. PhD Thesis, Univ Oxford, 196 pp

Bath AH (1974b) New isotopic age data on rocks from the Long Mynd, Shropshire. J Geol Soc Lond 130:567–574

Bath AH (1977) Experimental observation of exchange of Rb and Sr between clays and solution. 2nd Int Symp Water-Rock Inter and Int Assoc Geochem and Cosmochem, Strasbourg, IV, pp 244–249

Behairy AK, Chester R, Griffiths AJ, Johnson LR, Stoner JH (1975) The clay mineralogy of particulate material from some surface sea waters of the eastern Atlantic Ocean. Mar Geol 18:M45–M56

Berner RA, Holdren GR Jr (1977) Mechanism of feldspar weathering: some observational evidence. Geology 5:369–372

Berner RA, Holdren GR Jr (1979) Mechanism of feldspar weathering. II Observations of feldspars from soils. Geochim Cosmochim Acta 43:1173–1186

Bethke CM, Altaner SP (1986) Layer-by-layer mechanism of smectite illitization and application to a new rate law. Clays Clay Min 34:136–145

Birch GF, Willis JP, Rickard RS (1976) An electron microprobe study of glauconite from the continental margin of the west coast of South Africa. Mar Geol 22:271–283

Bird MI, Chivas AR (1988) Stable isotope evidence for low-temperature kaolinitic weathering and post-formational hydrogen-isotope exchange in Permian kaolinites. Isot Geosci 8:249–266

Biscaye PE (1965) Mineralogy of Recent deep-sea clay in the Atlantic Ocean and adjacent seas and oceans. Geol Soc Am Bull 76:803–832

Biscaye PE (1972) Strontium isotope composition and sediment transport in the Rio de la Plata estuary. Geol Soc Am Mem 133:349–357

Biscaye PE, Dasch EJ (1971) The rubidium, strontium, strontium-isotope system in deep-sea sediments: Argentine Basin. J Geophys Res 76:5087–5096

Biscaye PE, Chesselet R, Prospero JM (1974) Rb-Sr, $^{87}Sr/^{86}Sr$ isotope system as an index of provenance of continental dusts in the open Atlantic Ocean. J Rech Atmosph 5:819–829

Bischoff JL (1972) A ferroan nontronite from the Red Sea geothermal system. Clays Clay Min 20:217–223

Bjorkuum PA, Gjelsvik N (1986) An isochemic model for formation of authigenic kaolinite, K-feldspar and illite in sediments. J Sediment Petrol 58:506–511

Blatter CL (1974) Interaction of clay minerals with saline solutions at elevated temperatures. Proc 23rd Annu Clay Min Conf, Cleveland (Abstr), p 18

Blaxland AB (1974) Geochemistry and geochronology of chemical weathering, Butler Hill granite, Missouri. Geochim Cosmochim Acta 38:843–852

Bofinger VM, Compston W, Vernon MJ (1968) The application of acid leaching to the Rb-Sr dating of a middle Ordovician shale. Geochim Cosmochim Acta 32:823–833

Bofinger VM, Compston W, Gulson BL (1970) A Rb-Sr study of the lower Silurian State Circle Shale, Canberra, Australia. Geochim Cosmochim Acta 34:433–445

Boger PD, Faure G (1974) Strontium-isotope stratigraphy of a Red Sea core. Geology 2:181–183

Boger PD, Faure G (1976) Systematic variations of sialic and volcanic detritus in piston cores from the Red Sea. Geochim Cosmochim Acta 40:731–742

Boger PD, Boger JL, Jones LM, Faure G (1987) Effect of chemical weathering on the Rb-Sr date of feldspar in Neogene till, Mount Fleming, South Victoria Land, Antarctica. Isot Geosci 6:35–44

Boles JR, Franks SG (1979) Clay diagenesis in the Wilcox sandstones of southwest Texas – implications of smectite diagenesis on sandstone cementation. J Sediment Petrol 6;55–70

Bonhomme M (1962) Contribution à l'étude géochronologique de la plate-forme de l'Ouest africain. Thèse Doc ès-Sci. Univ Clermont-Ferrand, 62 pp

Bonhomme M, Prévôt L (1968) Application de la méthode rubidium-strontium à l'étude de l'âge radiométrique de quelques dépôts dévono-dinantiens du massif de la Bruche. Bull Serv Carte Géol Als Lorr 21:219–248

Bonhomme M, Lucas J, Millot G (1966a) Signification des déterminations isotopiques dans la géochronologie des sédiments. Actes du 151e Coll Int CNRS, Nancy 1965, pp 541–565

Bonhomme M, Cogné J, Leutwein F, Sonet J (1966b) Données nouvelles sur l'âge des séries rouges du golfe normanno-breton. C R Acad Sci Paris 262D:606–609

Bonhomme M, Vidal P, Cogné J (1968) Détermination de l'âge tectonique de la série ordovicienne et silurienne de l'Anse du Verya'ch (Presqu'île de Crozon, Finistère). Bull Serv Carte Géol Als Lorr 21:249–252

Bonhomme M, Clauer N, Cotillon P, Lucas J (1969) Datation rubidium-strontium de niveaux glauconieux du Crétacé de Haute Provence: mise en évidence d'une diagenèse. Bull Serv Carte Géol Als Lorr 22:235–247

Bonhomme MG, Millot G (1987) Diagenèse généralisée du Jurassique moyen (170–160 Ma) dans le bassin du Rhône inférieur jusqu'à la bordure des Cévennes (France). C R Acad Sci Paris 304 II:431–434

Bonhomme MG, Elsass F, Mosser C (1978) Argon isotopic geochemistry in clays – the influence of an inherited fraction. In: Zartmann RE (ed) 4th Int Conf Geochronol Cosmochronol and Isot Geol, US Geol Surv, Open File Rep 78-701:46–47

Bonhomme MG, Yerle JJ, Thiry M (1980) Datation K-Ar de fractions fines associées aux minéralisations. Le cas du bassin uranifère permo-houiller de Brousse-Broquiès (Aveyron). C R Acad Sci Paris 291D:121–124

Bonhomme MG, Bühmann D, Besnus Y (1983) Reliability of K-Ar dating of clays and silicifications associated with vein mineralizations in Western Europe. Geol Rdschau 72: 105–117

Bonnot-Courtois C (1981) Géochimie des terres rares dans les principaux milieux de formation et de sédimentation des argiles. Thèse Doc ès-Sci, Univ Paris-Sud, 217 pp

Bouquillon A, Chamley H (1986) Sédimentation et diagènese récentes dans l'éventail marin profond du Gange (Océan Indien). C R Acad Sci Paris 303D:1461–1466

Brannon JC, Podosek FA, McLimans RK (1992) Alleghenian age of the Upper Mississippi Valley zinc-lead deposit determined by Rb-Sr dating of sphalerite. Nature 356:509–511

Brass GW (1975) The effect of weathering on the distribution of strontium isotopes in weathering profiles. Geochim Cosmochim Acta 39:1647–1654

Brass GW (1976) The variation of the marine $^{87}Sr/^{86}Sr$ during Phanerozoic time: interpretation using a flux model. Geochim Cosmochim Acta 40:721–730

Brass GW, Turekian KK (1977) Comment on: "The strontium isotopic composition of seawater and seawater-oceanic crust interaction" by E.T.C. Spooner. Earth Planet Sci Lett 34:165–166

Braun JJ, Pagel M, Muller JP, Bilong P, Michard A, Guillet B (1990) Cerium anomalies in lateritic profiles. Geochim Cosmochim Acta 54:781–795

Bray CJ, Spooner ETC, Hall CM, York D, Bills TM, Krueger HW (1987) Laser probe $^{40}Ar/^{39}Ar$ and conventional K-Ar dating of illites associated with the McClean unconformity-related uranium deposits, north Saskatchewan, Canada. Can J Earth Sci 24:10–23

Brereton NR, Hooker PT, Miller JA (1976) Some conventional potassium-argon and $^{40}Ar/^{39}Ar$ age studies on glauconite. Geol Mag 113:329–340

Brigham RH, O'Neil JR (1985) Genesis and evolution of water in a two-type mica pluton: a hydrogen isotope study. Chem Geol 49:159–177

Brindley GW, Brown G (1984) (eds) Crystal structures of clay minerals and their X-ray identification. Min Soc Lond, Monogr 5, 495 pp

Briqueu L, Lancelot JR (1983) Sr isotopes, and K, Rb and Sr balance in sediments and igneous rocks from the subducted plate of the Vanuatu (New Hebrides) active margin. Geochim Cosmochim Acta 47:191–200

Brockamp O, Zuther M, Clauer N (1987) Epigenetic-hydrothermal origin of the sediment-hosted Mullenbach Uranium deposit, Baden-Baden, W Germany. Monogr Ser Mineral Deposits 27:87–98

Brongniart A (1807) Traité de minéralogie. Crapalet, Paris, 1, 490 pp

Brookins DG (1980) Geochronologic studies in the Grants mineral belt. N M Bur Mines Min Res Mem 38:52–58

Bros R, Stille P, Gauthier-Lafaye F, Weber F, Clauer N (1992) Sm-Nd isotopic dating of Proterozoic clay materials: an example from the Francevillian sedimentary series (Gabon). Earth Planet Sci Lett 113:207–218

Brousse R, Heintz E, Park F, Bellon H (1975) Gisement, faune et géochronologie du Puy Courny (Cantal, France). Géol Méditerr 2:135–142

Buatier M (1989) Genèse et évolution des argiles vertes hydrothermales océaniques: Les "Monts" du rift des Galapagos (Pacifique Equatorial). Thèse, Univ Strasbourg, 175 pp

Buatier M, Clauer N, Honnorez J, O'Neil JR (1988) A genetic model for hydrothermal Fe-rich clay minerals from Galapagos spreading center mounds: XRD, HRTEM, STEM and isotopic data. Annu Meet Geol Soc Am Abstr with Progr, Denver, Oct 31–Nov 3, A199

Burghele A, Zimmermann T, Clauer N, Kröner A (1984) Interpretation of $^{40}Ar/^{39}Ar$ and K-Ar dating of fine clay mineral fractions in Precambrian sediments. Terra Cognita 4:130

Burke WH, Denison RE, Hetherington EA, Koepnick RB, Nelson HF, Otto JB (1982) Variation of seawater $^{87}Sr/^{86}Sr$ throughout Phanerozoic time. Geology 10:516–519
Burley SD, Flisch M (1989) K-Ar chronology and the origin of illite in the Piper and Tartan Fields, Outer Moray Firth, U.K. North Sea. Clay Min 24:285–315
Burns AF, White JL (1963) The effect of potassium on the b-dimension of muscovite and dioctahedral soil micas. In: Rosenqvist IT, Graff-Petersen P (eds) Int Clay Conf. Pergamon Press, Oxford, pp 9–16
Burst JF (1958a) Mineral heterogeneity in "glauconite pellets". Am Min 43:481–497
Burst JF (1958b) "Glauconite" pellets: their mineral nature and applications to stratigraphic interpretations. Am Assoc Petr Geol Bull 42:310–337
Burst JF (1959) Post diagenetic clay mineral-environmental relationships in the Gulf Coast Eocene. Clays Clay Min 6:327–341
Busenberg E (1978) The products of the interaction of feldspars with aqueous solutions at 25 degrees C. Geochim Cosmochim Acta 42:1679–1686
Calvert CS, Buol SW, Weed SB (1980) Mineralogical characteristics and transformations of a vertical rock-saprolite-soil sequence in the North Carolina Piemont. II. Feldspar alteration products. Their transformations through the profile. Soil Sci Soc Am J 44: 1104–1112
Carrol D, Starkey HC (1960) Effect of sea-water on clay minerals. Clays Clay Min 7th Nat Conf. Pergamon Press, New York, pp 80–101
Chamley H (1989) Clay sedimentology. Springer, Berlin Heidelberg New York, 623 pp
Chaudhuri S (1976) The significance of rubidium-strontium age of sedimentary rock. Contrib Mineral Petrol 59:161–170
Chaudhuri S, Faure G (1967) Geochronology of the Keweenawan rocks of White Pine, Michigan. Econ Geol 62:1011–1033
Chaudhuri S, Brookins DG (1969) The isotopic age of the Flathead Sandstone (Middle Cambrian), Montana. J Sediment Petrol 39:364–366
Chaudhuri S, Brookins DG (1979) The Rb-Sr systematics in acid-leached clay minerals. Chem Geol 24:231–242
Chaudhuri S, Clauer N (1986) Fluctuations of isotopic composition of strontium in seawater during Phanerozoic Eon. Isot Geosci 5:293–304
Chaudhuri S, Cullers RL (1979) The distribution of rare-earth elements in deeply buried Gulf Coast sediments. Chem Geol 24:327–338
Chaudhuri S, Clauer N (1992a) Isotopic compositions of dissolved strontium and neodymium in continental surface and subsurface waters. In: Clauer N, Chaudhuri S (eds) Isotopic signatures and sedimentary records. Lecture Notes in Earth Sciences 43. Springer, Berlin Heidelberg New York, pp 467–495
Chaudhuri S, Clauer N (1992b) Signatures of radiogenic isotopes in deep subsurface waters in continents. In: Clauer N, Chaudhuri S (eds) Isotopic signatures and sedimentary records. Lecture Notes in Earth Sciences 43. Springer, Berlin Heidelberg New York, pp 497–529
Chaudhuri S, Clauer N (1993) Strontium isotopic compositions and potassium and rubidium contents of formation waters in sedimentary basins: clues to the origin of the solutes. Geochim Cosmochim Acta 57:429–437
Chaudhuri S, Broedel V, Clauer N (1987) Strontium isotopic evolution of oil-field waters from carbonate reservoir rocks in Bindley field, central Kansas, U.S.A. Geochim Cosmochim Acta 51:45–53
Chaudhuri S, Stille P, Clauer N (1992a) Sm-Nd isotopes in fine-grained clastic sedimentary materials: clues to sedimentary processes and recycling growth of the continental crust. In: Clauer N, Chaudhuri S (eds) Isotopic signatures and sedimentary records. Lecture Notes in Earth Sciences 43. Springer, Berlin Heidelberg New York, pp 287–319
Chaudhuri S, Robinson R, Clauer N, Jones LM (1992b) Evolution of formation waters in Early Pennsylvanian Morrowan sandstones in the Hugeton Embayement of the Anadarko Basin, Southern Midcontinent, U.S.A. Okla Geol Surv Circ 93:160–174

Chaudhuri S, Furlan S, Clauer N (1992c) The signature of water-rock interactions in the formation waters of sedimentary basins: some new evidence. In: Kharaka YK, Maest AS (eds) Proc 7th Int Symp on Water-Rock Inter, Balkema, Rotterdam, pp 907–910
Church TM, Velde B (1979) Geochemistry and origin of a deep-sea Pacific palygorskite deposit. Chem Geol 25:31–39
Clauer N (1970) Etude sédimentologique, géochimique et géochronologique des Schistes de Steige et de la Série de Villé. Thèse 3ème Cycle, Univ Strasbourg, 88 pp
Clauer N (1973) Utilisation de la méthode rubidium-strontium pour la datation de niveaux sédimentaires du Précambrien supérieur de l'Adrar mauritanien (Sahara occidental) et la mise en évidence de transformations précoces des minéraux argileux. Geochim Cosmochim Acta 37:2243–2255
Clauer N (1974) Utilisation de la méthode rubidium-strontium pour la datation d'une schistosité de sédiments peu métamorphisés: application au Précambrien II de la boutonnière de Bou Azzer-El Graara (Anti-Atlas). Earth Planet Sci Lett 22:404–412
Clauer N (1976) Géochimie isotopique du strontium des milieux sédimentaires. Application à la géochronologie de la couverture du craton ouest-africain. Sci Géol Mém, Strasbourg, 256 pp
Clauer N (1978) Behaviour of strontium and argon isotopes in biotites during a progressive natural weathering. In: Zartman RE (ed) 4th Int Conf Geochr Cosmochr and Isot Geol, US Geol Surv, Open-File Rep 78-701:68–71
Clauer N (1979a) A new approach to Rb-Sr dating of sedimentary rocks. In: Jäger E, Hunziker JC (eds) Lectures in isotope geology. Springer, Berlin Heidelberg New York, pp 30–51
Clauer N (1979b) Relationship between the isotopic composition of strontium in newly formed continental clay minerals and their source material. Chem Geol 27:115–124
Clauer N (1981a) Strontium and argon isotopes in naturally weathered biotites, muscovites, and feldspars. Chem Geol 31:325–334
Clauer N (1981b) 87Sr/86Sr ratios of the Barremian and Early Aptian seas. In: Thiede J, Vallier TL et al. (eds) Initial reports of the Deep Sea Drilling Project, 42. US Gov Printing Office, Washington, pp 781–783
Clauer N (1982a) Strontium isotopes of Tertiary phillipsites from the southern Pacific: timing of the geochemical evolution. J Sediment Petrol 52:1003–1009
Clauer N (1982b) The rubidium-strontium method applied to sediments: certitudes and uncertainties. In: Odin GS (ed) Numerical dating in stratigraphy. John Wiley, New York, pp 245–276
Clauer N, Bonhomme M (1970a) Homogénéisation isotopique du strontium entre les schistes de Steige et la série de Villé (Vosges) pendant la phase bretonne de l'orogenèse hercynienne. C R Acad Sci Paris 271D:1844–1847
Clauer N, Bonhomme M (1970b) Datations rubidium-strontium dans les schistes de Steige et la série de Villé (Vosges). Bull Serv Carte Géol Als Lorr 23:191–208
Clauer N, Chaudhuri S (1991) Evaluation of dating non-radioactive sediment-hosted ore deposits. In: Pagel M, Leroy JL (eds) Source, transport and deposition of metals. Balkema, Rotterdam, pp 377–380
Clauer N, Chaudhuri S (1992) Indirect dating of sediment-hosted ore deposits: promises and problems. In: Clauer N, Chaudhuri S (eds) Isotopic signatures and sedimentary records. Lecture Notes in Earth Sciences 43. Springer, Berlin Heidelberg New York, pp 361–388
Clauer N, Deynoux M (1987) New information on the probable isotopic age of the Late Proterozoic glaciation in West Africa. Precambrian Res 37:89–94
Clauer N, Hoffert M (1985) Sr isotopic constraints for the sedimentation rate of deep sea red clays in the southern Pacific ocean. In: Snelling NJ (ed) The chronology of the geological record. Geol Soc Lond Mem 10:290–296
Clauer N, Kröner A (1979) Strontium and argon isotopic homogenization of pelitic sediments during low-grade regional metamorphism: the Pan-African Upper Damara sequence of northern Namibia (southwest Africa). Earth Planet Sci Lett 43:117–131
Clauer N, Millot G (1978) Genèse des minéraux argileux et géochimie isotopique du strontium. In: Livre Jubilaire Jacques Flandrin. Doc Lab Géol Fac Sci Lyon HS 4:129–142

Clauer N, Olafsson J (1981) Icelandic thermal brines with a mantle Sr isotopic signature. Sci Géol Bull Strasbourg 34:243–245

Clauer N, Stille P (1993) Nd-Sr isotopic and rare-earth elementary constraints on the glauconitization process. VII Eur Uni Geosci, Strasbourg, April 4–8 1993, Terra Abstr 686–687

Clauer N, Tardy Y (1971) Distinction par la composition isotopique du strontium contenu dans les carbonates, entre le milieu continental des vieux socles cristallins et le milieu marin. C R Acad Sci Paris 273D:2191–2194

Clauer N, Hoffert M, Grimaud D, Millot G (1975) Composition isotopique du strontium d'eaux interstitielles extraites de sédiments récents: un argument en faveur de l'homogénéisation isotopique des minéraux argileux. Geochim Cosmochim Acta 39:1579–1582

Clauer N, Jeannette D, Tisserant D (1980) Datation isotopique de cristallisations successives d'un socle cristallin et cristallophyllien (Haute Moulouya, Moyen Maroc). Geol Rdschau 68:63–83

Clauer N, O'Neil JR, Bonnot-Courtois C (1982a) The effect of natural weathering on the chemical and isotopic compositions of biotites. Geochim Cosmochim Acta 46:1755–1762

Clauer N, Hoffert M, Karpoff AM (1982b) The Rb-Sr isotope system as an index of origin and diagenetic evolution of southern Pacific red clays. Geochim Cosmochim Acta 46:2659–2664

Clauer N, Caby R, Jeannette D, Trompette R (1982c) Geochronology of sedimentary and metasedimentary Precambrian rocks of the West African craton. Precambrian Res 18:53–71

Clauer N, Giblin P, Lucas J (1984) Sr and Ar isotope studies of detrital smectites from the Atlantic Ocean (DSDP, Legs 43, 48, and 50). Isot Geosci 2:141–151

Clauer N, Vidal P, Auvray B (1985a) Differential behavior of the Rb-Sr and K-Ar systems of spilitic flows and interbeded metasediments: the spilitic group of Erquy (Brittany, France). Paleomagnetic implications. Contrib Mineral Petrol 89:81–89

Clauer N, Ey F, Gauthier-Lafaye F (1985b) K-Ar dating of different rock types from the Cluff Lake uranium ore deposits (Saskatchewan-Canada). In: Lainé R, Alonso D, Svab M (eds) The Carswell structure uranium deposits, Saskatchewan. Geol Assoc Can Spec Pap 29:47–53

Clauer N, Muller JP, O'Neil JR (1989) Oxygen isotope signature of several generations of kaolinite in a laterite: geochemical implications. Int Clay Conf, Strasbourg, Abstr and Progr, 1 p

Clauer N, O'Neil JR, Bonnot-Courtois C, Holtzapffel T (1990) Morphological, chemical, and isotopic evidence for an early diagenetic evolution of detrital smectites in marine sediments. Clays Clay Min 38:33–46

Clauer N, Dallmeyer RD, Lécorché JP (1991) Age of the late Paleozoic tectonothermal activity in north-central Mauritanide, West Africa. Precambrian Res 49:97–105

Clauer N, Keppens E, Stille P (1992a) Sr isotopic constraints on the process of glauconitization. Geology 20:133–136

Clauer N, Stille P, Keppens E, O'Neil JR (1992b) Le mécanisme de la glauconitisation: apports de la géochimie isotopique du strontium, du néodyme et de l'oxygène de glauconies récentes. C R Acad Sci Paris 315 II:321–327

Clauer N, Savin SM, Chaudhuri S (1992c) Isotopic compositions of clay minerals as indicators of the timing and conditions of sedimentation and burial diagenesis. In: Clauer N, Chaudhuri S (eds) Isotopic signatures and sedimentary records, Lecture Notes in Earth Sciences 43. Springer, Berlin Heidelberg New York, pp 239–286

Clauer N, Cocker JD, Chaudhuri S (1992d) Isotopic dating of diagenetic illites in reservoir sandstones: influence of the investigator factor. In: Houseknecht DW, Pittman ED (eds) Origin, diagenesis, and petrophysics of clay minerals in sandstones. Soc Econ Paleontol Miner Spec Publ 47:5–12

Clauer N, Chaudhuri S, Kralik M, Bonnot-Courtois C (1993) Effects of experimental leaching on Rb-Sr and K-Ar isotopic systems and REE contents of diagenetic illite. Chem Geol 103:1–16

Clauer N, O'Neil JR, Furlan S (1994) Clay Minerals as records of temperature conditions and duration of thermal anomalies in the Paris Basin (France). Clay Min (in press)

Clauer N, Rais N, Schaltegger U, Piqué A K-Ar systematics of clay-to-mica minerals in a multi-stage low-grade metamorphic evolution. Chem Geol Isot Geosci Sect (accepted)
Clayton RN, Friedman I, Graf DL, Mayeda TK, Meents WF, Shimp F (1966) The origin of saline formation waters. I. Isotopic composition. J Geophys Res 71:3869–3882
Clayton RN, Rex RW, Syers JK, Jackson ML (1972) Oxygen isotope abundance in quartz from Pacific pelagic sediments. J Geophys Res 77:3907–3914
Cocker JD, Clauer N, Tsui TF, Swarbrick RE (1988) A diagenetic model for the northwest Hutton field. Conf Clay mineral diagenesis in hydrocarbon reservoirs and shales, Cambridge, UK, 1 p
Cole TG (1983) Oxygen isotope geothermometry and origin of smectite in the Atlantis II-Deep, Red Sea. Earth Planet Sci Lett 66:166–176
Collins AG (1975) Geochemistry of oilfield waters. Elsevier, New York, 496 pp
Compston W, Pidgeon RT (1962) Rb-Sr dating of shales by the total-rock method. J Geophys Res 67:3493–3502
Coplen TB, Hanshaw BB (1973) Ultrafiltration by a compacted clay membrane. I. Oxygen and hydrogen fractionation. Geochim Cosmochim Acta 37:2295–2310
Cordani UG, Kawashita K, Thomas-Filho A (1978) Applicability of the Rb-Sr method to shales and related rocks. Am Assoc Pet Geol Spec Publ 6:93–117
Cordani UG, Thomaz-Filho A, Brito-Neves BB, Kawashita K (1985) On the applicability of the Rb-Sr method to argillaceous sedimentary rocks: some examples from Precambrian sequences of Brazil. G Geol (Bologna) 47:253–280
Cormier RF (1956) Rubidium-strontium ages of glauconite and their application to the construction of a post-Precambrian time-scale. PhD Thesis, Massachussetts Institute of Technology, Cambridge
Courbe C, Velde B, Meunier A (1981) Weathering of glauconites: reversal of the glauconitization process in a soil in western France. Clay Min 16:231–244
Craig H (1961) Isotopic variations in meteoric waters. Science 133:1702–1703
Craig H, Gordon L (1965) Deuterium and oxygen-18 variations in the ocean and the marine atmosphere. Proc Conf on Stable isotopes in oceanographic studies and paleotemperatures, Spoleto 2:1–87
Cullers RL, Chaudhuri S, Arnold B, Moon L, Wolf CW (1975) Rare earth distributions in clay minerals and the clay-size fraction of the lower Permian Havensville and Eskridge shales of Kansas and Oklahoma. Geochim Cosmochim Acta 39:1691–1703
Curnelle R, Dubois P (1981) Evolution mésozoïque des grands bassins sédimentaires français: Bassins de Paris, d'Aquitaine et du Sud-Est. Bull Soc Géol Fr 8:529–546
Curtis CD (1983) A link between aluminium mobility and destruction of secondary porosity. Am Assoc Petr Geol Bull 67:380–384
Dallmeyer RD, Reuter A, Clauer N, Liewig N (1989) Chronology of Caledonian tectonothermal activity within the Gaissa and Lakefjord nappe complexes (Lower Allochthon), Finnmark, Norway. In: Gayer RA (ed) The caledonide geology of Scandinavia. Graham and Trotman, London, pp 9–26
Daniels EJ, Altaner SP (1990) Clay mineral authigenesis in coal and shale from the Anthracite region, Pennsylvania. Am Min 75:825–839
Daniels EJ, Aronson JL, Altaner SP, Clauer N (1994) Late Permian age of NH_4-bearing illite in anthracite from eastern Pennsylvania: temporal constraints on coalification in the Central Appalachians. Geol Soc Am Bull 106:760–766
Dasch EJ (1969) Strontium isotopes in weathering profiles, deep-sea sediments and sedimentary rocks. Geochim Cosmochim Acta 33:1521–1552
Dasch EJ, Dymond JR, Heath GR (1971) Isotopic analysis of metalliferous sediments from East Pacific Rise. Earth Planet Sci Lett 13:175–180
Davies G, Gledhill A, Hawkesworth C (1985) Upper crustal recycling in southern Britain: evidence from Nd and Sr isotopes. Earth Planet Sci Lett 75:1–12

Decarreau A (1983) Etude expérimentale de la cristallogenèse des smectites. Mesures des coefficients de partage smectite trioctaédrique – solution aqueuse pour les métaux M^{2+} de la première série de transition. Sci Géol Mém Strasbourg 74:184 pp

Degens ET, Epstein S (1962) Relationship between $^{18}O/^{16}O$ ratios in coexisting carbonates, cherts, and diatomites. Am Assoc Petrol Geol Bull 46:534–542

Del Nero M (1992) Modélisation thermodynamique et cinétique de la formation des bauxites et des cuirasses ferrugineuses. Thèse, Univ Strasbourg, 219 pp

Delvigne J, Martin H (1970) Analyse à la microsonde électronique de l'altération d'un plagioclase en kaolinite par l'intermédiaire d'une phase amorphe. Cah ORSTOM Sér Géol 2:259–295

Demir I (1988) Studies of smectite membrane behavior: electro-kinetic, osmotic, and isotopic fractionation processes at elevated pressures. Geochim Cosmochim Acta 52:727–737

DePaolo DJ (1986) Detailed record of the Neogene Sr isotopic evolution of seawater from DSDP site 5908. Geology 14:103–106

DePaolo DJ (1988) Neodymium isotope geochemistry. Springer, Berlin Heidelberg New York, 187 pp

DePaolo DJ, Wasserburg GJ (1976) Nd isotopic variations and petrogenetic models. Geophys Res Lett 3:249–252

Dethier DP (1986) Weathering rates and the chemical flux from catchments in the Pacific Northwest, USA. In: Coleman SM, Dethier DP (eds) Rates of chemical weathering of rocks and minerals, vol 21. Academic Press, San Diego, 503–530 pp

Deutrich T, Zwingmann H, Gaupp R, Clauer N (1993) Clay mineral diagenesis in Late Paleozoic sandstones of the NW German Basin. Terra Abstr, p 649

Dibble WE Jr, Tiller WA (1981) Non-equilibrium water/rock interactions. I. Model for inter-controlled reactions. Geochim Cosmochim Acta 45:79–92

Donnelly TW, Pritchard RA, Emmermann R, Puchett H (1980) The aging of oceanic crust: synthesis of the mineralogical and chemical results of Deep Sea Drilling Project, Legs 51 through 53. In: Initial Reports of the DSDP, 51–53, part 2, US Gov Printing Office, Washington, 1563–1577

Douglas LA, Fiessinger F (1971) Degradation of clay minerals by H_2O_2 treatments to oxidize organic matter. Clays Clay Min 19:67–68

Drever JI (1971) Early diagenesis of clay minerals, Rio America, Mexico. J Sediment Petrol 41:982–994

Droubi A, Vieillard P, Bourrié G, Fritz B, Tardy Y (1976) Etude théorique de l'altération des plagioclases. Bilans et conditions de stabilité des minéraux secondaires en fonction de la pression partielle de CO_2 et de la température (0 °C à 100 °C). Sci Géol Bull Strasbourg 29:45–62

Duddy IR (1980) Redistribution and fractionation of rare earth and other elements in a weathering profile. Chem Geol 30:363–381

Dunoyer de Segonzac G (1969) Les minéraux argileux dans la diagenèse. Passage au métamorphisme. Sci Géol Mém Strasbourg 29:320 pp

Dutton SP, Land LS (1985) Meteoric burial diagenesis of Pennsylvanian arkosic sandstones, southwestern Anadarko Basin, Texas. Am Assoc Petrol Geol Bull 69:22–38

Dymond J, Corliss JB, Heath GR, Field CW, Dasch EJ, Veeh HH (1973) Origin of metalliferous sediments from the Pacific Ocean. Geol Soc Am Bull 84:3355–3372

Dymond J, Biscaye PE, Rex RW (1974) Eolian origin of mica in Hawaiian soils. Geol Soc Am Bull 85:37–40

Eberl DD, Hower J (1976) Kinetics of illite formation. Geol Soc Am Bull 87:1326–1330

Eberl DD, Whitney G, Khoury H (1978) Hydrothermal reactivity of smectite. Am Min 63:401–405

Eberl DD, Srodon J, Northrop HR (1986) Potassium fixation in smectite by wetting and drying. In: Davis JA, Hayes KF (eds) Geochemical processes at mineral surfaces. Am Chem Soc Symp Ser 323:296–326

Eberl DD, Srodon J, Kralik M, Taylor BE, Peterman ZE (1990) Ostwald ripening of clays and metamorphic minerals. Science 248:474–477
Eggleton RA, Buseck PR (1980) High resolution electron microscopy of feldspar weathering. Clays Clay Min 28:173–179
Ehrenberg SN, Nadeau PH (1989) Formation of diagenetic illite in sandstones of the Garn Formation, Haltenbanken area, mid-Norwegian continental shelfs. Clay Min 24:233–253
Elderfield H, Greaves MJ (1981) Strontium isotope geochemistry of Icelandic geothermal systems and implications for sea water chemistry. Geochim Cosmochim Acta 45:2201–2212
Elderfield H, Gieskes JM, Baker PA, Oldfield RK, Hawkesworth CJ, Miller R (1982) $^{87}Sr/^{86}Sr$ and $^{18}O/^{16}O$ ratios, interstitial water chemistry and diagenesis in deep-sea carbonate sediments of the Ontong Java Plateau. Geochim Cosmochim Acta 46:2259–2268
Elderfield H, Upstill-Goddard R, Sholkovitz ER (1990) The rare earth elements in rivers, estuaries and coastal sea waters: processes affecting crustal input of elements to the ocean and their significance to the composition of sea water. Geochim Cosmochim Acta 54:971–991
Elliott WC, Aronson JL (1987) Alleghanian episode of K-bentonite illitization in the southern Appalanchian Basin. Geology 15:735–739
Elston DP, Bressler SL (1980) Paleomagnetic poles and polarity zonation from the middle Proterozoic Belt Supergroup, Montana and Idaho. J Geophys Res 85:339–355
Eslinger EV, Savin SM (1973) Oxygen isotope geothermometry of the burial metamorphic rocks of the Precambrian Belt Supergroup, Glacier National Park, Montana. Geol Soc Am Bull 84:2549–2560
Eslinger EV, Yeh HW (1981) Mineralogy, O^{18}/O^{16} and D/H ratios of clay-rich sediments from Deep Sea Drilling Project site 180, Aleutian Trench. Clays Clay Min 29:309–315
Eswaran H, Wong Chan Bin (1978) A study of deep weathering profile in granite in Peninsular Malaysia. III Alteration of feldspars. Soil Sci Soc Am J 42:154–158
Evernden JF, Curtis GH, Obradovich J, Kistler R (1961) On the evaluation of glauconite and illite for dating sedimentary rocks by the potassium-argon method. Geochim Cosmochim Acta 23:78–99
Ey F (1984) Un exemple de gisement d'uranium sous discordance: les minéralisations protérozoïques de Cluff Lake, Saskatchewan, Canada. Thèse 3ème Cycle, Univ Strasbourg, 171 pp
Fallick AE, Macaulay CI, Haszeldine RS (1993) Implications of linearly correlated oxygen and hydrogen isotopic compositions for kaolinite and illite in the Magnus Sandstone, North Sea. Clays Clay Min 2:184–190
Farmer VC, Russel JD, McHardy WJ, Newman ACD, Alhrichs JC, Rimsaite JYH (1971) Evidence for loss of protons and octahedral iron from oxidised biotites and vermiculites. Min Mag 38:121–137
Faure G (1982) The marine-strontium geochronometer. In: Odin GS (ed) Numerical dating in stratigraphy. John Wiley, New York, pp 73–79
Faure G (1986) Principles of isotope geology, 2nd edn. John Wiley, New York, 589 pp
Faure G, Barrett PJ (1973) Strontium isotope compositions of non-marine carbonate rocks from the Beacon Supergroup of the Trans-Antarctic Mountains. J Sediment Petrol 43:447–457
Faure G, Kovach J (1969) The age of the Gunflint Iron Formation of the Animikie series in Ontario, Canada. Geol Soc Am Bull 80:1725–1736
Faure G, Hurley PM, Fairbairn HW (1963) An estimate of the isotopic composition of strontium in rocks of the Precambrian shield of North America. J Geophys Res 68:2323–2329
Fisher GW, Lasaga AC (1981) Irreversible thermodynamics in petrology. In: Lasaga AC, Kirkpatrick RJ (eds) Kinetics of geochemical processes, Rev Miner 8, Miner Soc Am, Washington, pp 171–209
Fisher RV, Schmincke HU (1984) Pyroclastic rocks. Springer, Berlin Heidelberg New York, 472 pp

Fleet AJ (1984) Aqueous and sedimentary geochemistry of the rare earth elements. In: Henderson P (ed) Rare earth element geochemistry. Elsevier, Amsterdam, pp 343–373

Foland KA, Linder JS, Laskowski TE, Grant NK (1984) $^{40}Ar/^{39}Ar$ dating of glauconites: measured ^{39}Ar recoil loss from well-crystallized specimens. Chem Geol Isot Geosci Sect 2:241–264

Forbes P, Landais P, Pagel M, Meyer A (1987) Thermal evolution of the Guézouman Formation in the Akouta uranium deposit (Niger). Terra Cognita 7:343

Fordham AW (1973) The location of iron-55, strontium-85 and iodide-125 sorbed by kaolinite and dickite particles. Clays Clay Min 21:175–184

Foster WR, Custard HC (1982) Role of clay composition on smectite/illite diagenesis. Am Assoc Petr Geol Bull 66:1444

Fritz B, Jeannette D, Clauer N (1983) Brine geochemistry and rock petrography of a geothermal reservoir in deeply buried sandstones (Rhine graben). Terra Cognita 3:233

Fullagar PD, Bottino ML (1969) Rubidium-strontium age study of Middle Devonian Tioga bentonite. Southeast Geol 10:247–256

Fullagar PD, Raglan PC (1975) Chemical weathering and Rb-Sr whole-rock ages. Geochim Cosmochim Acta 39:1245–1252

Furlan S (1994) Transferts de matière au cours de la diagenèse d'enfouissement dans le bassin du delta de la Mahakam (Indonésie). Un nouveau concept pour le mécanisme de l'illitisation. Thèse, Univ Strasbourg, 190 pp

Furlan S, Clauer N, Chaudhuri S, Sommer F (1992) Paleo- and Recent diagenesis in sedimentary basins: the case of the Mahakam delta, Indonesia. In: Parnell J, Ruffell AH, Moles NR (eds) Geofluids '93, Contributions to an Int Conf on Fluid evolution, migration and interaction in rocks, pp 354–356

Furlan S, Clauer N, Chaudhuri S, Sommer F K-transfer during burial diagenesis in sediments beneath the Mahakam Delta (Kalimantan, Indonesia). Clays Clay Min (accepted)

Garrels RM (1984) Montmorillonite/Illite stability diagrams. Clays Clay Min 32:161–166

Garrels RM, Christ CL (1965) Solutions, minerals, and equilibria. Harper and Row, New York, 450 pp

Garrels RM, Mackenzie FT (1971) Evolution of sedimentary rocks. Norton, New York, 397 pp

Gaultier JP, Mamy J (1978) Evolution of exchange properties and crystallographic characteristics of biionic K-Ca montmorillonite submitted to alternate wetting and drying. Proc Int Clay Conf, Oxford, pp 167–175

Gebauer D, Grünenfelder M (1974) Rb-Sr whole-rock dating of late diagenetic to anchimetamorphic Paleozoic sediments in southern France (Montagne Noire). Contr Miner Petrol 47:113–130

Geyh MA, Schleicher H (1990) Absolute age determination. Physical and chemical dating methods and their application. Springer, Berlin Heidelberg New York, 503 pp

Ghosh PK (1972) Use of bentonites and glauconites in potassium 40/argon 40 dating in Gulf Coast stratigraphy. PhD Thesis, Univ Houston 136 pp

Gibbs R (1967) The geochemistry of the Amazon River Basin. Part I: The factors that control the salinity and the composition and concentration of suspended solids. Geol Soc Am Bull 78:1203–1232

Gieskes JM, Lawrence JR (1981) Alteration of vocanic matter in deep sea sediments: evidence from the chemical composition of interstitial waters from deep sea drilling cores. Geochim Cosmochim Acta 45:1687–1703

Gieskes JM, Elderfield H, Palmer MR (1986) Strontium and its isotopic composition in interstitial waters of marine carbonate sediments. Earth Planet Sci Lett 77:229–235

Girard JP, Savin SM, Aronson JL (1989) Diagenesis of the lower Cretaceous arkoses of the Angola margin: petrologic, K-Ar dating and $^{18}O/^{16}O$ evidence. J Sediment Petrol 59:519–538

Giresse P (1965) Observation sur la présence de "glauconie" actuelle dans les sédiments ferrugineux peu profonds du bassin gabonais. C R Acad Sci Paris 260:5597–5600

Giresse P, Odin GS (1973) Nature minéralogique et origine des glauconies du plateau continental du Gabon et du Congo. Sedimentology 20:457–488

Glasmann JR, Larter S, Briedis NA, Lundegard PD (1989a) Shale diagenesis in the Bergen High area, North Sea. Clays Clay Min 37:97–112

Glasmann JR, Lundegard PD, Clark RA, Penny BK, Collins ID (1989b) Geochemical .evidence for the history of diagenesis and fluid migration: Brent sandstone, Heather field, North Sea. Clay Min 24:255–284

Goldberg ED (1965) Minor elements in sea water. In: Riley JP, Skirrow G (eds) Chemical oceanography. Academic Press, New York, 163 pp

Goldberg ED, Kolde M, Schmitt RA, Smith RH (1963) Rare earths distribution in the marine environment. J Geophys Res 68:4209–4217

Goldich SS (1938) A study in rock-weathering. J Geol 46:17–58

Goldich SS, Gast PW (1966) Effects of weathering on the Rb-Sr and K-Ar ages of biotite from the Morton gneiss, Minnesota. Earth Planet Sci Lett 1:372–375

Goldich SS, Baadsgaard H, Edwards G, Weaver CE (1959) Investigations in radioactivity dating of sediments. Am Assoc Petr Geol Bull 43:654–662

Goldstein SJ, Jacobsen SB (1987) The Nd and Sr isotopic systematics of river-water dissolved material: implications for the sources of Nd and Sr in seawater. Chem Geol Isot Geosci Sect 66:245–272

Goldstein SJ, Jacobsen SB (1988) Nd and Sr isotopic systematics of river water suspended material: implications for crustal evolution. Earth Planet Sci Lett 87:249–265

Goldstein SL, O'Nions RK, Hamilton PJ (1984) A Sm-Nd isotopic study of atmospheric dusts and particulates from major river systems. Earth Planet Sci Lett 70:221–236

Goode ADT (1974) Oxidation of natural olivines. Nature 248:500–501

Gorokhov IM, Clauer N, Turchenko TL, Melnikov NN, Kutyavin EP, Pirrus E, Baskakov AV (1994) Rb-Sr systematics of Vendian-Cambrian claystones from East European platform: implications for a multi-stage illite evolution. Chem Geol 112:71–89

Gouvea Da Silva RB (1980) Migration des sels et des isotopes lourds à travers des colonnes de sediment non saturé sous climat aride. Thèse 3ème cycle, Univ Paris VI, 116 pp

Graf DL, Friedman I, Meents WF (1965) The origin of saline formation waters. II. Isotopic fractionation by shale in micropore systems. Illinois State Geol Surv Circ 393:32 pp

Grant NK, Laskowski TE, Foland KA (1984) Rb-Sr and K-Ar ages of Paleozoic glauconites from Ohio and Missouri, USA. Isot Geosci 2:217–239

Greene-Kelly R (1953) The identification of montmorillonite in clays. J Soil Sci 4:233–237

Greene-Kelly R (1955) Deshydratation of montmorillonite. Min Mag 30:604–615

Gregory RT, Taylor HP Jr (1981) An oxygen isotope profile in a section of Cretaceous oceanic crust, Smail Ophiolite, Oman: evidence for ^{18}O-buffering of the oceans by deep (>5 km) seawater-hydrothermal circulation at mid-ocean ridges. J Geophys Res 86:2737–2755

Grim RE (1968) Clay mineralogy, 2nd edn. McGraw-Hill, New York, 560 pp

Grim RE, Bray RH, Bradley WF (1937) The mica in argillaceous sediments. Am Min 22:813–829

Grousset F, Latouche C, Maillet N (1983) Clay-minerals: indicators of wind and current contributions to post-glacial sedimentation on the Azores-Iceland Ridge. Clay Min 18:65–75

Grousset FE, Chesselet R (1986) The Holocene sedimentary regime in the northern Mid-Atlantic Ridge region. Earth Planet Sci Lett 78:271–287

Grousset FE, Biscaye PE, Zindler A, Prospero J, Chester JD (1988) Neodymium isotopes as tracers in marine sediments and aerosols: North Atlantic. Earth Planet Sci Lett 87:367–378

Gruner JW (1934) The structure of vermiculites and their collapse by dehydration. Am Min 19:557–578

Gundogdu MN, Bonnot-Courtois C, Clauer N (1989) Isotopic and chemical signatures of sedimentary smectite and diagenetic clinoptilolite of a lacustrine Neogene basin near Bigadic, western Turkey. Applied Geochem 4:635–644

Guinier A (1964) Théorie et technique de la radiocristallographie. Dunod, Paris, 740 pp

Hall WE, Friedman I (1963) Composition of fluid inclusions, Cave-in-Rock fluorite district, Illinois, and Upper Mississippi Valley zinc-lead district. Econ Geol 58:886–911
Halter G, Sheppard SMF, Weber F, Clauer N, Pagel M (1987) Radiation-related retrograde hydrogen isotope and K-Ar exchange in clay minerals. Nature 330:638–641
Hamilton J, Fallick AE, MacIntyre RM, Elliot S (1987) Isotopic tracing of the provenance and diagenesis of lower Brent Group sands, North Sea. In: Brooks J, Glennie K (eds) Petroleum geology of north west Europe, Graham and Trotman, London, pp 939–949
Hamilton PJ, Kelley S, Fallick AE (1989) K-Ar dating of illite in hydrocarbon reservoirs. Clay Min 24:213–215
Hanor JS (1987) Origin and migration of subsurface sedimentary brines. Soc Econ Paleontol Miner, Short Course 21:247 pp
Harder H (1974) Illite mineral synthesis at surface temperatures. Chem Geol 14:241–254
Harder H (1976) Nontronite synthesis at low temperature. Chem Geol 18:169–180
Harper CT (1970) Graphic solution to the problem of $^{40}Ar^*$ loss from metamorphic minerals. Eclog Geol Helv 63:119–140
Harris WB, Fullagar PD (1989) Comparison of Rb-Sr and K-Ar dates of middle Eocene bentonite and glauconite, southeastern Atlantic Coastal plain. Geol Soc Am Bull 101:573–577
Hart SR, Staudigel H (1978) Oceanic crust: age of hydrothermal alteration. Geophys Res Lett 5:1009–1012
Hart SR, Staudigel H (1980) Ocean crust-sea water interactions: sites 417 and 418. In: Initial Reports of the Deep Sea Drilling Project, 51–53, part 2, pp 1169–1176
Hart SR, Staudigel H (1982) The control of alkalies and uranium sea-water by ocean crust. Earth Planet Sci Lett 58:202–212
Hart SR, Staudigel H (1986) Ocean crust vein mineral deposition: Rb/Sr ages, U-Th-Pb geochemistry, and duration of circulation at DSDP sites 261, 462 and 516. Geochim Cosmochim Acta 50:2751–2761
Hart SR, Tilton GR (1966) The isotope geochemistry of strontium and lead in Lake Superior sediments and water. In: The Earth beneath the continents. Am Geophys Union Geophys Mon Ser 10:127–137
Haskin LA, Haskin MA, Frey FA, Wildeman TR (1968) Relative and absolute terrestrial abundance of the rare earths. In: Ahrens LH (ed) Symposium on the origin and distribution of the elements. Pergamon, Oxford, pp 889–912
Hawkesworth CJ, Elderfield H (1978) The strontium isotopic composition of interstitial waters from sites 245 and 336 of the Deep Sea Drilling Project. Earth Planet Sci Lett 40:423–432
Hay RL, Lee M, Kolata DR, Mattthews DR, Morton JP (1988) Episodic potassic diagenesis of Ordovician tuffs in the Mississippi Valley area. Geology 16:743–747
Hayatsu A, Carmichael CM (1977) Removal of atmospheric argon contamination and the use and non-use of the K-Ar isochron method. Can J Sci 14:337–345
Haymon RM, Kastner M (1986) The formation of high temperature clay minerals from basalt alteration during hydrothermal discharge on the East Pacific Rise at 21 °N. Geochim Cosmochim Acta 50:1933–1939
Hearn PP Jr, Sutter JF (1985) Authigenic potassium feldspar in Cambrian carbonates: evidence of Alleghanian brine migration. Science 228:1529–1531
Hearn PP Jr, Sutter JF, Belkin HE (1987) Evidence for late-Paleozoic brine migration in Cambrian carbonate rocks of the central and southern Appalachians: implications for Mississippi Valley-type sulfide mineralization. Geochim Cosmochim Acta 51:1323–1334
Hedge CE (1978) Strontium isotopes in basalts from the Pacific ocean basin. Earth Planet Sci Lett 38:88–94
Helgeson HC, Mackenzie FT (1970) Silicate-sea water equilibria in the ocean system. Deep-Sea Res 17:877–892
Helgeson HC, Garrels RM, Mackenzie FT (1969) Evaluation of irreversible reactions in geochemical processes involving minerals and aqueous solutions. II. Applications. Geochim Cosmochim Acta 33:455–481

Helgeson HC, Brown TH, Nigrini A, Jones TA (1970) Calculation of mass transfer in geochemical processes involving aqueous solutions. Geochim Cosmochim Acta 34:569–592
Herzog LF, Pinson WH, Cormier RF (1958) Sediment age determination by Rb/Sr analysis of glauconite. Am Assoc Petrol Geol Bull 42:717–733
Hess J, Bender ML, Schilling JG (1986) Evolution of the ratio of strontium-87 to strontium-86 in seawater from Cretaceous to Present. Science 231:979–984
Hitchon B, Friedman I (1969) Geochemistry and origin of formation waters in the western Canada sedimentary basin. I. Stable isotopes of hydrogen and oxygen. Geochim Cosmochim Acta 33:1321–1349
Hoefs J (1980) Stable isotope geochemistry, 2nd edn. Springer, Berlin Heidelberg New York, 208 pp
Hoffert M (1980) Les "argiles rouges des grands fonds" dans le Pacifique Centre-Est. Authigenèse, transport, diagenèse. Sci Géol Mém, Strasbourg, 61:231 pp
Hoffert M, Karpoff AM, Clauer N, Schaaf A, Courtois C, Pautot G (1978) Néoformations et altérations dans trois facies volcanosédimentaires du Pacifique Sud. Oceanol Acta 1:187–202
Hoffman J, Hower J (1979) Clay mineral assemblages as low grade metamorphic geothermometers: application to the thrust faulted disturbed belt of Montana. In: Scholle PA, Schluger PS (eds) Aspects of diagenesis. Soc Econ Paleontol Mineral Spec Publ 26:55–79
Hoffman J, Hower J, Aronson JL (1976) Radiometric dating of time of thrusting in the disturbed belt of Montana. Geology 4:16–20
Hofmann AW, Hart SR, Hare PE (1972) Sr^{87}/Sr^{86} ratios of pore fluids from deep-sea cores. Rep Dir Geophys Lab Carnegie Inst, Washington, pp 563–564
Hofmann AW, Mahoney JW, Giletti BJ (1974) K-Ar and Rb-Sr data on the detrital and postdepositional history of Pennsylvania clay from Ohio and Pennsylvania. Geol Soc Am Bull 85:639–644
Hogdhal OT, Melson S, Bowen V (1968) Neutron activation analysis of lanthanide elements in sea water. Adv Chem 73:308–325
Hogg AJC, Peirson MJ, Fallick AE, Hamilton PJ, MacIntyre RM (1987) Clay mineral and isotope evidence for controls on reservoir properties of Brent Group sandstones, British North Sea. Terra Cognita 7:342
Honnorez J (1972) La palagonitisation, l'altération sous-marine du verre volcanique basique de Palagonia (Sicile). Vulkaninst. I. Friedloendler. Birkhäuser, Basel 9:131 pp
Howard JJ, Roy DM (1983) Development of layer charge and kinetics of smectite alteration. Clays Clay Min 33:81–88
Hower J (1961) Some factors concerning the nature and the origin of glauconite. Am Min 46:313–334
Hower J, Hurley PM, Pinson WH, Fairbairn HW (1963) The dependence of K-Ar age on the mineralogy of various particle size ranges in a shale. Geochim Cosmochim Acta 27:405–410
Hower J, Eslinger EV, Hower M, Perry EA (1976) Mechanism of burial metamorphism of argillaceous sediments. 1. Mineralogical and chemical evidence. Geol Soc Am Bull 87:725–737.
Huang WH, Keller WD (1970) Dissolution of rock forming minerals in organic acids. Am Min 55:2076–2094
Huff WD, Turkmenoglu AG (1981) Chemical characteristics and origin of Ordovician K-bentonites along the Cincinnati Arch. Clays Clay Min 29:113–123
Humphris SE, Thompson RN, Marriner GF (1980) The mineralogy and geochemistry of basalt weathering: holes 417A and 418A. In: Initial Reports of the DSDP Project, 51–53:1201–1213
Hunziker JC (1986) The evolution of illite to muscovite: An example of the behaviour of isotopes in low-grade metamorphic terrains. In: Deutsch S, Hofmann AW (eds) Isotopes in geology. Chem Geol 57:31–40

Hunziker JC, Frey M, Clauer N, Dallmeyer RD, Friedrichsen H, Flehmig W, Hochstrasser K, Roggwiller P, Schwander H (1986) The evolution of illite to muscovite: mineralogical and isotopic data from Glarus Alps, Switzerland. Contrib Miner Petrol 92:157–180
Hunziker JC, Frey M, Clauer N, Dallmeyer RD (1987) Reply to the comment on the evolution of illite to muscovite by J.R. Glasmann. Contr Miner Petrol 96:74–77
Huon S, Ruch P (1992) Mineralogical, K-Ar and $^{87}Sr/^{86}Sr$ isotope studies of Holocene and Late Glacial sediments in a deep-sea core from the northeast Atlantic Ocean. Mar Geol 107:275–282
Huon S, Piqué A, Clauer N (1987) Etude de l'orogenèse hercynienne au Maroc par la datation K-Ar de l'évolution métamorphique de schistes ardoisiers. Sci Géol Bull 40:273–284
Huon S, Jantschik R, Kübler B, Fontignie D (1991) Analyses K-Ar, Rb-Sr et minéralogiques des fractions argileuses de sédiments quaternaires, Atlantique N-E: résultats préliminaires. Schweiz Mineral Petrogr Mitt 71:275–280
Huon S, Cornée JJ, Piqué A, Rais N, Clauer N, Liewig N, Zayane R (1993) Mise en évidence au Maroc d'évènements thermiques d'âge triasico-liasique liés à l'ouverture de l'Atlantique. Bull Soc Géol Fr 164:165–176
Hurley PM, Cormier RF, Hower J, Fairbairn HW, Pinson WH (1960) Reliability of glauconite for age measurements by K-Ar and Rb-Sr methods. Am Assoc Petr Geol Bull 44:1793–1808
Hurley PM, Brookins DG, Pinson WH, Hart SR, Fairbairn HW (1961) K-Ar studies of Mississippi and other river sediments. Geol Soc Am Bull 72:1807–1816
Hurley PM, Heezen BC, Pinson WH, Fairbairn HW (1963) K-Ar age values in pelagic sediments of the North Atlantic. Geochim Cosmochim Acta 27:393–399
Hurst A, Irwin H (1982) Geochemical modelling of clay diagenesis in sandstones. Clays Clay Min 17:5–22
Ikpeama MOU, Boger PD, Faure G (1974) A study of strontium in core 119K, Discovery Deep, Red Sea. Chem Geol 13:11–22
Inoue A (1983) Potassium fixation by clay minerals during hydrothermal treatments. Clays Clay Min 31:81–91
Inoue A, Kohyama N, Kitagawa R (1987) Chemical and morphological evidence for the conversion of smectite to illite. Clays Clay Min 35:111–120
Jackson ML (1963) Interlayering of expansible layer silicates in soils by chemical weathering. Clays Clay Min 11:29–46
Jacobsen SB, Pimentel-Klose MR (1988a) Nd isotopic variations in Precambrian banded iron formations. Geophys Res Lett 15:393–396
Jacobsen SB, Pimentel-Klose MR (1988b) A Nd isotopic study of the Hamersley and Michipicoten banded iron formations: the source of REE and Fe in Archean oceans. Earth Planet Sci Lett 87:29–44
James AT, Baker DR (1976) Oxygen isotope exchange between illite and water at 22 °C. Geochim Cosmochim Acta 40:235–239
Jantschik R, Huon S (1992) Detrital silicates in northeast Atlantic deep-sea sediments during the Late Quaternary: mineralogical and K-Ar isotopic data. Eclog Geol Helv 85:195–212
Jeannette D, Liewig N, Mertz JD (1988) Les structures de la porosité de grès hydrothermalisés. Bull Minér III:613–623
Jennings S, Thompson GR (1986) Diagenesis of Plio-Peistocene sediments of the Colorado River delta, southern California. J. Sediment Petrol 56:89–98
Jones LM, Faure G (1969) The isotope composition of strontium and cation concentrations of Lake Vanda and Lake Bonney in southern Victoria Land, Antarctica. Ohio State Univ, Dep Geol, Ann Progr Rep, 82 pp
Jones LM, Faure G (1978) A study of strontium isotopes in lakes and surficial deposits of the ice-free valleys, southern Victoria Land, Antarctica. Chem Geol 22:107–120
Jourdan A, Thomas M, Brevart O, Robson P, Sommer F, Sullivan M (1987) Diagenesis as the control of the Brent sandstone reservoir properties in the greater Alwyn area (East Shetland basin). In: Brooks J, Glennie K (eds) Petroleum geology of northwest Europe. Graham and Trotman, London, pp 951–961

Juteau T, Noack Y, Whitechurch H, Courtois C (1979) Mineralogy and geochemistry of alteration products in holes 417A and 417D basement samples (Deep Sea Drilling Project Leg 51). In: Initial Reports of the DSDP Project, 51, part 2:1273–1279

Kastner M (1976) Diagenesis of basal sediments and basalts of sites 322 and 323, Leg 35, Bellingshausen Abyssal Plain. In: Hollister CD, Craddock C et al. (eds) Initial Reports of the DSDP, vol 35. US Gov Printing Office, Washington, pp 513–528

Kastner M, Gieskes JM (1976) Interstitial water profiles and sites of diagenetic reactions, Leg 35, DSDP, Bellingshausen Abyssal Plain. Earth Planet Sci Lett 33:11–20

Keith ML, Weber JN (1964) Carbon and oxygen isotopic composition of selected limestones and fossils. Geochim Cosmochim Acta 28:1787–1816

Keller WD (1978) Kaolinization of feldspar as displayed in scanning electron micrographs. Geology 6:184–188

Keller WD, Reynolds RC, Inoue A (1986) Morphology of clay minerals in the smectite-to-illite conversion series by scanning electron microscopy. Clays Clay Min 34:187–197

Kempe DRC (1974) The petrology of basalts, Leg 26, Deep Sea Drilling Project. In: Davies TA, Luyendeyh B et al. (eds) Initial Report of DSDP, vol 26. US Gov. Printing Office, Washington, pp 465–503

Keppens E, O'Neil JR (1984) Oxygen isotope variations in glauconies. Terra Cognita Spec Issue p 42

Keppens E, Pasteels P (1982) A comparison of rubidium-strontium and potassium-argon ages on glauconies. In: Odin GS (ed) Numerical dating in stratigraphy. John Wiley, New York, pp 225–239

Keto LS, Jacobsen SB (1987) Nd and Sr isotopic variations of early Paleozoic oceans. Earth Planet Sci Lett 84:27–41

Kettel D (1983) The East Groningen Massif – detection of an intrusive body by means of coalification. Geol Mijnbouw 6:203–210

Kharaka YK, Berry FAF (1980) Geochemistry of geopressured geothermal waters from the northern Gulf of Mexico and California basins. Proc 3rd Int Symp on Water-Rock Inter 3:95–96

Kharaka YK, Carothers WW (1986) Oxygen and hydrogen isotope geochemistry of deep basin brines. In: Fritz P, Fontes JC (eds) Handbook of environmental isotope geochemistry, vol 2. Elsevier, Amsterdam, pp 305–360

Kharaka YK, Thordsen JJ (1992) Stable isotope geochemistry and origin of waters in sedimentary basins. In: Clauer N, Chaudhuri S (eds) Isotopic signatures and sedimentary records. Lecture Notes in Earth Sciences 43. Springer, Berlin Heidelberg New York, pp 411–466

Kisch HJ (1980) Illite crystallinity and coal rank associated with lowest grade metamorphism of the Taveyanne greywacke in the Helvetic zone of the Swiss Alps. Eclog Geol Helv 73:753–777

.Kisch HJ (1983) Mineralogy and petrology of burial diagenesis (burial metamorphism) and incipient metamorphism in clastic rocks. In: Larsen G, Chillingar GV (eds) Diagenesis in sediments and sedimentary rocks, vol 2. Elsevier, Amsterdam, pp 289–493

Kish SA, Stein HJ (1979) The timing of ore mineralization, Viburnum Trend, southeast Missouri lead-zinc district: Rb-Sr glauconite dating. Annu Meet Geol Soc Am (Abstr with Prog) 11:458

Kittrick JA (1966) Forces involved in ion fixation by vermiculite. Soil Sci Soc Am Proc 30:801–803

Kittrick JA (1969) Interlayer forces in montmorillonite and vermiculite. Soil Sci Soc Am Proc 33:217–222

Kittrick JA (1971) Soil solution composition and stability of clay minerals. Soil Sci Soc Am Proc 35:450–454

Klay N, Jessberger EK (1984) $^{40}Ar/^{39}Ar$ dating of glauconites. Terra Cognita Spec Issue, p 22

Kligfield R, Hunziker JC, Dallmeyer RD, Schamel S (1986) Dating of deformation phases using K-Ar and $^{40}Ar/^{39}Ar$ techniques: results from northern Apennines. J Struct Geol 8:781–798

Knauth LP (1992) Origin and diagenesis of cherts: An isotopic perspective. In: Clauer N, Chaudhuri S (eds) Isotopic signatures and sedimentary records, Lecture Notes in Earth Sciences 43. Springer, Berlin Heidelberg New York, pp 123–152

Knauth LP, Beeunas MA (1986) Isotope geochemistry of fluid inclusions in Permian halite with implications for the isotopic history of ocean water and the origin of saline formation waters. Geochim Cosmochim Acta 50:419–433

Knauth LP, Epstein S (1976) Hydrogen and oxygen isotope ratios in silica from the JOIDES Deep Sea Drilling Project. Earth Planet Sci Lett 25:1–10

Knauth LP, Lowe DR (1978) Oxygen isotope geochemistry of cherts from Onverwacht Group (3.4 billion years), Transvaal, South Africa, with implications for secular variations in the isotopic composition of cherts. Earth Planet Sci Lett 41:209–222

Komarneni S, Jackson ML, Cole DR (1985) Oxygen isotope changes during mica alteration. Clays Clay Min 33:214–218

Kotzer TG, Kyser TK (1991) Retrograde alteration of clay minerals in uranium deposits: radiation catalyzed or simply low-temperature exchange? Chem Geol Isot Geosci Sect 86:307–321

Kovach J, Faure G (1977a) Strontium isotopic study of sediment from the Ross Sea. Antarct J 12:77–78

Kovach J, Faure G (1977b) Sources and abundance of volcanogenic sediment in piston cores from the Ross Sea, Antarctica. N Z J Geol Geophys 20:1017–1026

Kovach J, Faure G (1978) Use of strontium isotopes to study mixing of sediment derived from different sources: the Ross Sea, Antarctica. In: Zartman RE (ed) 4th Int Conf Geochr Cosmochr and Isot Geol, US Geol Surv, Open-File Rep 78–701, pp 230–232

Kralik M (1982) Rb-Sr age determinations on Precambrian carbonate rocks of the Carpentarian McArthur Basin, Northern Territories, Australia. Precambrian Res 18:157–170

Kralik M (1984) Effects of cation-exchange treatment and acid leaching on the Rb-Sr system of illite from Fithian, Illinois. Geochim Cosmochim Acta 48:527–533

Kristmannsdottir H (1977) Types of clay minerals in hydrothermally altered basaltic rocks, Reykjanes, Iceland. Jokull 26:30–39

Kröner A, Byerly GR, Lowe DR (1991) Chronology of Early Archean granite-greenstone evolution in the Barberton Mountain Land, South Africa, based on precise dating by single zircon evaporation. Earth Planet Sci Lett 103:41–54

Kubler B (1964) Les argiles indicateurs du métamorphisme. Rev Inst Fr Pétr 19:1093–1113

Kubler B (1966) La cristallinité de l'illite et les zones tout à fait supérieures du métamorphisme. Colloque sur les Etages Tectoniques. Univ Neuchatel, A la Baconnière, Neuchatel, Suisse, pp 105–122

Kulp JL, Engels J (1963) Discordances in K-Ar and Rb-Sr isotopic ages. In: Radioactive dating, Int Atomic Energy Agency, Vienna, pp 219–238

Kunk MJ, Brusewitz AM (1987) ^{39}Ar recoil in an I/S clay from the Ordovician "Big Bentonite Bed" at Kinnekulle, Sweden. 21st Annu Meet North-Central Section Geol Soc Am (Abstr with Prog) 19:230

Kyser TK (1987) Equilibrium fractionation factors for stable isotopes. In: Kyser TK (ed) Stable isotope geochemistry of low temperature processes. Short course handbook 13. Miner Assoc Can, Toronto, pp 1–84

La Iglesia A, Martin-Vivaldi JL, Lopez Aguayo F Jr (1976) Kaolinite crystallization at room temperature by homogeneous precipitation. III. Hydrolysis of feldspars. Clays Clay Min 24:36–42

Lancelot J, Vella V (1989) Datation U-Pb liasique de la pechblende de Rabejac. Mise en évidence d'une préconcentration uranifère permienne dans le bassin de Lodève (Hérault). Bull Soc Géol Fr 8:309–315

Land LS, Dutton SP (1978) Cementation of a Pennsylvanian deltaic sandstone – isotopic data. J Sediment Petrol 48:1167–1176

Lange S, Chaudhuri S, Clauer N (1983) Strontium isotopic evidence for the origin of barites and sulfides from the Mississippi Valley-type ore deposits in Southeast Missouri. Econ Geol 78:1255–1261

Langley KM (1978) Dating sediments by a K-Ar method. Nature 276:56–57

Lasaga AC (1984) Chemical kinetics of water-rock interactions. J Geophys Res 89:4009–4025

Lawrence JR (1979) $^{18}O/^{16}O$ of the silicate fraction of recent sediments used as a provenance indicator in the South Atlantic. Mar Geol 33:M1–M7

Lawrence JR (1989) The stable isotope geochemistry of deep-sea pore water. In: Fritz P, Fontes JC (eds) Handbook of environmental isotope geochemistry, vol 3. Elsevier, Amsterdam, pp 317–356

Lawrence JR, Drever JI (1981) Evidence for cold water circulation at DSDP Site 395: isotopes and chemistry of alteration products. J Geophys Res 86:5125–5133

Lawrence JR, Taylor HP Jr (1971) Deuterium and oxygen-18 correlation: clay minerals and hydroxides in Quaternary soils compared to meteoric water. Geochim Cosmochim Acta 35:993–1004

Lawrence JR, Taylor HP Jr (1972) Hydrogen and oxygen isotope systematics in weathering profiles. Geochim Cosmochim Acta 36:1377–1394

Lawrence JR, Drever JI, Anderson TF, Brueckner HK (1979) Importance of alteration of volcanic material in the sediments of Deep Sea Drilling site 323: chemistry, $^{18}O/^{16}O$ and $^{87}Sr/^{86}Sr$. Geochim Cosmochim Acta 43:573–588

Leach DL, Rowan EL (1986) Genetic link between Ouachita foldbelt tectonism and the Mississippi Valley-type lead-zinc deposits of the Ozark. Geology 14:931–935

Lee MJ, Brookins DG (1978) Rubidium-strontium minimum ages of sedimentation, uranium mineralization, and provenance, Morrison Formation (Upper Jurassic), Grants mineral belt, New Mexico. Am Assoc Petrol Geol Bull 62:1673–1683

Lee M, Aronson JL, Savin SM (1985) K-Ar dating of gas emplacement in Rotliegendes sandstones, Netherlands. Am Assoc Petrol Geol Bull 69:1381–1385

Lee M, Aronson JL, Savin SM (1989) Timing and conditions of Permian Rotliegendes sandstone diagenesis, southern North Sea: K/Ar and oxygen isotope data. Am Assoc Petrol Geol Bull 73:195–215

Le Guen M, Orgeval JJ, Lancelot J (1991) Lead isotope behaviour in a polyphased Pb-Zn ore deposit: les Malines (Cévennes, France). Min Depos 26:180–188

Lei Chou, Wollast R (1984) Study of the weathering of albite at room temperature and pressure with a fluidized bed reactor. Geochim Cosmochim Acta 48:2205–2217

Leprun JC (1979) Les cuirasses ferrugineuses des pays cristallins de l'Afrique Occidentale sèche. Genèse – transformation – dégradation. Sci Géol Mém Strasbourg 58:224 pp

Liewig N (1993) Datation isotopique d'illites diagénétiques de grès réservoirs à gaz, huile et eau du Nord-Ouest de l'Europe. Implications pétrogénétiques et géodynamiques. Thèse Doc ès-Sci, Univ Strasbourg, 238 pp

Liewig N, Clauer N, Sommer F (1987a) Rb-Sr and K-Ar dating of clay diagenesis in Jurassic sandstone reservoirs, North Sea. Am Assoc Petrol Geol Bull 71:1467–1474

Liewig N, Mossmann JR, Clauer N (1987b) Datation isotopique K-Ar d'argiles diagénétiques de réservoirs gréseux: mise en évidence d'anomalies thermiques du Lias inférieur en Europe Nord-Occidentale. C R Acad Sci Paris 304II:707–712

Lin FC, Clemency CV (1981) The kinetics of dissolution of muscovites at 25 °C and 1 atm CO_2 partial pressure. Geochim Cosmochim Acta 45:571–576

Lipson J (1958) Potassium-Argon dating of sedimentary rocks. Geol Soc Am Bull 69:137–150

Long LE, Agee WN Jr (1985) Rb-Sr dating of a paleosol, Llano uplift, Texas. Int Conf on. Isotopes in sedimentary cycle. Obernai, France, 1 p

Longstaffe FK (1983) Stable isotope studies of diagenesis in clastic rocks. Geoscience Can 10:43–58

Longstaffe FK (1986) Oxygen isotope studies of diagenesis in the basal Belly River sandstone, Pembina I-Pool, Alberta. J Sediment Petrol 56:78–88

Longstaffe FK (1989) Stable isotopes as tracers in clastic diagenesis. In: Hutcheon IE (ed) Burial diagenesis. Min Assoc Can, Short Course Ser 15:201–277

Longstaffe FK, Ayalon A (1987) Oxygen-isotope studies of clastic diagenesis in the Lower Cretaceous Viking Formation, Alberta: implications for the role of meteoric water. In: Marshall JD (ed) Diagenesis of sedimentary sequences. Geol Soc Am Spec Publ 36:277–296

Longstaffe FK, Ayalon A (1990) Hydrogen-isotope geochemistry of diagenetic clay minerals from Cretaceous sandstones, Alberta: evidence for exchange. Applied Geochem 5:657–668

Longstaffe FK, Tilley BJ, Ayalon A (1992) Controls on porewater evolution during sandstone diagenesis, Western Canada sedimentary basin: an oxygen isotope perspective. In: Houseknecht DW, Pittman ED (eds) Origin, diagenesis and petrophysics of clay minerals in sandstones. Soc Econ Paleontol Miner Spec Publ 47:13–34

Loucks RG, Debout DG, Galloway W (1977) Relationship of porosity formation and preservation to sandstone consolidation history – Gulf Coast Tertiary, Frio Formation. Bur Econ Geol Circ 77:109–120

Loveland PJ (1981) Weathering of a soil glauconite in southern England. Geoderma 29:35–54

Ludden JN, Thompson G (1978) The modification of the rare-earth-element abundance of oceanic basalts during low temperature weathering by sea water. Trans Am Geophys Union 59:408

MacEwan DMC, Ruiz Amil A, Brown G (1961) Interstratified clay minerals. In: Brown G (ed) The X-ray identification and crystal structures of clay minerals. Min Soc Lond, pp 392–445

Macfarlane AW, Holland H (1991) The timing of alkali metasomatism in paleosols. Can Miner 29:1043–1050

Mack LE, Awwiller DN (1990) Sm/Nd ratio as a diagenetic tracer, Paleogene Texas Gulf Coast. Am Assoc Petrol Geol Bull 74:711

MacKenzie FT, Garrels RM (1966) Silica-bicarbonate balance in the ocean and early diagenesis. J Sediment Petrol 36:1075–1084

MacKenzie RC, Wilson MJ, Mashhady AS (1984) Origin of palygorskite in some soils of the Arabian Peninsula. In: Singer A, Galan E (eds) Palygorskite-sepiolite: occurrence, genesis, and uses. Elsevier, Amsterdam, pp 177–186

Mackin JE (1986) Control of dissolved Al distributions in marine sediments by clay reconstitution reactions: experimental evidence leading to a unified theory. Geochim Cosmochim Acta 50:207–214

Manning LK, Frost CD, Branthaver JF (1991) A neodymium isotopic study of crude oils and source rocks: Potential applications for petroleum exploration. Chem Geol 91:125–138

Martineau F (1976) L'origine et l'histoire de la série spilitique d'Erquy. Arguments isotopiques (Sr, Ar) et géochimie des éléments en traces. Thèse 3ème cycle, Univ Rennes, 76 pp

Martin-Vivaldi JL, Cano-Ruiz J (1956) Sepiolite. II. Consideration on the mineralogical formula. CCM 4:173–176

Marvin RF, Wright JC, Walthall F (1965) K-Ar and Rb-Sr ages of biotite from the middle Jurassic part of the Carmel formation. Utah. US Geol Surv Prof Pap 525-B:104–107

Mattigod SV, Kittrick JA (1979) Aqueous solubility studies of muscovite: apparent non-stoichiometric solute activities at equilibrium. Soil Sci Soc Am J 43:180–187

McCubbin DG, Patton JL (1981) Burial diagenesis of illite/smectite, a kinetic model. Am Assoc Petrol Geol Bull 65:956

McCulloch MT, Wasserburg GJ (1978) Sm-Nd and Rb-Sr chronology of continental crust formation. Science 200:1003–1011

McDougall I, Polach HA, Stipp JJ (1969) Excess radiogenic argon in young subaerial basalts from the Auckland volcanic field, New Zeland. Geochim Cosmochim Acta 33:1485–1520

McHardy WJ, Wilson MJ, Tait JM (1982) Electron microscope and X-ray diffraction studies of filamentous illitic clays from sandstones of the Magnus field. Clay Min 17:23–39

McLennan SM (1988) Recycling of the continental crust. Pure Appl Geophys 128:683–724

McLennan SM (1989) Rare earth elements in sedimentary rocks: influence of provenance and sedimentary processes. Rev Min 21:169–200

McLennan SM, Taylor SR, McCulloch MT, Maynard JB (1990) Geochemical and Nd-Sr isotopic composition of deep-sea turbidites: crustal evolution and plate tectonic associations. Geochim Cosmochim Acta 54:2015–2050

McMurtry GM, Yeh HW (1981) Hydrothermal clay mineral formation of East Pacific Rise and Bauer Basin sediments. Chem Geol 32:189–205

McMurtry GM, Wang CH, Yeh HW (1983) Chemical and isotopic investigations on the origin of clay minerals from the Galapagos hydrothermal mounds field. Geochim Cosmochim Acta 47:475–490

McNutt RH, Frape SK, Dollar P (1987) A strontium, oxygen and hydrogen isotopic composition of brines, Michigan and Appalachian basins, Ontario and Michigan. Appl. Geochem 2:495–505

McPowell A (1979) A morphological classification of rock cleavage. Tectonophysics 58:21–34

McRea SG (1972) Glauconite. Earth Sci Rev 8:397–440

Mehegan JM, Robinson PT (1982) Secondary mineralization and hydrothermal alteration in the Reydarfjordur drill core, eastern Iceland. J Geophys Res B87:6511–6524

Melson WG, Thompson G (1973) Glassy abyssal basalts, Atlantic sea floor near St Pauli rocks: petrography and composition of secondary clay minerals. Geol Soc Am Bull 84: 703–716

Mendez Santizo J (1990) Diagenèse et circulations de fluides dans le gisement d'uranium de Lodève (Hérault). Thèse, Univ Strasbourg, 166 pp

Mendez Santizo J, Gauthier-Lafaye F, Liewig N, Clauer N, Weber F (1991) Existence d'un hydrothermalisme tardif dans le bassin de Lodève (Hérault). Arguments paléothermométriques et géochronologiques. C R Acad Sci Paris 312:739–745

Merrihue CM, Turner G (1966) Potassium-argon dating by activation with fast neutrons. J Geophys Res 71:2852–2857

Mevel C (1980) Mineralogy and geochemistry of secondary phases in low temperature-altered basalts from Deep Sea Drilling Project Legs 51, 52, and 53. In: Initial Reports of the DSDP Project, 51–53:1299–1312

Michard A (1989) Rare-earth element systematics in hydrothermal fluids. Geochim Cosmochim Acta 53:745–750

Michard A, Guriet P, Soudant M, Albarède F (1985) Nd isotopes in French Phanerozoic shales: external vs. internal aspects of crustal evolution. Geochim Cosmochim Acta 49:601–610

Michaud JG (1980) Gisements de Pb-Zn du Sud du Massif Central français (Cévennes-Montagne Noire) et caractéristiques géologiques de leur environnement. Bull Cent Rech Expl Prod Elf Aquitaine 3:335–380

Miller RG, O'Nions RK (1984) The provenance and crustal residence ages of British sediments in relation to palaeogeographic reconstruction. Earth Planet Sci Lett 68:459–470

Milliken KL, Land LS, Loucks RG (1981) History of burial diagenesis determined from isotope geochemistry, Frio Formation, Brazoria County, Texas. Am Assoc Petrol Geol Mem 37:434 pp

Milliman JD, Meade RH (1983) Worldwide delivery of river sediments to the oceans. J Geol 91:1–21

Millot G (1970) Geology of clays. Weathering, sedimentology, geochemistry. Springer, Berlin Heidelberg New York; Masson et Cie, Paris; Chapman & Hall, London, 429 pp

Mitchell JG, Taka AS (1984) Potassium and argon loss patterns in weathered micas: implications for detrital mineral studies with particular reference to the Triassic paleogeography of the British Isles. Sediment Geol 39:27–52

Mitchell JG, Penven MJ, Ineson PR, Miller JA (1988) Radiogenic argon and major-element loss from biotite during natural weathering: a geochemical approach to the interpretation of potassium-argon ages of detrital biotite. Chem Geol 72:111–126

Montigny R, Faure G (1969) Contribution au problème de l'homogénéisation isotopique du strontium des roches totales au cours du métamorphisme. Cas du Wisconsin Range, Antarctique. C R Acad Sci Paris 268D:1012–1015

Moorbath S, Stewart AD, Lawson DE, Williams GE (1967) Geochronological studies on the Torridonian sediments of north-west Scotland. Scott J Geol 3:389–412

Morton JP (1985a) Rb/Sr evidence for punctuated illite/smectite diagenesis in the Oligocene Frio Formation, Texas, Gulf Coast. Geol Soc Am Bull 96:1043–1049

Morton JP (1985b) Rb-Sr dating of diagenesis and source age of clays in Upper Devonian black shales of Texas. Geol Soc Am Bull 96:1043–1049

Morton JP, Long LE (1980) Rb-Sr dating of Paleozoic glauconite from the Llano region, Central texas. Geochim Cosmochim Acta 44:663–671

Mossmann JR (1991) K-Ar dating of authigenic illite-smectite clay material: application to complex mixtures of mixed-layer assemblages. Clay Min 26:189–198

Mossmann JR, Clauer N, Liewig N (1992) Dating thermal anomalies in sedimentary basins: the diagenetic history of clay minerals in the Triassic sandstones of the Paris Basin (France). Clay Min 27:211–226

Mosser C, Gense C (1979) Elements traces dans des kaolinites d'altération à Madagascar. Chem Geol 26:295–310

Muehlenbachs K (1986) Alteration of the oceanic crust and the ^{18}O history of seawater. In: Valley JW, Taylor HP Jr, O'Neil JR (eds) Stable isotopes in high temperature geological processes. Rev Miner 16:425–444

Muehlenbachs K, Clayton RN (1972) Oxygen isotope geochemistry of submarine greenstones. Can J Earth Sci 9:471–478

Muehlenbachs K, Clayton RN (1976) Oxygen isotope composition of the oceanic crust and its bearing on seawater. J Geophys Res 81:4365–4369

Muller JP (1987) Analyse pétrologique d'une formation latéritique meuble du Cameroun. Essai de traçage d'une différenciation supergène par les paragenèses minérales secondaires. Thèse Doc ès-Sci, Univ Paris VII, 174 pp

Muller JP, Bocquier G (1986) Dissolution of kaolinites and accumulation of iron oxides in lateritic ferruginous nodules. Mineralogical and microstructural transformations. Geoderma 37:113–136

Murata KJ, Friedman I, Gleason JD (1977) Oxygen isotope relations between diagenetic silica minerals in Monterey Shale, Temblor Range, California. Am J Sci 277:259–272

Murray J, Renard AF (1884) On the nomenclature, origin and distribution of deep-sea sediments. Proc R Soc Edinb 12:495–529

Nadeau PH, Wilson MJ, McHardy WJ, Tait JM (1984) Interstratified clays as fundamental particles. Science 225:923–925

Nadeau PH, Wilson MJ, McHardy WJ, Tait JM (1985) The conversion of smectite to illite during diagenesis: evidence from some illitic clays from bentonites and sandstones. Min Mag 49:393–400

Nägler TF, Schäffer HJ, Gebauer D (1992) A Sm-Nd isochron on pelites 1 Ga in excess of their depositional age and its possible significance. Geochim Cosmochim Acta 56:789–795

Nägler TF, Stille P, Chaudhuri S, Clauer N (1993) A Sr-Nd study on dissolved and suspended loads of Mississippi River waters with implications on global mass balance calculations. VII. Meet Eur Union Geosci, April 4–8 1993, Strasbourg, Terra Abstr, p 344

Nakai S, Halliday AN, Kesler SE, Jones HD (1990) Rb-Sr dating of sphalerites from Tennessee and the genesis of Mississippi Valley type ore deposits. Nature 346:354–357

Nesbitt HW (1979) Mobility and fractionation of rare earth elements during weathering of a granodiorite. Nature 279:206–210

Newman ACD (ed) (1987) Chemistry of clays and clay minerals. Miner Soc Mono 6, Min Soc, John Wiley, New York, 480 pp

Nier AO (1950) A redetermination of the relative abundances of the isotopes of carbon, nitrogen, oxygen, argon and potassium. Phys Rev 77:789–793

Obradovich JD (1965) Problems in the use of glauconite and related minerals for radioactive dating. PhD Thesis, Univ California Berkeley
Obradovich JD, Peterman ZE (1968) Geochronology of the Belt series, Montana. Can J Earth Sci 5:737–747
Odin GS (1975) Les glauconies: constitution, formation, âge. Thèse Doc-ès-Sci, Univ Paris VI, 250 pp
Odin GS, Bonhomme MG (1982) Argon behaviour in clays and glauconies during preheating experiments. In: Odin GS (ed) Numerical dating in stratigraphy. John Wiley, New York, pp 333–343
Odin GS, Dodson MH (1982) Zero isotopic age of glauconies. In: Odin GS (ed) Numerical dating in stratigraphy. John Wiley, New York, pp 277–305
Odin GS, Hunziker JC (1974) Etude isotopique de l'altération naturelle d'une formation à glauconie (méthode à l'argon). Contrib Miner Petrol 48:9–22
Odin GS, Hunziker JC (1982) Radiometric dating of the Albian-Cenomanian boundary. In: Odin GS (ed) Numerical dating in stratigraphy. John Wiley, New York, pp 537–556
Odin GS, Létolle R (1980) Glauconitization and phosphatization environments: a tentative comparison. In: Bentor YK (ed) Marine phosphorites. Soc Econ Paleontol Miner Spec Publ 29:227–237
Odin GS, Matter A (1981) De glauconarium origine. Sedimentology 28:611–641
Odin GS, Rex DC (1982) Potassium-argon dating of washed, leached, weathered and reworked glauconies. In: Odin GS (ed) Numerical dating in stratigraphy. John Wiley, New York, pp 363–385
Odin GS, Velde B, Bonhomme MG (1977) Radiogenic argon retention in glauconites as a function of mineral recrystallization. Earth Planet Sci Lett 37:154–158
Odin GS, Dodson MH, Hunziker JC, Kreuzer H (1979) Radiogenic argon in glauconies during their genesis. Int Geol Corr Progr Proj 133 Bull 6:7–8
Ohr M, Halliday AN, Peacor DR (1991) Sr and Nd isotopic evidence for punctuated clay diagenesis, Texas Gulf Coast. Earth Planet Sci Lett 105:110–126
O'Neil JR (1968) Hydrogen and oxygen isotopic fractionation between ice and water. J Phys Chem 72:3683
O'Neil JR, Kharaka YK (1976) Hydrogen and oxygen isotope exchange reactions between clay minerals and water. Geochim Cosmochim Acta 40:241–246
O'Neil JR, Clayton RN, Mayeda TK (1969) Oxygen isotope fractionation in divalent metal carbonates. J Chem Phys 51:5547–5558
O'Nions RK, Oxburgh ER, Hawkesworth CJ, MacIntyre RM (1973) New isotopic and stratigraphical evidence on the age of the Ingletonian: probable Cambrian of northern England. J Geol Soc Lond 129:445–452
O'Nions RK, Carter SR, Cohen RS, Evensen NM, Hamilton PJ (1978) Pb, Nd, and Sr isotopes in oceanic ferro-manganese deposits and ocean floor basalts. Nature 273:435–438
O'Nions RK, Hamilton PJ, Hooker PJ (1983) A Nd-isotope investigation of sediments related to crustal development in the British Isles. Earth Planet Sci Lett 63:229–240
Oxburgh ER, Turcotte DL (1974) Thermal gradients and regional metamorphism overthrust terrains with special reference to the eastern Alps. Schweiz Mineral Petrogr Mitt 54:641–662
Paces T (1973) Steady-state kinetics and equilibrium between ground water and granitic rock. Geochim Cosmochim Acta 37:2641–2663
Pagel M, Poty B, Sheppard SMF (1980) Contribution to some Saskatchewan uranium deposits mainly from fluid inclusion and isotopic data. In: Uranium in the Pine Creek Geosyncline, Int Atomic Energy Agency, Vienna, pp 639–654
Paquet H (1970) Evolution géochimique des minéraux argileux dans les altérations et les sols des climats méditerranéens et tropicaux à saisons contrastées. Sci Géol Mém Strasbourg 30:212 pp
Paquet H, Duplay J, Valleron-Blanc MM, Millot G (1987) Octahedral composition of individual particles in smectite-palygorskite and smectite-sepiolite assemblages. In: Schultz LG, Van

Olphen H, Mumpton FA (eds) Proc 8th Int Clay Conf Clay Min Soc, Bloomington, pp 73–77

Park YA, Pilkey OH (1981) Detrital mica: environmental significance of roundness and grain surface textures. J Sediment Petrol 51:113–120

Parron C, Nahon D (1980) Red bed genesis by lateritic weathering of glauconitic sediments. J Geol Soc 137:689–693

Pauling L (1930) The structure of the chlorites. Proc Natl Acad Sci USA 16:578–582

Perry EA (1974) Diagenesis and the K-Ar dating of shales and clay minerals. Geol Soc Am Bull 85:827–830

Perry EA, Hower J (1970) Burial diagenesis in Gulf Coast pelitic sediments. Clays Clay Min 18:165–177

Perry EA, Turekian KK (1974) The effects of diagenesis on the redistribution of strontium isotopes in shales. Geochim Cosmochim Acta 38:929–935

Perry EC (1967) The oxygen isotopic chemistry of ancient cherts. Earth Planet Sci Lett 3:62–66

Perry EC, Tan FC (1972) Significance of oxygen and carbon isotope variations in early Precambrian cherts and carbonate rocks of South Africa. Geol Soc Am Bull 83:664–683

Peterman ZE (1966) Rb-Sr dating in middle-Precambrian metasedimentary rocks of Minnesota. Geol Soc Am Bull 77:1031–1044

Peterman ZE, Hedge CE (1971) Related strontium isotopic and chemical variations in oceanic basalts. Geol Soc Am Bull 82:493–500

Peterman ZE, Hedge CE, Tourtelot HA (1970) Isotopic composition of sea water throughout Phanerozoic time. Geochim Cosmochim Acta 34:105–120

Petrovic R (1976) Rate control in feldspar dissolution. II. The protective effect of precipitates. Geochim Cosmochim Acta 40:1509–1522

Pevear DR (1992) Illite age analysis, a new tool for basin thermal history analysis. In: Kharaka YK, Maest AS (eds) Proc 7th Int Symp on Water-rock interaction. Balkema, Rotterdam, pp 1251–1254

Philippe S (1988) Systématique U-Pb et évolution comparée de minéralisations uranifères du bassin de l'Athabasca (Saskatchewan, Canada). Cas de gisements de la structure de Carswell et de Cigar Lake. Thèse, Univ Montpellier, 350 pp

Philippe S, Lancelot JR, Clauer N, Pacquet A (1993) Formation and evolution of the Cigar Lake uranium ore deposit based on U-Pb and K-Ar isotope systematics. Can J Earth Sci 30:720–730

Phillips FM, Bentley HW (1987) Isotopic fractionation during ion filtration. I. Theory. Geochim Cosmochim Acta 51:683–695

Piepgras DJ, Wasserburg GJ, Dasch EJ (1979) The isotopic composition of Nd in different ocean masses. Earth Planet Sci Lett 45:223–226

Piper DZ (1974) Rare earth elements in the sedimentary cycle: a summary. Chem Geol 14:285–304

Piqué A (1976) Front thermique syntectonique et mise en place du granite à Oulmès (Maroc central). Bull Soc Géol Fr 18:1233–1238

Piqué A (1981) Développement de la schistosité dans les grauwackes cambriennes de la Meseta côtière (Maroc). Sci Géol Bull Strasbourg 34:107–116

Polevaya NI, Murina GA, Kazakov GA (1961) Utilization of glauconite in absolute dating. In: Geochronology of rock systems. Ann New York Acad Sci 91:298–310

Pollastro RM (1985) Mineralogical and morphological evidence for the formation of illite at the expense of illite/smectite. Clays Clay Min 33:265–274

Pollastro RM (1989) Clay minerals as geothermometers and indicators of thermal maturity – application to basin history and hydrocarbon generation. Am Assoc Petrol Geol Bull 73:1171

Pomerol C (1974) Le Bassin de Paris. In: Debelmas J (ed) Géologie de la France. Doin, Paris

Posey HH, Stein HJ, Fullagar PD, Kish SA (1983) Rb-Sr isotopic analyses of Upper Cambrian glauconites, southern Missouri: implications for movement of Mississippi Valley-type ore fluids in the Ozark region. In: Kisvarsanyi G, Grant SK, Pratt WP, Koenig JW (eds) Int

Conf on Mississippi Valley-type lead-zinc deposits. Univ Missouri-La Rolla, Rolla, pp 166–172
Powers MC (1959) Adjustment of clays to chemical change and the concept of the equivalence level. Clays Clay Min 6:309–326
Powers MC (1967) Fluid-release mechanisms in compacting marine mudrocks and their importance in oil exploration. Am Assoc Petrol Geol Bull 51:1240–1254
Pritchard RG (1980) Alteration of basalts from Deep Sea Drilling Project Legs 51, 52, and 53, Holes 417A and 418A. In: Initial Reports of the DSDP Project, 51–53:1185–1192
Rai D, Lindsay WL (1975) A thermodynamic model for predicting the formation, stability, and weathering of common soil minerals. Soil Sci Soc Am Proc 39:991–996
Rais N (1992) Caractérisation minéralogique, cristallochimique et isotopique (K-Ar) d'un métamorphisme polyphasé de faible intensité. Exemple: les grauwackes cambriennes du Maroc Occidental. Thèse, Univ Brest, 193 pp
Rama Murthy V, Beiser E (1968) Strontium isotopes in ocean water and marine sediments. Geochim Cosmochim Acta 32:1121–1126
Ramseyer K, Boles JR (1986) Mixed-layer illite/smectite minerals in Tertiary sandstones and shales, San Joaquin basin, California. Clays Clay Min 34:115–124
Reuter A (1985) Korngrössenabhängigkeit von K-Ar Datierungen und Illit-Kristallinität anchizonaler Metapelite und assoziierter Metatuffe aus dem östlichen Rheinischen Schiefergebirge. Gött Arb Geol Paläontol 27:91 pp
Reuter A (1987) Implications of K-Ar ages of whole-rock and grain-size fractions of metapelites and intercalated metatuffs within an anchizonal terrane. Contrib Miner Petrol 97:105–115
Reuter A, Dallmeyer RD (1987a) $^{40}Ar/^{39}Ar$ age spectra of whole-rock and constituent grain-size fractions from anchizonal slates. Chem Geol 66:73–88
Reuter A, Dallmeyer RD (1987b) $^{40}Ar/^{39}Ar$ dating of cleavage formation in tuffs during anchizonal metamorphism. Contr Miner Petrol 97:352–360
Reuter A, Dallmeyer RD (1989) K-Ar and $^{40}Ar/^{39}Ar$ dating of cleavage formed during very low-grade metamorphism: a review. In: Daly JS, Cliff RA, Yardley BWD (eds) Evolution of metamorphic belts. Geol Soc Lond Spec Publ, Blackwell, Oxford, pp 161–172
Rex RW, Goldberg ED (1958) Quartz contents of pelagic sediments of the Pacific Ocean. Tellus 10:153–159
Reynolds RC (1967) Interstratified clay systems: calculation of the total one-dimensional diffraction function. Am Min 52:661–672
Reynolds RC (1980) Interstratified clay minerals. In: Brindley GW, Brown G (eds) Crystal structures of clay minerals and their X-ray diffraction identification. Min Soc Lond, pp 249–303
Reynolds RC (1985) NEWMOD © a computer program for the calculation of one-dimensional patterns of mixed-layer clays. RC Reynolds, 8 Brook Road, Hannover, NH 03755, USA
Reynolds RC, Hower J (1970) The nature of interlayering in mixed-layer illite-montmorillonite. Clays Clay Min 18:25–36
Rice S, Langmuir CH, Bender JF, Hanson GN, Bence AE (1980) Basalts from Deep Sea Drilling Project Holes 417A and 418D, fractionated melts of a light rare-earth-depleted source. In: Initial Reports of the DSDP Project, 51–53, part 2:1099–1111
Richardson SH, Hart SR, Staudigel H (1980) Vein mineral ages of old oceanic crust. J Geophys Res 58:7195–7200
Rinckenbach T (1988) Diagenèse minérale des sédiments pétrolifères du delta fossile de la Mahakam (Indonésie). Evolution minéralogique et isotopique des composants argileux et histoire thermique. Thèse, Univ Strasbourg, 209 pp
Roberson HE (1974) Early diagenesis: expansible soil clay-sea water reactions. J Sediment Petrol 44:441–449
Roberson HE, Lahann RW (1981) Smectite to illite conversion rates: effect of solution chemistry. Clays Clay Min 29:129–135
Robert M (1970) Etude expérimentale de la désagrégation du granite et de l'évolution des micas. Thèse Doc ès-Sci, Univ Paris, 194 pp

Robertson RHS (1961) Mineral use guide. Cleaver-Hume Press, London
Robinson AG, Coleman ML, Gluyas JG (1993) The age of illite cement growth, Village field area, southern North Sea: evidence from K-Ar ages and $^{18}O/^{16}O$ ratios. Am Assoc Petrol Geol Bull 77:68–80
Robinson PT, Flower MFJ, Swanson DA, Staudigel H (1979) Lithology and eruptive stratigraphy of Cretaceous oceanic crust, western Atlantic Ocean. In: Initial Reports of the DSDP Project, 53, part 2:1535–1555
Rodgers GP, Holland HD (1979) Weathering products within microcracks in feldspars. Geology 7:278–280
Ronov AB, Balashov YA, Migdisov AA (1967) Geochemistry of the rare earths in the sedimentary cycle. Geochem Int 4:1–17
Ross GJ (1968) Structural decomposition of an orthochlorite during its acid dissolution. Can Min 9:522–530
Ross CS, Hendricks SB (1945) Minerals of the montmorillonite group. US Geol Surv Prof Pap 205-B:23–79
Russell KL (1970) Geochemistry and halmyrolysis of clay minerals, Rio Ameca, Mexico. Geochim Cosmochim Acta 34:893–907
Salomons W, Hofman P, Boelens R, Mook WG (1975) The oxygen isotopic composition of the fraction less than 2 microns (clay fraction) in recent sediments from Western Europe. Mar Geol 18: M23–M28
Sardarov SS (1963) Preservation of radiogenic argon in glauconites. Geochem Int 10:937–944
Savin SM (1980) Oxygen and hydrogen isotope effects in low-temperature mineral-water interactions. In: Fritz P, Fontes JC (eds) Handbook of environmental isotope geochemistry, vol 1. Elsevier, Amsterdam, pp 283–328
Savin SM, Epstein S (1970a) The oxygen and hydrogen isotope geochemistry of clay minerals. Geochim Cosmochim Acta 34:25–42
Savin SM, Epstein S (1970b) The oxygen and hydrogen isotope geochemistry of ocean sediments and shales. Geochim Cosmochim Acta 34:43–63
Savin SM, Lee M (1988) Isotopic studies of phyllosilicates. In: Bailey SW (ed) Hydrous phyllosilicates (exclusive of micas). Rev Miner, Min Soc Am 19:189–223
Sayles FL, Mangelsdorf PC Jr (1977) The equilibration for clay minerals with seawater: exchange reactions. Geochim Cosmochim Acta 41:951–960
Sayles FL, Mangelsdorf PC Jr (1979) Cation-exchange characteristics of Amazon River suspended sediment and its reaction with seawater. Geochim Cosmochim Acta 43:767–780
Schaltegger U, Stille P, Rais N, Piqué A, Clauer N (1994) Nd and Sr isotopic dating of diagenesis and low-grade metamorphism of argillaceous sediments. Geochim Cosmochim Acta 58:1471–1481
Schultz LG (1969) Lithium and potassium adsorption, dehydroxylation temperature and structural water content of aluminous smectites. Clays Clay Min 17:115–149
Seyfried WE (1977) Sea water-basalt interaction from 25–300 °C and 1–500 bars. PhD Thesis, Univ California, Los Angeles, 242 pp
Shaffer NS, Faure G (1976) Regional variation of $^{87}Sr/^{86}Sr$ ratios and mineral compositions of sediment from the Ross Sea, Antarctica. Geol Soc Am Bull 87:1491–1500
Shaffiqullah M, Damon PE (1974) Evaluation of K-Ar isochron methods. Geochim Cosmochim Acta 38:1341–1358
Shemesh A, Kolodny Y, Luz B (1983) Oxygen isotope variations in phosphate of biogenic apatites. II. Phosphorite rocks. Earth Planet Sci Lett 64:405–416
Sheppard SMF (1986) Characterization and isotopic variations in natural waters. In: Valley JW, Taylor HP Jr, O'Neil JR (eds) Stable isotopes in high temperature geological processes. Rev Miner, Min Soc Am 16:165–183
Sholkovitz ER (1992) Chemical evolution of rare earth elements: fractionation between colloidal and solution phases of filtered river water. Earth Planet Sci Lett 114:77–84
Sholkovitz ER, Elderfield H (1988) The cycling of dissolved rare earth elements in Chesapeake Bay. Glob Bio-Geochem Cycl 2:157–176

Sigurgeirsson T (1962) Dating recent basalt by potassium-argon method. Rep Phys Dep Univ Iceland, 9 pp

Small JS (1993) Experimental determination of the rates of precipitation of authigenic illite and kaolinite in the presence of aqueous oxalate and comparison to the K/Ar ages of authigenic illite in reservoir sandstones. Clays Clay Min 41:191–208

Smalley PC, Nordaa A, Raheim A (1986) Geochronology and paleothermometry of Neogene sediments from the Voring Plateau using Sr, C and O isotopes. Earth Planet Sci Lett 78:368–378

Smith HS, O'Neil JR, Erlank AJ (1984) Oxygen isotope composition of minerals and rocks and chemical alteration patterns in pillow lava from the Barberton greenstone belt, South Africa. In: Kröner A, Hanson GN, Goodwin AM (eds) Archean geochemistry: the origin and evolution of Archean continental crust. Springer, Berlin Heidelberg New York, pp 115–137

Soubies F, Gout R (1987) Sur la cristallinité des biotites kaolinisées des sols ferrallitiques de la région d'Ambalavao. Cah ORSTOM Sér Pédol 23:111–121

Spiers GA, Dudas MJ, Muehlenbachs K, Pawluk S (1985) Isotopic evidence for clay mineral weathering and authigenesis in Cryoboralfs. Soil Sci Soc Am J 49:467–474

Spooner ETC (1976) The strontium isotopic composition of seawater and seawater-oceanic crust interaction. Earth Planet Sci Lett 31:167–174

Srodon J (1980) Precise identification of illite/smectite interstratifications by X-ray powder diffraction. Clays Clay Min 28:401–411

Srodon J (1981) X-ray identification of randomly interstratified illite/smectite in mixtures with discrete illite. Clay Min 16:297–304

Srodon J (1984) X-ray identification of illitic materials. Clays Clay Min 32:337–349

Srodon J, Eberl DD (1984) Illite. In: Bailey SW (ed) Micas. Rev Miner, Min Soc Am 13: pp 495–544

Srodon J, Morgan DJ, Eslinger EV, Eberl DD, Karlinger MR (1986) Chemistry of illite/smectite and end-member illite. Clays Clay Min 34:368–378

Stacey JS, Kramers JD (1975) Approximation of terrestrial lead isotope evolution by a two-stage model. Earth Planet Sci Lett 26:206–221

Stakes DS, O'Neil JR (1982) Mineralogy and stable isotope geochemistry of hydrothermally altered oceanic crust rocks. Earth Planet Sci Lett 57:285–304

Starinsky A, Bielski M, Lazar B, Steinitz G, Raab M (1983) Strontium isotope evidence on the history of oilfield brines, Mediterranean coastal plain, Israel. Geochim Cosmochim. Acta 47:687–695

Staudigel H, Hart SR (1983) Alteration of basaltic glass: mechanism and significance for the oceanic crust-sea water budget. Geochim Cosmochim Acta 47:337–350

Staudigel H, Hart SR (1985) Dating of crust hydrothermal alteration: strontium isotope ratios from hole 504 carbonates and reinterpretation of Sr isotope data from Deep Sea Drilling Project Sites 105, 332, 417 and 418. In: Initial Reports of the DSDP Project, 83:297–303

Staudigel H, Frey FA, Hart SR (1980) Incompatible trace-element geochemistry and $^{87}Sr/^{86}Sr$ in basalts and corresponding glasses and palagonites. In: Initial Reports of the DSDP Project, 51–53, part 2:1137–1143

Steiger RH, Jäger E (1977) Subcommission on geochronology: convention on the use of decay constants in geo- and cosmochronology. Earth Planet Sci Lett 36:359–362

Stille P (1992) Nd-Sr isotope evidence for dramatic changes of paleocurrents in the Atlantic Ocean during the past 80 m.y. Geology 20:387–390

Stille P, Clauer N (1986) Sm-Nd isochron-age and provenance of the argillites of the Gunflint Iron Formation in Ontario, Canada. Geochim Cosmochim Acta 50:1141–1146

Stille P, Clauer N (1994) The process of glauconitization. Chemical and isotopic evidence. Contr Miner Petrol 117:253–262

Stille P, Chaudhuri S, Kharaka YK, Clauer N (1992) Neodymium, strontium, oxygen and hydrogen isotope compositions of waters in present and past oceans: a review. In: Clauer N,

Chaudhuri S (eds) Isotopic signatures and sedimentary records. Lecture Notes in Earth Sciences 43. Springer, Berlin Heidelberg New York, pp 389–410
Stille P, Gauthier-Lafaye F, Bros R (1993) The Nd isotope system as a tool for petroleum research and exploration. Geochim Cosmochim Acta 5:4521–4525
Stordal MC, Wasserburg GJ (1986) Neodymium isotopic study of Baffin Bay water: sources of REE from very old terranes. Earth Planet Sci Lett 77:259–272
Stueber AM, Pushkar P, Hetherington EA (1984) A strontium isotopic study of Smackover brines and associated solids, southern Arkansas. Geochim Cosmochim Acta 48:1637–1649
Stueber AM, Pushkar P, Hetherington EA (1987) A strontium isotopic study of formation waters from the Illinois basin. Appl Geochem 2:477–494
Sucheki RK, Land LS (1983) Isotopic geochemistry of burial-metamorphosed volcanogenic sediments, Great Valley sequence, northern California. Geochim Cosmochim Acta 47:1487-1499
Sutter JF, Hartung JB (1984) Laser microprobe $^{40}Ar/^{39}Ar$ dating of mineral grains in situ. Scann Electr Micros, SEM Inc, Chicago, pp 1525–1529
Symons DTA, Sangster DF (1991) Paleomagnetic age of the central Missouri barite deposits and its genetic implication. Econ Geol 86:1–12
Tardy Y (1969) Géochimie des altérations. Etude des arênes et des eaux de quelques massifs cristallins d'Europe et d'Afrique. Mém Serv Carte Géol Als Lorr 31:199 pp
Tarutani T, Clayton RN, Mayeda TK (1969) The effect of polymorphism and magnesium substitution on oxygen isotope fractionation between calcium carbonate and water. Geochim Cosmochim Acta 33:987–996
Taylor HP Jr (1977) Water/rock interactions and the origin of H_2O in granitic batholiths. J Geol Soc Lond 133:509–558
Taylor HP Jr, Epstein S (1966) Deuterium-hydrogen ratios in coexisting minerals of metamorphic and igneous rocks. Trans Am Geophys Union Abstr 47:213
Tazaki K (1986) Observation of primitive clay precursors during microcline weathering. Contr Miner Petrol 92:86–88
Tazaki K, Fyfe WS, Heath GR (1986) Palygorskite formed on montmorillonite in North Pacific deep-sea sediments. Clay Sci 6:197–216
Tazaki K, Fyfe WS, Tsuji M, Katayama K (1987) TEM observations of the smectite-to-palygorskite transition in deep Pacific sediments. Appl Clay Sci 2:233–240
Teichmüller M, Teichmüller R, Weber K (1979) Inkohlung und Illit-Kristallinität Vergleichende Untersuchungen im Mesozoikum und Paläozoikum von Westfalen. Fortschr Geol Rheinl Westfalen 27:201–276
Tessier D (1984) Etude expérimentale de l'organisation des minéraux argileux. Hydratation, gonflement et structuration au cours de la dessication et de la réhumectation. Thèse Doc ès-Sci, Univ Paris VII, 361 pp
Thellier C (1984) Transfert d'une solution à travers un matériau pédologique lors d'une évaporation expérimentale. Bilans minéralogique, géochimique et isotopique. Thèse 3ème cycle, Univ Strasbourg, 130 pp
Thellier C, Clauer N (1989) Strontium isotopic evidence for soil-solution interactions during evaporation experiments. Chem Geol Isot Geosci Sect 73:299–306
Thellier C, Fritz B, Paquet H, Gac JY, Clauer N (1988) Chemical and mineralogical effects of saline water movement through a soil during evaporation. Soil Sci 146:22–29
Thompson CW (1874) Preliminary notes on the nature of the sea bottom procured by the soundings of HMS "Challenger" during her cruise in the Southern Sea in the early part of the year 1874. Proc R Soc Lond 23:32–49
Thompson GR, Hower J (1973) An explanation for low radiometric ages from glauconite. Geochim Cosmochim Acta 37:1473–1491
Thomson J, Carpenter MSN, Colley S, Wilson TRS, Elderfield H, Kennedy H (1984) Metal accumulation rates in northwest Atlantic pelagic sediments. Geochim Cosmochim Acta 48:1935–1948

Tilley BJ, Longstaffe FK (1989) Diagenesis and isotopic evolution of porewaters in the Alberta deep basin; the Falher Member and Cadomin Formation. Geochim Cosmochim Acta 53:2529–2546

Tisserant D, Odin GS (1979) Datation isotopique de glauconies miocènes d'Afrique Nord-Ouest. C R Somm Soc Géol Fr 4:188–190

Toulkeridis T, Clauer N, Stille P (1993) Pb-isotopic compositions and K-Ar dating of clay minerals associated with the Pb-Zn ores of Malines (Cévennes, France). Terra Abstr: 346

Toulkeridis T, Goldstein SL, Clauer N, Kröner A, Lowe DR (1994) Sm-Nd dating of Fig Tree clays: implications for the thermal history of the Barberton Greenstone Belt, South Africa. Geology 22:199–202

Touret O (1988) Structure des argiles hydratées. Thermodynamique de la déshydratation et de la compaction des smectites. Thèse, Univ Strasbourg, 172 pp

Trauth N (1977) Argiles évaporitiques dans la sédimentation carbonatée continentale et épicontinentale tertiaire. Bassins de Paris, de Mormoiron et de Salinelles (France), Jbel Ghassoul (Maroc). Sci Geol Mém Strasbourg 49:195 pp

Tsirambides AE (1986) Detrital and authigenic minerals in sediments from the western part of the Indian Ocean. Min Mag 50:69–74

Turpin L, Clauer N, Forbes P, Pagel M (1991) U-Pb, Sm-Nd and K-Ar systematics of the Akouta uranium deposit, Niger. Chem Geol Isot Geosci Sect 87:217–230

Valeton I (1958) Der Glaukonit und seine Begleitminerale aus dem Tertiär von Walsrode. Mitt Geol Staatsinst Hamburg, pp 88–131

Van Olphen H, Fripiat JJ (eds) (1979) Data handbook for clay minerals and other non-metallic minerals. Pergamon Press, Oxford, 346 pp

Veizer J (1989) Strontium isotopes in seawater through time. Annu Rev Earth Planet Sci 17:141–167

Veizer J (1992) Depositional and diagenetic history of limestones: stable and radiogenic isotopes. In: Clauer N, Chaudhuri S (eds) Isotopic signatures and sedimentary records. Lecture Notes in Earth Sciences 43. Springer, Berlin Heidelberg New York, pp 13–48

Veizer J, Compston W (1974) $^{87}Sr/^{86}Sr$ composition of sea-water during the Phanerozoic. Geochim Cosmochim Acta 38:1461–1484

Veizer J, Jansen SL (1979) Basement and sedimentary recycling and continental evolution. J Geol 87:341–370

Veizer J, Jansen SL (1985) Basement and sedimentary recycling. 2. Time dimension to global tectonics. J Geol 93:625–643

Veizer J, Compston W, Clauer N, Schidlowski M (1983) $^{87}Sr/^{86}Sr$ in Late Proterozoic carbonates: evidence for a "mantle" event at 900 Ma ago. Geochim Cosmochim Acta 47:295–302

Veizer J, Fritz P, Jones B (1986) Geochemistry of brachiopods: oxygen and carbon isotopic records of Paleozoic oceans. Geochim Cosmochim Acta 50:1679–1696

Velbel MA (1985) Geochemical mass balances and weathering rates in forested watersheds of the southern Blue Ridge. Am J Sci 285:904–930

Velde B (1965) Phengite micas: synthesis, stability, and natural occurrence. Am J Sci 263:886 913

Velde B (1977) Clays and clay minerals in natural and synthetic systems. Dev Sedimentol 21. Elsevier, Amsterdam, 218 pp

Velde B (1985) Clay minerals. A physico-chemical explanation of their occurrence. Dev Sedimentol 40. Elsevier, Amsterdam, 427 pp

Velde B, Bruzewitz AM (1981) Metasomatic and non-metasomatic low grade metamorphism in Ordovician bentonites in Sweden. Geochim Cosmochim Acta 46:447–452

Velde B, Meunier A (1987) Petrologic phase equilibria in natural clay systems. In: Newman ACD (ed) Chemistry of clays and clay minerals. Miner Soc Monogr 6. John Wiley, New York, pp 423–458

Velde B, Odin GS (1975) Further information related to the genesis of glauconite. Clays Clay Min 23:376–381

Venkatarathnam K, Biscaye PE, Ryan WBF (1972) Origin and dispersal of Holocene sediments in the eastern Mediterranean Sea. In: Stanley DJ (ed) The Mediterranean Sea, Dowden, Hutchinson and Ross, Stroudsburg, pp 455–469

Vidal P (1980) L'évolution polyorogénique du Massif Armoricain: apport de la géochronologie et de la géochimie isotopique du strontium. Mém Soc Géol Minér Bretagne 21:162 pp

Viets JG, Leach DL (1990) Genetic implications of regional and temporal trends in ore fluid geochemistry of Mississippi Valley-type deposits in the Ozark region. Econ Geol 85:842 2861

Wadleigh MA, Veizer J (1992) $^{18}O/^{16}O$ and $^{13}C/^{12}C$ in Lower Paleozoic articulate brachiopods: implications for the isotopic composition of seawater. Geochim Cosmochim Acta 56:431 443

Walker GF (1961) Vermiculite minerals. In: Brown G (ed) The X-ray identification and crystal structures of clay minerals, 2nd edn. Min Soc Lond, pp 297–324

Walker GF (1979) The decomposition of biotite in soil. Min Mag 28:693–703

Wasserburg GJ, Hayden RI, Jensen KJ (1956) Ar^{40}-K^{40} dating of igneous rocks and sediments. Geochim Cosmochim Acta 10:153–165

Wasserburg GJ, Jacobsen SB, DePaolo DJ, McCulloch MT, Wen T (1981) Precise determination of Sm/Nd ratios, Sm and Nd isotopic abundances in standard solutions. Geochim Cosmochim Acta 45:2311–2323

Weaver CE (1958) A discussion on the origin of clay minerals in sedimentary rocks. Clays Clay Minerals. Natl Acad Sci-Natl Res Council Publ 566, pp 159–173

Weaver CE (1959) The clay petrology of sediments. Proc 6th Natl Conf Clays and Clay Minerals. Natl Acad Sci-Natl Res Council Publ, pp 154–187

Weaver CE (1960) Possible uses of clay minerals in search of oil. Am Assoc Petrol Geol Bull 44:1505–1518

Weaver CE (1961) Clay minerals of the Ouachita structural belt and adjacent foreland. In: The Ouachita system. Bur Econ Geol Austin, pp 147–160

Weaver CE (1968) Electron microprobe study of kaolin. Clays Clay Min 16:187–189

Weaver CE, Beck KC (1971) Clay water diagenesis during burial: how mud becomes gneiss. Geol Soc Am Spec Pap 134:176 pp

Weaver CE, Broekstra BR (1984) Illite-Mica. In: Weaver CE (ed) Shale-slate metamorphism in southern Appalachians. Dev Petrol 10. Elsevier, Amsterdam, pp 67–97

Weaver CE, Pollard LD (1973) The chemistry of clay minerals. Elsevier, New York, 205 pp

Weaver CE, Wampler JM (1970) K-Ar illite burial. Geol Soc Am Bull 81:3423–3430

Weaver CE, Highsmith PB, Wampler JM (1984) Chlorite. In: Weaver CE (ed) Shale-slate metamorphism in Southern Appalachians. Dev Petrol 10. Elsevier, Amsterdam, pp 99–139

Weber JN (1965) Changes in the oxygen isotopic composition of seawater during the Phanerozoic evolution of the oceans. Geol Soc Am Spec Pap 82:218–219

Weber K (1972) Notes on determination of illite crystallinity. Neues Jahrb Geol Palaeontol Monatsh 267–276

Wedepohl KH (ed) (1978) Handbook of geochemistry. Springer, Berlin Heidelberg New York, 82-K-6, 92-K-3

White DE (1965) Saline waters of sedimentary rocks. In: Young Y, Galley GE (eds) Fluids in subsurface environments. Am Assoc Petrol Geol Mem 4:342–366

White WM (1979) Pb isotope geochemistry of the Galapagos Islands. Annu Rep Carnegie Inst Dep Terr Magn 331–335

Whitney G, Northrop HR (1987) Diagenesis and fluid flow in the San Juan Basin, New Mexico: regional zonation in the mineralogy and stable isotope composition of clay minerals in sandstone. Am J Sci 287:353–382

Whitney G, Northrop HR (1988) Experimental investigation of the smectite to illite reaction: dual reaction mechanisms and oxygen-isotope systematics. Am Min 73:77–90

Whitney PR, Hurley PM (1964) The problem of inherited radiogenic strontium in sedimentary age determinations. Geochim Cosmochim Acta 28:425–436

Willis KM, Stern RJ, Clauer N (1988) Age and geochemistry of Late Precambrian sediments of the Hammamat Series from the north-eastern desert of Egypt. Precambrian Res 42:173–187
Wilson MJ (1975) Chemical weathering of some primary rock-forming minerals. Soil Sci 119:349–355
Wilson MR, Kyser TK, Mehnert HH, Hoeve J (1987) Changes in the H-O-Ar isotope composition of clays during retrograde alteration. Geochim Cosmochim Acta 51:869–878
Wollast R (1967) Kinetics of alteration of K-feldspar in buffered solutions at low temperature. Geochim Cosmochim Acta 31:635–648
Worden JM, Compston W (1973) A Rb-Sr isotopic study of weathering in the Mertondale granite, Western Australia. Geochim Cosmochim Acta 37:2567–2576
Wu Y, Beales FWS (1981) A reconnaissance study by paleomagnetic methods of the age of mineralization along the Vibernum trend, southeast Missouri. Econ Geol 76:1879–1884
Yanase Y, Wampler JM, Dooley RE (1975) Recoil-induced loss of ^{39}Ar from glauconite and other minerals. Trans Am Geophys Union 56:472
Yau YC, Peacor DR, McDowell SD (1987) Smectite-to-illite reactions in Salton Sea shales; a transmission and analytical electron microscopy study. J Sediment Petrol 57:335–342
Yeh HW (1980) D/H ratios and late-stage dehydration of shales during burial. Geochim Cosmochim Acta 44:341–352
Yeh HW, Epstein S (1978) Hydrogen isotope exchange between clay minerals and sea water. Geochim Cosmochim Acta 42:140–143
Yeh HW, Eslinger EV (1986) Oxygen isotopes and the extent of diagenesis of clay minerals during sedimentation and burial in the sea. Clays Clay Min 34:403–406
Yeh HW, Savin SM (1976) The extent of oxygen isotope exchange between clay minerals and sea water. Geochim Cosmochim Acta 40:743–748
Yeh HW, Savin SM (1977) Mechanism of burial metamorphism of argillaceous sediments, 3. O-isotope evidence. Geol Soc Am Bull 88:1321–1330
Yoder HS, Eugster HP (1955) Synthetic and natural muscovites. Geochim Cosmochim Acta 8:225–280
York D, Masliwec H, Hall CM, Kuybida P, Kenyon WJ, Spooner ETC, Scott SD, Pye EG (1981) The direct dating of ore minerals. Ont Geol Surv Misc Pap 98:334–340
Zartman RE (1964) A geochronologic study of the Lone Grove pluton from the Llano uplift, Texas. J Petrol 5:359–408
Zartman RE, Brock M, Heyl AV, Thomas HH (1966) K-Ar and Rb-Sr ages of some alkalic intrusive rocks from central and eastern United States. Am J Sci 265:848–870
Zimmermann JL, Odin GS (1982) Kinetics of the release of argon and fluids from glauconies. In: Odin GS (ed) Numerical dating in stratigraphy. John Wiley, New York, pp 345–362
Zwingmann H, Deutrich T, Clauer N, Gaupp R (1993) Timing and tracing of clay authigenesis in Late Paleozoic reservoir sandstones of the NW German basin. Ann Meet Clay Min Soc, San Diego, 1 p

Subject Index

Printing: Mercedesdruck, Berlin
Binding: Buchbinderei Lüderitz & Bauer, Berlin